S0-AGV-204

Advances in
ORGANOMETALLIC CHEMISTRY

VOLUME 48

Advances in Organometallic Chemistry

EDITED BY

ROBERT WEST

DEPARTMENT OF CHEMISTRY
UNIVERSITY OF WISCONSIN
MADISON, WISCONSIN

ANTHONY F. HILL

AUSTRALIAN NATIONAL UNIVERSITY
RESEARCH SCHOOL OF CHEMISTRY
INSTITUTE OF ADVANCED STUDIES
CANBERRA, ACT, AUSTRALIA

FOUNDING EDITOR

F. GORDON A. STONE

VOLUME 48

ACADEMIC PRESS
A Division of Harcourt, Inc.

San Diego San Francisco New York Boston
London Sydney Tokyo

This book is printed on acid-free paper. 

Academic Press
A division of Harcourt, Inc.
525 B Street, Suite 1900, San Diego, California 92101-4495, USA
http://www.academicpress.com

Academic Press
Harcourt Place, 32 Jamestown Road, London NW1 7BY, UK
http://www.academicpress.com

International Standard Book Number: 0-12-031148-8

PRINTED IN THE UNITED STATES OF AMERICA
01 02 03 04 05 06 EB 9 8 7 6 5 4 3 2 1

Contents

Organoelement Chemistry of Main Group Porphyrin Complexes

PENELOPE J. BROTHERS

Contributors

Numbers in parentheses indicate the pages on which the authors' contributions begin.

VOLKER P. W. BÖHM (1), Department of Chemistry, University of North Carolina at Chapel Hill, Chapel Hill, North Carolina 27599

PENELOPE J. BROTHERS (289), Department of Chemistry, The University of Auckland, Auckland, New Zealand

MICHAEL I. BRUCE (71), Department of Chemistry, University of Adelaide, Adelaide, South Australia 5005, Australia

WOLFGANG A. HERRMANN (1), Anorganisch-chemisches Institut der Technischen Universität München, D-85747 Garching bei München, Germany

PAUL J. LOW (71), Department of Chemistry, University of Durham, Durham DH1 3LE, England

THOMAS WESKAMP (1), Symyx Technologies, Santa Clara, California 95051

ADVANCES IN ORGANOMETALLIC CHEMISTRY, VOL. 48

*Metal Complexes of Stable Carbenes**

WOLFGANG A. HERRMANN

Anorganisch-chemisches Institut der Technischen Universität München
D-85747 Garching bei München, Germany

THOMAS WESKAMP

Symyx Technologies
Santa Clara, California 95051

VOLKER P. W. BÖHM

Department of Chemistry
University of North Carolina at Chapel Hill
Chapel Hill, North Carolina 27599-3290

I
INTRODUCTION

A. *Historic Background*

Carbenes—molecules with a neutral dicoordinate carbon atom—have played an important role in organic chemistry ever since their first firm evidence of existence.

*Dedicated to Professor Henri Brunner on the occasion of his 65th birthday.

0065-3055/01 $35.00

However, despite the increasing interest in persistent intermediates since the days of Gomberg[1,2] and despite the fact that carbenes were introduced into organic chemistry by Doering and Hoffmann in the 1950s[3] and into organometallic chemistry by Fischer and Maasböl in the 1960s,[4] it was only in the late 1980s and early 1990s that the first carbenes were isolated [Eq. (1)].[5–8]

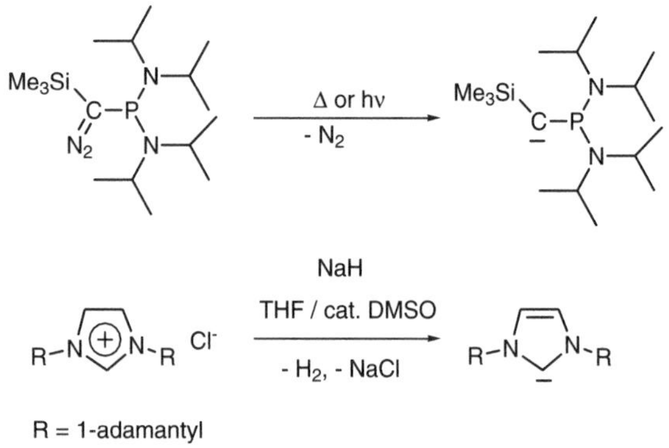

Me3Si C–P(N)(N) N2 —Δ or hν, - N2→ Me3Si C–P(N)(N)

(1)

R–N(+)N–R Cl⁻ —NaH, THF / cat. DMSO, - H2, - NaCl→ R–N N–R

R = 1-adamantyl

This discovery resulted in a revival of carbene chemistry, surprisingly more in organometallic chemistry than in organic chemistry.[9,10] One explanation for the interest of organometallic chemists in stable free carbenes might be the fact that the *metal complexes* **1** and **2** containing the subsequently isolated *N*-heterocyclic carbenes were prepared as early as in 1968 by Wanzlick and Schönherr and by Öfele.[11,12]

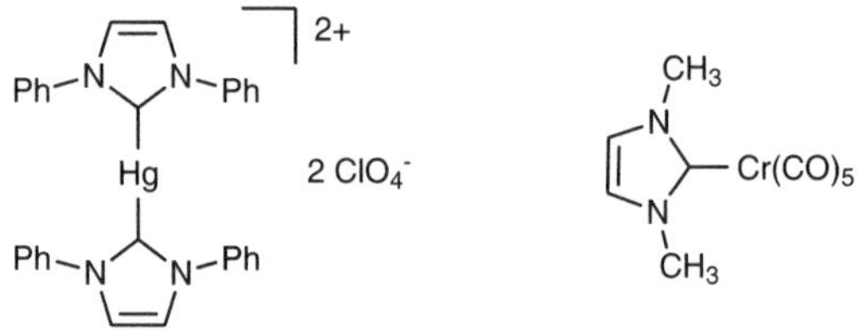

Ph–N N–Ph, Hg, Ph–N N–Ph, 2+, 2 ClO_4^-

CH_3, N, N, $Cr(CO)_5$, CH_3

1 **2**

That was only 4 years after the preparation of the first *Fischer*-type carbene complex **3**,[4] 6 years before the first *Schrock*-type carbene complex **4** was reported,[13] and more than 20 years before the isolation of stable imidazolin-2-ylidenes by Arduengo in 1991 [Eq. (1)].[7] Once attached to a metal, these *Wanzlick*- or *Arduengo*-carbenes have shown a reaction pattern completely different from that of the electrophilic *Fischer*- and nucleophilic *Schrock*-type carbene complexes.

CH_3–O–C(R)=W(CO)$_5$ H_3C–Ta(=CH_2)(Cp)$_2$

3 **4**

This article presents the principles known so far for the synthesis of metal complexes containing stable carbenes, including the preparation of the relevant carbene precursors. The use of some of these compounds in transition-metal-catalyzed reactions is discussed mainly for ruthenium-catalyzed olefin metathesis and palladium-/nickel-catalyzed coupling reactions of aryl halides, but other reactions will be touched upon as well. Chapters about the properties of metal– carbene complexes, their applications in materials science and medicinal chemistry, and their role in bioinorganic chemistry round the survey off. The focus of this review is on *N*-heterocyclic carbenes, in the following abbreviated as NHC and NHCs, respectively.

B. *Stable Carbenes*

Wanzlick *et al.* realized in the 1960s that the stability of carbenes should be increased by a special substitution pattern of the disubstituted carbon atom.[14–16] Substituents in the vicinal position that provide π-donor/σ-acceptor character to "fill" the p-orbital of the carbene carbon and stabilize the carbene lone pair by a negative inductive effect should reduce the electrophilicity of the singlet carbene and consequently reduce its reactivity (Scheme 1).

Based on this concept and the development of appropriate synthetic methods, many heteroatom-substituted carbenes have been isolated since the first successful attempts by Igau *et al.*[5] and by Arduengo *et al.*[7] The stability of carbenes was

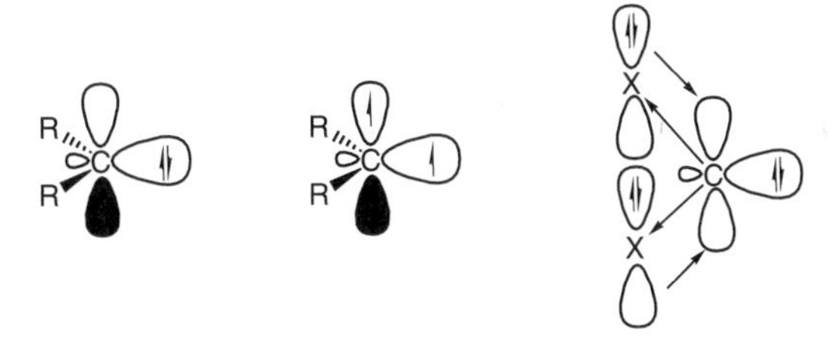

SCHEME 1. σ-Acceptors and π-donors vicinal to the carbene–carbon atom stabilize carbenes.

originally considered to be limited to cyclic diaminocarbenes (nitrogen provides good π-donor/σ-acceptor character) with steric bulk to prevent dimerization[17] and some aromatic character.[18] This holds true for imidazolin-2-ylidenes as well as for 1,2,4-triazolin-5-ylidenes. For this family of stable carbenes, many examples have been isolated so far, among them **5–13.**[9,19–21] **12** was even reported to be air stable.[22] Steric hindrance at the nitrogen substituents does not solely determine whether a carbene can be isolated: the 1,3-dimethylimidazolin-2-ylidene **5** can be distilled for its purification without significant decomposition.[23] However, steric parameters certainly influence the long-term stability of NHCs.[24,25]

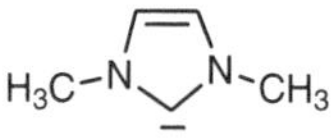

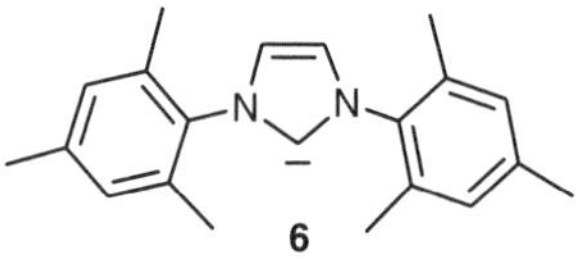

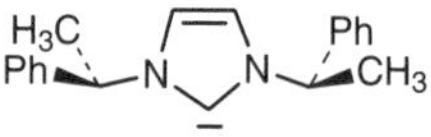

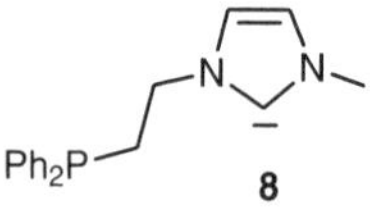

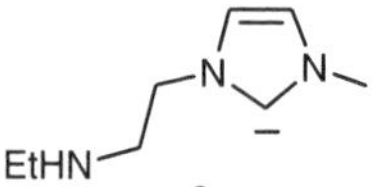

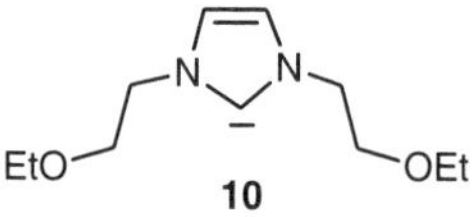

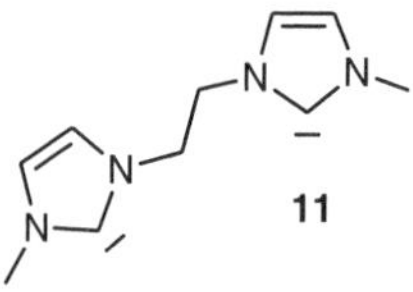

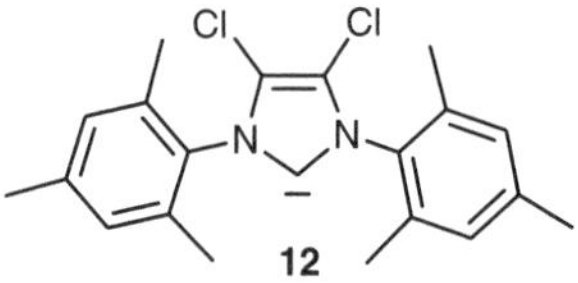

Later, imidazolidin-2-ylidenes such as **14,** a "saturated," more electron-rich and nonaromatic version of the imidazolin-2-ylidenes, were isolated.[25,26] Isolation of a six-membered tetrahydropyrimid-2-ylidene **15**[27,28] and of acyclic structures such as **16**[29,30] was a consequent extension since these compounds still possess two nitrogens vicinal to the carbene carbon, but lack the 6π-electron conjugation.

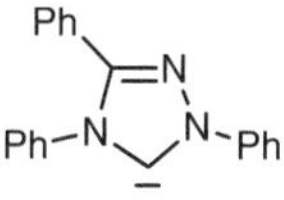

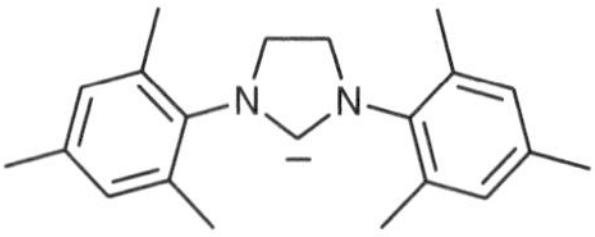

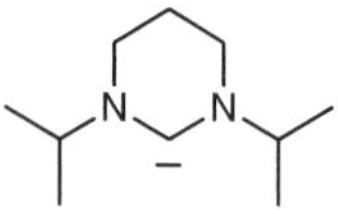

For all these compounds, the carbene carbon has two nitrogen substituents, which is in complete agreement with the consideration that strong π-donor substituents are an essential requirement for stable carbenes. However, one weaker

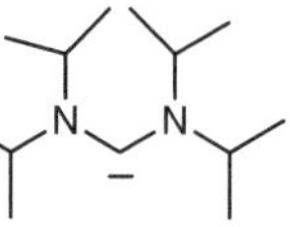

π-donor substituent, e.g., an alkoxy or alkylsulfido group, can be tolerated as was demonstrated for **17–19.**[31]

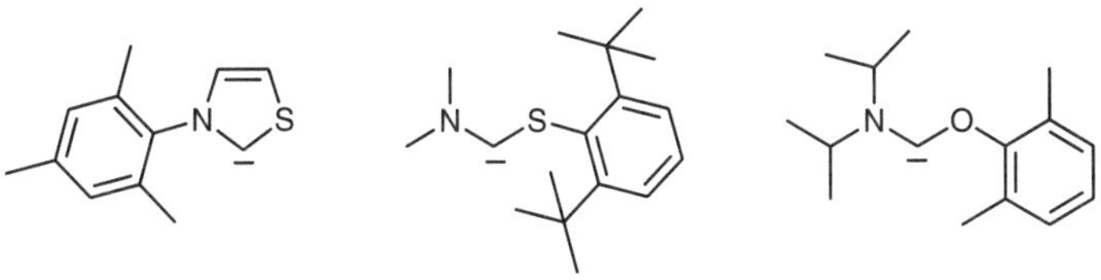

17 **18** **19**

C. *Precursors for Stable Carbenes*

1. *Precursors for NHCs with Unsaturated Backbone (Imidazolin-2-ylidenes and Benzimidazolin-2-ylidenes)*

In many cases the synthesis of NHC complexes starts from *N,N′*-disubstituted azolium salts. Imidazolium salts as precursors for imidazolin-2-ylidenes are generally accessible by two ways complementing each other: (i) nucleophilic substitution at the imidazole heterocycle or (ii) a multicomponent reaction building up the heterocycle with the appropriate substituents in a one-pot reaction.

Imidazolium salts that can be prepared by the first procedure, the alkylation of imidazole, are easy to obtain and often used for metal complex synthesis. Potassium imidazolide is reacted with the first equivalent of alkyl halide in toluene to give the 1-alkylimidazole.[32] Subsequent alkylation in 3-position is achieved by addition of another equivalent of alkyl halide [Eq. (2)].[33–35] A variant of this approach employs commercially available *N*-trimethylsilyl imidazole with 2 equiv of an alkyl chloride, under elimination of volatile Me_3SiCl.[36] The drawback of these simple routes is the fact that only primary alkyl halides can be reacted in satisfactory yields because secondary and tertiary alkyl halides give substantial amounts of elimination by-products.

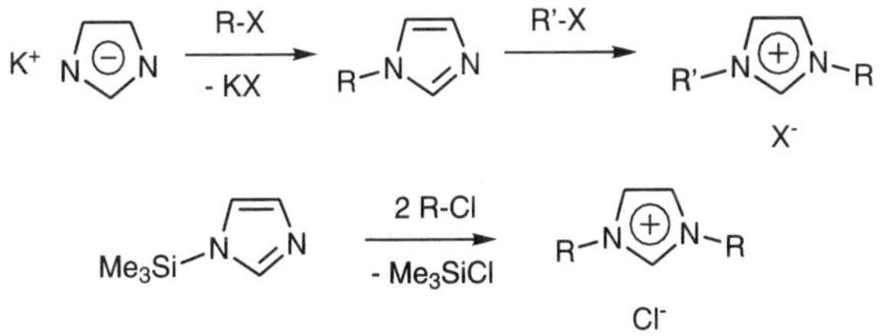

K+ R-X / - KX → R'-X → X-

Me3Si 2 R-Cl / - Me3SiCl → Cl-

(2)

In order to introduce other substituents at the 1- and 3-positions of the imidazolium salt the reaction of primary amines with glyoxal and formaldehyde in the presence of acid can be used [Eq. (3)].[20,37,38] Variation of the amine allows the preparation of imidazolium salt libraries which can be diversified by using different acids in order to change the anion of the imidazolium salt.[39] The use of chiral amines in this reaction results in the convenient generation of C_2-symmetric imidazolium salts.[21] It is possible to generate imidazolium salts with anilines that do not bear a *para*-substituent in a two-step sequence: synthesis of the bisimine in the first step and subsequent ring closure with formaldehyde and an acid.[40,41]

$$\mathrm{RNH_2} + \mathrm{OHC{-}CHO} + \mathrm{H_2C{=}O} + \mathrm{H_2NR} \xrightarrow[-3\,\mathrm{H_2O}]{\mathrm{HX}} \text{R–N}\overset{+}{\text{(im)}}\text{N–R}\ \mathrm{X^-} \qquad (3)$$

A method by Gridnev and Mihaltseva allows the combination of both strategies: (i) synthesis of the 1-alkylimidazole by a multicomponent reaction starting from glyoxal, formaldehyde, a primary amine and ammonium chloride, and (ii) subsequent alkylation by a primary alkyl halide to give the imidazolium salt [Eq. (4)].[42]

$$\mathrm{RNH_2} + \mathrm{OHC{-}CHO} + \mathrm{H_2C{=}O} + [\mathrm{NH_4}]\mathrm{Cl} \xrightarrow{\mathrm{H_3PO_4}} \text{R–N(im)N} \xrightarrow{\mathrm{R'{-}X}} \text{R–N}\overset{+}{\text{(im)}}\text{N–R'}\ \mathrm{X^-} \qquad (4)$$

Direct coupling of imidazole with aryl iodides in the presence of copper(I) triflate results in 1-aryl-imidazoles, which can be alkylated in a second step [Eq. (5)]. This route represents a variation of the Gridnev method.[43]

$$\mathrm{R{-}C_6H_4{-}I} + \mathrm{HN(im)N} \xrightarrow[-2\,\mathrm{HI}]{[\mathrm{Cu(OTf)}]_2} \mathrm{R{-}C_6H_4{-}N(im)N} \qquad (5)$$

The abstraction of a hydride is an additional route for the preparation of benzimidazolium salts: Treatment of 2,3-dihydro-1 *H*-benzimidazoles with tritylium tetrafluoroborate generates the benzimidazolium salt and triphenylmethane [Eq. (6)].[44]

$$\text{(dihydrobenzimidazole, N–R, N–R, CH}_2\text{)} \xrightarrow[-\,\mathrm{Ph_3CH}]{+\,[\mathrm{Ph_3C}][\mathrm{BF}]_4} \text{(benzimidazolium, N–R, N–R, C–H)}\ [\mathrm{BF}]_4^- \qquad (6)$$

The reaction of *N*-alkyl-*N*-formyl hydrazines with imidoyl chlorides gives 3,4-substituted 1-alkyl-4*H*-1,2,4-triazolium salts in a one-pot reaction.[45]

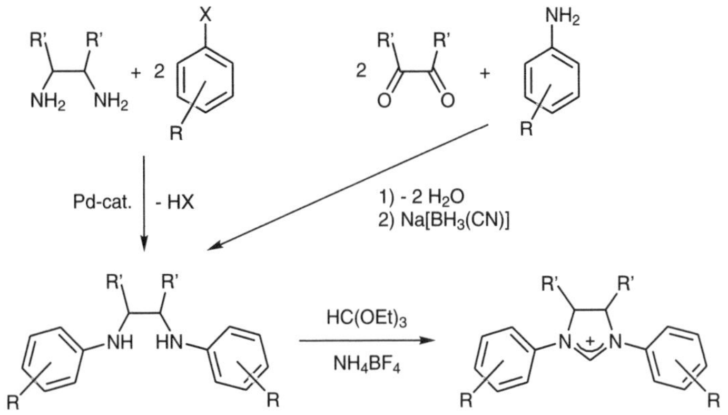

SCHEME 2. Synthesis of imidazolidinium salts.

2. *Precursors for NHCs with Saturated Backbone (Imidazolidin-2-ylidenes)*

The reaction of an *ortho*-ester, e.g., $HC(OEt)_3$, with a secondary bisamine in the presence of an ammonium salt yields imidazolidinium salts (Scheme 2).[46,47] The necessary secondary diamines can be generated by a classical condensation–reduction sequence or by applying the palladium-catalyzed Buchwald–Hartwig amination.[48] The latter reaction offers convenient access to imidazolidinium salts with chiral backbones starting from chiral diamines, a number of which are commercially available.[46]

3. *Precursors for Acyclic Carbenes*

Procedures for the synthesis of precursors for acyclic diamino-, aminoxy-, and aminothiocarbenes rely on the condensation of formamides with phenols, thiophenols, and amines, respectively (Scheme 3).[29–31]

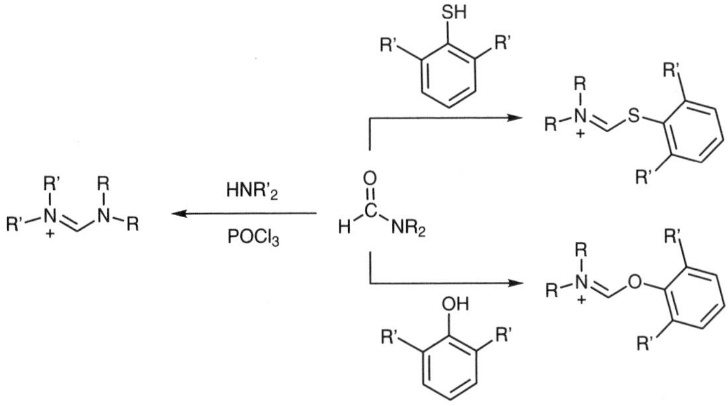

SCHEME 3. Preparation of precursors for acyclic carbenes.

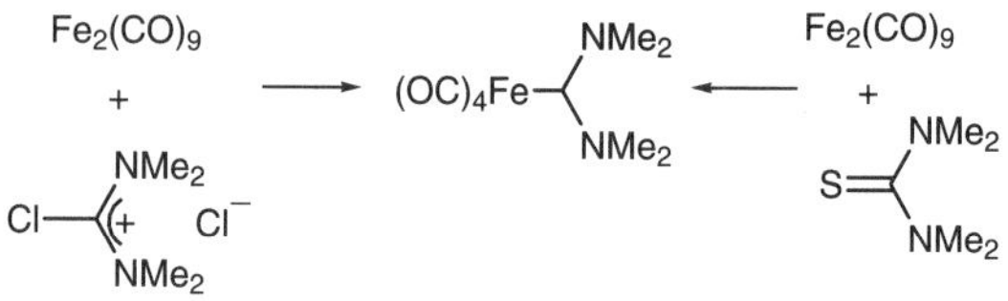

SCHEME 4. Synthetic routes to an iron complex containing an acyclic diaminocarbene.

Metal complexes of these carbenes have not been published yet, although complexes with $:C(NMe_2)_2$ are accessible, e.g., via the routes depicted in Scheme 4.[49–55]

II
METAL COMPLEXES

Metal complexes of stable carbenes are now known for almost all the metals of the periodic table. This chapter is divided into an overview of the NHC–main-group metal adducts that are generally synthesized by adding the free NHC to an appropriate metal precursor and a discussion of the various synthetic approaches toward transition metal complexes.

A. *Main-Group Metal Complexes*

The broader subject of the interaction of stable carbenes with main-group compounds has recently been reviewed.[56] Accordingly, the following discussion focuses on metallic elements of the s and p blocks. Dimeric NHC–alkali adducts have been characterized for lithium, sodium, and potassium. For imidazolin-2-ylidenes, alkoxy-bridged lithium dimer **20** and a lithium–cyclopentadienyl derivative **21** have been reported.[57] For tetrahydropyrimid-2-ylidenes, amido-bridged dimers **22** have been characterized for lithium, sodium, and potassium.[27,28] Since one of the synthetic approaches to stable NHCs involves the deprotonation of imidazolium cations with alkali metal bases, the interactions of alkali metal cations with NHCs are considered to be important for understanding the solution behavior of NHCs.

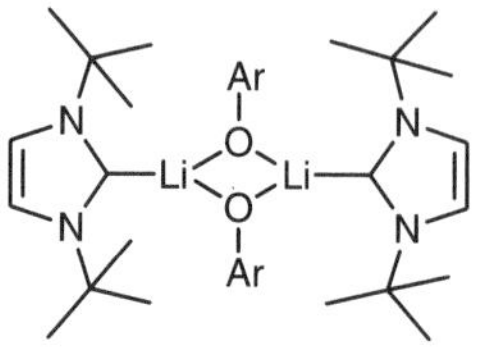

M = Li, Na, K

20 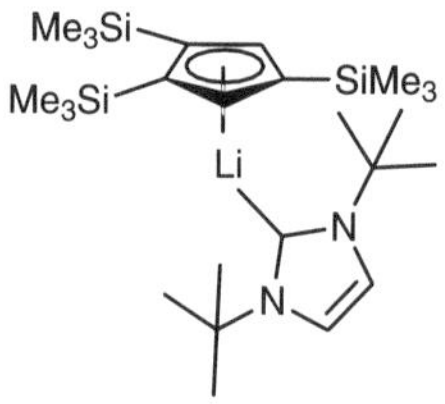 **21** 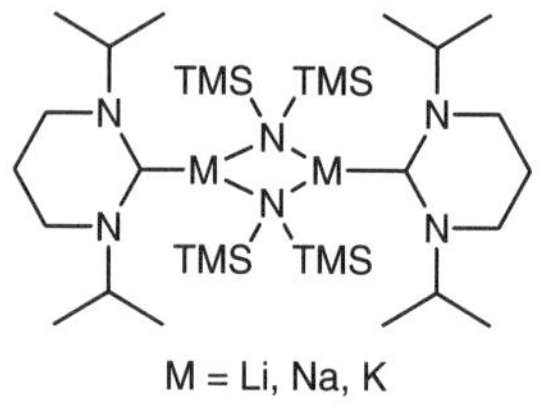 **22**

The fact that NHCs form stable compounds with beryllium, one of the hardest Lewis acids known and without p-electrons to "back donate," shows the nucleophilicity of these ligands. Reaction of 1,3-dimethylimidazolin-2-ylidene with polymeric $BeCl_2$ results in the formation of the neutral 2:1 adduct **23** or the cationic 3:1 adduct **24.**[58] The first NHC–alkaline earth metal complex to be isolated was the 1:1 adduct **25** with $MgEt_2$.[59] Whereas 1,3-dimesitylimidazolin-2-ylidene results in the formation of a dimeric compound, the application of sterically more demanding 1,3-(1-adamantyl)imidazolin-2-ylidene gives a monomeric adduct.[59]

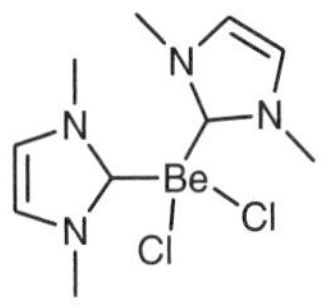

23 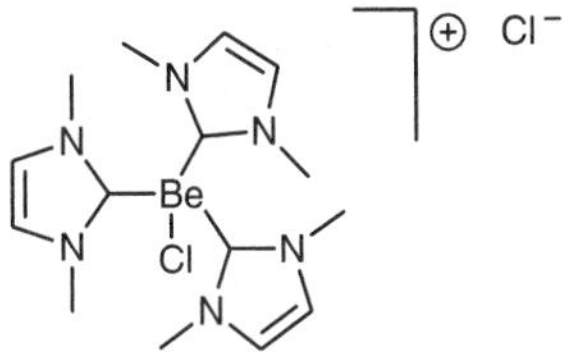 **24** 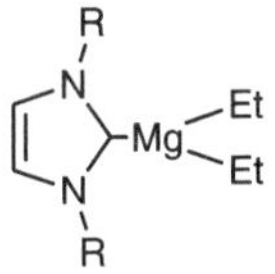 **25**

A variety of magnesium, calcium, strontium, and barium complexes **26–27** have been prepared starting with either the corresponding $[M\{N(SiMe_3)_2\}_2(thf)_2]$ or $[(\eta^5\text{-}C_5Me_5)_2M]$.[60,61] The resulting C_5Me_5 complexes **27** reveal different coordination modes for the C_5Me_5 ligand depending on the metal: In the case of Mg, one cyclopentadienyl ring is η^5-bound; the other is intermediate between η^3- and η^1-bound. The higher homologues, calcium and barium, feature two η^5-bound cyclopentadienyl rings. As expected, all these complexes show a nonlinear arrangement for the C_5Me_5-M-C_5Me_5 unit. With the heavier alkaline earth metals strontium and barium, it is also possible to isolate bis(NHC) complexes starting with $[M(\eta^5\text{-}C_5Me_5)_2]$.

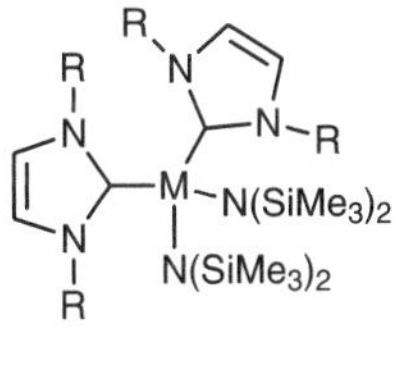

M = Ca, Sr, Ba

26

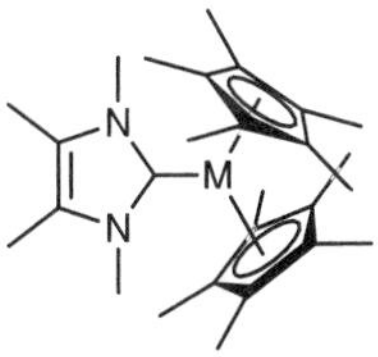

M = Mg, Ca, Sr, Ba

27

The metal–donor bonds are predominantly ionic and become more labile for calcium, strontium, and barium compared to beryllium and magnesium. The solubility and stability of the complexes decrease from calcium to barium. The 1:1 adducts of NHCs with BH_3 or BF_3 (**28** and **29**) are thermally stable and can be sublimed without decomposition.[62,63] This is in sharp contrast to the properties of conventional carbenes, which rely on a pronounced metal-to-ligand back donation and are, thus, not suited to forming adducts with electron-poor fragments such as boron(III). Other examples for boron(III) adducts of NHCs have been reported for BEt_3[64] and for a borabenzene derivative resulting in adduct **30.**[65]

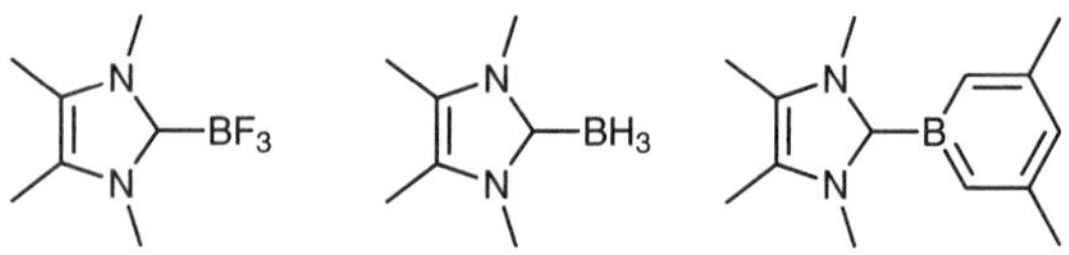

BF_3 BH_3 B

28 **29** **30**

NHC–group 13 metal adducts other than that for boron have been described for aluminum, gallium, and indium (**31–33**).[66–70]

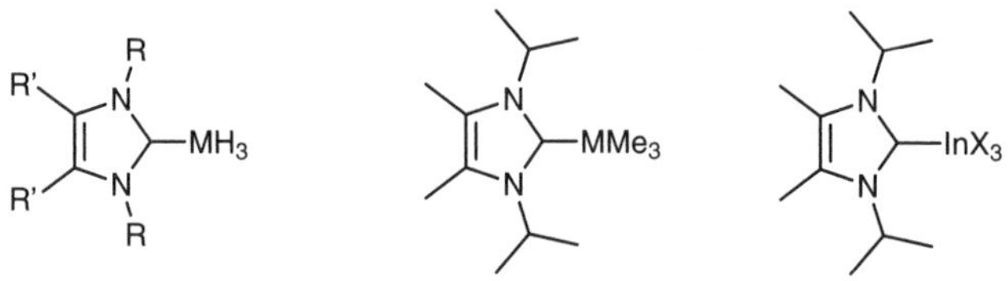

R R' R' R MH_3 MMe_3 InX_3

M = Al, Ga, In

31

M = Al, Ga

32

X = Cl, Br

33

Indium(III)–NHC complexes have been isolated in the form of their trihydrides **31** and trihalides **33.** For the halides, mono and bis(NHC) adducts are formed when InX_3 (X = Br, Cl) is exposed to the respective stoichiometric amounts of the NHC.[67] For the hydrides, a substitution reaction at $[InH_3(NMe_3)]$ or $Li[InH_4]$ gives the respective compounds.[68] The thermal stability of $[InH_3(NHC)]$ complexes was increased remarkably by using 1,3-dimesitylimidazolin-2-ylidene rather than less sterically demanding NHCs.[71] It was impossible to obtain bis(NHC) complexes with the IrH_3 fragment, presumably as a consequence of the reduced Lewis acidity of the InH_3 moiety compared to the indium halides, which form these 2:1 adducts readily. There are fewer data known for the hydride adducts **31** of aluminum and gallium. But the fact that these NHC adducts—again especially with sterically demanding NHCs—are reported to be of higher stability than adducts with other *Lewis* bases supports the postulation that NHCs are characterized by high nucleophilicity and *Lewis* basicity.[66] In general, the thermal stabilities of the NHC–group13-hydrides decrease from aluminum to indium. Trialkyl complexes **32** are known for aluminum and gallium.[70] NHC adducts with (semi-)metals of group **14** are known for silicon, germanium, tin, and lead.[72–74] The germanium(II) and tin(II) halides **34** and **35** are again prepared by an adduct formation starting from the free NHC and GeI_2 or $SnCl_2$, respectively. A tin alkyl adduct is formed by using bis(2,4,6-tri-*iso*-propylphenyl)tin(II) as the precursor. The structural data for all these compounds are consistent with the observations made for other NHC complexes: long M–C bonds suggesting single-bond character, which makes the compounds best described as Lewis acid–Lewis base adducts.

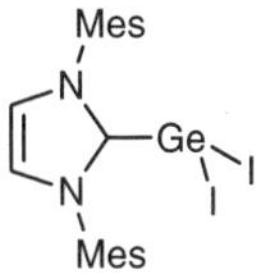

34 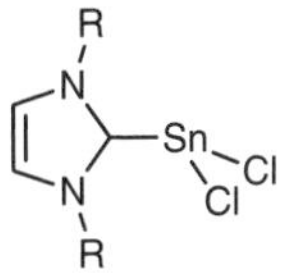 **35** 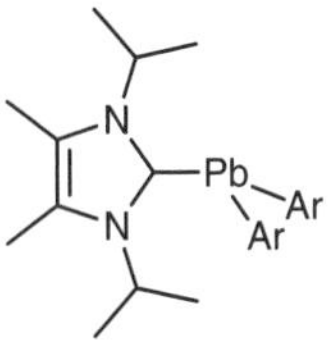 **36** 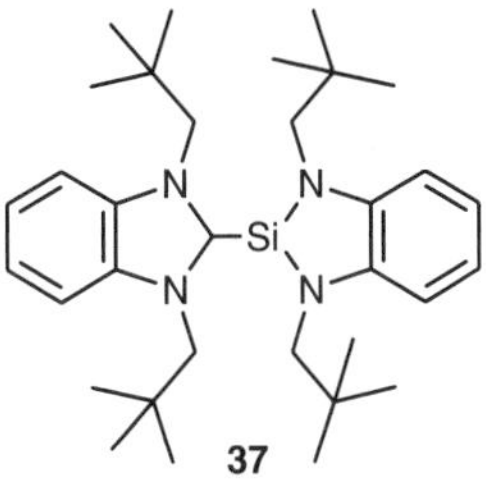 **37**

Plumbene–NHC complex **36** is generated by the reaction of an NHC with a bis(aryl)-lead(II) compound.[75] The NHC–silylene adduct **37** also features a long C–Si bond with significant $^{\delta+}$C–Si$^{\delta-}$ polarity.[76] Pentacoordinated silicon(IV) and tin(IV) compounds **38–40** can be generated when the precursor $SiCl_4$, Ph_2SiCl_2, Me_2SiCl_2, or Ph_2SnCl_2 is reacted with NHCs.[74,77]

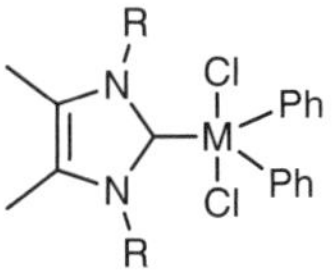

M = Si, Sn

38 **39** 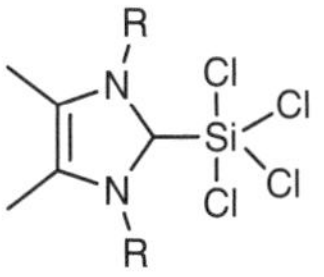 **40**

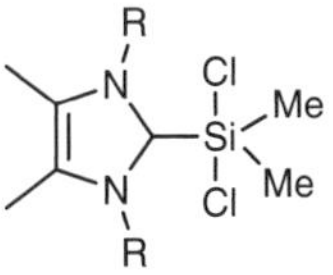

To our knowledge no (semi-)metal adducts of stable carbenes are known for groups 15 and 16 except **41–43.**[63,78,79] However, the oxidation of a 2-telluroimidazoline with iodine provides an interesting point of connection, effectively providing a TeI_2 adduct of a NHC.[80]

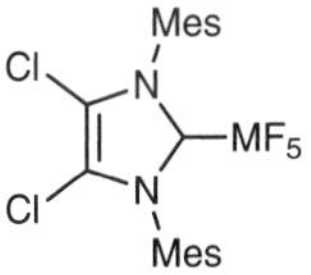

M = As, Sb R = Ph, C_6F_5

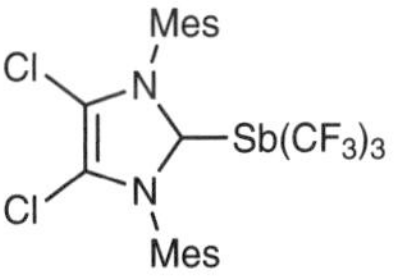

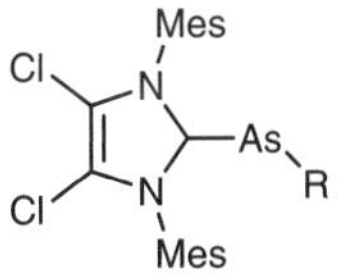

41 **42** **43**

B. *Transition Metal Complexes*

Since there are many more carbene–transition-metal complexes known than main-group adducts this chapter is organized by the synthetic methods that lead to the complexes rather than by the metal itself. Applications of these complexes to catalysis and materials science will be discussed in separate chapters of this article.

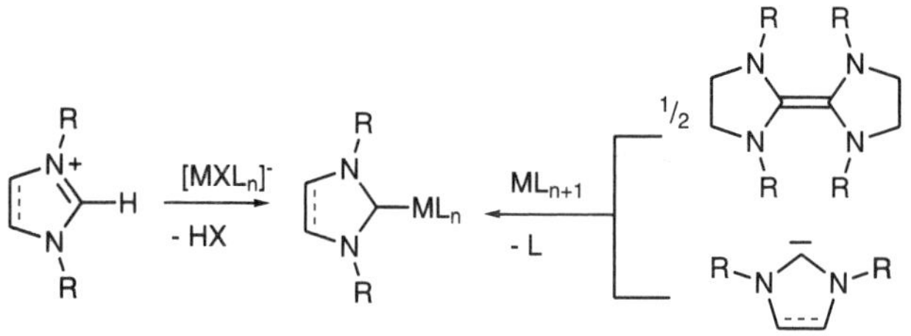

SCHEME 5. Major synthetic pathways for the generation of transition metal–NHC complexes.

The access to NHC complexes is mainly based on three routes: the *in situ* deprotonation of ligand precursors, the complexation of the free, preisolated NHCs, and the cleavage of electron-rich olefins (Scheme 5). A variety of other methods, mainly of importance in special cases, will be presented at the end of this chapter.

1. *In Situ Deprotonation of Ligand Precursors*

The *in situ* complexation of the ligand has the advantage of not having to prepare and isolate the free NHC. In cases where the carbene is hardly stable, not yet accessible at all, or difficult to handle, this approach offers the only chance to prepare the desired complex.

a. *Deprotonation by basic metallates.* Azolium cations can be deprotonated *in situ* by *Brönstedt* basic metallate anions in a formal redox and acid–base reaction. The metal of the base represents the ligand acceptor at the same time. Öfele prepared the first [(NHC)$Cr(CO)_5$] complexes by this method [Eq.(7)].[12,81] This route has also been used to prepare complexes of other metals and for imidazolium, benzimidazolium, pyrazolium, triazolium, and tetrazolium salts.[82–84] The limitation of this method is, however, the availability of the appropriate metallate which determines not only the nature and oxidation state of the central metal atom of the new complex but also its ligand environment.

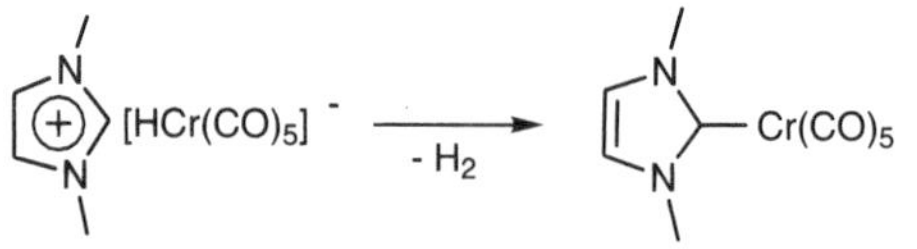

$$[HCr(CO)_5]^- \xrightarrow[-H_2]{} (NHC)Cr(CO)_5 \quad (7)$$

b. *Deprotonation by basic anions.* Brönstedt basic anions either on the metal precursor or on the azolium salt can form the desired ligand *in situ* by deprotonation. Commercially available metal acetates, acetylacetonates, or alkoxylates, which are also easy to prepare, have been used frequently. In the cases of coordinating counter-anions of the azolium salt, this anion is often incorporated into the new complex. To avoid this incorporation, perchlorate, hexafluorophosphate, or tetrafluoroborate have been used as the counterions of the azolium salts.

Wanzlick was the first to use an acetate salt in the synthesis of a mercury bis-NHC complex starting from mercury(II) diacetate [Eq. (8)].[11,85] There are other examples using the very same strategy.[86–89] Exchanging the anionic parts of the mercury precursor and the imidazolium salt, i.e., using $HgCl_2$ and imidazolium acetate, works as well.[90]

$Hg(OAc)_2$ + 2 [1,3-diphenylimidazolium] ClO_4^- → (- 2 AcOH) [Hg(NHC)$_2$]$^{2+}$ 2 ClO_4^- (8)

More than 25 years later, this method proved to be especially valuable for palladium(II) and nickel(II) complexes starting at the corresponding metal(II) diacetates and imidazolium or triazoliumsalts.[19,24,91–93] For palladium, it is possible to apply the *in situ* deprotonation method even without solvent,[20,24] but using THF or even better DMSO results in enhanced yields of the complexes.[92,93] In this respect, the crystal structure of an imidazolium tetrachloropalladadate salt can be seen as a model structure for the transition state of the deprotonation process.[94] Additionally, a variety of palladium and nickel complexes with methylene bridged, chelating bis(NHC)s were accessible only by this route [Eq.(9)][92,95–99] until these bidentate ligands were isolated as free dicarbenes.[100] The *in situ* deprotonation occurs as well when $Pd(OAc)_2$ is used as a catalyst in ionic liquid 1,3-dialkyl-imidazolium salts. Thus, in this type of solvent system, the respective palladium–NHC complexes are generated.[101,102] The chiral chelating NHC **44**—the only chiral bis(NHC) published so far—was attached to a palladium(II) center by this method.[103] The *in situ* deprotonation can be extended to other azolium salt precursors like benzimidazolium, benzothiazolium, or triazolium salts and the formation of their palladium complexes.[44,86,92,104,105] The first palladium(II) complexes containing such different NHC ligands as benzothiazolin-2-ylidene **45**[106] or a fluoroalkylated imidazolin-2-ylidenes **46**[107] were recently isolated by using this route. Since the

$Pd(OAc)_2$ + [methylene-bridged bis(imidazolium)] 2 I^- → (Δ, - HOAc) → (Δ, - HOAc) [chelating bis(NHC)]PdI_2 (9)

in situ approach seems to tolerate more functional groups in the azolium precursor than does the deprotonation of the azolium salt with NaH, it facilitates the use of bifunctional chelating ligands. Thus, e.g., a hemilabile pyridyl-functionalized NHC **47** has been attached to a palladium(II) center by this route.[108] In certain cases it does not seem to be necessary to use a basic anion to deprotonate benzimidazolium salts: $K_2[PtCl_4]$ and $Na_2[PtCl_6]$ are reported to perform the deprotonation as well.[109]

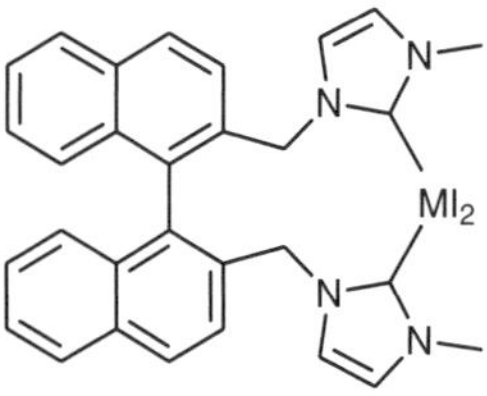

M = Pd, Ni

44

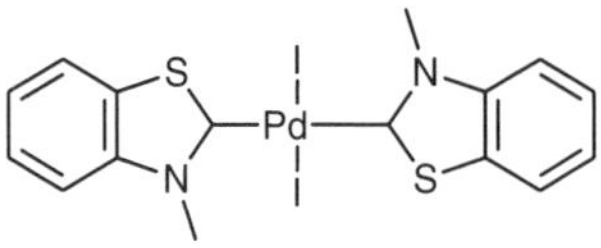

45

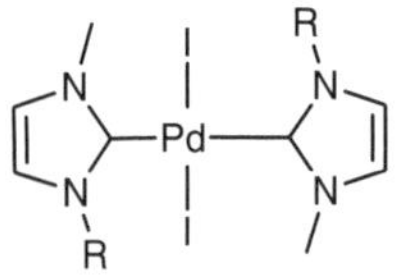

R = $CH_2CH_2C_6F_{13}$

46

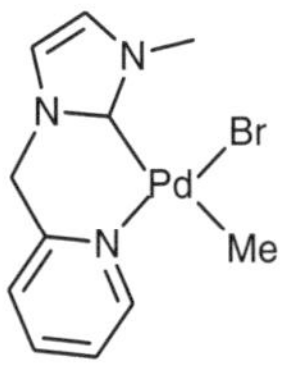

47

For rhodium(I) and iridium(I) compounds alkoxo ligands take over the role of the basic anion. Using μ-alkoxo complexes of (η^4-cod)rhodium(I) and iridium(I)—formed *in situ* by adding the μ-chloro bridged analogues to a solution of sodium alkoxide in the corresponding alcohol and azolium salts—leads to the desired NHC complexes even at room temperature [Eq. (10)].[19,110] Using imidazolium ethoxylates with $[(\eta^4\text{-cod})RhCl]_2$ provides an alternative way to the same complexes.[110] By this method, it is also possible to prepare benzimidazolin-2-ylidene complexes of rhodium(I).[110,111] Furthermore, an extension to triazolium and tetrazolium salts was shown to be possible.[112]

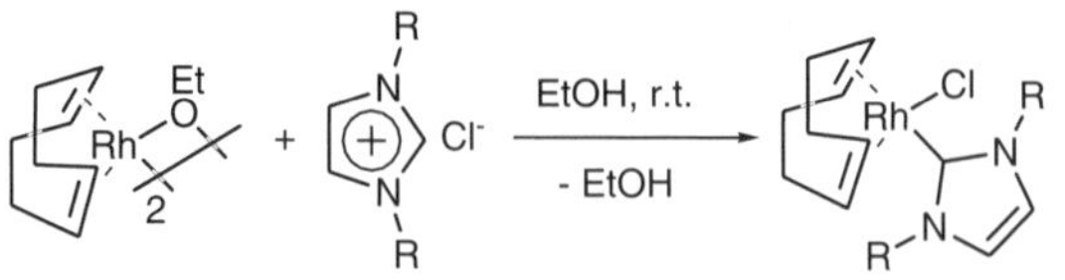

EtOH, r.t. / - EtOH (10)

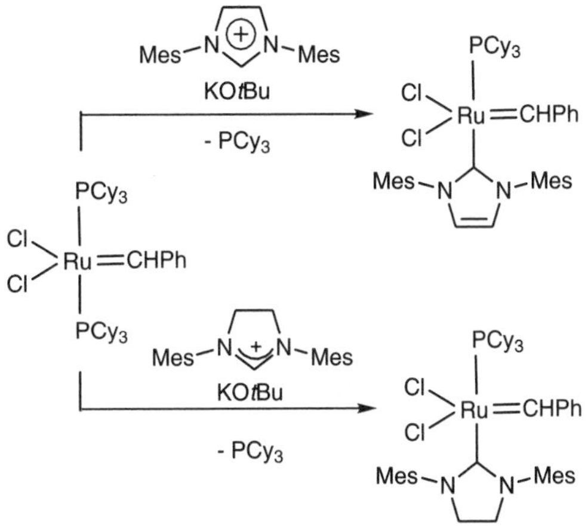

SCHEME 6. Synthesis of ruthenium–alkylidene complexes starting at the azolium salt without isolating the NHCs.

In situ deprotonation combined with a substitution of a phosphine ligand was reported as a convenient way for the synthesis of ruthenium–alkylidene complexes (Scheme 6).[46,113,114] For imidazolidin-2-ylidenes, this is the only way known to generate these complexes; for the imidazolin-2-ylidenes, it represents an alternative to phosphine exchange by the free NHC (*vide infra*). With the ruthenium(II) complex $[(\eta^5\text{-}C_5Me_5)Ru(OCH_3)]_2$ it is possible to react imidazolium salts under dimer cleavage and to isolate the stable 16-electron complex $[(\eta^5\text{-}C_5Me_5)Ru(NHC)Cl]$.[115]

The use of μ-hydroxo or μ-alkoxo bridged polynuclear complexes of chromium, molybdenum, tungsten, or rhenium in this route leads to the formation of monomeric bis(NHC) complexes, to the elimination of hydrogen, and to the partial oxidation of the metal [Eq.(11)].[116–118] Chelating and nonchelating imidazolium salts as well as benzimidazolium and tetrazolium salts can be used.

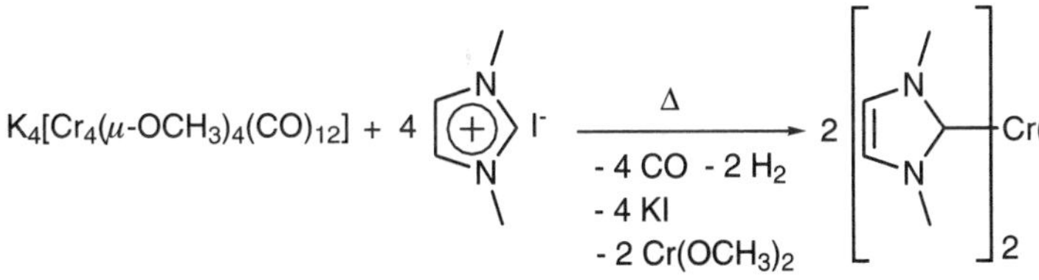

$$K_4[Cr_4(\mu\text{-}OCH_3)_4(CO)_{12}] + 4\ [\text{imidazolium}]^+ I^- \xrightarrow[\substack{-4\,CO\ -2\,H_2 \\ -4\,KI \\ -2\,Cr(OCH_3)_2}]{\Delta} 2\ [(\text{NHC})]_2Cr(CO)_4 \qquad (11)$$

Basic silver(I) oxide Ag_2O is a convenient precursor to silver(I) bis(NHC) complexes such as **48.**[40,119] The preparation proceeds even at room temperature. The cationic complex precipitates and is therefore easy to purify.

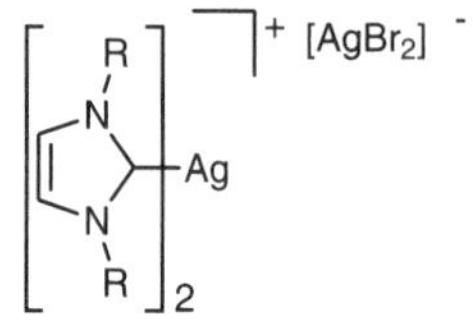

48

This complex represents a useful NHC transfer agent since it can be used as the NHC source for the preparation of NHC complexes of other metals (*vide infra*). Similar to Ag_2O, mercury(II) oxide HgO can be used to form Hg–NHC complexes in an ethanol/water mixture.[120]

Silver acetate provided access to a new class of compounds: ionic organometallic polymers based on 1,2,4-triazolin-3,5-diylidenes [Eq. (12)].[121,122] Using just 1 equiv of Ag(OAc) leads to the generation of a "conventional" bis(NHC)–Ag complex.

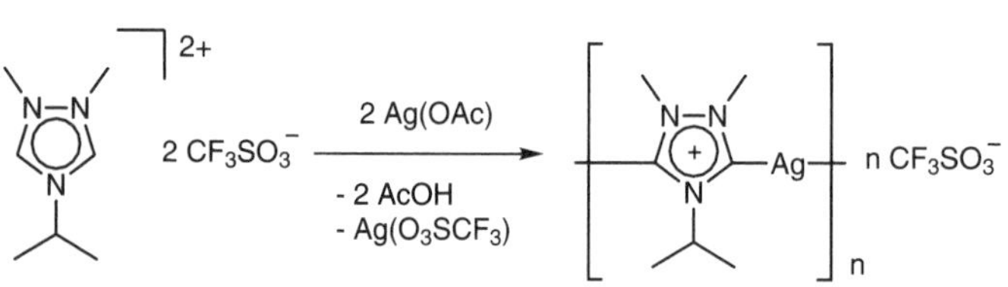

2 $CF_3SO_3^-$ —— 2 Ag(OAc), - 2 AcOH, - $Ag(O_3SCF_3)$ ——→ n $CF_3SO_3^-$ (12)

Loosely bound η^5-cyclopentadienyl anions can also serve as the base to deprotonate imidazolium salts. When chromocene is reacted with an imidazolium chloride in THF the metal precursor loses one molecule of cyclopentadiene to form the 14-electron complex [$(\eta^5$-$C_5H_5)Cr(NHC)Cl$] [Eq. (13)].[123] This complex can be further oxidized by $CHCl_3$ to give [$(\eta^5$-$C_5H_5)Cr(NHC)Cl_2$]. This route also works with nickelocene to generate the corresponding [$(\eta^5$-$C_5H_5)Ni(NHC)Cl$] complex.[124]

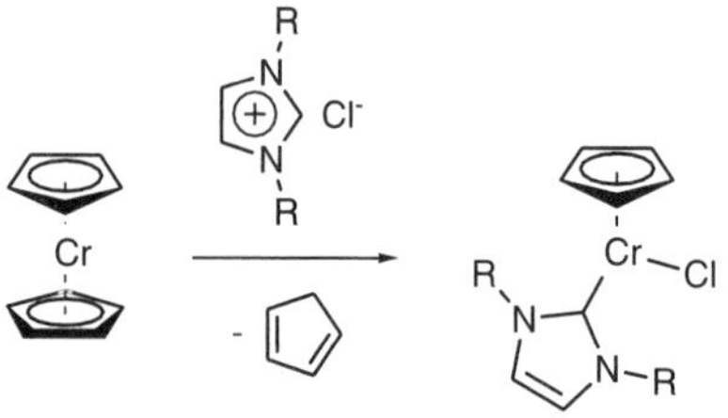

(13)

c. *Deprotonation by an external base.* The addition of an external base for the *in situ* deprotonation of the azolium salts can lead to products that differ from those obtained using basic anions. For example, potassium *tert*-butoxylate with an imidazolium perchlorate and 1 equiv of palladium(II) diacetate in the presence of sodium iodide form a dimeric mono(NHC) complex [Eq. (14)], whereas the reaction of palladium(II) diacetate alone results in the formation of monomeric bis(NHC) complexes (*vide supra*).[125] This method can also be used with triazolium salts.[125] The dimeric mono(NHC) complexes are valuable precursors for the introduction of other ligands by dimer cleavage, e.g., phosphines or different NHCs.[38,126,127] It is also possible to deprotonate imidazolium salts in the presence of $[(\eta^4\text{-cod})RhCl]_2$ with lithium *tert*-butoxylate in THF at room temperature.[38] Potassium *tert*-butoxylate and sodium hydride in THF can be used to coordinate NHCs to $Cr(CO)_6$ and to $W(CO)_6$ *in situ*.[128]

$Pd(OAc)_2$ + imidazolium ClO_4^- —(KO*t*Bu, NaI)→ (14)

Imidazolium salts can also be deprotonated with a phosphazene base at 0°C in THF and trapped, e.g., by $[(\eta^4\text{-cod})IrCl]_2$ [Eq. (15)].[129] The low temperature allows more sensitive imidazolium salts to be used, e.g., salts based on the antifungal drugs econazole or miconazole.

—($^1/_2$ $[Ir(cod)Cl_2]_2$, phosphazene base)→ Ir(cod)Cl (15)

By using phase transfer catalysts it is possible to generate NHC complexes with a diluted, aqueous sodium hydroxide solution. For example, benzimidazolium bromides and $[(Me_2S)AuCl]$ react in the biphasic system CH_2Cl_2/H_2O at room temperature in this way to give $[(NHC)_2Au]Br$ [Eq. (16)].[130] Analogously, $HgCl_2$ and benzimidazolium salts form NHC complexes via this route.[131]

$[Au(SMe_2)Cl]$ + benzimidazolium —(NaOH, $[NR_4]Br$, CH_2Cl_2 / H_2O, LiBr; $- SMe_2$)→ $[(NHC)_2Au]^+ Br^-$ (16)

$R = C_{12}H_{25}, C_{14}H_{29}, C_{16}H_{33}$

Triethylamine in THF can be used as the external base to deprotonate triazolium salts. The resulting NHCs were complexed *in situ*, e.g., to $[(\eta^6\text{-cymene})RuCl_2]_2$, $[(\eta^4\text{-cod})RhCl]_2$, and $[(\eta^5\text{-}C_5Me_5)RhCl_2]_2$.[132,133] Sodium carbonate in water/DMSO deprotonates imidazolium iodides in the presence of mercury(II) dichloride to give $[Hg(NHC)_2][HgI_3Cl]$.[134] A pyridine-functionalized imidazolium salt was deprotonated by lithium diisopropylamide (LDA) in THF and attached *in situ* to $[(\eta^4\text{-cod})Pd(Me)Br]$ [Eq.(17)].[135] After abstraction of the bromide anion with silver(I) a tetranuclear ring is formed.

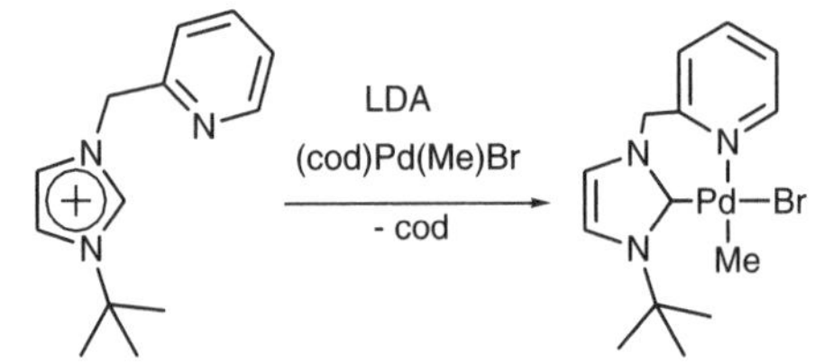

(17)

Addition of butyl lithium to a suspension of palladium(II) diiodide and methylene bridged bisimidazolium salts leads to the *in situ* formation and complexation of the NHC resulting in the cationic $[(\text{chelate})_2Pd]I_2$ in low yield.[136] Higher yields are obtained by deprotonation with palladium(II) acetate in DMSO (*vide supra*). By deprotonation with butyl lithium in THF it is also possible to prepare an NHC ligand analogous of *Trofimenko's* tris(pyrazolyl)borate.[137] Reaction with iron(II) chloride leads to the formation of a homoleptic hexa(NHC)iron(III) complex **49.** The same methodology works for benzimidazolium salts.[86]

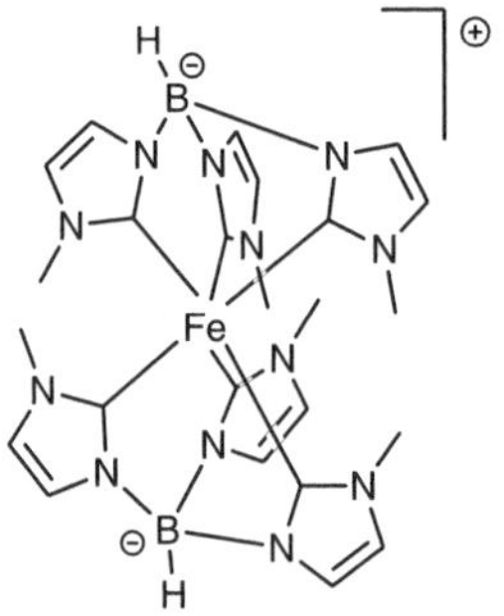

49

2. *Elimination of Small Molecules from Neutral Ligand Precursors*

The elimination of an alcohol from a neutral 2-alkoxy-1,2-dihydro-1*H*-imidazole leads to the formation of NHCs [Eq. (18)]. Upon heating, the elimination of alcohol forms the NHC, which in the case of imidazolin-2-ylidenes dimerizes to the corresponding tetraaminoethylene. This method was already used in 1961 to prepare imidazolin-2-ylidenes without, however, isolating them.[16]

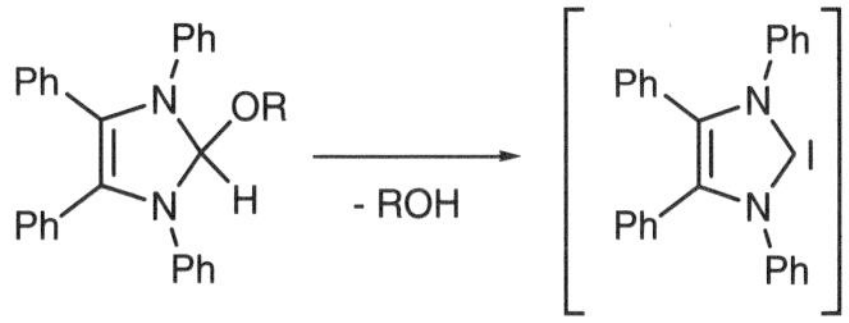

(18)

Imidazolidinium salts can also be transformed into the corresponding diamino *ortho*-esters by alkaline alkoxylate,[138] and upon alcohol elimination at elevated temperature the imidazolidin-2-ylidenes can be trapped.[46] The reaction of triazolium salts with sodium methanolate in methanol yields 5-methoxy-4,5-dihydro-1*H*-triazole which also eliminates methanol upon heating *in vacuo*. The resulting triazolin-5-ylidenes can either be isolated or trapped by an appropriate metal precursor [Eq. (19)].[138,139] Benzimidazolin-2-ylidenes are similarly accessible by this route.[138]

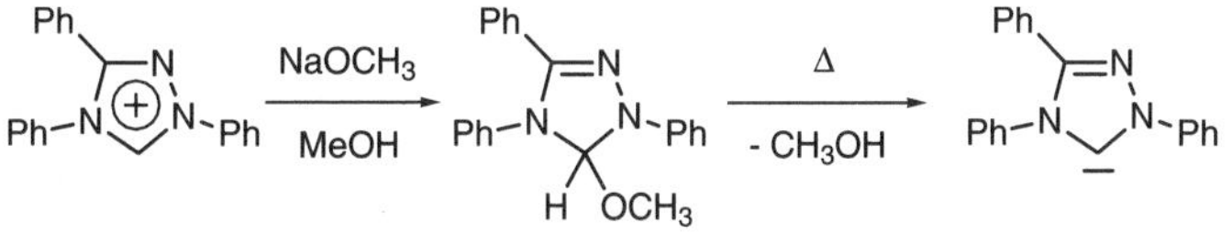

(19)

In a variation of this method, a dimethylamine adduct can be used in the same way as the methanol adduct described previously [Eq. (20)]. Nickel(II) and palladium(II) complexes with allyl-substituted NHCs are accessible by this route. These compounds cannot be prepared by the cleavage of an electron-rich olefin (*vide infra*) because of an amino *Claisen* rearrangement of the tetramino-substituted olefin.[140] However, $[(NHC)M(CO)_4]$ (M = Cr or Mo) were accessible via cleavage of electron-rich olefins with $[M(CO)_6]$ as the precursors but for the very same NHC.[141,142]

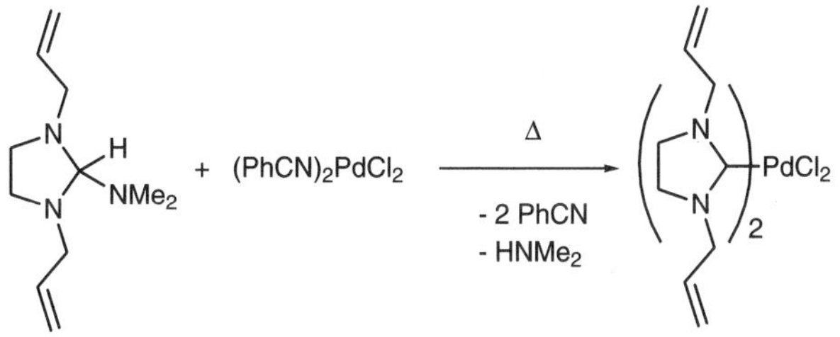

(20)

It is also possible to eliminate chloroform from trichloromethyl-substituted heterocycles. For example, *N,N'*-diphenyl-1,2-diaminoethane reacts with chloral to form 1,3-diphenyl-2-(trichloromethyl)imidazolidin which loses one molecule of chloroform upon heating [Eq. (21)]. The 1,3-diphenylimidazolidin-2-ylidene dimerizes spontaneously to form the corresponding electron-rich tetraaminoethylene.[14,16,143,144]

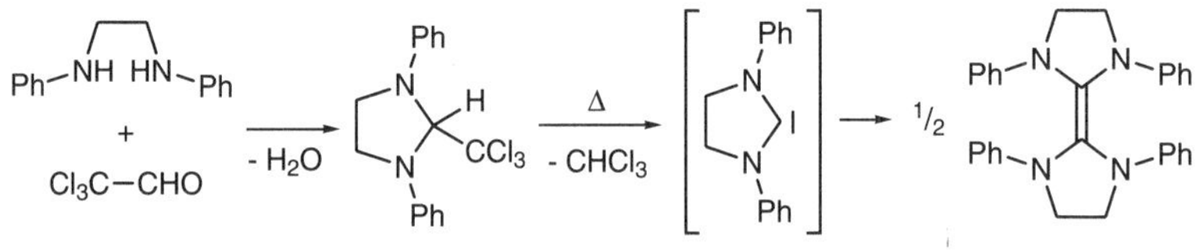

(21)

3. *Complexation of the Preformed, Free N-heterocyclic Carbenes*

Since the isolation of NHCs by *Arduengo* the direct application of these compounds has attracted much attention in complex synthesis.[7,9,10,145,146] The use of isolated NHCs has the advantage that a large variety of metal precursors without special requirements regarding the ligand sphere and the oxidation state can be used for the preparation of NHC complexes. Various methods have been developed to prepare the NHCs from suitable precursors. Azolium salts can be deprotonated by NaH and KO*t*Bu or dimsyl-anions ($DMSO^-$) in THF [Eq. (1)].[7,145] The generation of NHCs by NaH in a mixture of liquid ammonia and THF proves to be even higher yielding and applicable in a more general way [Eq. (22)].[19,20] In the case of *N,N′*-methylene bridged bisimidazolium salts the preparation of the free dicarbene is only possible by the use of potassium hexamethyldisilazide (KHMDS) in toluene.[100] With other methods, deprotonation occurs also at the methylene bridge.[93] Picolyl-functionalized NHCs can be prepared by deprotonation of the corresponding azolium salt precursor with LDA at low temperature.[135] These carbenes have not been isolated, but trapped as palladium(II) complexes.

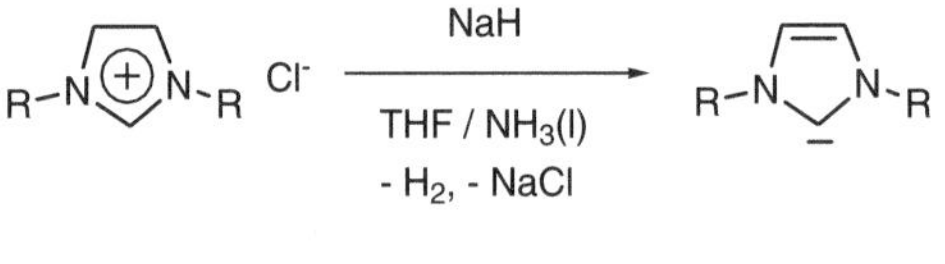

(22)

Cyclic thiourea derivatives like 1,3,4,5-tetramethylimidazole-2(3*H*)-thione—prepared by condensation of substituted thioureas with α-hydroxyketones—can be converted into the corresponding imidazolin-2-ylidene by desulfurization with sodium or potassium [Eq. (23)].[147] This method was used to prepare and isolate 1,3-bis-*neo*-pentylbenzimidazolin-2-ylidene with Na/K.[148] With LDA as the base it is also possible to generate free benzimidazolin-2-ylidenes in solution.[44]

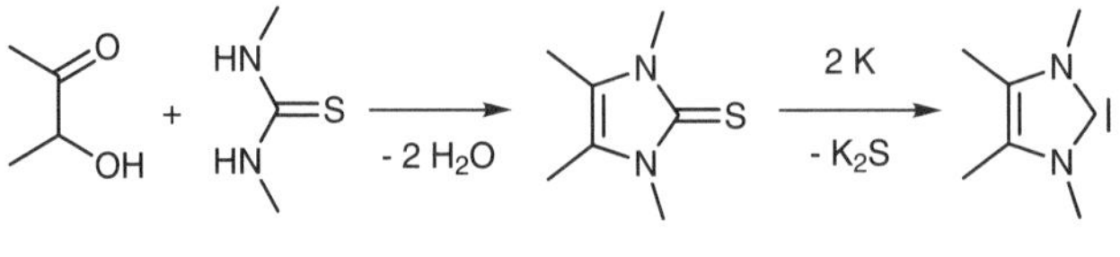

(23)

As mentioned earlier, triazolium salts can be converted into 5-methoxy-4,5-dihydro-1*H*-triazoles by reacting them with sodium methanolate in methanol. The heterocycles eliminate methanol upon heating *in vacuo* [Eq. (21)][138] and the formed triazolin-5-ylidenes can then be isolated.[145] The same method works with imidazolium and benzimidazolium salts.[46,138]

a. *Cleavage of dimeric complexes.* Nucleophilic NHCs can cleave dimeric complexes with bridging ligands like halides, carbon monoxide, or acetonitrile. Examples for this type of complex formation are the reactions of $[(\eta^4\text{-cod})MCl]_2$ or $[(\eta^5\text{-}C_5Me_5)MCl_2]_2$ (M = Rh, Ir) with free NHCs [Eq. (24)].[19–21,149,150] By using less sterically demanding NHCs it is also possible to incorporate two NHC ligands into a then cationic rhodium complex.[19] Cleaving $[Rh(CO)_2Cl]_2$ with NHCs leads to the formation of a bis-ligated complex $(NHC)_2Rh(CO)Cl$.[19] For the cleavage of $[Rh(coe)_2Cl]_2$ with 2 equiv of 1,3-dimesitylimidazolin-2-ylidene, an intramolecular C–H activation is reported to form an *ortho*-metalated NHC complex.[151] A similar phenomenon has been observed for iridium–NHC complexes.[150] Dimer cleavage and incorporation of just one NHC occur also with $[(\eta^6\text{-cymene})RuCl_2]_2$[19,20,152] and $[Os(CO)_3Cl_2]_2$.[19] Higher nuclear clusters can also be cleaved, e.g., $[(\eta^5\text{-}C_5Me_5)RuCl]_4$ is monomerized to $[(\eta^5\text{-}C_5Me_5)Ru(NHC)Cl]$ by free NHC.[115,153]

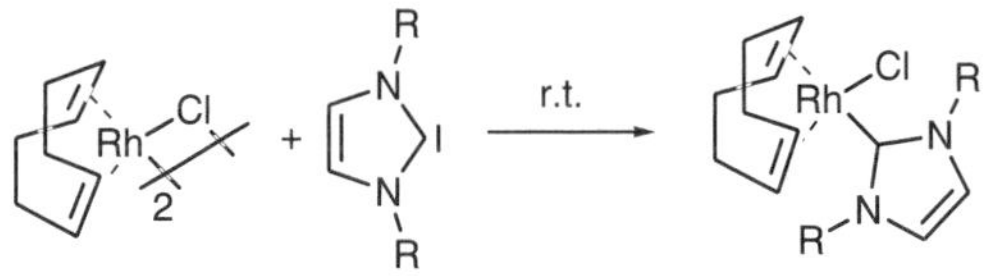

(24)

It is possible to break up bridging chlorides in $TiCl_4$ with NHCs leading to $[(NHC)TiCl_4]$.[154] The coordination of just one NHC is complementary to the reaction of the solvent adduct $[(thf)_2TiCl_4]$ with two NHC ligands (*vide infra*). Analogously, the mono(NHC) complexes are obtained from $[Y\{N(SiMe_3)_2\}_3]$ and $[La\{N(SiMe_3)_2\}_3]$ in hexane.[155] Iron(II) dihalides form $[(NHC)_2FeX_2]$ complexes in toluene upon treatment with the free NHC at 80°C.[156] $[(tmp)_2Yb]_n$ was cleaved by one NHC (tmp = tetramethylphospholyl) [Eq. (25)].[157]

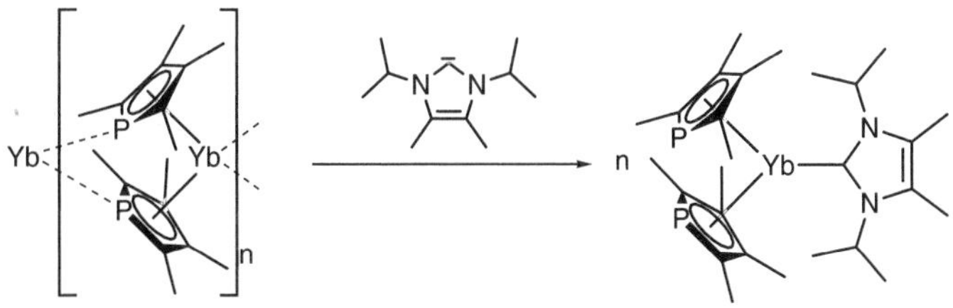

(25)

b. *Exchange of phosphine ligands.* Phosphines and other labile ligands (*vide infra*) can be exchanged for NHCs. As most phosphines are easily exchangeable even below room temperature this method represents an important means of NHC complex preparation. In certain cases it has been found that a sequential exchange of phosphines can lead to the clean formation of mixed phosphine/NHC complexes. On the olefin metathesis catalyst $[RuCl_2(PR_3)_2(=CHPh)]$ both phosphines can be exchanged for various NHCs without affecting the benzylidene moiety [Eq. (26)].[158] Using more bulky NHCs leads to the exchange of only one of the phosphines resulting in a mixed phosphine/NHC complex.[153,159–161] The exchange reaction proceeds as well with $[(\eta^5\text{-}C_5Me_5)Ru(PCy_3)Cl]$.[153] Exchanging triphenylphosphine on $[(Ph_3P)_3RuCl_2]$ with an excess of NHC results in the formation of $[(NHC)_4RuCl_2]$.[162]

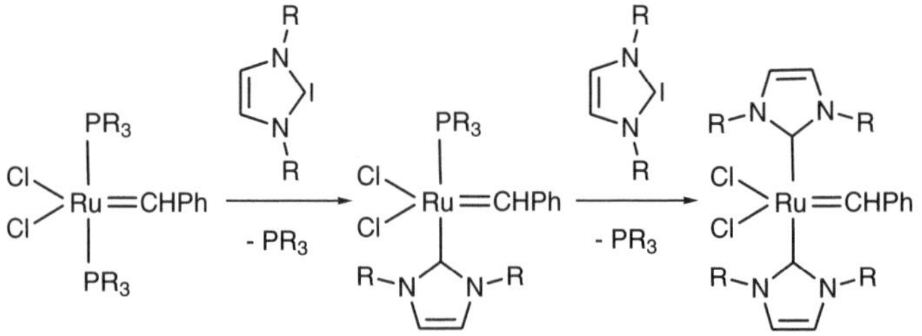

(26)

The ligand exchange procedure allows the formation of palladium(0) complexes of various NHCs. Starting from bis(tri-*ortho*-tolylphosphine)palladium(0) quantitative ligand exchange provides the formation of bis(NHC)palladium(0) complexes [Eq. (27)].[163] Again, by using sterically more demanding NHCs like 1,3-diadamantylimidazolin-2-ylidene the exchange of just one of the phosphines occurs.[164] Tri-*ortho*-tolylphosphine can be exchanged in a clean reaction, but not triphenylphosphine or tricyclohexylphosphine.

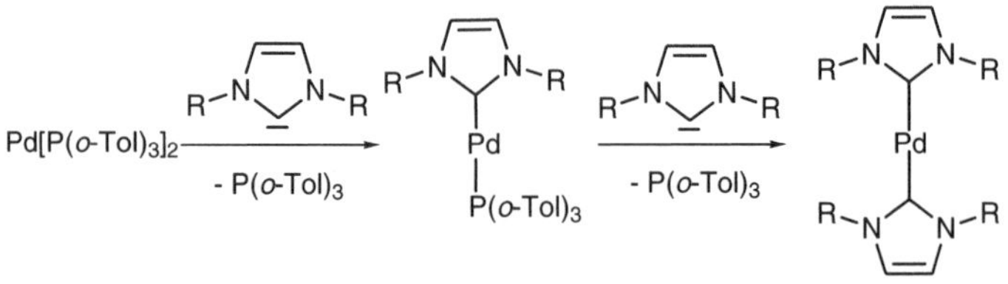

(27)

Nickel complexes can also be prepared by phosphine exchange. The triphenylphosphines in $[(Ph_3P)_2NiCl_2]$ can be completely exchanged for NHC ligands.[91] Substitution of trimethylphosphine in $[(Me_3P)_2NiCl_2]$ is also possible.[100] The reaction with chelating NHCs yields either the monocationic $[(Me_3P)Ni(chelate)Cl]Cl$ or the dicationic $[Ni(chelate)_2]Cl_2$ depending on the reaction conditions.

c. *Exchange of other ligands.* In carbonyl complexes like $Cr(CO)_6$, $Mo(CO)_6$, $W(CO)_6$, $Fe(CO)_5$, or $Ni(CO)_4$ one or two carbon monoxide molecules can be thermally substituted by NHC ligands [Eq. (28)].[20,21,128,165,166] Further substitution requires photolysis conditions.[167]

$$\mathrm{Fe(CO)_5} + \text{NHC (imidazolin-2-ylidene, N-R)} \xrightarrow{-\,\mathrm{CO}} \text{(NHC)}\mathrm{-Fe(CO)_4} \tag{28}$$

Exchange of coordinated solvent molecules like THF in $[(\eta^5\text{-}C_5Me_5)_2M(thf)]$ ($M = Sm$, Yb) by free NHC leads to the mono imidazolin-2-ylidene complexes.[168–171] The same exchange methodology works with $[(thf)_{3.25}ErCl_3]$ to coordinate three NHCs and with $[(thf)_2Y\{N(SiHMe_2)_2\}_3]$ resulting in different degrees of exchange depending on the stoichiometry of the reaction.[155] In the latter complex a β-agostic interaction of the ytterbium metal center with the Si–H bond is observed.

$[(thf)_2NiCl_2]$ is a valuable precursor to the bis(NHC) complexes of nickel(II).[91] The bis(thf)tetrachloro complexes of titanium, zirconium, hafnium, niobium, and tantalum allow the exchange of both solvent molecules for NHC ligands.[172] In $[(thf)W(CO)_5]$ the solvent molecule can be exchanged for an NHC ligand selectively.[44] The same selectivity is observed in $[(\eta^5\text{-}C_5H_5)Cr(thf)Cl_2]$ and $[(\eta^5\text{-}C_5Me_5)Cr(thf)Cl_2]$.[173] In $[Cr(mes)_2(thf)_3]$ just two solvent molecules are displaced by NHCs because of the interference with the sterically demanding mesityl ligands.[174] All three acetonitrile molecules can be exchanged by NHCs in $[(CH_3CN)_3M(CO)_3]$ ($M = Cr$, Mo, W) to give *fac*-$[(NHC)_3M(CO)_3]$.[128] Metals in high oxidation states can also be subjected to the replacement of solvent molecules. For example, $[(thf)_2MO_2Cl_2]$ with $M = Cr$ or Mo have been shown to exchange THF molecules for 1,3-dimethylimidazolin-2-ylidenes.[175] Reacting silver(I) and copper(I) triflates in THF with free NHCs can be regarded as well as an exchange of solvent molecules but without the preisolation of the THF complexes.[176] The same holds true for the reactions of $ZnEt_2$, $Zn(\eta^5\text{-}C_5Me_5)_2$, $CdMe_2$,[59,61,177] and for the bridged cyclopentadienyl complex $[(Cp')_2Sm(allyl)]$ with free NHC in THF.[171] A similar reactivity is observed for $HgCl_2$.[178]

Amines have also been exchanged by NHC, e.g., TMEDA can be replaced by two NHCs in $[(tmeda)_2VCl_2]$,[128] and pyridine by one NHC in complexes of chromium, molybdenum, and tungsten.[172]

Olefins like 1,5-cyclooctadiene can be subjected to ligand exchange if no other ligands can be replaced (*vide supra*). Bis(NHC) complexes of nickel(0) and platinum(0) have been prepared from $Ni(\eta^4\text{-cod})_2$ and $Pt(\eta^4\text{-cod})_2$, respectively.[127,179] No side reaction of the free NHC with the olefin such as cyclopropanation has been observed. The corresponding palladium(0) complexes cannot be prepared by this route but can be obtained by exchange of phosphine ligands [Eq. (27)].[163] The exchange of η^4-1,5-cyclooctadiene ligands on palladium(0) is possible if an

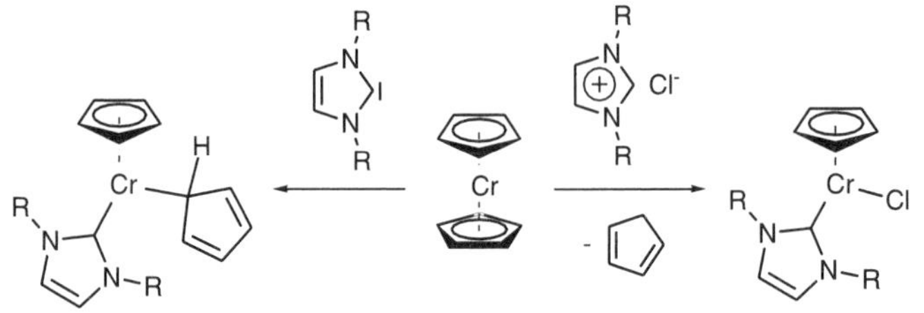

SCHEME 7. Reaction of chromocene with imidazolin-2-ylidenes and imidazolium salts.

electron-deficient alkene like tetracyanoethylene or maleic anhydride is present.[180] But even this reaction does not lead to homoleptic palladium(0)–NHC complexes because the acceptor olefin is not replaced. Starting at $[(\eta^4\text{-cod})Pd(CH_3)Cl]$ the olefin can be exchanged by one or two NHCs to form $[(NHC)Pd(CH_3)Cl]_2$ or $[(NHC)_2Pd(CH_3)Cl]$, respectively.[181,182] Attempts to exchange dibenzylideneacetone ligands from $Pd_2(dba)_3$ by free NHCs have failed in the isolation of defined palladium(0) complexes although catalytically active species are formed (*vide infra*).[24,181] Both cyclooctene ligands can be exchanged on $[(coe)_2RhCl]_2$ by two NHCs even if the bulky 1,3-dimesitylimidazolin-2-ylidene is used.[151]

In certain cases it is also possible to replace anionic ligands by neutral NHC ligands. Nickelocene and chromocene are capable of coordinating one imidazolin-2-ylidene ligand by shifting one of the cyclopentadienyl anions from η^5- to η^1-coordination (Scheme 7).[183] Under certain conditions it is possible to proceed in this reaction with another equivalent of NHC to remove a cyclopentadienyl ligand completely to form, e.g., the cationic $[(\eta^5\text{-}C_5H_5)Ni(NHC)_2][C_5H_5]$ complex with a noncoordinating cyclopentadiene anion.[183] This complements the deprotonation of imidazolium salts by cyclopentadienyl anions (*vide supra*).[123,124]

4. *Cleavage of Electron-Rich Olefins*

Electron-rich olefins are nucleophilic and therefore subject to thermal cleavage by various electrophilic transition metal complexes. As the formation of tetraaminoethylenes, i.e., enetetramines, is possible by different methods, various precursors to imidazolidin-2-ylidene complexes are readily available.[184] Dimerization of nonstable NHCs such as imidazolidin-2-ylidenes is one of the routes used to obtain these electron-rich olefins [Eq. (29)].[144] The existence of an equilibrium between free NHC monomers and the olefinic dimer was proven only recently for benzimidazolin-2-ylidenes.[17,185,186] In addition to the previously mentioned methods it is possible to deprotonate imidazolidinium salts with *Grignard* reagents in order to prepare tetraaminoethylenes.[143] The isolation of stable imidazolidin-2-ylidenes was achieved by deprotonation of the imidazolidinium salt with potassium hydride in THF.[26]

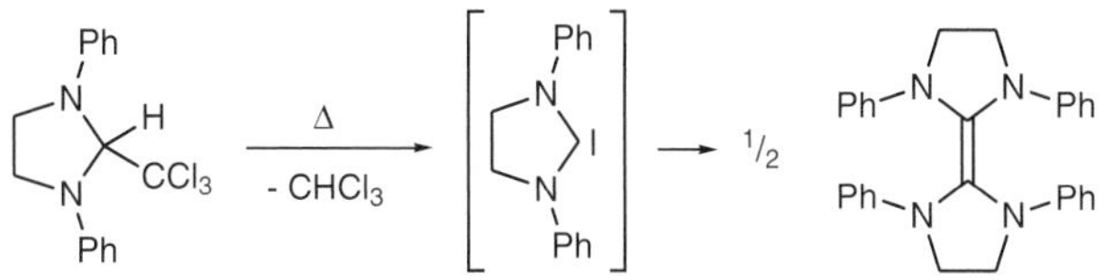

(29)

Heating tetraaminoethylenes in refluxing toluene in the presence of metal precursors yields the corresponding NHC complexes.[187–189] Metal carbonyls of manganese, chromium, iron, ruthenium, osmium, cobalt, or nickel are the most common precursors in this reaction.[187,190–197] Generally one or two carbon monoxide molecules are replaced by imidazolidin-2-ylidene ligands [Eq. (30)]. Higher substitution is nevertheless possible in certain cases.[198] For $Ru_3(CO)_{12}$ the exchange of one of the carbonyl ligands occurs without disruption of the cluster structure.[190] The selective substitution of solvent molecules in the presence of carbonyls is also possible, e.g., in $[(CH_3CN)M(CO)_5]$ with M = Cr, Mo, W.[199] The exchange of phosphine ligands works, e.g., in the Wilkinson catalyst $[(Ph_3P)_3RhCl]$, in $[(Ph_3P)_3RuCl_2]$ or $[(Et_3P)_2PtCl_2]$[198,200–205] and dimeric complexes can be cleaved as well, e.g., $[(\eta^4\text{-cod})RhCl]_2$[198,203,206–209] Dimeric cyclometallated palladium(II) complexes can be cleaved thermally by enetetramines resulting in the formation of monomeric NHC complexes.[210]

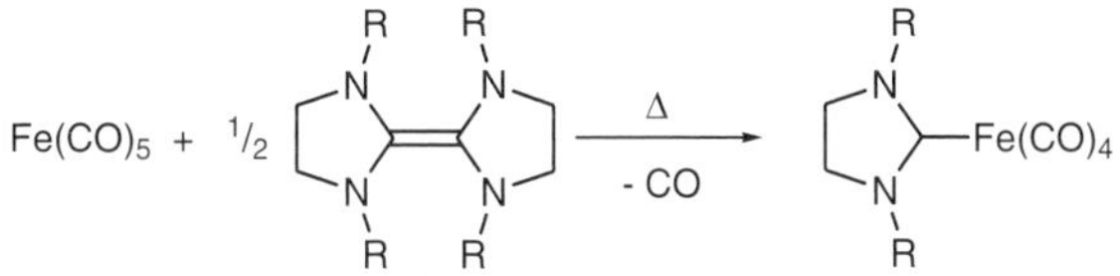

(30)

A special type of reaction is observed with the platinum(IV) complex $[PtI(Me)_3]$ which cleaves the *N,N,N′,N′*-tetraphenyltetraaminoethylene under reduction to form the dimeric cyclometallated mono(NHC) complex of platinum(II) iodide [Eq. (31)]. Cyclometallation with the same ligand is also observed for ruthenium.[211,212] Additional cyclometallations with various substituents of NHCs have been reported for ruthenium(II),[132,213] rhodium(III),[132] iridium(I),[150,208] palladium(II),[214–216] and platinum(II).[217] In the case of iridium, alkyl groups can be activated twice.[150] In rare cases like for nickel(II) μ-bridging NHCs have been obtained.[190]

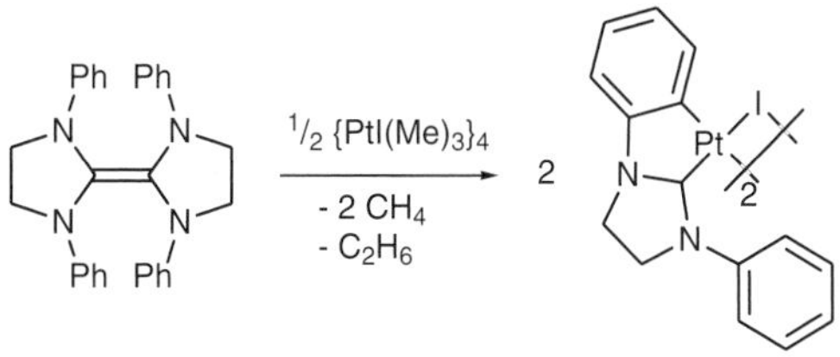

(31)

Benzimidazolin-2-ylidene complexes are conveniently prepared from the corresponding enetetramines. In refluxing toluene the olefin is attacked, e.g., by $[(\eta^4\text{-cod})RhCl]_2$ to form the mono(NHC) or the bis(NHC) complex depending on the reaction conditions.[218]

6. *Other Methods*

In addition to the previously mentioned most common methods some less frequently used reactions have also led to NHC complexes. In all these cases certain requirements for the metal, the complex precursor or the NHC itself, have limited the approach and have prevented broader applications.

a. *Vapor phase synthesis.* In the case of sublimable NHCs the vapor deposition method can lead to the formation of desired complexes. In the case of 1,3-di-*tert.*-butylimidazolin-2-ylidene and group 10 metals this method was successfully applied in the preparation of homoleptic bis(1,3-di-*tert.*-butylimidazolin-2-ylidene)metal(0) complexes of nickel, palladium, and platinum.[219]

b. *The metalla-Ugi reaction.* The isolobal analogy of oxygen and a d^6-transition metal complex fragment ML_5 is the idea behind the application of the *Ugi*-4-component condensation for hydantoins in transition metal complex synthesis:[220] Anionic cyano complexes are reacted with an aldehyde, an isocyanide, and an ammonium salt to form the desired NHC complexes [Eq. (32)] This principle has been demonstrated for various metals, e.g., chromium and tungsten.[221] It is also possible to start the reaction sequence from a defined isocyanide complex, e.g., with platinum or gold as the metal.[54,222] The α-H atom of the isocyanide is attacked by an amine base; subsequent cycloaddition of a dipolarophile to the ylid-structure gives the desired imidazolin-2-ylidene complex. If the isocyanide ligand is replaced by an *N*-isocyanimine ligand this reaction sequence leads to triazolin-5-ylidene complexes.[223]

$$\begin{array}{c} t\text{Bu–N}\equiv\text{C} \\ \text{PhCHO} \quad + \quad [(\text{NC})\text{Cr}(\text{CO})_5][\text{NEt}_4] \\ [\text{PhNH}_3]\,\text{Cl} \end{array} \xrightarrow[\substack{-\,\text{H}_2\text{O} \\ -\,[\text{NEt}_4]\text{Cl}}]{} \text{imidazolin-2-ylidene–Cr(CO)}_5 \text{ (4-NH}t\text{Bu, 5-Ph, 1-H, 3-Ph)} \qquad (32)$$

In a three-component reaction, a cationic platinum isocyanide complex $[(Ph_3P)_2Pt(CNR)Cl][BF_4]$ is reacted with a β-bromoamine and butyl lithium to give an imidazoldin-2-ylidene complex.[224] This transformation can be a two-component reaction if the isocyanide ligand contains already the necessary amine functionality. This was shown for chromium, molybdenum, tungsten, and rhenium carbonyls.[225]

c. *Ligand transfer reactions.* NHC ligands can be transferred via an intermolecular way from one metal center to another. The first examples of this reaction

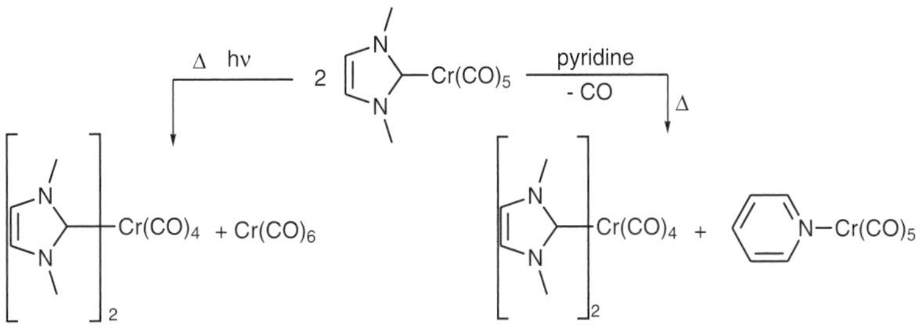

SCHEME 8. Disproportionation of [(NHC)Cr(CO)$_5$].

were found in a disproportionation reaction of [(NHC)Cr(CO)$_5$]. The complex was heated under photolysis conditions to form [(NHC)$_2$Cr(CO)$_4$] and Cr(CO)$_6$ (Scheme 8).[81] The same reaction proceeds for analogous molybdenum and tungsten compounds.[226] Thermally induced disproportionation of [(NHC)Cr(CO)$_5$] complexes occurs only in the presence of other donor ligands like pyridine or tricyclohexylphosphine (Scheme 8).[227] Again, further substitution requires photolysis conditions.[81,128,227]

NHCs have been successfully transferred from chromium, molybdenum, and tungsten carbonyls to rhodium(I), palladium(II), platinum(II), copper(I), silver(I), and gold(I).[228,229] Reacting [(NHC)W(CO)$_5$] with [(PhCN)$_2$PdCl$_2$] results in mono- or bisligated complexes depending on the reaction conditions. With [(PhCN)$_2$PtCl$_2$] the complex [(NHC)Pt(CO)Cl$_2$] is obtained indicating that a carbonyl transfer is also possible by this method. [(Me$_2$S)AuCl] leads to the formation of the cationic [(NHC)$_2$Au]Cl and dimeric [Rh(CO)$_2$Cl]$_2$ gives [(NHC)$_2$ Rh(CO)Cl].

Cationic bis(NHC)silver(I) complexes can be used to transfer both NHC ligands to [(CH$_3$CN)$_2$PdCl$_2$] or [(Me$_2$S)AuCl] in CH$_2$Cl$_2$ at ambient temperature [Eq. (33)].[199] The silver(I)–NHC complex does not have to be isolated prior to the NHC transfer reaction which makes this method more convenient.[230] The silver(I) halide which forms during the transfer can be filtered off the product solution and can be recycled. Under phase transfer conditions this reaction can be run catalytically in silver (*vide supra*). In certain cases it has been advantageous to use a mixture of CH$_2$Cl$_2$ and ethanol as the solvent in this reaction.[231] This procedure has been extended to the preparation of a variety of palladium complexes with donor-functionalized NHCs.[232] Due to its convenience and the broad range of complexes accessible, this method is very likely to gain an important role as a synthetic route to NHC complexes.

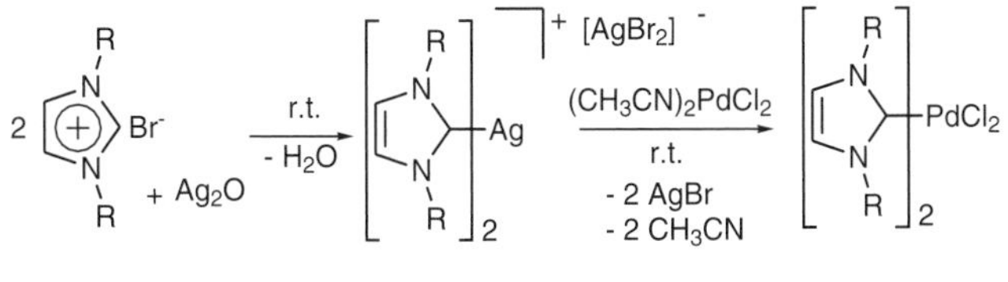

(33)

One example is known for the transfer of an NHC from a main-group element to a transition metal: The NHC adduct of BH_3 was reported to transfer the carbene to $[Mn(CO)_5Br]$.[64]

d. *Rare examples.* Stoichiometric oxidation of chromium(0) and iron(0) NHC complexes with $Ag[BF_4]$ was used to prepare paramagnetic chromium(I) and iron(I) complexes.[233,234] The electrochemical oxidation of chromium(0), molybdenum(0), and tungsten(0) NHC complexes is possible. Starting at *trans*-$[(NHC)_2M(CO)_4]$ the cationic *cis*-complexes are formed.[235,236] The oxidative addition of a suitable substrate equals a two-electron oxidation. With gold(I) complexes of the type $[(NHC)_2Au]^+$ the oxidative addition of dihalogenes such as Br_2 leads to the corresponding gold(III)–NHC complexes.[237] Reduction pathways are not established yet, e.g., palladium(II) complexes of imidazolin-2-ylidenes have failed to be reduced without affecting the integrity of the ligand environment.[238]

2-Lithioimidazoles—prepared from 1-substituted imidazoles with butyl lithium at low temperatures—can be transmetallated by, e.g., [(L)AuCl] (L = Me_2S, PPh_3 tetrahydrothiophene) and subsequently be quenched by HCl to give a 3-hydroimidazolin-2-ylidene complex [Eq. (34)].[239–241] Subsequent alkylation with methyl triflate is a way to the preparation of more stable 1,3-disubstituted imidazolin-2-ylidenes.[240,241] The same reaction sequence is applicable to the formation of copper(I), chromium(0), molybdenum(0), and tungsten(0) complexes and the use of thioazolyl, benzothiazolyl, triazolyl, and pyrazolyl ligands.[242–244] Depending on the reactants and the conditions, mono- and bis(NHC) complexes can be formed.

2 (1-R-imidazole) —BuLi→ 2 (2-Li-1-R-imidazole) —1) $(Me_3S)AuCl$; 2) HCl→ $[(NHC)_2Au]^+ Cl^-$ (34)

The reaction of thiourea derivatives with a metal complex to form NHC complexes is a combination of the NHC formation from thioureas with potassium or sodium [Eq. (23)] and the cleavage of electron rich olefins. For example, a 10-S-3-tetraazapentalene derivative is cleaved by $Pd(PPh_3)_4$ and $[(Ph_3P)_3RhCl]$, respectively [Eq. (35)].[245] Other substitution patterns in the carbene precursor, including selenium instead of sulfur can also be used.[246–248]

S=(tetraazapentalene)=S + $Pd(PPh_3)_4$ —(− S=PPh_3, − 2 PPh_3)→ PPh_3–Pd(S)(S) complex (35)

A variation of the thermal elimination of an alcohol from the neutral 2-alkoxy-1,2-dihydro-1*H*-imidazole is the preformation of a chelate *vic*-bisamine complex which is subsequently attacked by an *ortho*-ester to form the desired NHC complex. This principle has been shown with nickel and platinum [Eq. (36)].[249–251] The reverse sequence is also possible as was demonstrated by the attack of a *vic*-bisamine on a tungsten carbonyl cation $[(PhC{\equiv}CPh)_3W(CO)]^+$.[252]

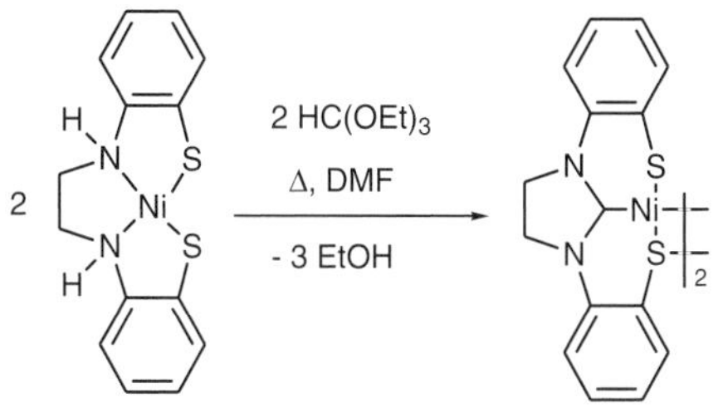

(36)

Certain transition metal complexes can serve as templates for the synthesis of chelating NHC ligands. For example, 1-phenylphosphole complexes of palladium(II) are attacked in a *Diels–Alder* reaction by 1-vinylimidazole.[253] If 1,2-dichloroethane is used as the solvent the imidazole is alkylated *in situ* and then subjected to a spontaneous carbometallation reaction [Eq. (37)].

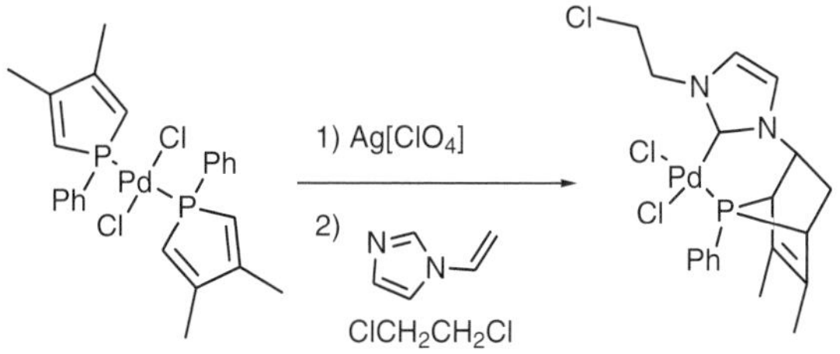

(37)

7. *Bimetallic Complexes*

NHC ligands have been used to bridge different metals resulting in homobimetallic systems. Examples exist for palladium(II),[254] rhodium(I),[19] as well as chromium(0).[234] Homobimetallic ruthenium(II) systems have been shown to be superior catalysts in cyclopropanation reactions compared to their monometallic analogues (*vide infra*).[203]

A supramolecular, helical structure was obtained from bridging bis(NHC) ligands and $Hg(OAc)_2$ in acetonitrile.[89]

Heterobimetallic systems have not been investigated extensively. An astonishing result was, however, observed when treating osmium(VIII) oxide with a chromium(0)–NHC complex: OsO_4 added stoichiometrically to the C=C double bond of the NHC instead of oxidizing chromium(0) [Eq. (38)].[255]

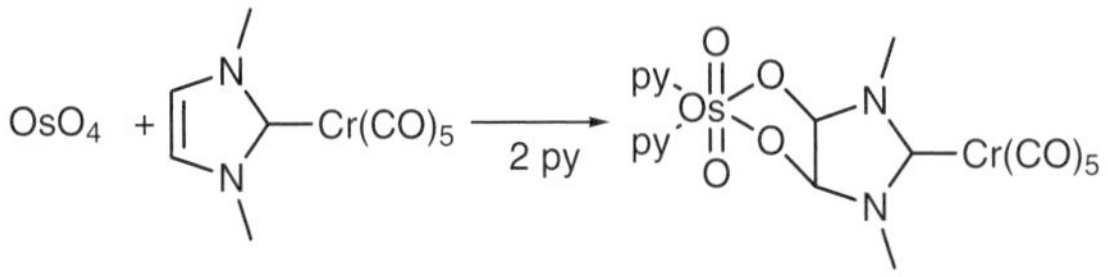

(38)

Another example for a bimetallic NHC complex is the combination of a ruthenium hydride fragment with an ytterbium NHC complex. The NHC serves partly as the "hydrogen trap" [Eq. (39)].[157]

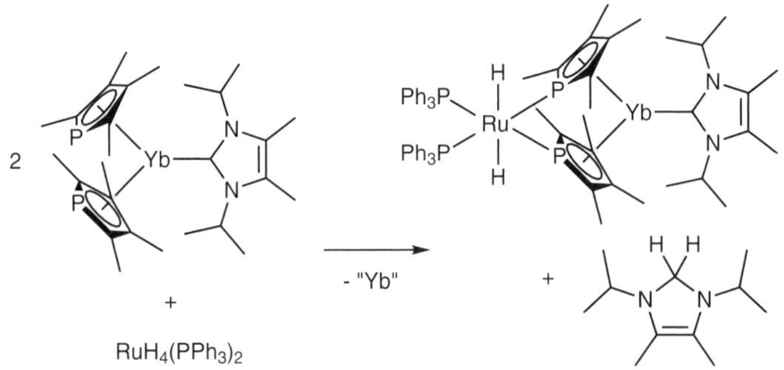

(39)

Ferrocenyl-substituted NHC can also be considered as sources for heterobimetallic complexes. They have been attached to tungsten(0), mercury(II), and palladium(II) complexes. Cyclovoltametric studies show no interaction of the metal centers with each other.

A special application of bimetallic ruthenium complexes was found in the olefin metathesis reaction (*vide infra*).[159,160] The two metal centers were closely attached to one another through μ-halide anions. The labile assembly was the key feature to the formation of highly active catalysts.

8. *Summary*

The survey on published methods for generating NHC metal complexes shows that a broad variety of different approaches exists. In general, the preparation by cleavage of dimeric metal precursors or exchange of other ligands with free NHC is the most convenient and general approach. In most examples the maximum number of NHC ligands on the metal is achieved by this method. Nevertheless, the necessity to prepare the free NHCs is a limitation.

In cases where the free NHC cannot be synthesized the complex formation has to be accomplished *in situ* from a ligand precursor, e.g., the imidazolium salt in the case of imidazolin-2-ylidenes. By this method, it is often possible to prepare complexes which do not have the maximum number of NHC ligands attached to the metal center.

Cleavage of tetraaminoethylenes has been most frequently used in the preparation of imidazolidin-2-ylidene complexes. The ligand exchange reaction, i.e., ligand transfer from one metal to another, seems to be a simple alternative for the preparation of complexes, which have not yet been accessible. In particular, silver(I)–NHC complexes are convenient transfer agents. This method is likely to become more important for NHC complex synthesis in the future.

Rather exotic methods like vapor phase, multicomponent, and template syntheses are considered to be important in cases where the other ways failed. Generally, the main purpose of their application was to prove fundamental principles.

Low and medium oxidation states like 0, +I, +II, and +III seem to be stabilized best by NHCs. Difficulties in preparing and isolating the desired compounds occur with higher and lower oxidation states, although extreme cases such as complexes of rhenium(VII) are known. In summary, almost every transition metal has been used to prepare NHC complexes by one of the routes described previously.

C. *Properties of NHC Ligands*

The properties of NHC ligands will be discussed before presenting the applications of NHC complexes, e.g., in homogeneous catalysis, since the profound understanding of these properties certainly helped in predicting their behavior in these applications. The results obtained from various analytical methods can be summarized as follows.

The character of the metal–carbon bond is described best as a strong σ-bond. The length of the metal–carbon bond is usually in the range of a typical metal–hydrocarbyl single bond and π-backbonding is considered to be negligible. In IR spectroscopy, the CO stretching frequencies in metal–carbonyl complexes represent a sensitive probe for electronic ligand properties. $[Fe(CO)_4L]$ and $[Cr(CO)_5L]$ with L representing different carbenes have been investigated (Table I).[256]

The electron density induced at the metal center increases in the order :C(OR)R < :C(NHR)R < $:C(NHR)_2$ $\approx$ imidazolidin-2-ylidene $\approx$ imidazolin-2-ylidene. According to these data the difference between diamino-substituted carbenes whether cyclic or acyclic, aromatic or nonaromatic, seems to be quite small.

TABLE I

THE CARBONYL ABSORPTION FREQUENCIES [cm^{-1}] IN $[Cr(CO)_5L]$ AND $[Fe(CO)_4L]$

Complex	L	$A_1^{(1)}$ (cm^{-1})	$A_2^{(2)}$ (cm^{-1})	E (cm^{-1})
$[Cr(CO)_5L]$	$:C(OCH_3)(C_6H_5)$	1963	2062	1952
	$:C(CH_3)(NHC_6H_5)$	1937	2057	1937
	$:C(NH_2)(C_6H_5)$	1921	2058	1944
	Imidazolidin-2-ylidene	—	2061	1928
	Imidazolin-2-ylidene	1897	2056	1925
$[Fe(CO)_4L]$	$:C(NHCH_3)_2$	1967	2046	1931, 1925
	Imidazolidin-2-ylidene	1963	2044	1929
	Imidazolin-2-ylidene	1897	2056	1938

TABLE II

THE CARBONYL ABSORPTION FREQUENCIES [cm^{-1}] IN *fac*-[$Mo(CO)_3L_3$] AND *trans*-[$RhL_2(CO)Cl$]

Complex	L	Absorption frequency (cm^{-1})
fac-[$Mo(CO)_3L_3$]	PPh_3	1835, 1934
	CH_3CN	1783, 1915
	Imidazolin-2-ylidene	1764, 1881
	Pyridine	1746, 1888
trans-[$Rh(CO)L_2Cl$]	1,3-Dimethylimidazolin-2-ylidene	1924
	1,3-Dicyclohexyl-imidazolin-2-ylidene	1929
	PCy_3	1939
	PMe_3	1957
	PPh_3	1983
	$P(C_6F_5)_3$	2003
	$P(OPh)_3$	2018

The comparison of NHCs with various other monodentate ligands such as phosphines and amines on a [$MoL_3(CO)_3$],[256,257] a *trans*-[$RhL_2(CO)X$],[110] and various other [$M(CO)_nL_m$][82] complexes shows the significantly increased donor capacity relative to phosphines, even to trialkylphosphines (Table II).[82,110,258] The π-acceptor capability of NHCs is on the order of those of nitriles and pyridine.

IR data have also been used to compare the *trans*-influence of NHCs and phosphines.[258] Supported by NMR spectroscopy the *trans*-effect on platinum centers was found to decrease in the order $PEt_3 \approx$ imidazolidin-2-ylidene > benzothiazolin-2-ylidene. Similar results were reported with gold(I): benzimidazolin-2-ylidene $\approx PR_3 > NR_3$, halides.[231]

Calorimetric measurements in solution support these observations: A relative enthalpy scale for different NHCs coordinated to ruthenium(II) [Eq. (40)] shows that NHCs have a higher donor capability than the best phosphine donor ligands. The only exception is the sterically extremely demanding 1,3-di(1-adamantyl)imidazolin-2-ylidene (Table III).[153,259] The enthalpies for the formation of sulfur adducts indicate as well that basic trialkylphosphines are comparable with NHCs.[260]

$^{1}/_{4}$ [$(C_5Me_5)RuCl]_4$ + R–N N–R ⟶ Cl Ru R N N R (40)

It has been shown for several metals that an exchange of phosphines versus NHCs proceeds rapidly and without the need for the NHC to be present in excess

TABLE III
REACTION ENTHALPIES ΔH_{rxn} [kcal/mol] FOR THE REACTION ACCORDING TO Eq. (40)

NHC	$-\Delta H_{rxn}$ (kcal/mol)
1,3-Dicyclohexylimidazolin-2-ylidene	85.0
1,3-Dimesitylimidazolin-2-ylidene	62.6
1,3-Di(1-adamantyl)imidazolin-2-ylidene)	27.4
PCy_3	41.9
$P\textit{i}Pr_3$	37.4

to drive potential equilibria (*vide supra*). This is experimental evidence for the increased donor capacity of NHCs compared to phosphines. As mentioned earlier, x-ray structures for NHC complexes with metals from all over the periodic table reveal a metal–carbon bond that is exceptionally long for carbenes. The fundamental difference between a typical alkylidene moiety and an NHC as a ligand is mirrored in the X-ray structure of $[RuCl_2(NHC)_2(=CHC_6H_4Cl)]$ where two types of carbenes are attached to the same metal center (Fig. 1).[158] The ruthenium–carbon

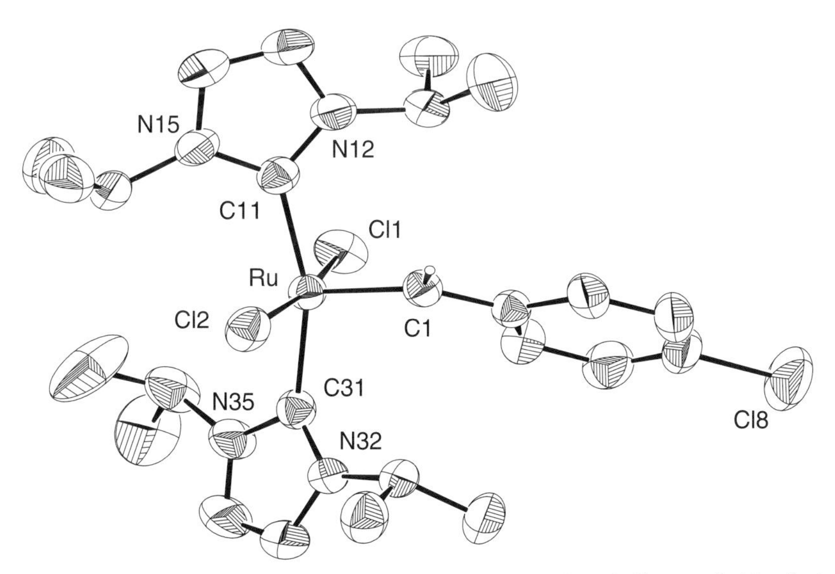

FIG. 1. Platon plot of $[RuCl_2(NHC)_2(=CHC_6H_4Cl)]$ (NHC = 1,3-diisopropylimidazolin-2-ylidene). Thermal ellipsoids are at 50% probability. All hydrogen atoms except at C1 are omitted for greater clarity. Selected bond distances [Å] and angles [°]: Ru–Cl1 2.3995(9), Ru–Cl2 2.3921(8), Ru–C1 1.821(3), Ru–C11 2.107(3), Ru–C31 2.115(3), Cl1–Ru–Cl2 170.32(3), Cl1–Ru–C1 98.37(10), Cl1–Ru–C11 88.26(8), Cl1–Ru–C31 94.76(8), Cl2–Ru–C1 91.14(10), Cl2–Ru–C11 89.21(8), Cl2–Ru–C31 85.14(8), C1–Ru–C11 94.48(12), C1–Ru–C31 100.92(13), C11–Ru–C31 163.68(11).

bond of the *Schrock* carbene—generally written as a double bond (σ-donor and π-acceptor)—has a bond length of 1.821(3) Å, whereas the Ru–C bond length of the NHC (2.107(3) Å and 2.115(3) Å) justifies its representation as a single bond (σ-donor and virtually no π-acceptor).

Only in homoleptic $M(L)_2$ (L = 1,3-dimesitylimidazolin-2-ylidene) of zero-valent nickel and platinum significantly shorter metal–carbon bonds for NHCs and, thus, metal-to-ligand back donation can be observed.[179] The Ni–C bond length is about 0.15 Å shorter than in $[Ni(CO)_2(L)_2]$ (L = 1,3-dimesitylimidazolin-2-ylidene) which cannot be explained exclusively by the change of the coordination number.

The differences between structural parameters for free NHCs and metal-bound NHCs are very small. Generally, only the NCN angle is affected by the coordination and is in average increased by about 2 degrees. The dependence of the ring parameters of the NHC on the coordinated metal follows the general rule that metal–carbon bonds become stronger when going from 3d to 4d and 5d transition metals. This effect results in an increase in the N–C–N angle and a decrease in the N–C bond length.

A first approach to quantify steric parameters of NHCs based on X-ray data in analogy to the *Tolman* angle for phosphines has been published recently.[259] According to this simple model NHCs can be considered as "fences" requiring parameters for their length and height. In contrast to phosphines, two different angles have to be introduced depending on the perspective from which the NHC is viewed.

The position of different carbene ligands relative to the residual metal complex fragment can be strongly influenced by packing effects: For $[(NHC)W(CO)_5]$ with only slightly different NHC ligands, three different rotamers with an eclipsed, a staggered and an intermediate arrangement have been structurally characterized.[149]

On mixed ligated complexes of palladium(II) and platinum(II), the *cis* coordination of the NHC and a phosphine ligand is thermodynamically favored.[38,127,261,262] Thermal isomerization was reported for *trans*-$[(NHC)_2Cr(CO)_4]$ (M = Cr, Mo) to the *cis*-complex.[83,263] For $[(NHC)_2Mo(CO)_4]$ this proceeds even in the solid state.[264] A barrier of rotation due to a double-bond character of the metal–NHC bond could not be determined so far. This is in agreement with the single-bond character of the metal–carbon bond. Barriers of rotation determined so far are due to steric hindrance.[155,158,265]

In ^{13}C-NMR spectra, the signals for the carbene carbon are usually shifted upfield by about 20–30 ppm upon complexation of the free NHC to a transition metal. ^{53}Cr-NMR data of $[LCr(CO)_5]$ complexes underline that NHC are a special case of carbene ligands because of their lack of π-acceptor ability.[266] Photoreactions of metal complexes containing NHCs by laser flash and continuous photolysis show that NHCs are quite inert ligands in photolysis reactions.[267,268] He I and He II photoelectron spectra of platinum(0)- and palladium(0) bis(imidazolin-2-ylidene)

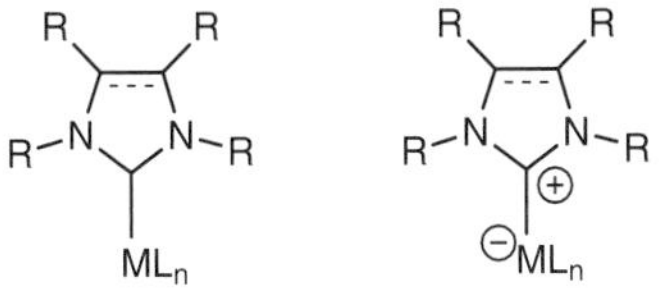

SCHEME 9. Generally accepted representations of the metal–carbon bond of metal coordinated NHCs.

complexes reveal as well little evidence for π-bonding as the NHC π-orbitals are essentially unperturbated upon complexation.[269]

In summary, the metal–carbon bond of an NHC is significantly different from a "real" metal–carbene bond both of the *Fischer*- or *Schrock*-type. Thus, the representations of the metal–carbon bond according to Scheme 9 is now generally accepted.

As a result, the reactivity of these metal–NHC compounds is also unique. They prove to be rather resistant toward attack by nucleophiles or electrophiles at the divalent carbon atom. Additionally, theoretical calculations and experimental investigations agree that the ligand dissociation energy for an NHC is higher than for a phosphine.[95,159,270–272]

III

APPLICATIONS IN HOMOGENEOUS CATALYSIS

Metal–carbene complexes with *Fischer*- or *Schrock*-type carbenes are very useful reagents for the transfer of CR_2-moieties (R = H, alkyl, aryl, alkoxy, amino) resulting in the application of these compounds as catalysts in reactions such as cyclopropanation or olefin metathesis[273] as well as in stoichiometric transformations like the Dötz reaction.[274] However, due to their high reactivity, these types of carbenes can usually not be considered as an addition to the supporting ligand portfolio for homogeneous catalysis. Since NHCs show a completely different reactivity than *Fischer*- or *Schrock*-carbenes and—as outlined previously—fulfill a lot of the criteria for a ligand in homogeneous catalysis, they have been applied as directing ligands in various catalytic transformations since the 1970s, extensively, however, just for the last few years.[9,10]

The first examples of the application of NHCs as directing ligands were published by Nile and Lappert.[275–279] These results showed the applicability of NHCs as ligands in homogeneous catalysis in principle and were therefore pioneering. However, it was still to be shown that NHCs represent a real alternative to known ligands. Thus, superior properties of these ligands in selected transition metal-catalyzed reactions were to be demonstrated. The comparison of spectroscopic and crystallographic data of complexes containing NHCs with metal compounds containing other ligands reveal that ligand properties of NHCs are comparable to those of trialkylphosphines. More precisely, spectroscopic data assign them an even higher basicity and donor capability.[82,110,256] Keeping in mind that even triarylphosphines and trialkylphosphines are not readily interchangeable in most transition metal catalyzed processes, NHCs are less likely to be an alternative in reactions where triarylphosphines perform best, but are a more promising alternative where trialkylphosphines have proven superior. Additionally, due to their strong metal–carbon bond, NHCs have to be considered as typical directing or innocent spectator ligands and are therefore unlikely to be good substitutes for coordinatively labile ligands, which are actively involved in the catalytic cycle by dissociation equilibria.[95,159,270]

The recent applications of NHCs in ruthenium-catalyzed olefin metathesis and palladium/nickel-catalyzed coupling reactions show the value of such a profound understanding about ligand properties before using them for specific catalytic transformations.

A. *Olefin Metathesis*

Together with Schrock's molybdenum–imido compound **50**[280] the ruthenium–phosphine complexes **51** and especially **52** developed by Grubbs[281–284] proved to be an outstanding achievement in the development of molecular catalysts for olefin metathesis reactions (Scheme 10).

Although the first application of NHCs as ligands for the metathesis reaction were reported in 1978 with [(NHC)Mo(CO)$_4$] as catalyst and EtAlCl$_2$ as the cocatalyst, the breakthrough for these ligands in this reaction was achieved only recently

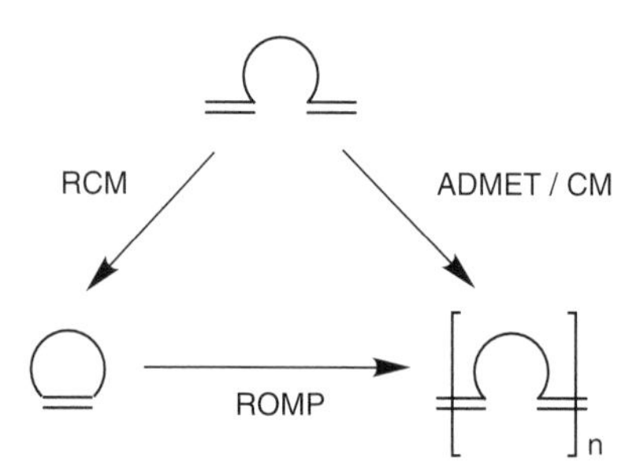

SCHEME 10. Olefin metathesis: RCM (ring closing metathesis), ROMP (ring opening metahesis polymerization), ADMET (acyclic diene metathesis), CM (cross metathesis).

and is based on an evaluation of the catalytic performance of **51** and **52.**

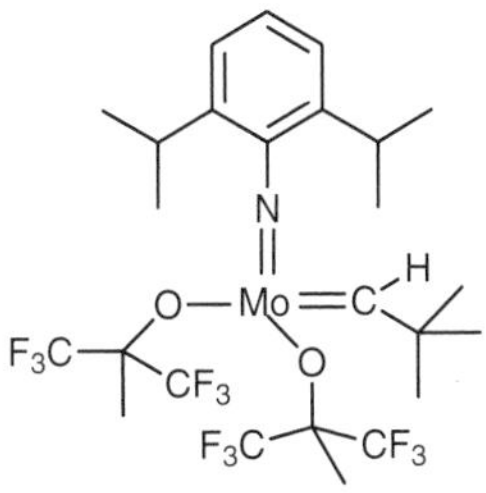

50 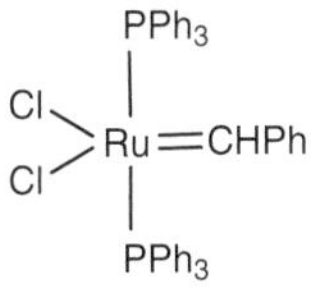 **51** 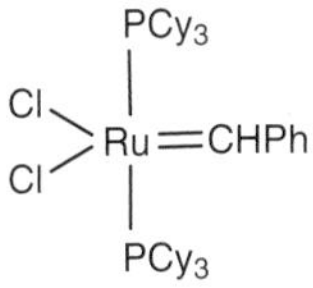 **52**

1. *Imidazolin-2-ylidenes as the Directing Ligands*

A closer look on the history of the development of catalyst **52** shows that this class of compounds was to some degree predestined for the application of NHCs. Complex **51** containing triphenylphosphines is an active catalyst for olefin metathesis.[281,283] However, the substitution of the triphenylphosphines by more electron-donating and sterically more demanding tricyclohexylphosphines is accompanied by a significantly increased stability and catalytic performance.[282–284] Thus, complexes of type **53**[158,285] can be seen as a logical development with respect to the phosphine complexes **51** and **52.**

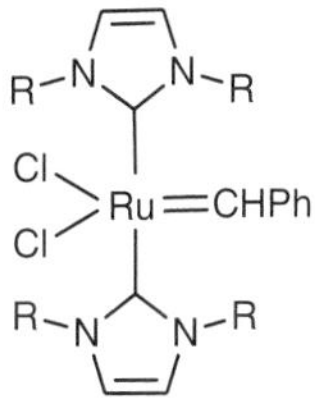

53

Although NHCs have to be considered as even more electron-donating than trialkylphosphines, the catalytic performance of **53** turned out to be "just" comparable to **52,** but not significantly better. It depends on the substrate whether the one or the other catalyst performs slightly better in ring opening metathesis polymerization (ROMP).

A breakthrough in catalytic metathesis applications was achieved with the second generation of ruthenium–NHC–alkylidene complexes: In **54, 55,** and **56** NHCs are combined with coordinatively more labile ligands such as phosphines or organometallic fragments on the ruthenium center.[153,159–161]

54 55 56

The significant increase in catalytic performance is shown by a comparison of the bimetallic compounds **55** and **56,** NHC/phosphine complex **54,** dicarbene complex **53,** and diphosphine complex **52** in ROMP of 1,5-cyclooctadiene (Fig. 2).[159]

This development is a good example for the fine-tuning of a tailor-made catalyst based on mechanistic considerations and theoretical calculations: The mechanistic scheme for **52** postulates the dissociation of a phosphine ligand as the key step in the dominant reaction pathway (Scheme 11).[286]

Thus, the phosphine ligand in **52**—as in many other phosphine containing catalysts—has to be considered as a good compromise for two quite opposite requirements: (i) it provides enough electron density to the metal for the generation and stability of the catalytically active fragment **57** and (ii) the ruthenium–phosphorous bond energy is not too high to suppress the dissociative pathway.

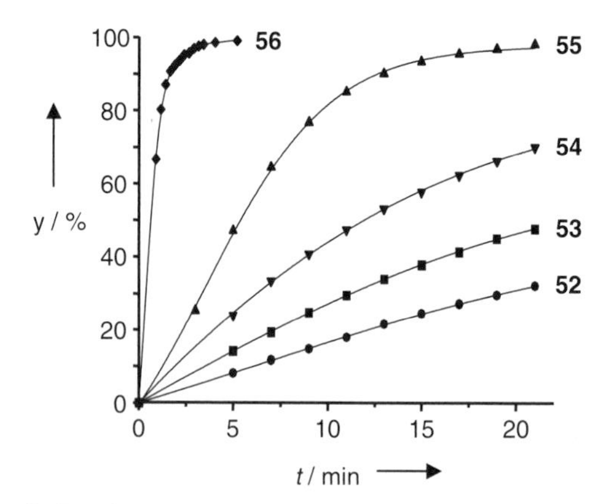

FIG. 2. ROMP of 1,5-cyclooctadiene (y = yield of polyoctadienamer). NMR-monitored comparison of catalysts **52, 53, 54, 55,** and **56** ($T = 25°C$; 1.70 μmol catalyst in 0.55 ml of CD_2Cl_2; [1,5-cyclooctadiene]/[catalyst] = 250:1).

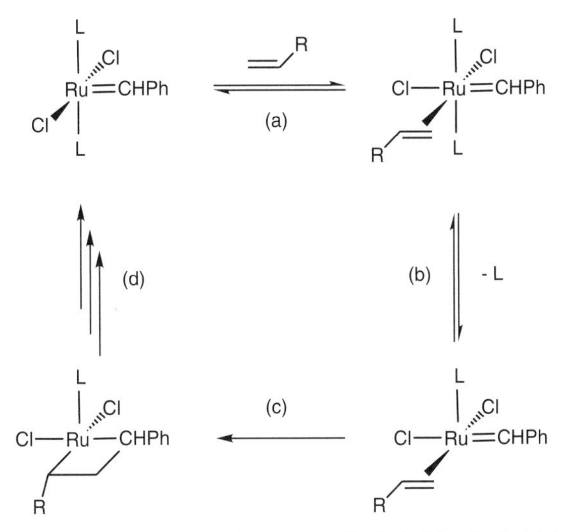

SCHEME 11. Dominant dissociative reaction pathway that is decisive for the high catalytic performance of **52.**

NHCs are certainly not the best choice for requirement (ii), but seem to be superior with respect to (i). This has been quantified by theoretical calculations: The dissociation energies of NHCs and phosphines for ruthenium alkylidene model compounds by density functional (DFT) methods according to Eq. (41) demonstrate that the ligand dissociation energies ascend in the series $PH_3 < PMe_3 <$ NHC (Table IV).[159]

TABLE IV
CALCULATED LIGAND DISSOCIATION ENERGIES ΔE [kcal mol^{-1}] FOR THE MODEL COMPOUNDS ACCORDING TO Eq. (41).[a]

Model compound	ΔE for PH_3	ΔE for PMe_3	ΔE for NHC
$L^1 = L^2 = PH_3$	18.2 (19.4)	—	—
$L^1 = L^2 = PMe_3$	—	27.0 (25.8)	—
$L^1 = L^2 =$ NHC[b]	—	—	45.0 (42.2)
$L^1 = PH_3$; $L^2 =$ NHC[b]	18.7 (15.8)	—	46.9 (49.7)
$L^1 = PMe_3$; $L^2 =$ NHC[b]	—	26.0 (24.9)	42.0 (43.4)

[a] Ligand dissociation energies without ethylene coordination are given in parentheses.
[b] NHC is modeled by 1,3-dihydroimidazolin-2-ylidene.

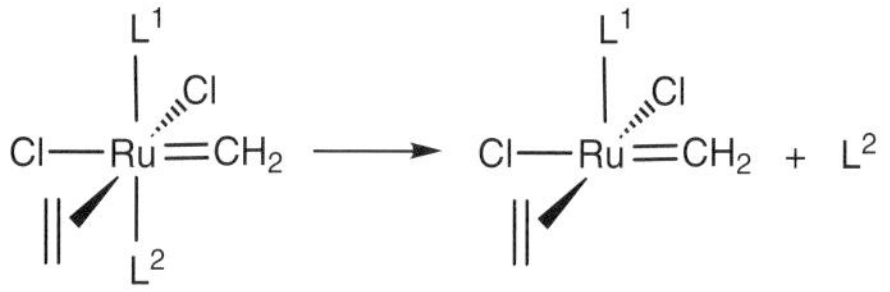

$$Cl_2Ru(=CH_2)(L^1)(L^2)(\eta^2\text{-olefin}) \longrightarrow Cl_2Ru(=CH_2)(L^1)(\eta^2\text{-olefin}) + L^2 \qquad (41)$$

As a consequence of the higher coordination energy, the dicarbene complexes **53** disfavor a dissociative pathway similar to that of **52.** A mixed NHC/phosphine complex of type **54,** however, reveals a phosphine dissociation energy in the same order of magnitude as **52.** Therefore, **54** is able to populate the dissociative pathway just as readily as **52.** In contrast to **52,** however, a phosphine-free species **58** is to be considered as the key intermediate in the catalytic cycle.

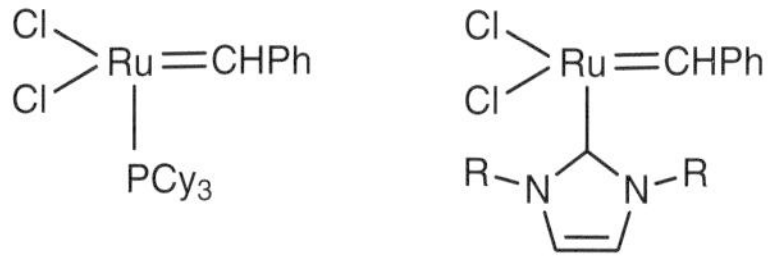

$Cl_2Ru{=}CHPh(PCy_3)$ **57** $\qquad$ $Cl_2Ru{=}CHPh(\text{NHC}, R{-}N\ N{-}R)$ **58**

The increased activity of **54** shows that intermediate **58** is more active than its phosphine analogue **57.** This is confirmed by the experimental results and the DFT calculations for the bimetallic derivative **56.**[159]

Within the NHCs, increased bulk of the substituents at the nitrogen leads to higher activities in a number of metathesis applications. Thus, the 1,3-mesityl-substituted NHC[153,161] is advantageous for some applications compared to cyclohexyl or other CHR_2 substituents at the nitrogen.[287] Thus, it is the combination of an NHC with coordinatively more labile ligands on the ruthenium center, which allows NHCs to develop their full potential in this class of catalysts that has found extensive applications in metathesis reactions.[161,288–302]

2. *Imidazolidin-2-ylidenes as the Directing Ligands*

Based on these findings, another starting point for modifications besides modifying the labile "coligand" is the NHC ligand itself, especially since there are many more carbenes known and accessible. One of the many possible modifications turned out to be very successful and can—based on its catalytic activity and catalyst lifetime—be considered as the third generation of NHC catalysts in olefin metathesis. The recipe for success is to use NHCs with saturated backbones, i.e., imidazolidin-2-ylidenes.[46]

The resulting complexes **59** display catalytic activities in ROMP that exceed even the performance of molybdenum-imido system **50** (Fig. 3), which has been considered as intrinsically more active than late-transition metal systems, but also more sensitive toward polar functionalities, water, and air. Thus, to combine "molybdenum-like" activity with "ruthenium-like" functional group tolerance has been a goal that has been targeted coming from both the molybdenum and the ruthenium side.

PCy3
Cl
Ru=CHPh
Cl
R–N N–R
R' R'

59

Although discovered quite recently, catalyst **59** has already found many applications in ROMP, RCM, and cross-metathesis.[291,303–313] The higher efficiency of NHCs with saturated backbones compared to their unsaturated analogues can be explained by their higher basicity. However, the numbers available for the basicity of various stable cyclic diaminocarbenes are not that different. Another starting point for an explanation is the stability of the free NHCs.[18] At least for 1,3-disubstituted imidazolidin-ylidene and 1,3-disubstituted imidazolin-ylidene, it seems that there is a correlation between the instability of the free carbene (Scheme 12) and the stability and catalytic performance of the corresponding ruthenium alkylidene complex.[303,314] Thus, the stability of the catalytically active fragment is higher the less stable the free carbene is, since its dissociation tendency is limited due to thermodynamic reasons.

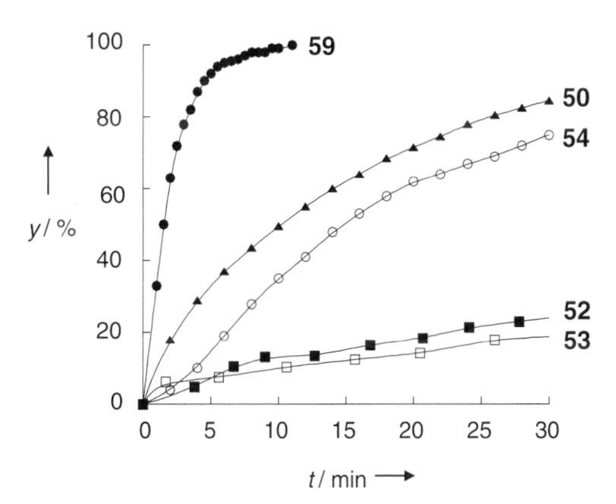

FIG. 3. ROMP of 1,5-cyclooctadiene (y = yield of polyoctadienamer). NMR-monitored comparison of catalysts **50, 52, 53, 54,** and **59** ($T = 20°C$; $[M]_0 = 0.5$ mM for **50, 54,** and **59;** $[M]_0 = 3.0$ mM for **52** and **53;** [1,5-cyclooctadiene]/[catalyst] = 300:1). This figure was provided by R. H. Grubbs and C. Bielawski.

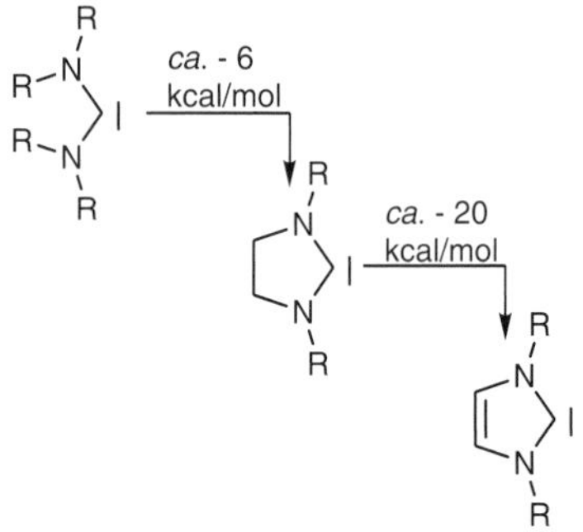

SCHEME 12. Calculated stabilization effects in diaminocarbenes.

B. *Heck-Type Reactions*

Other successful examples of catalysts containing NHC ligands are found in palladium- and nickel-catalyzed carbon–carbon bond formations. The catalyst development with these metals has focused in particular on *Heck*-type reactions, especially the *Mizoroki–Heck* reaction itself [Eq. (42)][315–318] and various cross coupling reactions [Eq. (43)], e.g., the *Suzuki–Miyaura* reaction ([M] = $B(OH)_2$)[319–322] and the *Kumada–Corriu* reaction ([M] = MgBr).[323–325] Related reactions like the *Sonogashira* coupling [Eq. (44)][326–329] and the *Buchwald–Hartwig* amination [Eq. (45)][330–333] were also successfully performed with these types of catalysts.

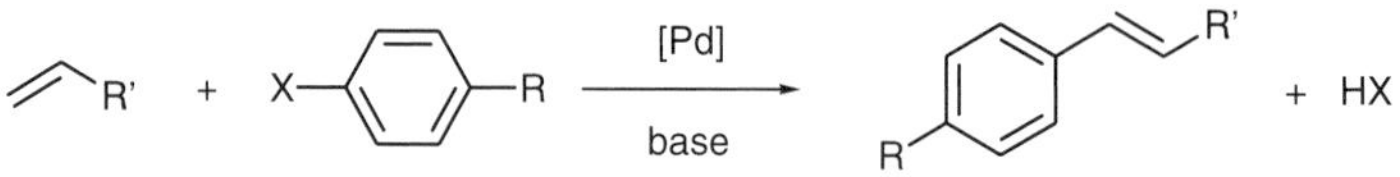

(42)

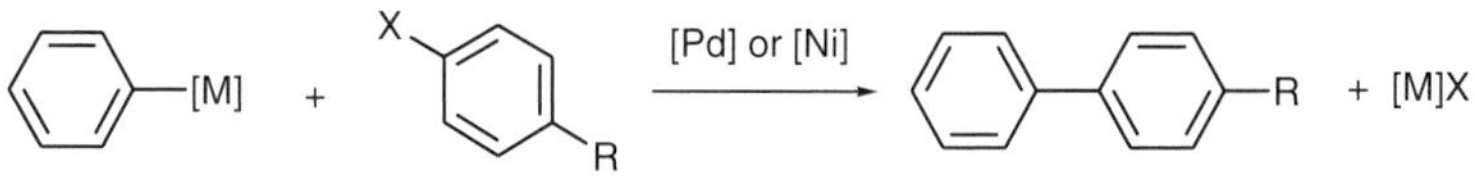

(43)

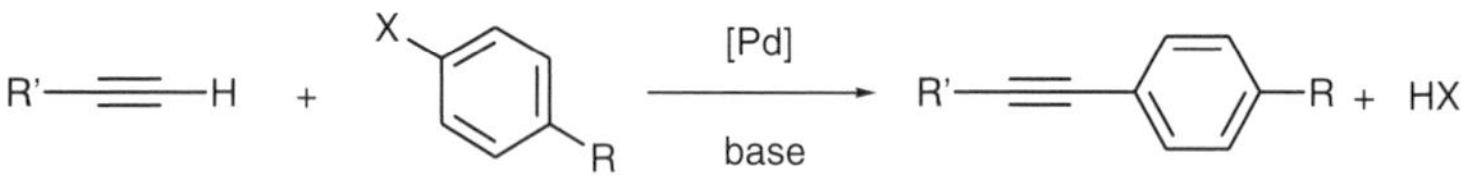

(44)

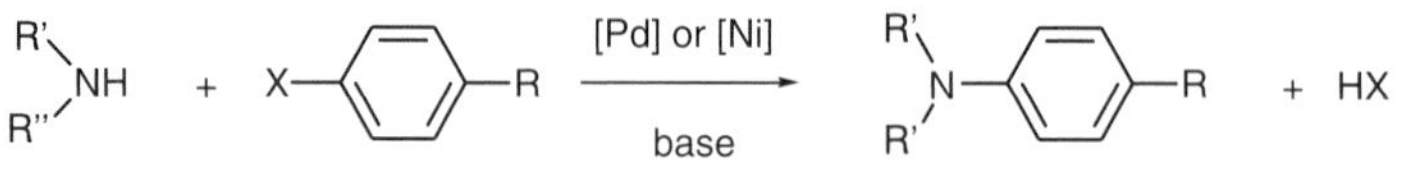

(45)

[M] = $B(OH)_2$, SnR_3, MgX, ZnX, $Si(OR)_3$ X = I, Br, Cl, OSO_2CF_3, N_2^+, C(O)Cl

The mechanistic similarity of these reactions is based on the identity of the first step of the catalytic cycle: the oxidative addition of the aryl halide to a palladium(0) species [Eq. (46)].

Electron-donating ligands like trialkyl phosphines as well as NHCs facilitate the oxidative addition of aryl halides; making the ligands sterically demanding has shown in many cases an additional accelerating effect on catalysis, which is mostly accounted to influences on subsequent reaction steps. Consequently, the development of highly active palladium catalysts for the activation of aryl chlorides in *Heck*-type reactions has focused on using sterically demanding, basic, monodentate phosphines and NHCs.[334]

$$C_6H_5\text{–X} + [Pd]^0 \longrightarrow C_6H_5\text{–}[Pd]^{II}\text{–X} \qquad (46)$$

1. *Palladium(II) Pre-catalysts*

Because of their convenient preparation from palladium(II) salts and stable NHC-precursors (*vide supra*), palladium(II) complexes were first examined as potential catalysts for *Heck*-type reactions. Due to the high thermal stability, temperatures up to 150°C can be used to activate even less reactive substrates, like, e.g., aryl chlorides. Immobilization of such catalysts has been shown recently (*vide infra*).[95]

In the *Mizoroki–Heck* reaction aryl bromides and activated aryl chlorides could be employed with moderate turnovers. This holds true for both the complexes of monodentate NHC[24,106,181] such as **60** as well as the complexes of chelating ones such as **61.**[95,96,135,232] An increase in activity for catalysts like **60** was achieved using $[NBu_4]Br$ as the solvent.[101,335] Additionally, it was shown that the *Suzuki–Miyaura* reaction, the *Sonogashira* coupling and the *Buchwald–Hartwig* amination were also catalyzed by defined palladium(II) complexes of chelating NHCs employing less reactive aryl bromides and aryl chlorides.[98,232,336]

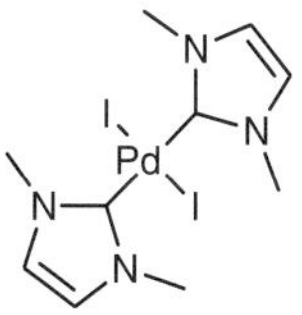

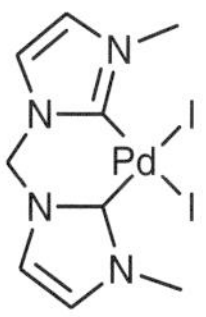

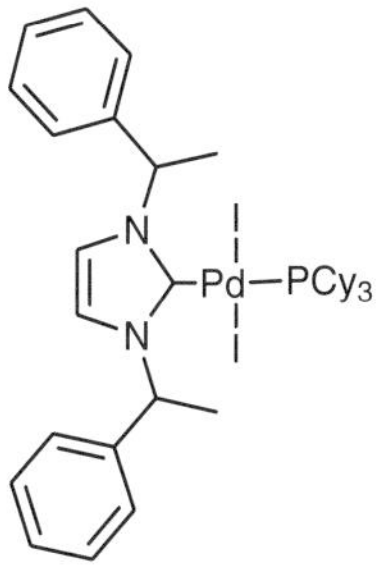

Following the strategy that has been very efficient for olefin metathesis, palladium(II) complexes such as **62** containing both phosphine and NHC ligands were used in catalysis.[126,127] Increased activities in the *Mizoroki–Heck* and in the

Suzuki–Miyaura as well as moderate activity in the *Stille* reaction ([M] = SnR_3) were observed. In contrast to bis(NHC) complexes, inactivity in the *Sonogashira* reaction was due to increased activity in the homocoupling of alkynes [Eq. (47)], an undesired side reaction.

$$\mathrm{H{-}C{\equiv}C{-}R} \xrightarrow[\text{(base)}]{\text{[Pd] / (CuI)}} \mathrm{R{-}C{\equiv}C{-}C{\equiv}C{-}R} \; + \; \mathrm{R{-}C{\equiv}C{-}CH{=}CH{-}R} \qquad (47)$$

2. *Well-Defined Palladium(0) Catalysts*

The mechanistic requirement for a palladium(0) complex regarding the initiation of the catalytic cycle in *Heck*-type reactions suggest the preparation of suitable palladium(0) complexes of NHC. In contrast to simple synthetic routes to nickel(0)- and platinum(0) complexes starting from free NHCs and Ni(η^4-cod)$_2$ or Pt(η^4-cod)$_2$,[179] the synthesis of the corresponding palladium(0) complexes is not possible by this method due to the instability of Pd(η^4-cod)$_2$.[337] Heteroleptic palladium(0) complexes containing NHC and an electron deficient olefin for stabilization were synthesized and successfully used as catalysts in the *Mizoroki–Heck* reaction of aryl iodides.[180] Since it has been demonstrated that the coordination of olefins to palladium(0) catalysts generally slows down the reaction rate of the oxidative addition,[338] the synthesis of homoleptic palladium(0)–NHC complexes was expected to furnish more active catalysts. The vapor phase deposition of palladium metal in the presence of a sublimable free NHC like 1,3-di-*tert*-butylimidazolin-2-ylidene allows the isolation of complex **63** (Scheme 13).[219] As this method is limited to sublimable, thermally stable NHC, a more general route to this type of complex is desirable for the evaluation of ligand effects in catalysis. Ligand exchange of free NHC on bis[tri(*ortho*-tolyl)phosphine]palladium(0) leads to the clean formation of the desired complexes (Scheme 10).[163] In the case of sterically demanding NHCs the isolation of the intermediate mixed phosphine–NHC complexes **64** is possible.[339]

63 **64**

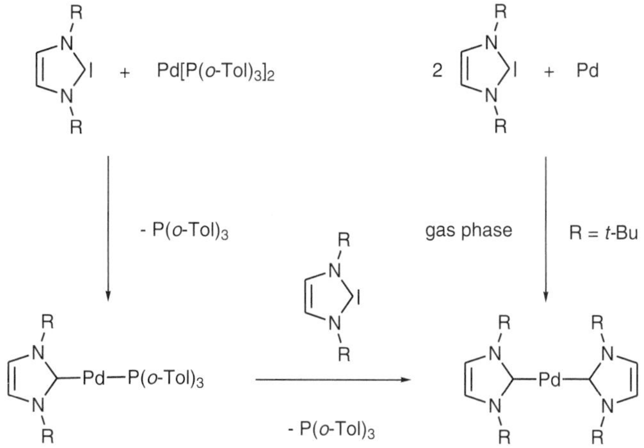

SCHEME 13. Synthesis of palladium(0)–NHC complexes.

Application of the complexes **63** in the *Mizoroki–Heck* reaction did not reveal higher activity than the previously examined palladium(II) complexes. However, in the *Suzuki–Miyaura* reaction, a drastically increased activity was observed with complex **63.** Catalysis starts without a measurable induction period at mild temperatures accompanied by an extraordinarily high turnover frequency (TOF) of 552 [mol product $\times$ mol $Pd^{-1} \times h^{-1}$] at the start of the reaction for the coupling of *p*-chlorotoluene and phenyl boronic acid [Eq. (48)].[163]

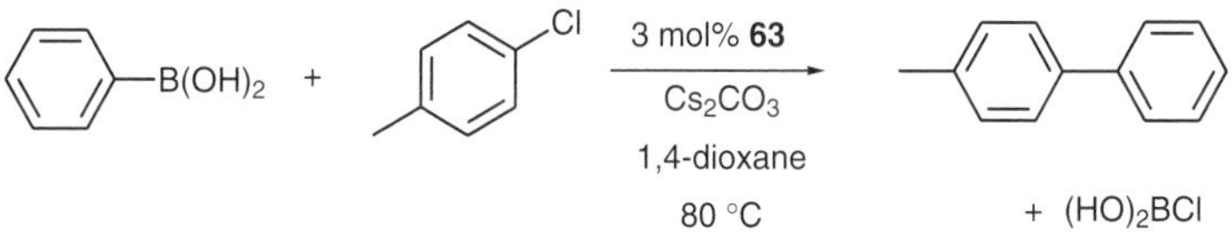

$$PhB(OH)_2 + p\text{-}ClC_6H_4CH_3 \xrightarrow[\text{1,4-dioxane, 80 °C}]{\text{3 mol\% } \mathbf{63},\ Cs_2CO_3} p\text{-}CH_3C_6H_4\text{–}Ph + (HO)_2BCl \quad (48)$$

In sharp contrast to the observations with mixed palladium(II) complexes, the mixed palladium(0) complex **64** showed inferior activity in all tested reactions as compared to **63,** rendering this catalyst useless for the activation of aryl chlorides.

For nickel(0) complexes prepared from $Ni(\eta^4\text{-cod})_2$ and an excess of the free NHC,[179] it was shown that they exhibit outstanding catalytic activity in the *Kumada–Corriu* reaction at room temperature toward unreactive substrates like aryl chlorides and even aryl fluorides.[127,340] Again, an essential element of these catalysts is the need for sterically demanding NHC ligands as observed for the palladium catalysts.

3. *In Situ Systems*

Significant progress regarding more convenient application of these systems was made by the observation that highly active catalysts can be formed directly in the reaction mixture from imidazolium salts and commercially available metal sources. This *in situ* formation of an NHC–metal complex can be achieved by the addition of an excess of a base, which is typically present as a stoichiometric reagent in most of the *Heck*-type reactions. The use of such *in situ* systems consisting of imidazolium salts and palladium(0) or palladium(II) precursors has proven highly efficient for the *Suzuki–Miyaura* reaction,[341,342] the *Kumada–Corriu* reaction,[41] the cross-coupling of phenyltrimethoxysilane,[343] and the *Buchwald–Hartwig* amination,[344] even enabling the conversion of aryl chlorides (using **65** and **66**). For amination, *in situ* systems based on imidazolidinium salts **67** and palladium(0) were used successfully for the transformation of aryl chlorides under very mild conditions.[345] For the *Kumada–Corriu* reaction, nickel salts combined with imidazolium salts **65** and **66** give a very active catalyst system for aryl chlorides and even aryl fluorides.[39,127] In all cases, sterically very demanding NHCs perform best in terms of activity of the catalyst systems. The inactivity of *in situ* systems in the *Sonogashira* reaction is attributed to homocoupling of alkynes [Eq. (47)].[346]

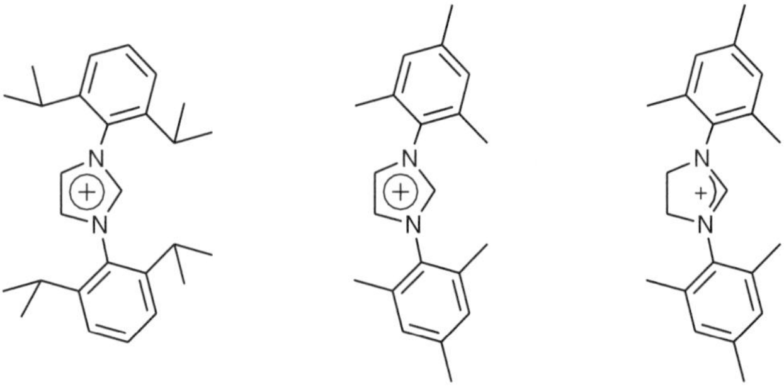

65 **66** **67**

The *Mizoroki–Heck* reaction in liquid imidazolium salts as the solvent is a special case of an *in situ* system: Under the reaction conditions NHC complexes of palladium are formed as the active catalyst from the solvent and the ligand-free palladium precursor.[102] In general, ionic liquids are novel reaction media for homogeneous catalysis. They allow easy separation of product and catalyst after the reaction.[347,348]

4. *Mechanism*

Studies regarding the nature of the catalytically active species for NHC complexes in *Heck*-type reactions have focused on the *Mizoroki–Heck* reaction and have consistently revealed a palladium(0) species as the active catalyst. The induction period is shortened upon addition of a reducing agent,[24] and postulated intermediates of the reaction were isolated and characterized as well as employed in stoichiometric and catalytic reactions.[180,181] Theoretical studies using DFT calculations showed the mechanism for NHC complexes to most likely be in agreement with phosphine chemistry.[349]

Catalysis experiments in the *Suzuki–Miyaura* and the *Kumada–Corriu* reaction suggest monoligated, 12-electron complexes of palladium(0) and nickel(0) to be the catalytically active species.[127,345] Comparison of the activity of bis(NHC) and mono(NHC) complexes bearing potentially chelating pyridine residues on the nitrogen ring atoms shows that the mono-ligation furnishes more active catalysts in the *Mizoroki–Heck,* the *Suzuki–Miyaura,* and the *Sonogashira* reaction.[232] Analogous observations in more detailed studies of phosphine systems support this view.[350]

C. *Other Reactions*

NHCs have as well been tested in a variety of other reactions. In most of these transformations, they were not found to be real alternatives to other ligand classes, but it has to be taken into account that only a small number of structurally different stable carbenes have been investigated so far. One of the main problems applying NHCs as ligands in homogeneous catalysis is the fact that these ligands form a metal–carbon bond, which can interfere in the catalytic process. In migratory insertion steps of the desired reaction there is a possible competition of this step with the insertion into the metal–NHC bond. This unwanted side reaction breaks the metal–NHC bond irreversibly and is therefore a major pathway of catalyst deactivation. In general, this side reaction becomes important for reactions with slow insertion steps, for applications at high temperatures, and for chemistry employing high-pressure gas-phase reagents. Observations of this decomposition pathway have been barely published so far.[351]

Historically, NHC complexes were investigated for the first time as catalysts and discussed as catalytic intermediates in the dismutation of electron rich tetraaminoethylenes [Eq. (49)].[278] Mixtures of two differently substituted olefins were reacted in the presence of rhodium(I) complexes and the products obtained showed "mixed" substitution patterns. Starting from *Wilkinson*'s catalyst [$(Ph_3P)_3RhCl$], NHC complexes are formed as intermediates which could be isolated and used as even more active catalysts. In this first example, however, the NHC actively participates in the reaction and can thus not be considered to be an innocent spectator ligand.

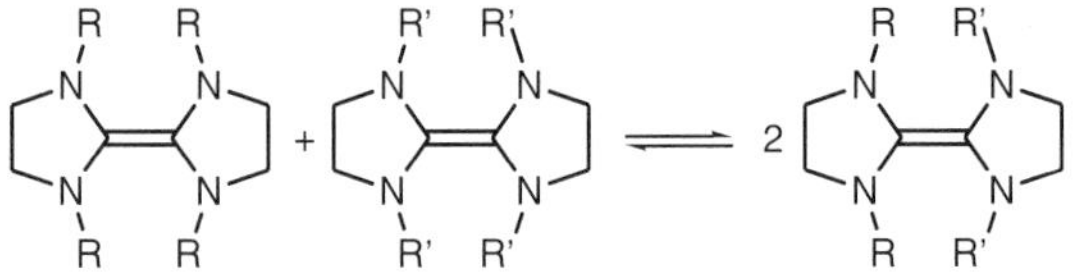

(49)

1. *Hydrosilylation*

Rhodium(I) and ruthenium(II) complexes containing NHCs have been applied in hydrosilylation reactions with alkenes, alkynes, and ketones. Rhodium(I) complexes with imidazolidin-2-ylidene ligands such as [RhCl(η^4-cod)(NHC)], [RhCl(PPh_3)$_2$(NHC)], and [RhCl(CO)(PPh_3)(NHC)] have been reported to lead to highly selective anti-*Markovnikov* addition of silanes to terminal olefins [Eq. (50)].[275,276]

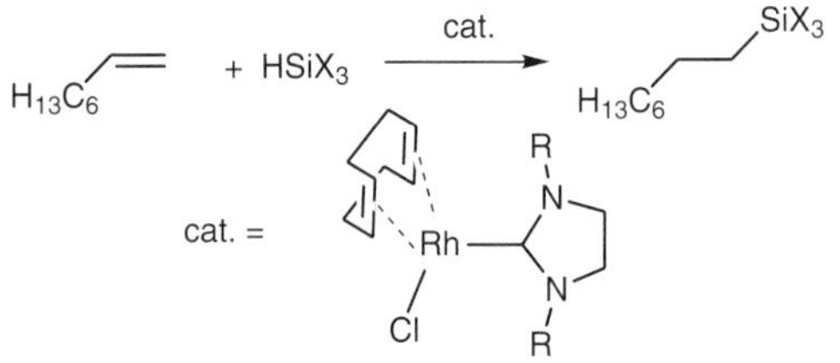

$$H_{13}C_6\text{CH=CH}_2 + HSiX_3 \xrightarrow{\text{cat.}} H_{13}C_6\text{CH}_2\text{CH}_2SiX_3 \qquad (50)$$

cat. = [RhCl(η^4-cod)(NHC)]

Wilkinson's catalyst [RhCl(PPh_3)$_3$], a standard catalyst for this reaction, is reported to give lower yields with less regioselectivity in these reactions. Conjugated dienes gave mixtures of 1,4- and 1,2-addition products in the presence of rhodium–NHC systems, whereas [RhCl(PPh_3)$_3$] leads to selective 1,4-addition.

In the hydrosilylation of alkynes the product distribution depends on the reaction conditions.[277] With [RhCl(η^4-cod)(NHC)] and [RhCl(PPh_3)$_2$(NHC)] silane addition results in mixtures of the *cis* and *trans* isomers. Irradiation increases the reaction rates for these reactions. However, no details about the fate of the NHC ligand under these conditions are reported.

Hydrosilylation of ketones was investigated with rhodium(I)–and ruthenium(II)–NHC systems such as [RhCl(PPh_3)$_2$(NHC)] and [$RuCl_2$(NHC)$_4$].[275,276] The rhodium catalysts are efficient catalysts for this transformation with the exception of [RhCl(CO)(NHC)$_2$]. Presumably, two NHCs ligands are too many as discussed for olefin metathesis. The dependence of catalytic activity on the nature of the nitrogen substituent of the NHC demonstrates the influence of the NHC ligand in the catalytic cycle. The best results for ketone hydrosilylation were achieved with 1,3-diphenylimidazolidin-2-ylidene. Asymmetric hydrosilylation is especially interesting for substrates that cannot be reduced with H_2 in high enantiomeric excess. The usefulness of chiral C_2-symmetric easy-to-prepare imidazolin-2-ylidenes for the transfer of optical information in catalytic reactions could be shown by the hydrosilylation of acetophenone with diphenylsilane. Monodentate NHC ligands are able to generate about 30% ee [Eq. (51)].[21] Similar enantiomeric excess can be achieved by appropriate triazolin-2-ylidene complexes of rhodium(I).[352,353] With more sophisticated chiral NHCs the enantiomeric excess can be increased to about 70%.[162]

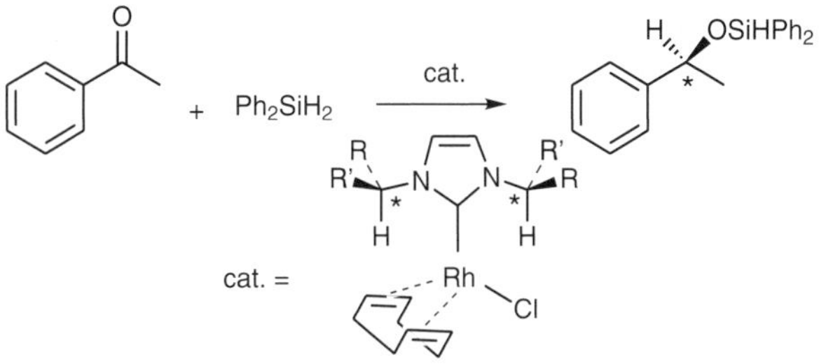

(51)

2. *Hydroformylation*

Rhodium(I) complexes with 1,3-dimethylimidazolin-2-ylidene ligands were used in the hydroformylation of olefins.[167] However, the activity and selectivity toward formation of branched versus linear aldehyde cannot compete with rhodium–phosphine systems.[354,355] Similar catalyst systems with the sterically more demanding 1,3-dimesitylimidazolin-2-ylidene give higher branched/linear ratios for vinyl arenes (95:5), but the turnover frequency is still low compared to established systems [Eq. (52)].[356]

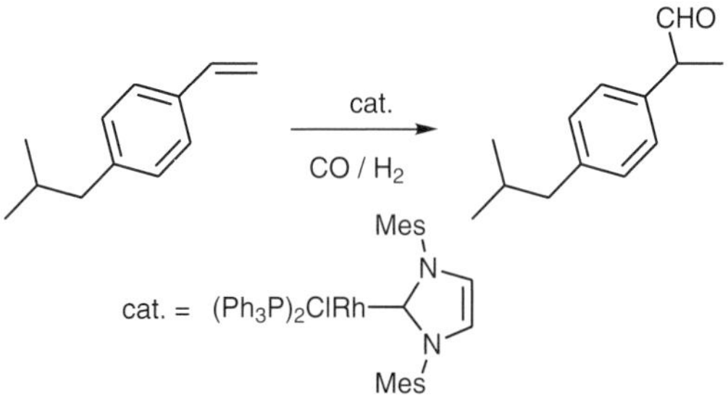

(52)

3. *Hydrogenation*

Hydrogenations have not been looked at very intensively so far.[162,276] The results for rhodium complexes containing NHCs with saturated and unsaturated backbones applied to the hydrogenation of dehydroamino acids are far from being comparable to the best phosphine systems. As for hydroformylation, the poor performance illustrates that due to their strong σ-donor capability NHCs do not exhibit the catalyst properties required in these reactions. Usually rhodium complexes containing π-accepting phosphines exhibit the highest activities in these transformations. More basic alkyl-substituted phosphines are, however, reported to be advantageous for certain substrates.[357]

4. *Polymerization*

Although polymerization plays an important role in today's organometallic catalysis, almost no reports exist about the application of NHCs in this reaction. Ethylene polymerization catalyzed by a chromium–NHC complex was reported but the efficiency of this catalyst is only moderate.[173] After activation with MAO the catalysts produces highly branched PE which contains low- and high-molecular-weight fractions. Molybdenum(0) complexes of 1,3-bismethylimidazolin-2-ylidene are reported for the polymerization of methacrylate[279] and tungsten for the polymerization of diphenylacetylene.[252]

Cationic palladium(II) systems with chelating NHCs catalyze the copolymerization of ethylene and carbon monoxide under mild conditions and at low pressure to give strictly alternating polyketone with high molecular weight.[97] Because of the high molecular weight the amount of precatalyst actually participating in the polymerization reaction is considered to be very low. However, when Pd–Me complexes containing NHC ligands are exposed to CO, decomposition of the catalyst is fast and might serve as an explanation for the observed polymer properties.[351]

Ruthenium(II)–NHC systems can be used for atom transfer radical polymerization (ATRP).[358] Generally, similar results as for the analogous phosphine complexes are obtained. For the ATRP of styrene and methyl methacrylate (MMA) $[(NHC)_2FeBr_2]$ was found to rival copper(I)-based systems and to yield poly (MMA) with low polydispersities.[156] Polymerizations based on olefin metathesis that are catalyzed by ruthenium–NHC complexes are discussed separately (*vide supra*).

5. *Cyclopropanation of Olefins*

Rhodium(I) and ruthenium(II) complexes containing NHCs with hemilabile ether moieties were successfully applied as catalysts for the cyclopropanation of olefins with diazoalkanes [Eq. (53)].[203]

+ N_2CHCO_2Et —cat.→ CO_2Et

cat. = (R, N, N, OMe) $-RhClL_2$; (R, N, N, OMe) $-RuCl_2$(cymene) (53)

6. *Furan Synthesis*

Ruthenium(II) systems containing imidazol-2-ylidene or imidazolidin-2-ylidene have been used to catalyze the synthesis of 2,3-dimethylfuran starting at (*Z*)-3-methylpent-2-en-4-yn-1-ol [Eq. (54)].[162,359] The activity of the catalyst strongly depends on the nature of the NHC ligand. Benzimidazolin-2-ylidenes give the best results for this transformation.[162,360] Similar systems have also been used for olefin metathesis reactions.[152,167]

OH —cat.→ O

cat. = (benzimidazolin-2-ylidene, N-R, N-R)—$RuCl_2$(cymene) (54)

7. *Alkyne Dimerization*

The 16-electron ruthenium(II) complexes [(η^5-C_5Me_5)Ru(NHC)Cl] with sterically demanding NHCs catalyze the carbon–carbon coupling of terminal alkynes HC≡CR (R = Ph, $SiMe_3$, *t*Bu, *p*-Tol) under mild conditions. The product selectivity strongly depends on the substituent R.[115]

D. *Heterogenization*

The difficult recycling of catalysts from the reaction mixture represents a major drawback of homogeneous systems compared to heterogeneous catalysts. Many concepts have been developed to overcome this disadvantage of homogeneous catalysts.[354,355,361] One of them is the immobilization of homogeneous catalysts on a solid support via an organic linker. However, the "weak point" of this approach is usually the link between the organic part (polymeric support, linker, and ligand) and the metal, i.e., generally the metal–ligand bond. If that bond is weak, the metal will leach out and the catalytic activity will decrease with the number of runs (Scheme 14). The high dissociation energy of the NHC ligand should suppress catalyst leaching much more efficiently than an attachment via more labile ligands such as phosphines.[95,159,270]

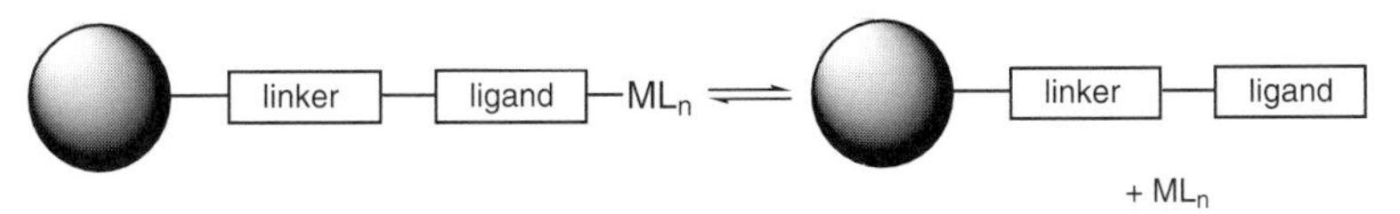

SCHEME 14. The leaching problem.

Catalytic applications have been reported for three systems so far: a rhodium complex on a *Merrifield* resin for hydrosilylation,[362] a palladium compound **68** for the *Mizoroki–Heck* coupling on a *Wang* resin,[95] and a ruthenium complex **69** for olefin metathesis again on a *Merrifield* polystyrene resin.[363] The first reaction published to be catalyzed by an immobilized NHC catalyst was the hydrosilylation of acetophenone.[362] The triazolium salt precursor was immobilized on a *Merrifield* resin via a dihydropyran linker and the rhodium(I) catalyst was then prepared in THF/NEt_3 from $[(\mu^4\text{-cod})RhCl]_2$. A chiral triazolium salt gave up to 24% ee and 80% yield for an asymmetric hydrosilylation reaction. Recycling was shown to be possible but leaching was not quantified. Immobilized imidazolin-2-ylidene and benzimidazolin-2-ylidene complexes have been applied as well for hydrosilylation.[111,364] For the palladium-catalyzed *Mizoroki–Heck* reaction, the effect of the strong metal–NHC bond was increased by the use of chelating bis(NHC) ligands.[95] These well-defined complexes were attached to a *Wang*-resin by etherification via a bromoalkyl-substituted nitrogen of the NHC to form **68.** High reaction temperatures are necessary for catalyst activity with aryl bromides, but the recycling of the catalyst is possible for many times without significant loss of activity. Leaching was quantified and is mainly an issue in the first run probably during the formation of the catalytically active species. The overall leaching observed is much lower than that for related phosphine systems. For ruthenium-catalyzed olefin metathesis, a precursor for the 1,3-dimesitylimidazolidin-2-ylidene was successfully attached to a *Merrifield* polystyrene resin before the metal complex **69** was generated on the resin.[363] The generated species on the support mirrors the latest generation of highly active ruthenium catalysts with NHCs.[46] Various metathesis reactions have been performed by using 5 mol% of the support-bound catalyst. However, more profound data with respect to the influence of different resins, the recycling potential, the long-term catalytic activity, and the metal leaching will be necessary for final conclusions about the usefulness of this technique.

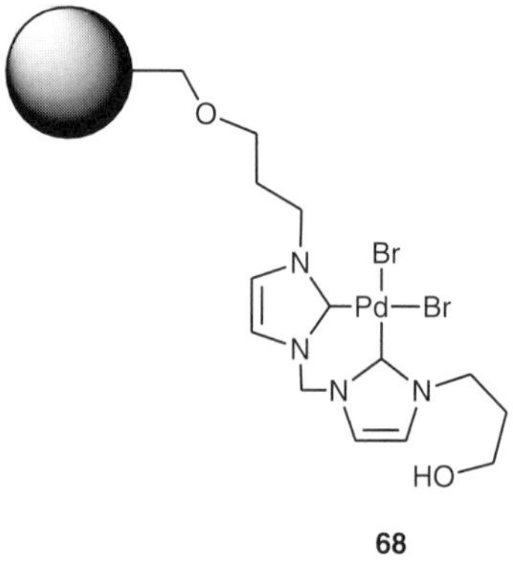

O N N N N Br Pd Br HO

68

O N N Cl Cl Ru=CHPh PCy_3

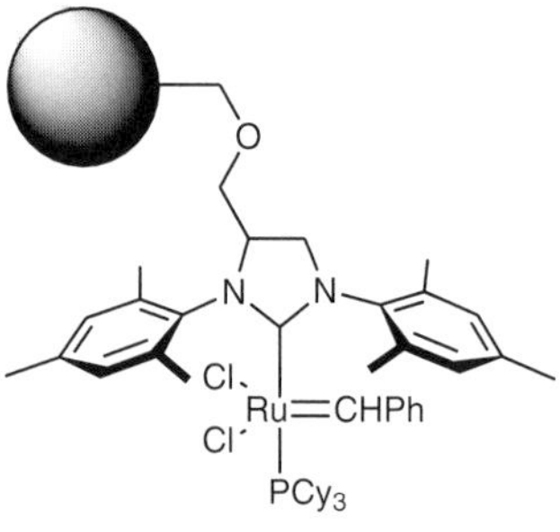

69

Another recycling strategy for homogeneous catalysts is biphasic catalysis. One phase contains the starting material and the product, respectively; the other phase contains the catalyst, which can ideally be recycled by simple decantation. The catalyst has generally to be modified to ensure its solubility in the desired medium and to avoid its leaching into the product phase. Examples of homogeneously catalyzed reactions exist mainly for water as the "catalyst supporting phase."[355,361,365] Fluorous solvents that are immiscible with standard organic solvents,[366] as well as supercritical CO_2,[367] or ionic liquids[347,348] represent alternative approaches. All of these concepts have not yet been shown to work for NHC complexes although appropriate modified NHCs and complexes have been prepared. Water-soluble complexes bearing imidazolin-2-ylidenes with charged side chains, e.g., **70,** have been prepared and were shown to be stable in water and other protic solvents rendering them potentially useful as biphasic catalysts in water.[38,167,368,369]

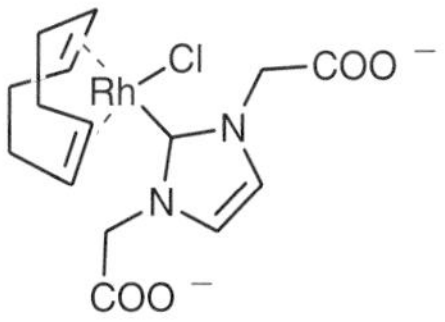

70

Palladium(II) complexes of imidazolin-2-ylidenes bearing perfluorinated side chains **46** were prepared and suggested to be useful in flourous biphasic catalysis as well as in supercritical CO_2.[107] However, catalytic applications have not been reported yet.

E. *Asymmetric Catalysis*

Asymmetric homogeneous catalysis generally requires chiral ligands. Approaches to chiral NHCs have focused on the generation of chiral centers either in the 4- and 5-position of imidazolidinium salts **71** or in the α-position of the nitrogen substituents for imidazolium salts **72.**

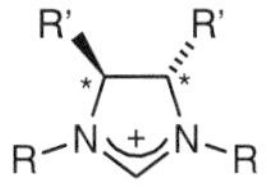

71  **72**

Several complexes containing these types of ligands including triazol-based analogues have been reported over the last 2 decades.[21,46,125,132,133,149,364,370] However, applications in asymmetric catalysis are so far limited to enantioselective hydrosilylation (*vide supra*)[21,162,362] and an enantioselective *Mizoroki–Heck* reaction[125] as well as to monodentate NHC ligands. Monodentate ligands are known to be less effective than bi- or oligodentate chiral ligands in many asymmetric transformations. The extension of NHC-based chiral systems toward derivatives containing "more" chiral information, i.e., chiral modules that have been applied successfully in many transformations (e.g., binaphthyl-, ferrocenyl-, or salen-based structures), has just started.[103] Due to the strong metal–carbene bond, the chiral information should be efficiently anchored to the metal center and not suffer from being "diluted" by dissociation equilibria. Again, it might be necessary to combine NHC moieties with other functionalities such as phosphines to tune both enantiomeric excess and catalytic performance.

F. *High-Throughput Screening*

Ligand screening has become increasingly important for the rapid discovery of homogeneous catalysts.[371] However, ligand synthesis itself is often the limiting step. Thus, a simple synthetic route to an entire ligand class including numerous electronic and steric variations is highly desirable. Furthermore, this ligand class should be easy to handle, i.e., not sensitive to air and moisture, and stable enough to be stored for long periods of time. Free carbenes certainly do not fulfil these criteria. However, their azolium salt precursors represent a very robust class of compounds that is conveniently accessible. The techniques established for the generation of *in situ* catalyst systems (*vide infra*) open up the possibility that libraries of azolium salts can be used directly and deprotonated under reaction conditions. This circumvents the necessity to prepare a library of free NHCs.[39,41,344]

Up to now, NHCs or their azolium precursors, respectively, have been screened only in palladium- or nickel-catalyzed *Heck*–type reactions. This is probably based on the fact that the generation of catalytically relevant palladium and nickel complexes starting directly at the azolium salt had been known before (*vide supra*).

For *Heck*-type reactions, the following screening approaches have been reported: For the *Kumada–Corriu* reaction a ligand library with phosphines and imidazolium was screened against various palladium and nickel precursors. Aryl chlorides were coupled with aryl *Grignard* reagents at ambient temperature; the screening for the desired product was performed by ^{19}F-NMR.[39] Sterically demanding imidazolium salts **65** and **66** were identified to be most active in combination with a nickel precursor. In the *Sonogashira* coupling of phenylacetylene and aryl bromides at room temperature the same library was tested applying a color assay for product determination.[346] However, the azolium salts did not "light up" in this reaction. Instead $P(tBu)_3$ was identified as the best ligand. A different library consisting of phosphines and imidazolidinium salts in combination with a palladium source was examined in the *Buchwald–Hartwig* amination.[372] Imidazolidinium salt **67** combined with a palladium precursor turned out to work best. Again, the catalysts for amine formation at room temperature were identified by employing a color assay.

The *in situ* preparation of a ruthenium–alkylidene catalyst for olefin metathesis is the first step for extending this high-throughput approach toward other catalytic transformations and opens up the way to the screening of azolium salt libraries for olefin metathesis reactions.[113]

IV

OTHER APPLICATIONS

A. *Materials Science*

The development of new materials can also take advantage of the strong metal–ligand bond, which allows the development of high temperature resistant applications. Thus, NHC containing compounds can help to overcome decomposition of metal containing liquid crystals—so-called metallomesogens (MLC)—at the clearing point.[373] Additionally, many functionalized NHC derivatives can be easily designed considering electronic and steric effects of the ligands.

Liquid–crystalline cationic gold(I) bis(NHC) complexes were synthesized under phase-transfer catalysis conditions from a benzimidazolium salt with C_{12}, C_{14}, or C_{16} alkyl chains on the nitrogen atoms and [(Me_2S)AuCl] [Eq. (16)].[130] A lamellar β-mesophase is formed by the unusual interdigitation of the long alkyl residues. This mesophase is the result of a compromise between aromatic π interactions, hydrophobic interactions of alkyl chains, and hydrogen-bonding forces between the anions and C–H hydrogen atoms. In a similar manner, thermally stable, mesomorphic liquid–crystalline palladium(II) complexes of imidazolin-2-ylidenes and benzimidazolin-2-ylidenes were prepared by carbene transfer from silver(I) precursors.[230]

Gold(I)–NHC complexes of benzimidazolin-2-ylidenes [(NHC)AuX] (X = Cl, Br, I, thiophenolate, phenylacetylide) are luminous in acetonitrile solution and in the solid state with long lifetimes at room temperature. Multiple emissions have been observed for different NHCs and anions X. For 1,3-dimethylimidazolin-2-ylidene, crystalline samples show gold(I)–gold(I) and ring π–π intermolecular interactions that explain multiple emissions originating from NHC ^{3}IL and gold(I)–gold(I) ^{3}MC transitions.[231]

A helical homobimetallic mercury(II) complex with a bridging bis(NHC) ligand serves as a starting point for a supramolecular assembly.[89] Also tetrameric cyclic palladium(II) complexes have been obtained with bridging NHC–pyridine ligands.[135]

1,2,4-Triazolin-3,5-diylidenes were reported to form infinite chain-like structures with silver(I) resulting in ionic organometallic polymers [Eq. (12)].[121,122]

B. *Bioinorganic Chemistry*

Bioinorganic chemists have been attracted by the complex formations of NHC because the imidazolin-2-ylidene motif is encountered frequently in living organisms. The imidazole moiety is part of the purin bases in both DNA and RNA as well as the amino acid histidine which appears in proteins and enzymes and is in many cases considered to play a decisive role within the catalytically active center.[374] The possible formation of NHC complexes under physiological conditions or *in vivo* has been addressed by investigation of *N*-confused caffeine **73** or purine **74** complexes.

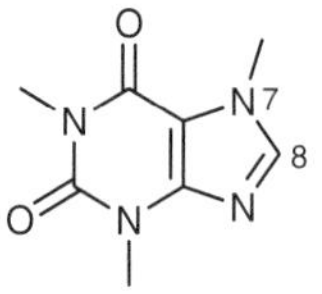

73 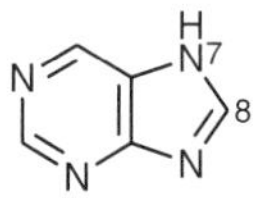 **74**

When mixing $[(H_3N)_4RuCl_2]$ and caffeine, the formation of a C(8) carbene ruthenium(II) complex was observed in addition to coordination via nitrogen N(7).[375,376] The *N*-bound ligand isomerizes to give the NHC in the presence of diluted acid.

Using mercury(II) as the metal to coordinate to purine derivatives also showed that spontaneous NHC complex formation is observed even if the N(7) position is not alkylated.[377–379] Depending on the pH of the solution, $[MeHg(NO_3)]$ coordinates sequentially to N(7), N(1), and finally to C(8). This type of metal-induced activation of the carbene carbon was also observed for gold(I) complexes.[380,381] The formation of NHC complexes starting at alkylated purine base salts with mercury(II) acetate is also possible.[88] The observation of C–H activation and the formation of NHC complexes of heavy metals with purines upon coordination of the metal to the vicinal nitrogen atom gives an alternative explanation for the mutagenicity of these metals, especially in the case of organomercurials. It is noteworthy that this reaction sequence does not require alkylation of both nitrogen atoms vicinal to the carbene carbon atom, although this facilitates the NHC complex formation. However, with none of the nitrogen atoms being alkylated, only *N*-coordination is observed. In the case of unalkylated species, the activation of the ligand to form the C-bound NHC complexes is achieved by either protonation of the nitrogen atom or by coordination of a *Lewis* acid. A similar activation sequence has been observed with borane coordination to an imidazole-nitrogen atom.[64]

Porphyrins and their metal complexes, respectively, occur as active sites in a number of enzymes. *N*-confused porphyrins such as **75** have been discussed regarding their analogy to NHCs and their potential importance in porphyrin chemistry.[382–385] Isolation of the first metal complexes containing palladium, silver, and copper was reported recently.[386,387] Macrocyclic imidazolylboranes **76** represent an interesting variation of inverted porphyrins. However, no metal complexes have been reported yet.[388]

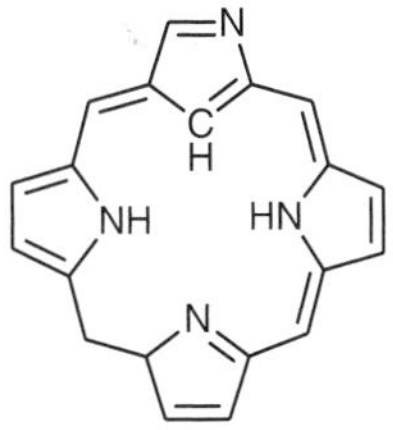

75 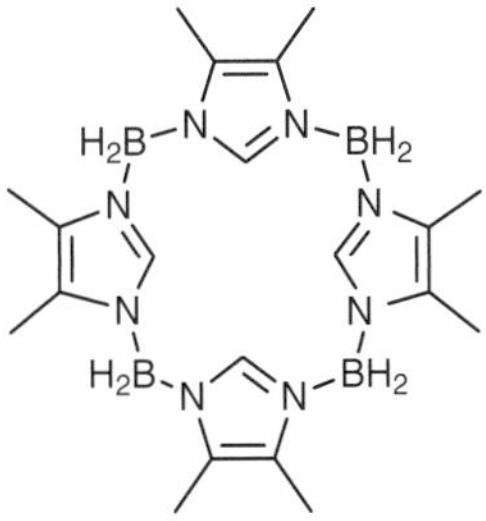 **76**

C. *Medicinal Chemistry*

Possible pharmaceutical properties of NHC metal complexes have been investigated. Due to the strong metal–ligand bond, no relevant dissociation equilibria under physiological conditions are to be expected which renders the complexes useful for delivering tailor made structures to the cells *in vivo.* Rhodium(I) compunds such as $[(\eta^4\text{-cod})Rh(NHC)Cl]$ and ruthenium(II) complexes like $[(\eta^6\text{-cymene})Ru(NHC)Cl_2]$ with imidazolidin-2-ylidene ligands were evaluated for their *in vitro* antimicrobial activity against the *Gram*-positive *Enterococcus faecalis* and *Staphylococcus aureus,* and the *Gram*-negative *Escherichia coli* and *Pseudomonas aeruginosa.* Different compounds were found to inhibit the growth of the *Gram*-positive bacteria with the minimum inhibitory concentrations (MIC) being as low as 5 μg/ml. The *Gram*-negative bacteria were inhibited by three rhodium(I) compounds with substantially higher MIC (200–400 μg/ml).[389] Extending the study to further rhodium(I) and ruthenium(II) complexes of benzimidazolin-2-ylidenes and to the antimicrobial activity against 98 *Staphylococcus* strains an inhibitory effect for some of the complexes was observed.[390] The four most effective compounds of these studies were tested for their serum MIC values and side effects on hepatic and renal functions on *Wistar* rats in order to determine whether they can be used for therapeutic purposes. However, none of the tested compounds showed antimicrobial activity at their serum concentrations.[391] In another series of experiments, the ruthenium(II) complexes bearing imidazolidin-2-ylidene ligands showed more pronounced antimicrobial activity against *Gram*-positive bacteria and fungi than complexes containing nitrogen donors as ligands.[392] Hydrophobic substituents on the NHC made those compounds significantly more effective.

V

SUMMARY

Metal complexes of stable carbenes—or more precisely metal complexes of carbenes that are now known to be stable—have developed from laboratory curiosities to widely used compounds. The basis for this development was laid by Wanzlick's and Öfele's discoveries in the 1960s, the recent revival has certainly been driven by Arduengo's first isolation of an *N*-heterocyclic carbene in 1991, and the result is a permanently increasing number of synthetic routes towards precursors for stable carbenes, towards stable carbenes themselves, and their metal complexes. Simultaneously with their accessibility, the applicability of these compounds to various fields such as homogeneous catalysis, materials science, medicinal and bioinorganic chemistry has been evaluated.

Although a relatively broad variety of stable carbenes is now accessible, cyclic diaminocarbenes (*N*-heterocyclic carbenes, NHC) have been used almost exclusively for the generation of metal coordination compounds. The free NHCs are usually prepared by the deprotonation of the corresponding azolium salt precursors. Those are readily accessible following relatively simple condensation–cyclization protocols, often in form of multicomponent one-pot reactions. Other routes to get to carbenes such as cleavage of tetraaminoethylenes complement the deprotonation route. Complexes of stable carbenes are known for most of the metals of the periodic table of elements. The most general approach for their preparation is via the free carbene that can either be isolated before the complexation or trapped *in situ* by the appropriate metal precursor. In cases where the free carbene cannot be synthesized the complex formation has to be accomplished *in situ* from a ligand precursor, e.g., the azolium salt. The metal–carbon bond of an NHC is significantly different from a "real" metal–carbene bond when *Fischer*- or *Schrock*-type carbenes are involved and it is described best as a strong σ-bond. The length of this bond is usually in the range of a typical metal-hydrocarbyl single bond and π-backbonding is considered to be negligible.

As a result, the reactivity of these metal–NHC compounds is also unique. They prove to be rather resistant towards attack of nucleophiles or electrophiles at the divalent carbon atom. Additionally, theoretical calculations and experimental investigations agree that the ligand dissociation energy for an NHC is higher than for a phosphine.

Promising applications for metal–NHC compounds in materials science and medicinal chemistry are based on the strong metal–carbon bond and the high donor capability of the NHC. The most extensive investigations have been carried out in the field of homogeneous catalysis. Here, NHCs have to be considered as typical directing or innocent spectator ligands, best comparable to trialkylphosphines. The recent successful applications of NHCs in ruthenium-catalyzed olefin metathesis and palladium/nickel-catalyzed coupling reactions prove this concept. The interest of bioinorganic chemists in metal–NHC complexes has its roots in their similarity to "confused" purine or porphyrine complexes that might play a role in enzyme catalyzed processes.

ACKNOWLEDGMENTS

The authors thank the *Alexander von Humboldt-Stiftung* (fellowships for T. Weskamp and V. P. W. Böhm) and the *Fonds der Chemischen Industrie* (studentships for T. Weskamp and V. P. W. Böhm). Special thanks goes to U. Tracht (Symyx Technologies) and to our pioneer in NHC chemistry, K. Öfele, for helpful discussions.

REFERENCES

(1) Gomberg, M. *Chem. Ber.* **1900,** *33,* 3150.
(2) Griller, D.; Ingold, K. U. *Acc. Chem. Res.* **1976,** *9,* 13.
(3) Doering, W. v. E.; Hoffmann, A. K. *J. Am. Chem. Soc.* **1954,** *76,* 6162.
(4) Fischer, E. O.; Maasböl, A. *Angew. Chem. Int. Ed. Engl.* **1964,** *3,* 580.
(5) Igau, A.; Grutzmacher, H.; Baceiredo, A.; Bertand, G. *J. Am. Chem. Soc.* **1988,** *110,* 6463.
(6) Bourissou, D.; Bertand, G. *Adv. Organomet. Chem.* **1999,** *44,* 175.
(7) Arduengo, A. J., III; Harlow, R. L.; Kline, M. *J. Am. Chem. Soc.* **1991,** *113,* 361.
(8) Arduengo, A. J., III; *Acc. Chem. Res.* **1999,** *32,* 913.
(9) Herrmann, W. A.; Köcher, C. *Angew. Chem. Int. Ed. Engl.* **1997,** *36,* 2163.
(10) Bourissou, D.; Guerret, O.; Gabbaï, F. P.; Bertrand, G. *Chem. Rev.* **2000,** *100,* 39.
(11) Wanzlick, H.-W.; Schönherr, H.-J. *Angew. Chem. Int. Ed. Engl.* **1968,** *7,* 141.
(12) Öfele, K. *J. Organomet. Chem.* **1968,** *12,* P42.
(13) Schrock, R. R. *J. Am. Chem. Soc.* **1974,** *96,* 6796.
(14) Wanzlick, H.-W.; Esser, F.; Kleiner, H.-J. *Chem. Ber.* **1963,** *96,* 1208.
(15) Wanzlick, H.-W. *Angew. Chem. Int. Ed. Engl.* **1962,** *1,* 75.
(16) Wanzlick, H.-W.; Kleiner, H.-J. *Angew. Chem.* **1961,** *73,* 493.
(17) Hahn, F. E.; Wittenbecher, L.; Le Van, D.; Fröhlich, R. *Angew. Chem. Int. Ed.* **2000,** *39,* 541.
(18) Heinemann, C.; Müller, T.; Apeloig, Y.; Schwarz, H. *J. Am. Chem. Soc.* **1996,** *118,* 2023.
(19) Herrmann, W. A.; Elison, M.; Fischer, J.; Köcher, C.; Artus, G. R. J. *Chem. Eur. J.* **1996,** *2,* 772.
(20) Herrmann, W. A.; Köcher, C.; Gooßen, L. J.; Artus, G. R. J. *Chem. Eur. J.* **1996,** *2,* 1627.
(21) Herrmann, W. A.; Gooßen, L. J.; Köcher, C.; Artus, G. R. J. *Angew. Chem. Int. Ed. Engl.* **1996,** *35,* 2805.
(22) Arduengo, A. J., III; Davidson, F.; Dias, H. V. R.; Goerlich, J. R.; Khasnis, D.; Marshall, W. J.; Prakasha, T. K. *J. Am. Chem. Soc.* **1997,** *119,* 12742.
(23) Fischer, J. Dissertation, Technische Universität München (Germany), 1996.
(24) Herrmann, W. A.; Elison, M.; Fischer, J.; Köcher, C.; Artus, G. R. J. *Angew. Chem. Int. Ed. Engl.* **1995,** *34,* 2371.
(25) Denk, M. K.; Thadani, A.; Hatano, K.; Lough, A. J. *Angew. Chem. Int. Ed. Engl.* **1997,** *36,* 2607.
(26) Arduengo, A. J., III; Goerlich, J. R.; Marshall, W. J. *J. Am. Chem. Soc.* **1995,** *117,* 11027.
(27) Alder, R. W.; Blake, M. E.; Bortolotti, C.; Bufali, S.; Butts, C. P.; Linehan, E.; Oliva, J. M.; Orpen, A. G.; Quayle, M. *Chem. Commun.* **1999,** 241.
(28) Alder, R. W.; Blake, M. E.; Bortolotti, C.; Bufali, S.; Butts, C. P.; Linehan, E.; Oliva, J. M.; Orpen, A. G.; Quayle, M. *Chem. Commun.* **1999,** 1049.

(29) Alder, R. W.; Allen, P. R.; Murray, M.; Orpen, A. G. *Angew. Chem. Int. Ed. Engl.* **1996,** *35,* 1121.
(30) Alder, R. W.; Blake, M. E. *Chem. Commun.* **1997,** 1513.
(31) Alder, R. W.; Butts, C. P.; Orpen, A. G. *J. Am. Chem. Soc.* **1998,** *120,* 11526.
(32) Fournari, P.; de Cointet, P.; Laviron, E. *Bull. Soc. Chim. Fr.* **1968,** 2438.
(33) Chan, B. K. M.; Chan, N.-H.; Grimmett, M. R. *Aust. J. Chem.* **1977,** *30,* 2005.
(34) Haque, M. R.; Rasmussen, M. *Tetrahedron* **1994,** *50,* 5535.
(35) Grimmett, M. R. *Imidazole and Benzimidazole Synthesis;* Academic Press: London, 1997.
(36) Harlow, K. J.; Hill, A. F.; Welton, T. *Synthesis* **1996,** 697.
(37) Arduengo, A. J., III; US 5077414, 1991.
(38) Herrmann, W. A.; Gooßen, L. J.; Spiegler, M. *J. Organomet. Chem.* **1997,** *547,* 357.
(39) Böhm, V. P. W.; Weskamp, T.; Gstöttmayr, C. W. K.; Herrmann, W. A. *Angew. Chem. Int. Ed.* **2000,** *39,* 1602.
(40) Bildstein, B.; Malaun, M.; Kopacka, H.; Wurst, K.; Mitterböck, M.; Ongania, K.-H.; Opromolla, G.; Zanello, P. *Organometallics* **1999,** *18,* 4325.
(41) Huang, J.; Nolan, S. P. *J. Am. Chem. Soc.* **1999,** *121,* 9889.
(42) Gridnev, A. A.; Mihaltseva, I. M. *Synth. Commun.* **1994,** *24,* 1547.
(43) Kiyomori, A.; Marcoux, J.-F.; Buchwald, S. L. *Tetrahedron Lett.* **1999,** *40,* 2657.
(44) Bildstein, B.; Malaun, M.; Kopacka, H.; Ongania, K.-H.; Wurst, K. *J. Organomet. Chem.* **1999,** *572,* 177.
(45) Teles, J. H.; Breuer, K.; Enders, D.; Gielen, H. *Synth. Commun.* **1999,** *29,* 1.
(46) Scholl, M.; Ding, S.; Lee, C. W.; Grubbs, R. H. *Org. Lett.* **1999,** *1,* 953.
(47) Saba, S.; Brescia, A. M.; Kaloustian, M. K. *Tetrahedron Lett.* **1991,** *32,* 5031.
(48) Cabanal-Duvillard, I.; Mangeney, P. *Tetrahedron Lett.* **1999,** *40,* 3877.
(49) Pebler, J.; Petz, W. *Z. Naturfosch. B* **1977,** *32,* 1431.
(50) Petz, W. *Angew. Chem. Int. Ed. Engl.* **1975,** *14,* 367.
(51) Petz, W. *J. Organomet. Chem.* **1979,** *172,* 415.
(52) Bodensieck, U.; Stoeckli-Evans, H.; Süss-Fink, G. *J. Organomet. Chem.* **1992,** *433,* 149.
(53) Azam, K. A.; Hossain, M. A.; Hursthouse, M. B.; Kabir, S. E.; Malik, K. M. A.; Vahrenkamp, H. *J. Organomet. Chem.* **1998,** *555,* 285.
(54) Fehlhammer, W. P.; Finck, W. *J. Organomet. Chem.* **1991,** *414,* 261.
(55) Fischer, E. O.; Reitmeier, R. *Z. Naturforsch. B* **1983,** *38,* 582.
(56) Carmalt, C. J.; Cowley, A. H. *Adv. Inorg. Chem.* **2000,** *50,* 1.
(57) Arduengo, A. J., III; Tamm, M.; Calabrese, J. C.; Davidson, F.; Marshall, W. J. *Chem. Lett.* **1999,** 1021.
(58) Herrmann, W. A.; Runte, O.; Artus, G. *J. Organomet. Chem.* **1995,** *501,* C1.
(59) Arduengo, A. J., III; Dias, H. V. R.; Davidson, F.; Harlow, R. L. *J. Organomet. Chem.* **1993,** *462,* 13.
(60) Runte, O. Dissertation, Technische Universität München (Germany), 1997.
(61) Arduengo, A. J., III; Davidson, F.; Krafczyk, R.; Marshall, W. J.; Tamm, M. *Organometallics* **1998,** *17,* 3375.
(62) Kuhn, N.; Henkel, G.; Kratz, T.; Kreutzberg, J.; Boese, R.; Maulitz, A. H. *Chem. Ber.* **1993,** *126,* 2041.
(63) Arduengo, A. J., III; Davidson, F.; Krafczyk, R.; Marshall, W. J.; Schmutzler, R. *Monatsh. Chem.* **2000,** *131,* 251.
(64) Wacker, A.; Pritzkow, H.; Siebert, W. *Eur. J. Inorg. Chem.* **1998,** 843.
(65) Zheng, X.; Herberich, G. E. *Organometallics* **2000,** *19,* 3751.
(66) Arduengo, A. J., III; Dias, H. V. R.; Calabrese, J. C.; Davidson, F. *J. Am. Chem. Soc.* **1992,** *114,* 9724.

(67) Black, S. J.; Hibbs, D. E.; Hursthouse, M. B.; Jones, C.; Malik, K. M. A.; Smithies, N. A. *J. Chem. Soc. Dalton Trans.* **1997,** 4313.
(68) Hibbs, D. E.; Hursthouse, M. B.; Jones, C.; Smithies, N. A. *Chem. Commun.* **1998,** 869.
(69) Francis, M. D.; Hibbs, D. E.; Hursthouse, M. B.; Jones, C.; Smithies, N. A. *J. Chem. Soc. Dalton Trans.* **1998,** 3249.
(70) Li, X.-W.; Su, J.; Robinson, G. H. *Chem. Commun.* **1996,** 2683.
(71) Abernethy, C. D.; Cole, M. L.; Jones, C. *Organometallics* **2000,** *19,* 4852.
(72) Arduengo, A. J., III; Dias, H. V. R.; Calabrese, J. C.; Davidson, F. *Inorg. Chem.* **1993,** *32,* 1541.
(73) Schäfer, A.; Weidenbruch, M.; Saak, W.; Pohl, S. *J. Chem. Soc. Chem. Commun.* **1995,** 1157.
(74) Kuhn, N.; Kratz, T.; Bläser, D.; Boese, R. *Chem. Ber.* **1995,** *128,* 245.
(75) Stabenow, F.; Saak, W.; Weidenbruch, M. *Chem. Commun.* **1999,** 1131.
(76) Boesveld, W. M.; Gehrhus, B.; Hitchcock, P. B.; Lappert, M. F.; Rague-Schleyer, P. v. *Chem. Commun.* **1999,** 755.
(77) Harrison, P. G.; Idowu, O. A. *Inorg. Chim. Acta* **1984,** *81,* 213.
(78) Arduengo, A. J., III; Calabrese, J. C.; Cowley, A. H.; Dias, H. V. R.; Goerlich, J. R.; Marshall, W. J.; Riegel, B. *Inorg. Chem.* **1997,** *36,* 2151.
(79) Arduengo, A. J., III; Krafczyk, R.; Schmutzler, R.; Mahler, W.; Marshall, W. J. *Z. Anorg. Allg. Chem.* **1999,** *625,* 1813.
(80) Kuhn, N.; Kratz, T.; Henkel, G. *Z. Naturforsch. B* **1996,** *51,* 295.
(81) Öfele, K.; Herberhold, M. *Angew. Chem. Int. Ed. Engl.* **1970,** *9,* 739.
(82) Öfele, K.; Kreiter, C. G. *Chem. Ber.* **1972,** *105,* 529.
(83) Öfele, K.; Roos, E.; Herberhold, M. *Z. Naturforsch. B* **1976,** *31,* 1070.
(84) Huttner, G.; Gartzke, W. *Chem. Ber.* **1972,** *105,* 2714.
(85) Luger, P.; Ruban, G. *Acta Crystallogr. Sect. B* **1971,** *27,* 2276.
(86) Bildstein, B.; Malaun, M.; Kopacka, H.; Ongania, K.-H.; Wurst, K. *J. Organomet. Chem.* **1998,** *552,* 45.
(87) Arduengo, A. J., III; Harlow, R. L.; Marshall, W. J.; Prakasha, T. K. *Heteroat. Chem.* **1996,** *7,* 421.
(88) Beck, W.; Kottmair, N. *Chem. Ber.* **1976,** *109,* 970.
(89) Chen, J. C. C.; Lin, I. J. B. *J. Chem. Soc. Dalton Trans.* **2000,** 839.
(90) Schönherr, H.-J.; Wanzlick, H.-W. *Chem. Ber.* **1970,** *103,* 1037.
(91) Herrmann, W. A.; Gerstberger, G.; Spiegler, M. *Organometallics* **1997,** *16,* 2209.
(92) Herrmann, W. A.; Schwarz, J.; Gardiner, M. G.; Spiegler, M. *J. Organomet. Chem.* **1999,** *575,* 80.
(93) Herrmann, W. A.; Schwarz, J.; Gardiner, M. G. *Organometallics* **1999,** *18,* 4082.
(94) Ortwerth, M. F.; Wyzlic, M. J.; baughman, R. G. *Acta Crystallogr. Sect. C* **1998,** *54,* 1594.
(95) Schwarz, J.; Böhm, V. P. W.; Gardiner, M. G.; Grosche, M.; Herrmann, W. A.; Hieringer, W.; Raudaschl-Sieber, G. *Chem. Eur. J.* **2000,** *6,* 1773.
(96) Herrmann, W. A.; Reisinger, C.-P.; Spiegler, M. *J. Organomet. Chem.* **1998,** *557,* 93.
(97) Gardiner, M. G.; Herrmann, W. A.; Reisinger, C.-P.; Schwarz, J.; Spiegler, M. *J. Organomet. Chem.* **1999,** *572,* 239.
(98) Herrmann, W. A.; Böhm, V. P. W.; Reisinger, C.-P. *J. Organomet. Chem.* **1999,** *576,* 23.
(99) Bertrand, G.; Diéz-Barra, E.; Fernández-Baeza, J.; Gornitzka, H.; Moreno, A.; Otero, A.; Rodríguez-Curiel, R. I.; Tejeda, J. *Eur. J. Inorg. Chem.* **1999,** 1965.
(100) Douthwaite, R. E.; Haüssinger, D.; Green, M. L. H.; Silcock, P. J.; Gomes, P. T.; Martins, A. M.; Danopoulos, A. A. *Organometallics* **1999,** *18,* 4584.
(101) Böhm, V. P. W.; Herrmann, W. A. *Chem. Eur. J.* **2000,** *6,* 1017.
(102) Xu, L.; Chen, W.; Xiao, J. *Organometallics* **2000,** *19,* 1123.
(103) Clyne, D. S.; Jin, J.; Genest, E.; Gallucci, J. C.; RajanBabu, T. V. *Org. Lett.* **2000,** *2,* 1125.

(104) Herrmann, W. A.; Fischer, J.; Öfele, K.; Artus, G. R. J. *J. Organomet. Chem.* **1997,** *530,* 259.
(105) Hahn, F. E.; Foth, M. *J. Organomet. Chem.* **1999,** *585,* 241.
(106) Caló, V.; Del Sole, R.; Nacci, A.; Schingaro, E.; Scordari, F. *Eur. J. Org. Chem.* **2000,** 869.
(107) Xu, L.; Chen, W.; Bickley, J. F.; Steiner, A.; Xiao, J. *J. Organomet. Chem.* **2000,** *598,* 409.
(108) Chen, J. C. C.; Lin, I. J. B. *Organometallics* **2000,** *19,* 5113.
(109) Demidov, V. N.; Kukushkin, Y. N.; Vedeneeva, L. N.; Belyaev, A. N. *J. Gen. Chem. USSR* **1988,** *58,* 652.
(110) Köcher, C.; Herrmann, W. A. *J. Organomet. Chem.* **1997,** *532,* 261.
(111) Mühlhofer, M. Diplomarbeit, Technische Universität München (Germany),1999.
(112) Prinz, M. Diplomarbeit, Technische Universität München (Germany), 1996.
(113) Morgan, J. P.; Grubbs, R. H. *Org. Lett.* **2000,** *2,* 3153.
(114) Jafarpour, L.; Nolan, S. P. *Organometallics* **2000,** *19,* 2055.
(115) Baratta, W.; Herrmann, W. A.; Rigo, P.; Schwarz, J. *J. Organomet. Chem.* **2000,** *593–594,* 489.
(116) Ackermann, K.; Hofmann, P.; Köhler, F. H.; Kratzer, H.; Krist, H.; Öfele, K.; Schmidt, H. R. *Z. Naturforsch. B* **1983,** *38,* 1313.
(117) Herrmann, W. A.; Mihalios, D.; Öfele, K.; Kiprof, P.; Belmedjahed, F. *Chem. Ber.* **1992,** *125,* 1795.
(118) Öfele, K.; Herrmann, W. A.; Mihalios, D.; Elison, M.; Herdtweck, E.; Priermeier, T.; Kiprof, P. *J. Organomet. Chem.* **1995,** *498,* 1.
(119) Wang, H. M. J.; Lin, I. J. B. *Organometallics* **1998,** *17,* 972.
(120) Norris, A. R.; Buncel, E.; Taylor, S. E. *J. Inorg. Biochem.* **1982,** *16,* 279.
(121) Guerret, O.; Solé, S.; Gornitzka, H.; Teichert, M.; Trinquier, G.; Bertrand, G. *J. Am. Chem. Soc.* **1997,** *119,* 6668.
(122) Guerret, O.; Solé, S.; Gornitzka, H.; Trinquier, G.; Bertrand, G. *J. Organomet. Chem.* **2000,** *600,* 112.
(123) Voges, M. H.; Rømming, C.; Tilset, M. *Organometallics* **1999,** *18,* 529.
(124) Abernethy, C. D.; Cowley, A. H.; Jones, R. A. *J. Organomet. Chem.* **2000,** *596,* 3.
(125) Enders, D.; Gielen, H.; Raabe, G.; Runsink, J.; Teles, J. H. *Chem. Ber.* **1996,** *129,* 1483.
(126) Weskamp, T.; Böhm, V. P. W.; Herrmann, W. A. *J. Organomet. Chem.* **1999,** *585,* 348.
(127) Herrmann, W. A.; Böhm, V. P. W.; Gstöttmayr, C. W. K.; Grosche, M.; Reisinger, C. P.; Weskamp, T. *J. Organomet. Chem.* **2001,** *617–618,* 618.
(128) Öfele, K.; Herrmann, W. A.; Mihalios, D.; Elison, M.; Herdtweck, E.; Scherer, W.; Mink, J. *J. Organomet. Chem.* **1993,** *459,* 177.
(129) Davis, J. H.; Lake, C. M.; Bernard, M. A. *Inorg. Chem.* **1998,** *37,* 5412.
(130) Lee, K. M.; Lee, C. K.; Lin, I. J. B. *Angew. Chem. Int. Ed. Engl.* **1997,** *36,* 1850.
(131) Cooksey, C. J.; Dodd, D.; Johnson, M. D. *J. Chem. Soc. B* **1971,** 1380.
(132) Enders, D.; Gielen, H.; Raabe, G.; Runsink, J.; Teles, J. H. *Chem. Ber./Recueil* **1997,** *130,* 1253.
(133) Enders, D.; Gielen, H.; Runsink, J.; Breuer, K.; Brode, S.; Boehn, K. *Eur. J. Inorg. Chem.* **1998,** 913.
(134) Clark, A. M.; Oliver, A. G.; Rickard, C. E. F.; Wright, L. J.; Roper, W. R. *Acta Crystallogr. Sect. C* **2000,** *56,* 26.
(135) Tulloch, A. A. D.; Danopoulos, A. A.; Tooze, R. P.; Cafferkey, S. M.; Kleinhenz, S.; Hursthouse, M. B. *Chem. Commun.* **2000,** 1247.
(136) Fehlhammer, W. P.; Bliss, T.; Kernbach, U.; Brüdgam, I. *J. Organomet. Chem.* **1995,** *490,* 149.
(137) Kernbach, U.; Ramm, M.; Luger, P.; Fehlhammer, W. P. *Angew. Chem. Int. Ed. Engl.* **1996,** *35,* 310.
(138) Teles, J. H.; Melder, J.-P.; Ebel, K.; Schneider, R.; Gehrer, E.; Harder, W.; Brode, S.; Enders, D.; Breuer, K.; Raabe, G. *Helv. Chim. Acta* **1996,** *79,* 61.
(139) Enders, D.; Breuer, K.; Raabe, G.; Runsink, J.; Teles, J. H.; Melder, J.-P.; Ebel, K.; Brode, S. *Angew. Chem. Int. Ed. Engl.* **1995,** *34,* 1021.

(140) Chamizo, J. A.; Morgado, J.; Bernés, S. *Transition Met. Chem.* **2000,** *25,* 161.
(141) Chamizo, J. A.; Hitchcock, P. B.; Jasim, H. A.; Lappert, M. F. *J. Organomet. Chem.* **1993,** *451,* 89.
(142) Chamizo, J. A.; Morgado, J.; Álvarez, C.; Toscano, R. A. *Transition Met. Chem.* **1995,** *20,* 508.
(143) Lemal, D. M.; Kawano, K. I. *J. Am. Chem. Soc.* **1962,** *84,* 1761.
(144) Wanzlick, H.-W.; Schikora, E. *Angew. Chem.* **1960,** *72,* 494.
(145) Arduengo, A. J., III; Dias, H. V. R.; Harlow, R. L.; Kline, M. *J. Am. Chem. Soc.* **1992,** *114,* 5530.
(146) Weskamp, T.; Böhm, V. P. W.; Herrmann, W. A. *J. Organomet. Chem.* **2000,** *600,* 12.
(147) Kuhn, N.; Kratz, T. *Synthesis* **1993,** 561.
(148) Hahn, F. E.; Wittenbecher, L.; Boese, R.; Bläser, D. *Chem. Eur. J.* **1999,** *5,* 1931.
(149) Herrmann, W. A.; Goossen, L. J.; Artus, G. R. J.; Köcher, C. *Organometallics* **1997,** *16,* 2472.
(150) Prinz, M.; Grosche, M.; Herdtweck, E.; Herrmann, W. A. *Organometallics* **2000,** *19,* 1692.
(151) Huang, J.; Stevens, E. D.; Nolan, S. P. *Organometallics* **2000,** *19,* 1194.
(152) Jafarpour, L.; Schanz, H.-J.; Stevens, E. D.; Nolan, S. P. *Organometallics* **1999,** *18,* 3760.
(153) Huang, J.; Stevens, E. D.; Nolan, S. P.; Petersen, J. L. *J. Am. Chem. Soc.* **1999,** *121,* 2674.
(154) Kuhn, N.; Kratz, T.; Bläser, D.; Boese, R. *Inorg. Chim. Acta* **1995,** *238,* 179.
(155) Herrmann, W. A.; Munck, F. C.; Artus, G. R. J.; Runte, O.; Anwander, R. *Organometallics* **1997,** *16,* 682.
(156) Louie, J.; Grubbs, R. H. *Chem. Commun.* **2000,** 1479.
(157) Desmurs, P.; Dormond, A.; Nief, F.; Baudry, D. *Bull. Soc. Chim. Fr.* **1997,** *134,* 683.
(158) Weskamp, T.; Schattenmann, W. C.; Spiegler, M.; Herrmann, W. A. *Angew. Chem. Int. Ed.* **1998,** *37,* 2490.
(159) Weskamp, T.; Kohl, F. J.; Hieringer, W.; Gleich, D.; Herrmann, W. A. *Angew. Chem. Int. Ed.* **1999,** *38,* 2416.
(160) Weskamp, T.; Kohl, F. J.; Herrmann, W. A. J. *Organomet. Chem.* **1999,** *582,* 362.
(161) Scholl, M.; Trnka, T. M.; Morgan, J. P.; Grubbs, R. H. *Tetrahedron Lett.* **1999,** *40,* 2247.
(162) Steinbeck, M. Dissertation, Technische Universität München (Germany), 1998.
(163) Böhm, V. P. W.; Gstöttmayr, C. W. K.; Weskamp, T.; Herrmann, W. A. *J. Organomet. Chem.* **2000,** *595,* 186.
(164) Gstöttmayr, C. W. K.; Böhm, V. P. W.; Weskamp, T.; Herrmann, W. A. unpublished results.
(165) Kuhn, N.; Kratz, T.; Boese, R.; Bläser, D. *J. Organomet. Chem.* **1994,** *470,* C8.
(166) Kuhn, N.; Kratz, T.; Boese, R.; Bläser, D. *J. Organomet. Chem.* **1994,** *479,* C32.
(167) Köcher, C. Dissertation, Technische Universität München (Germany), 1997.
(168) Schumann, H.; Glanz, M.; Winterfeld, J.; Hemling, H.; Kuhn, N.; Kratz, T. *Angew. Chem. Int. Ed. Engl.* **1994,** *33,* 1733.
(169) Schumann, H.; Glanz, M.; Winterfeld, J.; Hemling, H.; Kuhn, N.; Kratz, T. *Chem. Ber.* **1994,** *127,* 2369.
(170) Arduengo, A. J., III; Tamm, M.; McLain, S. J.; Calabrese, C. J.; Davidson, F.; Marshall, W. J. *J. Am. Chem. Soc.* **1994,** *116,* 7927.
(171) Baudry-Barbier, D.; Andre, N.; Dormond, A.; Pardes, C.; Richard, P.; Visseaux, M.; Zhu, C. J. *Eur. J. Inorg. Chem.* **1998,** 1721.
(172) Herrmann, W. A.; Öfele, K.; Elison, M.; Kühn, F. E.; Roesky, P. W. *J. Organomet. Chem.* **1994,** *480,* C7.
(173) Döhring, A.; Göhre, J.; Jolly, P. W.; Kryger, B.; Rust, J.; Verhovnik, G. P. J. *Organometallics* **2000,** *19,* 388.
(174) Danopoulos, A. A.; Hankin, D. M.; Wilkinson, G.; Cafferkey, S. M.; Sweet, T. K. N.; Hursthouse, M. B. *Polyhedron* **1997,** *16,* 3879.
(175) Herrmann, W. A.; Lobmaier, G. M.; Elison, M. *J. Organomet. Chem.* **1996,** *520,* 231.
(176) Arduengo, A. J., III; Dias, H. V. R.; Calabrese, J. C.; Davidson, F. *Organometallics* **1993,** *12,* 3405.

(177) Arduengo, A. J., III; Goerlich, J. R.; Davidson, F.; Marshall, W. J. Z. *Naturforsch. B* **1999,** *54,* 1350.
(178) Faust, R.; Göbelt, B. *Chem. Commun.* **2000,** 919.
(179) Arduengo, A. J., III; Gamper, S. F.; Calabrese, J. C.; Davidson, F. *J. Am. Chem. Soc.* **1994,** *116,* 4391.
(180) McGuinness, D. S.; Cavell, K. J.; Skelton, B. W.; White, A. H. *Organometallics* **1999,** *18,* 1596.
(181) McGuinness, D. S.; Green, M. J.; Cavell, K. J.; Skelton, B. W.; White, A. H. *J. Organomet. Chem.* **1998,** *565,* 165.
(182) Green, M. J.; Cavell, K. J.; Skelton, B. W.; White, A. H. *J. Organomet. Chem.* **1998,** *554,* 175.
(183) Abernethy, C. D.; Clyburne, J. A. C.; Cowley, A. H.; Jones, R. A. *J. Am. Chem. Soc.* **1999,** *121,* 2329.
(184) Winberg, H. E.; Carnahan, J. E.; Coffman, D. D.; Brown, M. *J. Am. Chem. Soc.* **1965,** *87,* 2055.
(185) Liu, Y.; Lindner, P. E.; Lemal, D. M. *J. Am. Chem. Soc.* **1999,** *121,* 10626.
(186) Böhm, V. P. W.; Herrmann, W. A. *Angew. Chem. Int. Ed.* **2000,** *39,* 4036.
(187) Lappert, M. F. *J. Organomet. Chem.* **1988,** *358,* 185.
(188) Çetinkaya, B.; Dixneuf, P.; Lappert, M. F. *J. Chem. Soc. Chem. Commun.* **1973,** 206.
(189) Çetinkaya, B.; Dixneuf, P.; Lappert, M. F. *J. Chem. Soc. Dalton Trans.* **1974,** 1827.
(190) Lappert, M. F.; Pye, P. L. *J. Chem Soc. Dalton Trans.* **1977,** 2172.
(191) Hitchcock, P. B.; Lappert, M. F.; Pye, P. L. *J. Chem. Soc. Dalton Trans.* **1978,** 826.
(192) Hitchcock, P. B.; Lappert, M. F.; Thomas, S. A.; Thorne, A. J.; Carty, A. J.; Taylor, N. J. *J. Organomet. Chem.* **1986,** *315,* 27.
(193) Macomber, D. W.; Rogers, R. D. *Organometallics* **1985,** *4,* 1485.
(194) Delgado, S.; Moreno, C.; Macazaga, M. J. *Polyhedron* **1991,** *10,* 725.
(195) Hitchcock, P. B.; Lappert, M. F.; Pye, P. L. *J. Chem. Soc. Dalton Trans.* **1977,** 2160.
(196) Lappert, M. F.; Pye, P. L. *J. Chem. Soc. Dalton Trans.* **1977,** 1283.
(197) Lappert, M. F.; Pye, P. L.; McLaughlin, G. M. *J. Chem. Soc. Dalton Trans.* **1977,** 1272.
(198) Coleman, A. W.; Hitchcock, P. B.; Lappert, M. F.; Maskell, R. K.; Müller, J. H. *J. Organomet. Chem.* **1985,** *296,* 173.
(199) Lappert, M. F.; Martin, T. R.; McLaughlin, G. M. *J. Chem. Soc. Chem. Commun.* **1980,** 635.
(200) Hitchcock, P. B.; Lappert, M. F.; Pye, P. L. *J. Chem. Soc. Dalton Trans.* **1978,** 826.
(201) Manjolovic-Muir, L.; Muir, K. W. *J. Chem. Soc. Dalton Trans.* **1974,** 2427.
(202) Cardin, D. J.; Çetinkaya, B.; Çetinkaya, E.; Lappert, M. F. *J. Chem. Soc. Dalton Trans.* **1973,** 594.
(203) Çetinkaya, B.; Özdemir, I.; Dixneuf, P. H. *J. Organomet. Chem.* **1997,** *534,* 153.
(204) Lappert, M. F.; Pye, P. L. *J. Chem. Soc. Dalton Trans.* **1978,** 837.
(205) Lappert, M. F.; Pye, P. L. *J. Chem. Soc. Chem. Commun.* **1976,** 644.
(206) Çetinkaya, B.; Hitchcock, P. B.; Lappert, M. F.; Shaw, D. B.; Spyropoulos, K.; Warhurst, N. J. W. *J. Organomet. Chem.* **1993,** *459,* 311.
(207) Doyle, M. J.; Lappert, M. F.; Pye, P. L.; Terreros, P. *J. Chem. Soc. Dalton Trans.* **1984,** 2355.
(208) Hitchcock, P. B.; Lappert, M. F.; Terreros, P. *J. Organomet. Chem.* **1982,** *239,* C26.
(209) Hitchcock, P. B.; Lappert, M. F.; Terreros, P.; Wainwright, K. P. *J. Chem. Soc. Chem. Commun.* **1980,** 1180.
(210) Hiraki, K.; Onishi, M.; Sewaki, K.; Sugino, K. *Bull. Chem. Soc. Jpn.* **1978,** *51,* 2548.
(211) Hitchcock, P. B.; Lappert, M. F.; Pye, P. L.; Thomas, S. *J. Chem. Soc. Dalton Trans.* **1979,** 1929.
(212) Hitchcock, P. B.; Lappert, M. F.; Pye, P. L. *J. Chem. Soc. Chem. Commun.* **1977,** 196.
(213) Owen, M. A.; Pye, P. L.; Piggott, B.; Caparelli, M. V. *J. Organomet. Chem.* **1992,** *434,* 351.
(214) Hiraki, K.; Onishi, M.; Sugino, K. *J. Organomet. Chem.* **1979,** *171,* C50.
(215) Hiraki, K.; Sugino, K. *J. Organomet. Chem.* **1980,** *201,* 469.
(216) Hiraki, K.; Sugino, K.; Onishi, M. *Bull. Chem. Soc. Jpn.* **1980,** *53,* 1976.
(217) Hiraki, K.; Onishi, M.; Ohnuma, K.; Sugino, K. *J. Organomet. Chem.* **1981,** *216,* 413.

(218) Çetinkaya, E.; Hitchcock, P. B.; Küçükbay, H.; Lappert, M. F.; Al-Juaid, S. *J. Organomet. Chem.* **1994,** *481,* 89.
(219) Arnold, P. L.; Cloke, F. G. N.; Geldbach, T.; Hitchcock, P. B. *Organometallics* **1999,** *18,* 3228.
(220) Ugi, I.; Offermanns, K. *Isonitrile Chemistry;* Academic Press:London, 1971.
(221) Rieger, D.; Lotz, S. D.; Kernbach, U.; Schröder, S.; André, C.; Fehlhammer, W. P. *Inorg. Chim. Acta* **1994,** *222,* 275.
(222) Fehlhammer, W. P.; Bartel, K.; Völkl, A.; Achatz, D. *Z. Naturforsch. B* **1982,** *37,* 1044.
(223) Weinberger, B.; Degel, F.; Fehlhammer, W. P. *Chem. Ber.* **1985,** *118,* 51.
(224) Michelin, R. A.; Zanotto, L.; Braga, D.; Sabatino, P.; Angelici, R. J. *Inorg. Chem.* **1988,** *27,* 93.
(225) Liu, C.-Y.; Chen, D.-Y.; Lee, G.-H.; Peng, S.-M.; Liu, S.-T. *Organometallics* **1996,** *15,* 1055.
(226) Kreiter, C. G.; Öfele, K.; Wieser, G. W. *Chem. Ber.* **1976,** *109,* 1749.
(227) Öfele, K.; Herberhold, M. *Z. Naturforsch. B* **1973,** *28,* 306.
(228) Liu, S.-T.; Hsieh, T.-Y.; Lee, G.-H.; Peng, S.-M. *Organometallics* **1998,** *17,* 993.
(229) Ku, R.-Z.; Huang, J.-C.; Cho, J.-Y.; Kiang, F.-M.; Reddy, K. R.; Chen, Y.-C.; Lee, K.-J.; Lee, J.-H.; Lee, G.-H.; Peng, S.-M.; Liu, S.-T. *Organometallics* **1999,** *18,* 2145.
(230) Lee, C. K.; Chen, J. C. C.; Lee, K. M.; Liu, C. W.; Lin, I. J. B. *Chem. Mater.* **1999,** *11,* 1237.
(231) Wang, H. M. J.; Chen, C. Y. L.; Lin, I. J. B. *Organometallics* **1999,** *18,* 1216.
(232) McGuinness, D. S.; Cavell, K. J. *Organometallics* **2000,** *19,* 741.
(233) Lappert, M. F.; MacQuitty, J. J.; Pye, P. L. *J. Chem. Soc. Dalton Trans.* **1981,** 1583.
(234) Lappert, M. F.; McCabe, R. W.; MacQuitty, J. J.; Pye, P. L.; Riley, P. I. *J. Chem. Soc. Dalton Trans.* **1980,** 90.
(235) Rieke, R. D.; Kojima, H.; Saji, T.; Rechberger, P.; Öfele, K. *Organometallics* **1988,** *7,* 749.
(236) Rieke, R. D.; Kojima, H.; Öfele, K. *J. Am. Chem. Soc.* **1976,** *98,* 6735.
(237) Raubenheimer, H. G.; Olivier, P. J.; Lindeque, L.; Desmet, M.; Hrusak, J.; Kruger, G. J. *J. Organomet. Chem.* **1997,** *544,* 91.
(238) Böhm, V. P. W.; Herrmann, W. A. unpublished results.
(239) Bonati, F.; Burini, A.; Pietroni, B. R.; Bovio, B. *J. Organomet. Chem.* **1989,** *375,* 147.
(240) Raubenheimer, H. G.; Lindeque, L.; Cronje, S. *J. Organomet. Chem.* **1996,** *511,* 177.
(241) Kruger, G. J.; Olivier, P. J.; Lindeque, L.; Raubenheimer, H. G. *Acta Crystallogr. Sect. C* **1995,** *51,* 1814.
(242) Raubenheimer, H. G.; Stander, Y.; Marais, E. K.; Thompson, C.; Kruger, G. J.; Cronje, S.; Deetlefs, M. *J. Organomet. Chem.* **1999,** *590,* 158.
(243) Raubenheimer, H. G.; Cronje, S.; Olivier, P. J. *J. Chem. Soc. Dalton Trans.* **1995,** 313.
(244) Raubenheimer, H. G.; Cronje, S.; van Rooyen, P. H.; Olivier, P. J.; Toerien, J. G. *Angew. Chem. Int. Ed. Engl.* **1994,** *33,* 672.
(245) Matsumura, N.; Kawano, J.-I.; Fukunishi, N.; Inoue, H. *J. Am. Chem. Soc.* **1995,** *117,* 3623.
(246) Yasui, M.; Yoshida, S.; Kakuma, S.; Shimamoto, S.; Matsumura, N.; Iwasaki, F. *Bull. Chem. Soc. Jpn.* **1996,** *69,* 2739.
(247) Iwasaki, F.; Manabe, N.; Yasui, M.; Matsumura, N.; Kamiya, N.; Iwasaki, H. *Bull. Chem. Soc. Jpn.* **1996,** *69,* 2749.
(248) Iwasaki, F.; Nishiyama, H.; Yasui, M.; Kusamiya, M.; Matsumura, N. *Bull. Chem. Soc. Jpn.* **1997,** *70,* 1277.
(249) Sellmann, D.; Prechtel, W.; Knoch, F.; Moll, M. *Organometallics* **1992,** *11,* 2346.
(250) Sellmann, D.; Prechtel, W.; Knoch, F.; Moll, M. *Inorg. Chem.* **1993,** *32,* 538.
(251) Sellmann, D.; Allmann, C.; Heinemann, F.; Knoch, F.; Sutter, J. *J. Organomet. Chem.* **1997,** *541,* 291.
(252) Ku, R.-Z.; Chen, D.-Y.; Lee, G.-H.; Peng, S.-M.; Liu, S.-T. *Angew. Chem. Int. Ed. Engl.* **1997,** *36,* 2631.
(253) Lang, H.; Vittal, J. J.; Leung, P.-H. *J. Chem. Soc. Dalton Trans.* **1998,** 2109.
(254) Herrmann, W. A.; Gooßen, L. J.; Spiegler, M. *Organometallics* **1998,** *17,* 2162.

(255) Herrmann, W. A.; Roesky, P. W.; Elison, M.; Artus, G.; Öfele, K. *Organometallics* **1995,** *14,* 1085.
(256) Elison, M. Dissertation, Technische Universität München (Germany), 1996.
(257) Elschenbroich, C.; Salzer, A. *Organometallchemie;* Teubner: Stuttgart, 1993.
(258) Cardin, D. J.; Çetinkaya, B.; Lappert, M. F. *J. Organomet. Chem.* **1974,** *72,* 139.
(259) Huang, J.; Schanz, H.-J.; Stevens, E. D.; Nolan, S. P. *Organometallics* **1999,** *18,* 2370.
(260) Huang, J.; Schanz, H.-J.; Stevens, E. D.; Nolan, S. P.; Capps, K. B.; Bauer, A.; Hoff, C. D. *Inorg. Chem.* **2000,** *39,* 1042.
(261) Cardin, D. J.; Çetinkaya, B.; Çetinkaya, E.; Lappert, M. F.; Manjolovic-Muir, L. J.; Muir, K. W. *J. Organomet. Chem.* **1972,** *44,* C59.
(262) Çetinkaya, B.; Çetinkaya, E.; Lappert, M. F. *J. Chem. Soc. Dalton Trans.* **1973,** 906.
(263) Lappert, M. F.; Pye, P. L.; Rogers, A. J.; McLaughlin, G. M. *J. Chem. Soc. Dalton Trans.* **1981,** 701.
(264) Scheidsteger, O.; Huttner, G.; Bejenke, V.; Gartzke, W. Z. *Naturforsch. B* **1983,** *38,* 1598.
(265) Doyle, M. J.; Lappert, M. F. *J. Chem. Soc. Chem. Commun.* **1974,** 679.
(266) Hafner, A.; Hegedus, L. S.; deWeck, G.; Hawkins, B.; Dötz, K. H. *J. Am. Chem. Soc.* **1988,** *110,* 8413.
(267) Oishi, S.; Tokumaru, K. J. *Phys. Org. Chem.* **1989,** *2,* 323.
(268) Hegedus, L. S.; Schwindt, M. A.; DeLombaert, S.; Imwinkelried, R. *J. Am. Chem. Soc.* **1990,** *112,* 2264.
(269) Green, J. C.; Scurr, R. G.; Arnold, P. L.; Cloke, F. G. N. *Chem. Commun.* **1997,** 1963.
(270) Boehme, C.; Frenking, G. *Organometallics* **1998,** *17,* 5801.
(271) Green, J. C.; Surr, R. G.; Arnold, P. L.; Cloke, F. G. N. *Chem. Commun.* **1997,** 1963.
(272) Alder, R. W.; Blake, M. E.; Oliva, J. M. *J. Phys. Chem. A* **1999,** *103,* 11200.
(273) Zaragoza-Dörwald, F. *Metal Carbenes in Organic Synthesis;* Wiley-VCH: Weinheim, 1998.
(274) Dötz, K. H.; Tomuschat, P. *Chem. Soc. Rev.* **1999,** *28,* 187.
(275) Hill, J. E.; Nile, T. A. *J. Organomet. Chem* **1977,** *137,* 293.
(276) Lappert, M. F. In *Transition Metal Chemistry;* Müller, A.; Diemann, E., Eds.; Verlag Chemie: Heidelberg, 1981.
(277) Lappert, M. F.; Maskell, R. K. *J. Organomet. Chem.* **1984,** *264,* 217.
(278) Cardin, D. J.; Doyle, M. J.; Lappert, M. F. *J. Chem. Soc. Chem. Commun.* **1972,** 927.
(279) Hill, J. E.; Nile, T. A. *Transition Met. Chem.* **1978,** *3,* 315.
(280) Schrock, R. R.; Feldman, J.; Cannizzo, L. F.; Grubbs, R. H. *Macromolecules* **1987,** *20,* 1169.
(281) Nguyen, S. T.; Johnson, L. K.; Grubbs, R. H. *J. Am. Chem. Soc.* **1992,** *114,* 3974.
(282) Nguyen, S. T.; Grubbs, R. H.; Ziller, J. W. *J. Am. Chem. Soc.* **1993,** *115,* 9858.
(283) Schwab, P.; France, M. B.; Ziller, J. W.; Grubbs, R. H. *Angew. Chem. Int. Ed.* **1995,** *34,* 2039.
(284) Schwab, P.; Grubbs, R. H.; Ziller, J. W. *J. Am. Chem. Soc.* **1996,** *118,* 100.
(285) Weskamp, T.; Schattenmann, W. C.; Spiegler, M.; Herrmann, W. A. *Angew. Chem. Int. Ed.* **1999,** *38,* 262.
(286) Dias, E. L.; Nguyen, S. T.; Grubbs, R. H. *J. Am. Chem. Soc.* **1997,** *119,* 3887.
(287) Denk, K. Diplomarbeit, Technische Universität München (Germany), 1999.
(288) Ackermann, L.; Fürstner, A.; Weskamp, T.; Kohl, F. J.; Herrmann, W. A. *Tetrahedron Lett.* **1999,** *40,* 4787.
(289) Frenzel, U.; Weskamp, T.; Kohl, F. J.; Schattenmann, W. C.; Nuyken, O.; Herrmann, W. A. *J. Organomet. Chem.* **1999,** *586,* 263.
(290) Hamilton, J. G.; Frenzel, U.; Kohl, F. J.; Weskamp, T.; Rooney, J. J.; Herrmann, W. A.; Nuyken, O. *J. Organomet. Chem.* **2000,** *606,* 8.
(291) Fürstner, A.; Thiel, O. R.; Blanda, G. *Org. Lett.* **2000,** *2,* 3731.
(292) Fürstner, A.; Thiel, O. R.; Ackermann, L.; Schanz, H.-J.; Nolan, S. P. *J. Org. Chem.* **2000,** *65,* 2204.

(293) Fürstner, A.; Thiel, O. R.; Kindler, N.; Bartkowska, B. *J. Org. Chem.* **2000,** *65,* 7990.
(294) Jafarpour, L.; Huang, J.; Stevens, E. D.; Nolan, S. P. *Organometallics* **1999,** *18,* 5416.
(295) Huang, J.; Schanz, H.-J.; Stevens, E. D.; Nolan, S. P. *Organometallics* **1999,** *18,* 5375.
(296) Schanz, H.-J.; Jafarpour, L.; Stevens, E. D.; Nolan, S. P. *Organometallics* **1999,** *18,* 5187.
(297) Ackermann, L.; ElTom, D.; Fürstner, A. *Tetrahedron* **2000,** *56,* 2195.
(298) Benningshof, J. C. J.; Blaauw, R. H.; van Ginkel, A. E.; Rutjes, F. P. J. T.; Fraanje, J.; Goubitz, K.; Schenk, H.; Hiemstra, H. *Chem. Commun.* **2000,** 1465.
(299) Bourgeois, D.; Mahuteau, J.; Pancrazi, A.; Nolan, S. P.; Prunet, J. *Synthesis* **2000,** 869.
(300) Briot, A.; Bujard, M.; Gouverneur, V.; Nolan, S. P.; Mioskowski, C. *Org. Lett.* **2000,** *2,* 1517.
(301) Itoh, T.; Mitsukura, K.; Ishida, N.; Uneyama, K. *Org. Lett.* **2000,** *2,* 1431.
(302) Frenzel, U.; Nuyken, O.; Kohl, F. J.; Schattenmann, W. C.; Weskamp, T.; Herrmann, W. A. *Polym. Mater. Sci. Eng.* **1999,** *80,* 135.
(303) Bielawski, C. W.; Grubbs, R. H. *Angew. Chem. Int. Ed.* **2000,** *39,* 2903.
(304) Chatterjee, A. K.; Grubbs, R. H. *Org. Lett.* **1999,** *1,* 1751.
(305) Chatterjee, A. K.; Morgan, J. P.; Scholl, M.; Grubbs, R. H. *J. Am. Chem. Soc.* **2000,** *122,* 3783.
(306) Lee, C. W.; Grubbs, R. H. *Org. Lett.* **2000,** *2,* 2145.
(307) Garber, S. B.; Kingsbury, J. S.; Gray, B. L.; Hoveyda, A. H. *J. Am. Chem. Soc.* **2000,** *122,* 8168.
(308) Hyldtoft, L.; Madsen, R. *J. Am. Chem. Soc.* **2000,** *122,* 8444.
(309) Efremov, I.; Paquette, L. A. *J. Am. Chem. Soc.* **2000,** *122,* 9324.
(310) Limanto, J.; Snapper, M. L. *J. Am. Chem. Soc.* **2000,** *122,* 8071.
(311) Maynard, H. D.; Okada, S. Y.; Grubbs, R. H. *Macromolecules* **2000,** *33,* 6239.
(312) Smulik, J. A.; Diver, S. T. *Org. Lett.* **2000,** *2,* 2271.
(313) Wright, D. L.; Schulte, J. P.; Page, M. A. *Org. Lett.* **2000,** *2,* 1847.
(314) Ulman, M.; Grubbs, R. H. *J. Org. Chem.* **1999,** *64,* 7202.
(315) Mizoroki, T.; Mori, K.; Ozaki, A. *Bull. Chem. Soc. Jpn.* **1971,** *44,* 581.
(316) Heck, R. F.; Nolley, J. P. *J. Org. Chem.* **1972,** 2320.
(317) Shibasaki, M.; Vogl, E. M. *J. Organomet. Chem.* **1999,** *576,* 1.
(318) de Meijere, A.; Meyer, F. E. *Angew. Chem. Int. Ed. Engl.* **1994,** *33,* 2379.
(319) Miyaura, N.; Suzuki, A. *J. Chem. Soc., Chem. Commun.* **1979,** 866.
(320) Miyaura, N.; Yanagi, T.; Suzuki, A. *Synth. Commun.* **1981,** *11,* 513.
(321) Stanforth, S. P. *Tetrahedron* **1998,** *54,* 263.
(322) Suzuki, A. *J. Organomet. Chem.* **1999,** *576,* 147.
(323) Tamao, K.; Sumitani, K.; Kumada, M. *J. Am. Chem. Soc.* **1972,** *94,* 4374.
(324) Corriu, R. J. P.; Masse, J. P. *J. Chem. Soc. Chem. Commun.* **1972,** 144.
(325) Kumada, M. *Pure Appl. Chem.* **1980,** *52,* 669.
(326) Sonogashira, K.; Tohda, Y.; Hagihara, N. *Tetrahedron Lett.* **1975,** 4467.
(327) Cassar, L. *J. Organomet. Chem.* **1975,** *93,* 253.
(328) Dieck, H. A.; Heck, F. R. *J. Organomet. Chem.* **1975,** *93,* 259.
(329) Sonogashira, K. In *Metal-catalyzed Cross-coupling Reactions;* Diederich, F.; Stang, P. J., Eds.; Wiley-VCH: Weinheim, 1998.
(330) Louie, J.; Hartwig, J. F. *Tetrahedron Lett.* **1995,** *36,* 3609.
(331) Guram, A. S.; Rennels, R. A.; Buchwald, S. L. *Angew. Chem. Int. Ed. Engl.* **1995,** *34,* 1348.
(332) Hartwig, J. F. *Acc. Chem. Res.* **1998,** *31,* 852.
(333) Yang, B. H.; Buchwald, S. L. *J. Organomet. Chem.* **1999,** *576,* 125.
(334) Stürmer, R. *Angew. Chem. Int. Ed.* **1999,** *38,* 3307.
(335) Caló, V.; Nacci, A.; Lopez, L.; Mannarini, N. *Tetrahedron Lett.* **2000,** *41,* 8973.
(336) Herrmann, W. A.; Reisinger, C.-P.; Spiegler, M. *J. Organomet. Chem.* **1998,** *557,* 93.
(337) Atkins, R. M.; Mackenzie, R.; Timms, P. L.; Turney, T. W. *J. Chem. Soc., Chem. Commun.* **1975,** 764.
(338) Amatore, C.; Fuxa, A.; Jutand, A. *Chem. Eur. J.* **2000,** *6,* 1474.

(339) Gstöttmayr, C. W. K. Diplomarbeit. Technische Universität München (Germany), 1999.
(340) Böhm, V. P. W.; Gstöttmayr, C. W. K.; Weskamp, T.; Herrmann, W. A. unpublished results.
(341) Zhang, C.; Huang, J.; Trudell, M. L.; Nolan, S. P. *J. Org. Chem.* **1999,** *64,* 3804.
(342) Zhang, C.; Trudell, M. L. *Tetrahedron Lett.* **2000,** *41,* 595.
(343) Lee, H. M.; Nolan, S. P. *Org. Lett.* **2000,** *2,* 2053.
(344) Huang, J.; Grasa, G.; Nolan, S. P. *Org. Lett.* **1999,** *1,* 1307.
(345) Stauffer, S. R.; Lee, S.; Stambuli, J. P.; Hauck, S. I.; Hartwig, J. F. *Org. Lett.* **2000,** *2,* 1423.
(346) Böhm, V. P. W.; Herrmann, W. A. *Eur. J. Org. Chem.* **2000,** 3679.
(347) Welton, T. *Chem. Rev.* **1999,** *99,* 2071.
(348) Wasserscheid, P.; Keim, W. *Angew. Chem. Int. Ed.* **2000,** *39,* 3772.
(349) Albert, K.; Gisdakis, P.; Rösch, N. *Organometallics* **1998,** *17,* 1608.
(350) Hartwig, J. F.; Paul, F. *J. Am. Chem. Soc.* **1995,** *117,* 5373.
(351) McGuinness, D. S.; Cavell, K. J. *Organometallics* **2000,** *19,* 4918.
(352) Enders, D.; Gielen, H.; Breuer, K. *Tetrahedron Asymm.* **1997,** *8,* 3571.
(353) Enders, D.; Breuer, K.; Teles, J. H.; Ebel, K. J. *Prakt. Chem.* **1997,** *339,* 397.
(354) Cornils, B.; Herrmann, W. A. *Applied Homogeneous Catalysis with Organometallic Compounds;* VCH: Weinheim, 1996.
(355) Cornils, B.; Herrmann, W. A. *Aqueous-Phase Organometallic Catalysis;* VCH: Weinheim, 1998.
(356) Chen, A. C.; Ren, L.; Decken, A.; Crudden, C. M. *Organometallics* **2000,** *19,* 3459.
(357) Schrock, R. R.; Osborn, J. A. *J. Chem. Soc. Chem. Commun.* **1970,** 567.
(358) Simal, F.; Delaude, L.; Jan, D.; Demonceau, A.; Noels, A. F. *Polym. Prepr.* **1999,** *40(2),* 336.
(359) Çetinkaya, B.; Özdemir, I.; Bruneau, C.; Dixneuf, P. H. *J. Mol. Catal. A* **1997,** *118,* L1.
(360) Küçükbay, H.; Çetinkaya, B.; Guesmi, S.; Dixneuf, P. H. *Organometallics* **1996,** *15,* 2434.
(361) Cornils, B. J. *Mol. Catal. A* **1999,** *143,* 1.
(362) Enders, D.; Gielen, H.; Breuer, K. *Mol. Online* **1998,** *2,* 105.
(363) Schürer, S. C.; Gessler, S.; Buschmann, N.; Blechert, S. *Angew. Chem. Int. Ed.* **2000,** *39,* 3898.
(364) Gooßen, L. J. Dissertation, Technische Universität München (Germany), 1997.
(365) Joó, F.; Kathó, A. J. *Mol. Catal. A* **1997,** *116,* 3.
(366) Fish, R. H. *Chem. Eur. J.* **1999,** *5,* 1677.
(367) Stewart, I. H.; Hutchings, G. J.; Derouane, E. G. *Curr. Top. Catal.* **1999,** *2,* 17.
(368) Herrmann, W. A.; Elison, M.; Fischer, J.; Köcher, C.; Öfele, K. DE 4447066, 1996.
(369) Herrmann, W. A.; Elison, M.; Fischer, J.; Köcher, C.; Öfele, K. US 5728839, 1998.
(370) Coleman, A. W.; Hitchcock, P. B.; Lappert, M. F.; Maskell, R. K.; Müller, J. H. *J. Organomet. Chem.* **1983,** *250,* C9.
(371) Jandeleit, B.; Schäfer, D. J.; Powers, T. S.; Turner, H. W.; Weinberg, W. H. *Angew. Chem. Int. Ed.* **1999,** *38,* 2494.
(372) Hartwig, J. F. ACS National Meeting, San Francisco, March 2000, INOR 1.
(373) Serrano, J. L. *Metallomesogens;* VCH: Weinheim, 1996.
(374) Copeland, R. A. *Enzymes;* Wiley: New York, 2000.
(375) Clarke, M. J.; Taube, H. *J. Am. Chem. Soc.* **1975,** *97,* 1397.
(376) Krentzien, H. J.; Clarke, M. J.; Taube, H. *Bioinorg. Chem.* **1975,** *4,* 143.
(377) Buncel, E.; Norris, A. R.; Racz, W. J.; Taylor, S. E. *J. Chem. Soc., Chem. Commun.* **1979,** 562.
(378) Buncel, E.; Norris, A. R.; Racz, W. J.; Taylor, S. E. *Inorg. Chem.* **1981,** *20,* 98.
(379) Norris, A. R.; Buncel, E.; Taylor, S. E. *J. Inorg. Biochem.* **1982,** *16,* 279.
(380) Bovio, B.; Burini, A.; Pietroni, B. R. *J. Organomet. Chem.* **1993,** *452,* 287.
(381) Bonati, F.; Burini, A.; Pietroni, B. R.; Bovio, B. *J. Organomet. Chem.* **1991,** *408,* 271.
(382) Ghosh, A. *Angew. Chem. Int. Ed. Engl.* **1995,** *34,* 1028.
(383) Chmielewski, P. J.; Latos-Grazynski, L.; Rachlewicz, K.; Glowiak, T. *Angew. Chem. Int. Ed. Engl.* **1994,** *33,* 779.
(384) Furuta, H.; Asano, T.; Ogawa, T. *J. Am. Chem. Soc.* **1994,** *116,* 767.

(385) Sessler, J. L. *Angew. Chem. Int. Ed. Engl.* **1994,** *33,* 1348.
(386) Furuta, H.; Maeda, H.; Osuka, A. *J. Am. Chem. Soc.* **2000,** *122,* 803.
(387) Furuta, H.; Maeda, H.; Osuka, A.; Yasutake, M.; Shinmyozu, T.; Ishikawa, Y. *Chem. Commun.* **2000,** 1143.
(388) Weiss, A.; Pritzkow, H.; Siebert, W. *Angew. Chem. Int. Ed.* **2000,** *39,* 547.
(389) Çetinkaya, B.; Çetinkaya, E.; Küçükbay, H.; Durmaz, R. *Arzneim.-Forsch.* **1996,** *46,* 821.
(390) Durmaz, R.; Küçükbay, H.; Çetinkaya, E.; Çetinkaya, B.; Turk, B. *J. Med. Sci.* **1997,** *27,* 59.
(391) Durmaz, R.; Koroglu, M.; Küçükbay, H.; Temel, I.; Ozer, M. K.; Refiq, M.; Çetinkaya, E.; Çetinkaya, B.; Yologlu, S. *Arzneim.-Forsch.* **1998,** *48,* 1179.
(392) Çetinkaya, B.; Özdemir, I.; Binbasioglu, B.; Durmaz, R.; Gunal, S. *Arzneim.-Forsch.* **1999,** *49,* 538.

ADVANCES IN ORGANOMETALLIC CHEMISTRY, VOL. 48

Transition Metal Chemistry of 1,3-Diynes, Poly-ynes, and Related Compounds

PAUL J. LOW

Department of Chemistry
University of Durham
Durham DH1 3LE, England

MICHAEL I. BRUCE

Department of Chemistry
University of Adelaide
Adelaide, South Australia 5005, Australia

0065-3055/01 $35.00

I

INTRODUCTION

The chemistry of metal complexes featuring alkyne and alkynyl (acetylide) ligands has been an area of immense interest for decades. Even the simplest examples of these, the mononuclear metal acetylide complexes $L_nMC{\equiv}CR$, are now so numerous and the extent of their reaction chemistry is so diverse as to defy efforts at a comprehensive review.[1–3] The utility of these complexes is well documented. Some metal alkynyl complexes have been used as intermediates in preparative organic chemistry and together with derived polymeric materials, many have useful physical properties including liquid crystallinity[4–6] and nonlinear optical behaviour.[7] The structural properties of the M−C≡C moiety have been used in the construction of remarkable supramolecular architectures based upon squares, boxes, and other geometries.[8]

Coordination of metal centers to the alkyne π-system has led to the development of alkyne protecting group strategies[9] and to the stabilization of reactive intermediates such as propargylium ions.[10] The geometrical and electronic reorganization of the alkyne which occurs following complexation has resulted in new synthetic strategies, like the Pauson–Khand reaction.[11] The multifaceted coordination modes available to alkynes through combinations of σ and π bonding, coupled with the propensity to bridge two or more metal atoms, have led to a plethora of polynuclear and cluster compounds.

The presence of multiple C≡C moieties in di- and poly-ynes hints at a vast potential for expansion of this chemistry. Recent developments in the chemistry and material properties of these complexes have led to a surge of interest in the chemistry of these highly unsaturated ligands. Some of these results have been the subject of several recent short reviews and highlight articles.[12–14]

Poly-ynyl moieties linking a variety of redox-active metal centers are also efficient carriers of electronic charge.[15–17] The potential of these systems to function as "molecular wires"[18–21] and rectifiers[22,23] prompted many investigations.[24–33] In a recent review of the role of the Fe(dppe)Cp* fragment in such complexes, it was concluded that the terminal metal–ligand fragments play a crucial role in determining the degree of electronic communication along the wire.[34] Long-range photo-induced electron transfer in several oligomeric alkyne complexes has been demonstrated.[35–37]

Introduction of two or more terminal diyne units, either at a metal center or, for example, on an η-ring complex, leads to the possibility of generating metal-containing polymers. Many examples of such polymers derived from alkynes have been described, but studies of similar systems derived from di- or poly-ynes are in their infancy.[37,38] The role of transition metals in polymeric π-conjugated organic frameworks, including organometallic σ-alkynyl polymers and polymeric systems containing π-bonded metal–ligand fragments has recently been reviewed.[38] Further development of the chemistry of σ-ethynyl, σ-diynyl, σ-poly-ynyl, and ethynyl-substituted η-C_4, C_5 or C_6 ligand–metal complexes should also afford novel molecular architectures, including linear, star-shaped, and spherical aggregates.[40–53]

The presence of multiple C≡C moieties affords the potential to assemble novel arrays of metal centers by combinations of σ- and π-bonding interactions.[54–57] However, examples of multimetallic complexes containing poly-yne and poly-yndiyl ligands are limited, although preliminary reports suggest that there is much scope for the preparation of new structural types.

In this article we have tried to draw together a comprehensive review of this area and in general, when discussing the various complexes that have been reported, we take a Periodic Group approach. Complexes with σ-bonded diynyl and poly-ynyl ligands are presented, together with a description of their reaction chemistry and routine spectroscopic properties, followed by a description of the various η^2-bonded systems featuring one or more metal centers. Experimental and theoretical work leading to a description of the MC- and CC-bonding interactions in di- and poly-ynyl systems is presented in Section X.D. Numerous reactions of diynes and poly-ynes with metal species that lead to products distinct from those previously discussed are described in Section VII, as are C—C bond forming reactions which lead to poly-ynes. Other ligands containing diynyl groups are summarized in Section IX. The preparative chemistry and electronic structures (where possible) of metalladiynes are addressed (Section XI), and a survey of novel materials obtained from diyne complexes completes the review. Even given the relatively recent surge in interest in these poly-yne complexes, we have found it necessary to be selective in our coverage, which extends to the end of 1999.

II

DIYNYL COMPLEXES, $M\{(C{\equiv}C)_2R\}_n$ ($n = 1, 2$)

A. *Synthetic Methods*

While in principle diynyl complexes may be formed either by reactions of preformed diyne reagents with metal centers or via coupling reactions of metal alkynyls with other alkynyl reagents, only the first procedure has been explored in any depth to date. The latter route is more commonly used to prepare longer chain analogues (see following). The key step in the preparative methods is therefore the formation of the bond between the metal center and the poly-yne ligand. There are several successful synthetic approaches which are generally comparable with those used to prepare the analogous alkynyl complexes. However, the thermal sensitivity of some 1,3-diynes may limit the application of some of the traditional methods.

Examples of diynyl complexes featuring elements from most groups of the transition metals are now known. Table I summarizes the diynyl complexes known to us at the end of 1999, and provides an indication of the preparative method employed for individual complexes. The most widely utilized preparative methods are (a) reactions of diyne anions with metal–halide complexes (Method A), (b) Cu(I)-catalyzed reactions of 1,3-diynes with metal–halide complexes (Method B), (c) oxidative addition of 1,3-diynes to suitable electron-deficient metal complexes (Method C), (d) metal exchange and coupling reactions involving Group 14 diynyl reagents (Method D), (e) deprotonation of alkyne or vinylidene intermediates (Method E). These methodologies will be discussed in turn, followed by a description of the various applications by Group.

TABLE I

TRANSITION METAL DIYNYL COMPLEXES $\{L_nM\}(C{\equiv}CC{\equiv}CR)_n$

ML_x	R	*n*	Synthetic Method[a]	$\nu(C{\equiv}C)$	$\delta C(J)^b$ C_α, C_β, C_γ, C_δ	Reference
$TiClCp^{Si}_2$	Et	1	A	2193 m, 2032 vs	136.0, 121.7, 90.2, 65.8	98
$TiClCp^{Si}_2$	Fc	1	A	2175, 2028	143.4, 117.5, 87.2, 66.0	97
$Ti(C{\equiv}CRc)Cp^{Si}_2$	Fc	1	A	2173, 2058, 2023	n.d.	97
$Ti(C{\equiv}CFc)Cp^{Si}_2$	Fc	1	A	2173, 2054, 2024	147.2, 116.5, 86.7, 66.2	97
$TiCp_2$	Fc	2	A	2172, 2021	n.d.	97
$TiCp^{Si}_2$	Fc	2	A	2171, 2021	149.4, 119.7, 89.5, 66.0	97
$TiCp^{Si}_2$	Fc; Rc	2	A	2173, 2021	n.d.	97
$TiCp^{Si}_2$	$SiMe_3$	2	A	2001 s, 1994 vs	143.1, 118.2, 96.3, 88.6	97, 99
$TiCp^{Si}_2$	Et	2	A	2189 m, 2023 s	141.3, 118.3, 92.5, 65.8	98, 99
$Zr\{2\text{-}N(SiMe_3)\text{-}4\text{-}MeC_5H_4N\}_3$	$SiMe_3$	1	A	n.d.	143.6, 92.2, 90.4, 80.6	100
$Cr(CO)_5$	$C(NEt_2)C(Me){=}C(NMe_2)_2$	1	F	2149 w, 2000 w	175.3, 102.2/94.7/87.3 or 55.8	102
$Cr(CO)_5$	$C(NMe_2)_2$	1	A	2140 w, 1998 w	174.1, 99.5/95.7/49.6	101
Mo(CO)(dppe)Cp	$SiMe_3$	1	A	2185 w, 2065 w	137.0 (J_{CP} 32), 100.2/87.9/70.5	107
Mo(CO)(dppe)Cp	H	1	B, G	1981 w, 1968 w	136.7 (J_{CP} 21), 94.5/71.9/59.9	107
$Mo(CO)_3Cp$	H	1	B	2145 m	110.46, 87.25, 70.27, 62.10	104, 109
$Mo(CO)_2(C_7H_7)$	$SiMe_3$	1	A	2117, 2161	114.2, 95.9, 90.8, 74.8	108
$Mo(CO)_2(C_7H_7)$	H	1	A, G	2127	111.6, 94.4, 71.1, 57.1	108
$Mo(CO)(PMe_3)(\eta\text{-}C_7H_7)$	H	1	A	2114	125.6 (J_{CP} 26.2), 93.7, 71.7 (J_{CP} 4.4 Hz), 56.2 (J_{CP} 2.5)	108
$Mo(CO)_3Cp$	Ph	1	H	2183 m	110.92, C_β n.d., 76.42/72.59	104
$W(O)_2Cp^*$	Ph	1	I	n.d.	n.d.	112, 257
$W(O_2)OCp^*$	Ph	1	I	n.d.	n.d.	112, 257
$W(CO)_3Cp^*$	Ph	1	B	n.d.	n.d.	106, 112
$W(CO)_3Cp$	H	1	B	2145 m	110.52, 71.60 (J_{CW} 88), 70.13 (J_{CP} 94), 63.30	104
$W(CO)_3Cp$	$SiMe_3$	1	J	2174 w, 2127 w	111.74, 110.86, 90.04, 73.59	104
$W(CO)_3Cp$	$P(O)Ph_2$	1	J	2138 m	110.57 (J_{CW} 16), 91.32; C_γ, C_δ n.d.	104
$W(CO)_3Cp$	Ph	1	H	2183 m, 2059 m	111.03 (J_{CW} 22); C_β n.d.; 76.19/73.78	104
$W(CO)_3Cp$	$C_6H_4Me\text{-}4$	1	H	2180 m	n.d.	104
$W(CO)_3Cp$	$C_6H_4OMe\text{-}4$	1	H	2174 w	n.d.	104
$W(CO)_3Cp$	$C_6H_4CO_2Me\text{-}4$	1	H	2179 m	n.d.	104
cis-$W(CO)_2(PPh_3)Cp$	Ph	1	I	2172 m	113.27 (J_{CW} 8), 100.14, 72.75/72.2	104
$W(CO)_5$	$C(NEt_2)C(Me){=}C(NMe_2)_2$	1	F	2148 w, 2001 w	175.2, 102.5/93.9/87.4 or 58.3	102
$W(CO)_5$	$C(NMe_2)_2$	1	A	2140 w, 1999 w	153.8 (J_{CW} 102.6), 94.6 (J_{CW} 25.6), 99.5/52.1	101
$Re(NO)(PPh_3)Cp^*$	$SiMe_3$	1	E	2119 m, 2097 m	105.8 (J_{CP} 15.9), 112.3, 93.5 (J_{CP} 2.7), 80.6	87

(continued)

TABLE I (*continued*)

ML_x	R	*n*	Synthetic Method[a]	$\nu(C{\equiv}C)$	$\delta C(J)^b$ C_α, C_β, C_γ, C_δ	Reference
$Re(NO)(PPh_3)Cp^*$	H	1	G	2113 s, 1975 w	102.1 (J_{CP} 15.9), 110.8, 72.4, 65.2	87
$Re(NO)(PPh_3)Cp^*$	Me	1	J	2194 m, 2029 m	96.8 (J_{CP} 17.3), 111.6, 69.1 (J_{CP} 3.1), 71.9	31, 87
$Re(NO)(PPh_3)Cp^*$	Cu	1	J	n.d.	n.d.	144
$Re(NO)(PPh_3)Cp^*$	C(OMe)=[Mn $(CO)_2(\eta\text{-}C_5Cl_5)$]	1	J	2057 m, 2046 m	167.1 (J_{CP} 14.8), 127.4/116.3/82.4	31, 114
$Re(NO)(PPh_3)Cp^*$	C(OMe)=[Mn $(CO)_2(\eta\text{-}C_5Br_5)$]	1	J	2054 m	169.7 (J_{CP} 11.4), 127.6/116.7/82.6	31
$Re(NO)(PPh_3)Cp^*$	C(OMe)=[Mn $(CO)_2(\eta\text{-}C_5H_4Cl)$]	1	J	2064 m	155.6 (J_{CP} 15.3), 118.3	31
$Re(NO)(PPh_3)Cp^*$	C(OMe)=[Fe$(CO)_4$]	1	J	2018 vs	183.6 (J_{CP} 14.1), 127.4/119.6/88.1	31
$Re(CO)_3(Bu^t{}_2bpy)$	H	1	A	n.d.	n.d.	80
$Re(CO)_3(Bu^t{}_2bpy)$	Ph	1	A	n.d.	n.d.	80
$Fe(CO)_2(\eta\text{-}C_9H_7)$	Bu^n	1	A	n.d.	n.d.	555
$Fe(CO)[P(OMe_3)]Cp$	Bu^n	1	I	n.d.	n.d.	555
$Fe(CO)_2Cp^*$	$SiMe_3$	1	A	2171 w, 2120 w	106.3, 96.5, 94.7, 69.5	27
$Fe(CO)_2(\eta\text{-}C_5Ph_5)$	$SiMe_3$	1	A	2185 s, 2129 s	99.8, 99.4, 92.6, 71.1	27
$Fe(dppe)Cp^*$	$SiMe_3$	1	I	2165 w, 2090 s, 1980 m	142.2 (J_{CP} 38), 102.3 (J_{CP} 2), 96.2 (J_{CP} 3), 64.7	27
$Fe(CO)_2Cp^*$	H	1	G	2142 m	102.9, 95.0 (J_{CH} 6), 73.4 (J_{CH} 51), 54.1 (J_{CH} 216)	27
$Fe(CO)_2(\eta\text{-}C_5Ph_5)$	H	1	G	2141 w	94.1, 99.7 (J_{CH} 5), 72.8 (J_{CH} 51), 55.9 (J_{CH} 259)	27
$Fe(dppe)Cp^*$	H	1	G	2099 s, 1958 w	136.6 (J_{CP} 38), 100.7 (J_{CH} 5), 75.1 (J_{CH} 50, J_{CP} 3), 50.5 (J_{CH} 248)	27
$Fe(CO)_2Cp$	$SiMe_3$	1	A	n.d.	n.d.	24, 148
$Fe(CO)_2Cp$	nBu	1	D	2203 vw	97.4/83.8/67.9/67.1 (n.a.)	148
$Fe(CO)_2Cp$	Ph	1	D	2184 m	97.3/96.2/77.3/64.8 (n.a.)	148
$Fe(CO)(PPh_3)Cp$	$SiMe_3$	1	I	n.d.	n.d.	24, 148
$Fe(CO)_2Cp$	H	1	B, G	2174 w, 2127 w	96.80, 89.85, 70.88, 54.35 (n.a.)	24, 83, 104, 157, 505
$Fe(CO)_2Cp^*$	$SiMe_3$	1	A	2178, 2125	n.d.	115
$Fe(CO)_2Cp^*$	H	1	G	2141	106.4, 92.8 (J_{CH} 7), 71.9 (J_{CH} 50), 53.5 (J_{CH} 252),	115
$Fe(CO)_2Cp$	$C(NMe_2)W(CO)_5$	1	A	2140 w, 2055 w	128.52, 97.66, 118.65, 66.17	66
$Fe(CO)(PPh_3)Cp$	H	1	I	2170 w	109.4 (J_{CP} 39.4), 99.1, 72.1, 54.3	24
$Ru(PPh_3)_2Cp$	H	1	A	1971 m	116.4 (J_{CP} 24.6), 94.4/73.9, 128.4	62
$Ru(PPh_3)_2Cp$	Ph	1	E	2162 s, 2025 s	80.47 (J_{CP} 11.0), 102.40/95.64/62.87	119
$Ru(CO)_2Cp$	$C(NMe_2)$=[$W(CO)_5$]	1	D	2145 w, 2060 w	119.24, 92.30, 117.99, 67.81	66
trans-$RuCl(dppe)_2$	Ph	1	E	2154, 2018	131.67 (J_{CP} 15), 81.22 (J_{CP} 2), 62.73 (J_{CP} 1), 95.53	95

TABLE I (*continued*)

ML_x	R	*n*	Synthetic Method[a]	ν(C≡C)	δC(*J*)[b] C_α, C_β, C_γ, C_δ	Reference
trans-RuCl(dppe)$_2$	Fc	1	E	2163 s, 2018 s	n.d.	123
trans-Ru(dppe)$_2$	Fc	2	E	2171 s, 2017 s	n.d.	123
trans-Ru(C≡CFc)(dppe)$_2$	Fc	1	E	2169 s, 2063 s, 2024 s	133.6 (J_{CP} 15), 98.8, 76.7, 69.6	123
trans- Ru(C≡C $C_6H_4NO_2$-4)(dppe)$_2$	Fc	1	E	2170 s, 2052 s, 2026 s	129.7 (J_{CP} 15), 119.4 (J_{CP} 4), 100.6, 62.0	123
cis-Ru(dppm)$_2$	$SiMe_3$	2	A	2108 s	97.05/95.05/69.71	122
trans-Ru(dppm)$_2$	$SiMe_3$	2	A	2107 s	98.22/95.63/67.99	122
trans-RuCl(dppm)$_2$	$SiMe_3$	1	A	2107 s	96.11/95.95/67.11	122
trans-RuCl(dppm)$_2$	$CPh_2(OSiMe_3)$	1	D, E	2176 s, 2024 s	125.97 (J_{CP} 15.1), 93.48, 80.24/63.79	85, 86
trans-Ru(dppm)$_2$	$CPh_2(OSiMe_3)$	2	E	2175 s, 2020 s	131.18 (J_{CP} 15.2), 96.42, 80.27/64.96	85, 86
trans-Ru(C≡CPh)(dppm)$_2$	$CPh_2(OSiMe_3)$	1	E	2178 s, 2077 m, 2028 m	132.91 (J_{CP} 14.4), 96.26, 80.52/64.55	125
trans-Ru(C≡CBu)(dppm)$_2$	$CPh_2(OSiMe_3)$	1	E	2175 s, 2098 m, 2020 m	C_αn.d., 95.25 (J_{CP} 1.2), 80.81/63.71	125
trans-RuCl(dppe)$_2$	$CPh_2(OSiMe_3)$	1	E	2177 s, 2029 s	127.10 (J_{CP} 15.2), 94.07, 79.90, 68.03	124
$Ru(CO)_2(PEt_3)_2$	$SiMe_3$	2	A	2165 m, 2121 m	103.9 (J_{CP} 12.5), 93.2, 92.4, 70.8	117, 118
$Ru(CO)_2(PEt_3)_2$	H	2	G	2137 m	101.3 (J_{CP} 12.0), 91.7, 72.1, 54.5	117, 118
$RuCl(PMe_3)(\eta^6$-$C_6Me_6)$	CPh_2OSiMe_3	1	E	2186 s, 2035 m	121.11 (J_{CP} 39.5), 86.38/78.26/76.75	89, 90, 120
$RuCl(PMe_2Ph)(\eta^6$-$C_6Me_6)$	CPh_2OSiMe_3	1	E	2185 s, 2033 m	121.11 (J_{CP} 39.5), 78.40 (J_{CP} 3.6), 86.73/76.80	89, 90, 120
$RuCl(PMePh_2)(\eta^6$-$C_6Me_6)$	CPh_2OSiMe_3	1	E	2189 s, 2038 m	120.22 (J_{CP} 37.2), 78.30 (J_{CP} 3), 87.83/76.68	120
$RuCl(PMePh_2)(\eta^6$-$C_6Me_6)$	$[C(C_6H_4NMe_2$-$4)_2]^+$	1	E	2085, 1996	180.11 (J_{CP} 36.1), 121.15 (J_{CP} 3), 100.13, 76.98	89
$OsHCl(P^iPr_3)_2$	CPh_2OH	1	C	2059	80.7 (J_{CP} 11.0), 104.1, 78.5/75.0/74.5	557
$RhHCl(PPr^i_3)_2$	CPh_2OSiMe_3	1	C	2187 s, 2041 s	109.76 (J_{RhC} 50.9, J_{CP} 16.5), 88.24 (J_{RhC} 11.4, J_{CP} 2.2), 78.80 (J_{RhC} 3.8, J_{CP} 1.9), 77.17	127
$RhHCl(py)(PPr^i_3)_2$	CPh_2OSiMe_3	1	I	2187 s, 2041 s	109.76 (J_{CRh} 50.9, J_{CP} 16.5), 88.24 (J_{CRh} 11.4 J_{CP} 2.2), 78.80 (J_{CRh} 3.8, J_{CP} 1.9), 77.17	127
$Rh(CO)(PPr^i_3)_2$	$SiMe_3$	1	D	2150, 2105	121.25 (J_{CRh} 43.3, J_{CP} 22.1), 103.0 (J_{RhC} 13.1, J_{CP} 2.5), 93.2/77.2	126
$Rh(CO)(PPr^i_3)_2$	Ph	1	I	2150	126.33 (J_{CRh} 43.4, J_{CP} 21.6), 102.22 (J_{CRh} 12.7, J_{CP} 3.2), 79.15, 71.95 (J_{CP} 2.1)	71
$RhH(CO)(PPr^i_3)_2$	Ph	2	I	n.d.	n.d.	71
$RhH(py)(PPr^i_3)_2$	Ph	2	I	2140, 2010	n.d.	71

(*continued*)

TABLE I (*continued*)

ML_x	R	*n*	Synthetic Method[a]	$\nu(C{\equiv}C)$	$\delta C(J)^b$ C_α, C_β, C_γ, C_δ	Reference
$RhH(PPr^i{}_3)_2$	Ph	2	C, E	2140, 2010	n.d.	71
$RhHCl(PPr^i{}_3)_2$	Ph	1	C	n.d.	n.d.	71
$RhHCl(py)(PPr^i{}_3)_2$	Ph	1	I	2165, 2050	n.d.	71
$IrHCl(PPr^i)_2$	$SiMe_3$	1	L	2180, 2110	80.0 (J_{CP} 11.2), 94.4/92.5, 75.5	79
$IrHCl(py)(PPr^i)_2$	$SiMe_3$	1	I	2180, 2115	85.1 (J_{CP} 11.4), 95.6, 85.0/71.4	79
$IrHCl(PPr^i)_2$	CPh_2OH	1	L	2180, 2020	82.65 (J_{CP} 11.3), 76.68 (J_{CP} 1.8), 89.86 (J_{CP} 1.2), 72.56 or 65.67	72
$IrHCl(py)(PPr^i)_2$	CPh_2OH	1	I	2188, 2043	n.d.	72
$Ni(PPh_3)Cp$	H	1	A, B	2138	85.9 (J_{CP} 44), 71.5 (J_{CP} 3), 99.5, 66.1	128, 129
$Ni(PEt_3)_2$	H	2	A	n.d.	n.d.	65
trans-$Ni(PBu_3)_2$	H	2	B	n.d.	n.d.	132
trans-$Ni(PEt_3)_2$	Me	2	A	n.d.	n.d.	65
trans-$Pd(PBu_3)_2$	H	2	A	n.d.	n.d.	130
trans-$Pt(PEt_3)_2$	Me	2	B	n.d.	n.d.	65
trans-$Pt(PEt_3)_2$	$SiMe_3$	2	A	n.d.	102.7 (J_{PtC} 1268, J_{PC} 14.8), 93.6 (J_{PtC} 284), 94.1 (J_{PtC} 35), 76.6	131
trans-$Pt(PEt_3)_2$	H	2	G	n.d.	n.d.	131
trans-$Pt(tol)\{P(tol)_3\}_2$	H	1	B	2143 s	111.2 (J_{CP} 13.8, J_{CPt} 650), 96.0 (J_{CH} 5.2, J_{CPt} 225), 73.6 (J_{CH} 50.9), 58.8 (J_{CH} 249.6)	147
trans-$Pt(tol)(PPh_3)_2$	H	1	B	2143 s	110.7, 96.3, 73.7, 59.2	147
cis-$Pt(PBu_3)_2$	H	2	B	n.d.	n.d.	68
trans-$Pt(PBu_3)_2$	H	2	B	2142 s	99.7 (J_{PtC} 1001; J_{PC} 30), 92.3 (J_{PtC} 285), 73.4, 60.4	68, 131, 464
trans-$Pt(PBu_3)_2$	$SiMe_3$	2	A	n.d.	102.9 (J_{PtC} 1003, J_{PC} 14.8), 93.3 (J_{PtC} 305), 94.1, 76.5	131
Pt(dppe)	H	2	B	2147 s	93.7 m, 77.21 m, 71.75, 61.74	193
Pt(dcype)	$SiMe_3$	2	B	2128 s, 2181 s	n.d.	464
Pt(dcype)	H	2	B, G	2082 s, 2146 s	99.9 (J_{PtC} 1090; J_{PC} 15, 135), 93.3 (J_{PtP} 308; J_{PC} 0.1, 32), 72.6 (J_{PtC} 36), 60.7	464
$Pt(PMe_2Ph)_2$	Ph	2	B	n.d.	n.d.	154
Pt(dppp)	H	2	B	2151 s	72.0, 64.4 (C_α, C_β n.d.)	133
$Pt(PEt_3)_2$	H	2	B	2147 s	94.7, 80.9, 72.0, 60.9	133
Pt(dppe)	Me	2	J	2148 s, 2082 m	96.9, 93.1, 71.8, 61.7	133
Pt(dppe)	H, $SiMe_3$	2	J	2149 s, 2089 m	n.d.	133
Pt(dppe)	H, $Au(PPh_3)$	2	J	2140 m, 2084 m	n.d.	133
Pt(dppe)	$SiMe_3$	2	J	n.d.	n.d.	133
Pt(dppe)	Ph	2	H	n.d.	n.d.	133
Cu	Me	1	A	n.d.	n.d.	135
Cu	$SiMe_3$	1	A	n.d.	n.d.	136

TABLE I (*continued*)

ML_x	R	*n*	Synthetic Method[a]	$\nu(C{\equiv}C)$	$\delta C(J)^b$ C_α, C_β, C_γ, C_δ	Reference
Au[c]	H	2	A	2141 s	83.56, 71.84, 56.73 (C_α n.d.)	110
$Au_2(\mu$-dppm)	H	1	A	2140, 2080	89.7, 84.1, 71.4, 65.4	110
$Au(PPh_3)$	H	1	A	2148 s	129.55, 128.79, 69.39, 60.39	
Au	H, $Cu_3(\mu\text{-I})_3(\mu\text{-dppm})_3$	2	B	n.d.	n.d.	110
$[Zn(NPMe_3)]_4$	$SiMe_3$	1	A	n.d.	108.84/92.46/90.95/75.48	139
$[Zn(NPEt_3)]_4$	$SiMe_3$	1	A	n.d.	111.36/92.83/90.73/76.67	139
ZnCl	$SiMe_3$	1	A	n.d.	n.d.	138
Hg	Ph	2	A/B	n.d.	n.d.	135
Hg	Me	2	A/B	n.d.	n.d.	135
Hg	$C(NMe_2)=[W(CO)_5]$	2	A	Not observed	138.39, 87.52, 111.53, 71.76	66

[a]Methods: A—reactions of diyne anions with metal–halide complexes; B—Cu(I)-catalyzed reactions of 1,3-diynes with metal–halide complexes; C—oxidative addition of 1,3-diynes to suitable electron-deficient metal complexes; D—metal-exchange and coupling reactions involving Group 14 diynyl reagents; E—deprotonation of alkyne or vinylidene intermediates; F—addition of electrophilic reagent to cumulenylidene complexes; G—desilylation of corresponding $SiMe_3$ complexes; H—Sonogashira coupling protocols; I—Ligand modification about metal center of diynyl complexes; J—deprotonation and quenching of diynyl complexes; K—deprotonation of diyne complexes; L—elimination of H_2 and oxidative addition reactions.

[b]n.d. not determined or not recorded.

[c][ppn][$Au(C{\equiv}CC{\equiv}CH)_2$].

1. *Syntheses from Diacetylide Anions and Metal–Halide Complexes*

The acetylenic hydrogens of 1,3-diynes are about as acidic as those of 1-alkynes and diynes are therefore readily metallated.[58,59] All told, the various diynyl anions $RC{\equiv}CC{\equiv}C^-$ have been successfully employed in the preparation of both diynyl and bis(butadiynyl) metal complexes featuring metals from across the Periodic Table. Treatment of $HC{\equiv}CC{\equiv}CR$ (R = H, alkyl, trialkylsilyl, aryl) with one equivalent of BuLi affords $LiC{\equiv}CC{\equiv}CR$,[60] while Grignard reagents are made in the conventional manner. Thus, treatment of $HC{\equiv}CC{\equiv}CR$ (R=H, alkyl, trialkylsilyl, aryl) with 1 equiv of EtMgBr affords $BrMgC{\equiv}CC{\equiv}CR$; metal exchange reactions between $MgBr_2$ and the diynyllithium have also been used.[59] However, as a result of the tendency of the mono-deprotonated derivatives of buta-1,3-diyne to disproportionate to the insoluble dianions, reactions involving these reagents are rather sensitive to the conditions employed. Nevertheless, the mono-Grignard $BrMgC{\equiv}CC{\equiv}CH$ and mono-lithiated buta-1,3-diyne have been used to synthesize diynyl complexes of nickel[61] and ruthenium,[62] respectively.

The problems of disproportionation, coupled with the thermal sensitivity of 1,3-diynes, which often polymerize in a rapid, highly exothermic, and occasionally explosive manner in the pure state, have led most workers to examine alternate methods of generating 1,3-diynyl anions. The reaction of *Z*-1-methoxy-1-buten-3-yne with 2 equiv of BuLi gives LiC≡CC≡CH and LiOMe.[63] Substituted derivatives may be obtained following sequential treatment of the same *cis*-HC≡CCH=CH(OMe) with 1 equiv of BuLi and RCH_2Cl, which affords *cis*-RCH_2C≡CCH=CH(OMe). These substituted methoxy-ene-ynes yield LiC≡CC≡CCH_2R upon treatment with 2 equiv of LDA. Quenching with aqueous NH_4Cl or silylation give RCH_2C≡CC≡CR′ (R′ = H, $SiMe_3$, respectively).[64]

An excess of $NaNH_2$ in liquid NH_3 reacts directly with $MeCCl_2CH_2CCl_2Me$ to give NaC≡CC≡CMe, which was found to displace chloride from $MCl_2(PEt_3)_2$ to give *trans*-M(C≡CC≡CMe)$_2$(PEt_3)$_2$ (M = Ni, Pt).[65]

The $SiMe_3$-protected diyne synthon LiC≡CC≡$CSiMe_3$ is readily obtained by treating the doubly protected analogue Me_3SiC≡CC≡$CSiMe_3$ with MeLi.LiBr; 1–10 mol% of the KF.18-crown-6 complex has also been used to achieve this transformation.[59,60] The relative nucleophilicity of Li(C≡C)$_n$$SiMe_3$ decreases with increases in chain length, and only LiC≡$CSiMe_3$ reacts with $M(CO)_6$ (M = Cr, W).[66] However, the diynyl reagent is still sufficiently nucleophilic to displace halide from a variety of transition metal complexes (see Table I). This property, together with the simplicity of preparing LiC≡CC≡$CSiMe_3$ from the air-stable crystalline diyne Me_3SiC≡CC≡$CSiMe_3$ has resulted in this reagent becoming the most widely used diacetylide anion in the preparation of diynyl complexes.

2. *Cu(I)-Catalyzed Reactions of 1,3-Diynes with Metal–Halide Complexes*

The use of CuI to catalyze reactions between transition metal halides and 1-alkynes has considerable precedence in organic chemistry.[67] Copper(I)-catalyzed reactions between HC≡CC≡CH and a variety of Group 10 complexes in amine solvents under mild conditions were initially reported in 1977, when the *cis*- and *trans*-isomers of Pt(C≡CC≡CH)$_2$(PBu_3)$_2$ were obtained from the corresponding dihalides and HC≡CC≡CH with retention of configuration at the metal center.[68] More recently, the present authors and others have applied this reaction to the preparation of diynyl complexes featuring metals across the Periodic Table.[69] Similar reactions have also been used to prepare a large number of polymeric diynyl Group 10 metal complexes.[38]

While the mechanisms of these reactions have not been investigated in detail, it is likely that an intermediate copper(I) alkynyl is formed, which undergoes an alkynyl–halide exchange with the metal halide, resulting in the formation of the transition metal σ-alkynyl complex and a Cu(I) halide, which completes the catalytic cycle (Scheme 1).[68,70]

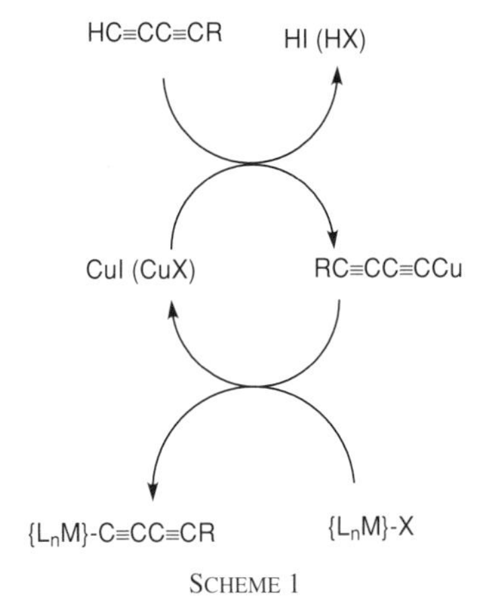

SCHEME 1

3. *Oxidative Addition of Terminal Diynes to Electron-Deficient Metal Centers*

In a manner analogous to the reactions of 1-alkynes, 1,3-diynes oxidatively add to electronically unsaturated metal centers. Thus $\{RhCl(PPr^i_3)_2\}_n$ and HC≡CC≡CPh gave the Rh(III)-diynyl compound $RhHCl(C{\equiv}CC{\equiv}CPh)(PPr^i_3)_2$, while the acetato complex $Rh(\eta^2\text{-}O_2CMe)(PPr^i_3)_2$ reacted with 2 equiv of phenylbutadiyne in the presence of Na_2CO_3 to give the five-cordinate hydridorhodium(III) complex $RhH(C{\equiv}CC{\equiv}CPh)_2(PPr^i_3)_2$.[71] A five-coordinate Ir(III) complex *trans*-$IrHCl\{C{\equiv}CC{\equiv}CCPh_2(OH)\}(PPr^i_3)_2$ was similarly obtained prepared from $IrH_2Cl(PPr^i_3)_2$ and $HC{\equiv}CC{\equiv}CCPh_2(OH)$.[72]

4. *Metal Exchange and Coupling Reactions Involving Group 14 Diynyl Reagents*

Group 14 reagents are widely used in preparative organic poly-ynyl chemistry.[3] Typical examples include the use of trialkylsilyl moieties as protecting and stabilizing groups in poly-yne synthesis[73] while tetravalent tin reagents have proven to be of remarkable utility in the construction of carbon—carbon bonds.[67,74,75]

As previously mentioned, the stable crystalline buta-1,3-diyne synthon $Me_3SiC{\equiv}CC{\equiv}CSiMe_3$ is readily and selectively mono-desilylated by MeLi.LiBr to generate the useful reagent $LiC{\equiv}CC{\equiv}CSiMe_3$, which is sufficiently nucleophilic to displace labile ligands from a wide variety of transition metal substrates. The silylated diynyl complexes so prepared are readily proto-desilylated by standard methods, such as treatment with $[NBu_4]F$ or KOH/MeOH to give the corresponding terminal diynyl complexes. These procedures have the advantage of avoiding the preparation of terminal diyne reagents.

Fluoride-catalyzed desilylation of trimethylsilyl-substituted diynes in the presence of metal halides affords σ-diynyl complexes in high yields in reactions that are considered to proceed via the intermediate vinylidene or silyl-vinylidene.[76–79] Thus, treatment of $ReCl(CO)_3(Bu^t{}_2\text{-bpy})$ with $Me_3SiC{\equiv}CC{\equiv}CSiMe_3$ in the presence of KF and AgOTf in refluxing MeOH for 24 h gave $Re(C{\equiv}CC{\equiv}CH)(CO)_3(Bu^t{}_2\text{-bpy})$.[80] Similar reactions with $RuCl(PPh_3)_2Cp$[81] and $RuCl_2(PMeR_2)(\eta^6\text{-}C_6Me_6)$ (R = Me, Ph)[82] in the presence of other halide abstracting reagents gave diynyl complexes in good yield.

The Stille reaction, in which an aryl-, alkenyl-, or alkynyl-stannane is cross-coupled with an aryl or vinyl halide, pseudo-halide, or arenediazonium salt in the presence of a Pd(0) catalyst, is a common method for the preparation of new carbon—carbon bonds.[75] Similarly, reaction of $FeI(CO)_2Cp$ with $Me_3SnC{\equiv}CC{\equiv}CH$ in the presence of $PdCl_2(NCMe)_2$ (5%) gives $Fe(C{\equiv}CC{\equiv}CH)(CO)_2Cp$.[83] Alkynyl complexes of $M(CO)_nCp$ (M = Ru, Mo, W) have also been prepared using Stille-type reactions suggesting that diynyl complexes involving these metals might also be prepared using this approach.[84]

Organostannanes can be used for the preparation of metal complexes containing poly-ynyl ligands from reactions that proceed without a catalyst. Thus the reaction of *cis*-$RuCl_2(dppm)_2$ with $Bu_3SnC{\equiv}CC{\equiv}CCPh_2(OSiMe_3)$ in the presence of $NaPF_6$ resulted in formation of *trans*-$RuCl\{C{\equiv}CC{\equiv}CCPh_2(OSiMe_3)\}(dppm)_2$. Presumably the reaction is driven by the elimination of $SnClBu_3$.[85,86]

5. *Deprotonation of Alkyne or Vinylidene Complexes*

Deprotonation ($KOBu^t$) of the η^2-alkyne complex $[Re(\eta^2\text{-}HC_2C{\equiv}CSiMe_3)(NO)(PPh_3)Cp^*]BF_4$ affords $Re(C{\equiv}CC{\equiv}CSiMe_3)(NO)(PPh_3)Cp^*$ in excellent yield.[87]

One of the most general methods of preparing transition metal compounds featuring acetylide ligands is by deprotonation of vinylidene complexes, which are conveniently obtained from reactions of 1-alkynes with a wide variety of transition metal complexes *via* a 1,2-H shift.[88] The nature of the products obtained from reactions of 1,3-diynes with metal complexes species depends on the electronic nature of the metal–ligand group. The reaction of $RuCl_2(L)(\eta^6\text{-}C_6Me_6)$ (L = PMe_3, PMe_2Ph, PPh_3) with $HC{\equiv}CC{\equiv}CCPh_2(OSiMe_3)$ in the presence of both a halide abstracting agent ($NaPF_6$) and a strong base ($NHPr^i{}_2$) afforded the diynyl complexes $RuCl\{C{\equiv}CC{\equiv}CCPh_2(OSiMe_3)\}(L)(\eta^6\text{-}C_6Me_6)$,[82,89,90] presumably via an ethynylvinylidene intermediate. With the more electron-rich fragments [$Ru(PPh_3)_2Cp$, $RuCl(dppm)_2$, or $RuCl(dppe)_2$], 1,3-diynes tend to undergo a 1,4-H shift affording highly reactive butatrienylidene complexes. In contrast to the arene examples, the intermediates derived from these more electron-rich species are often complicated by nucleophilic addition reactions involving the carbon chain (Section VII.B.2).[91–95]

B. *Diynyl Complexes by Group*

1. *Titanium and Zirconium*

Mono- and bis-diynyl complexes of titanium have been prepared from the reactions of titanocene dichlorides with $LiC{\equiv}CC{\equiv}CR$ in the appropriate stoichiometry. While mono(alkynyl)titanocene chloride complexes are known to disproportionate readily in solution to give the corresponding bis(alkynyl)- and dichloro-titanocenes this reaction is suppressed in the presence of silyl-substituted cyclopentadienyl ligands[96] and complexes such as $TiCl(C{\equiv}CC{\equiv}CR)Cp^{Si}{}_2$ may be readily prepared. Stepwise replacement of chloride in $TiCl_2Cp^{Si}{}_2$ using $LiC{\equiv}CC{\equiv}CR$ (R = Et, $SiMe_3$, Fc, Rc) afforded $TiCl_{2\text{-}n}(C{\equiv}CC{\equiv}CR)_nCp^{Si}{}_2$ ($n = 1, 2$), of which the ruthenocene complexes are sensitive to light. The metallocene monoalkynyls react further with $Li(C{\equiv}C)_mMc$ ($m = 1$, Mc = Fc, Rc; $m = 2$, M = Rc) to give the mixed compounds $Ti(C{\equiv}CFc)(C{\equiv}CC{\equiv}CFc)Cp^{Si}{}_2$ and $Ti(C{\equiv}CC{\equiv}CFc)\{(C{\equiv}C)_mRc\}Cp^{Si}{}_2$ ($m = 1, 2$).[97–99]

The reaction of $ZrCl\{2\text{-}N(SiMe_3)\text{-}4\text{-}MeC_5H_4N\}_3$ with $LiC{\equiv}CC{\equiv}CSiMe_3$ gives a structurally characterized mono-diynyl complex **1**.[100]

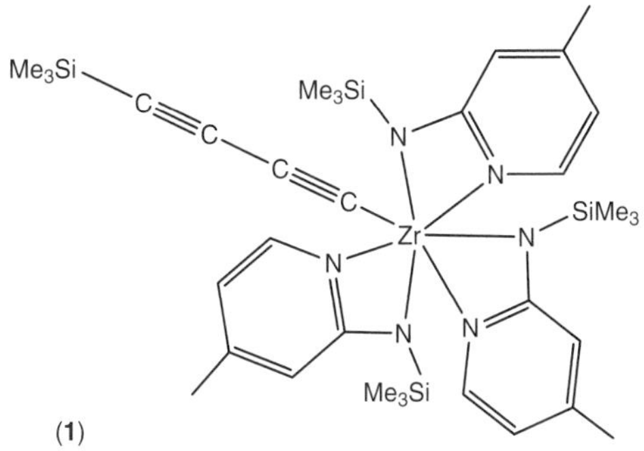

(**1**)

2. *Chromium, Molybdenum, and Tungsten*

Reactions of $M(CO)_5(thf)$ (M = Cr, W) with $LiC{\equiv}CC{\equiv}CC(NMe_2)_3$, generated from $Me_3SiC{\equiv}CC{\equiv}CC(NMe_2)_3$ and LiBu *in situ,* gave $[M\{C{\equiv}CC{\equiv}CC(NMe_2)_3\}(CO)_5]^-$ which were not isolated, but treated with $BF_3 \cdot OEt_2$ to abstract one NMe_2 group with formation of the orange pentatetraenylidenes $M\{{=}C{=}C{=}C{=}C{=}C(NMe_2)_2\}(CO)_5$. Of the resonance structures **A–F** for these formally cumulated species, the zwitterionic diynyl forms **D–F** better represent the structural data.[101] Both pentatetraenylidene complexes $M\{{=}C{=}C{=}C{=}C{=}C(NMe_2)_2\}(CO)_5$ (M = Cr, W) were susceptible to [2 + 2] cycloaddition reactions of the $C_\delta{=}C_\varepsilon$ double bond with the ynamine $MeC{\equiv}CNEt_2$, to give, after ring-opening,

A-F

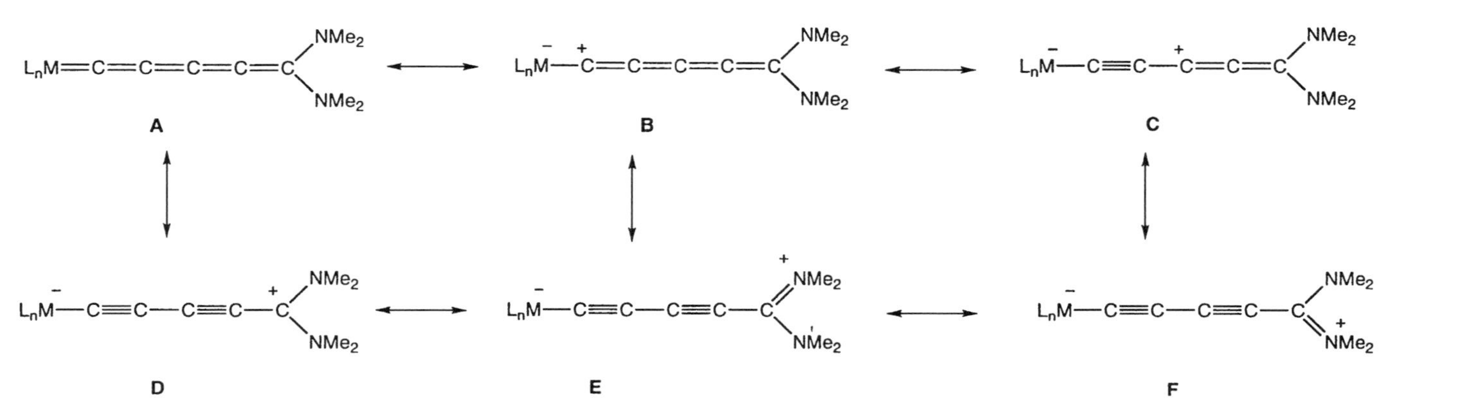

M{=C=C=C=C=C(NEt$_2$)CMe=C(NMe$_2$)$_2$}(CO)$_5$. Again, the IR and structural data are consistent with a significant contribution from the diynyl tautomers in these complexes.[102,103]

Although diynyl–chromium complexes are rare, similar Mo and W complexes have been extensively investigated. In general, the metal—diynyl bonds in these complexes are generated from Cu(I)-catalyzed reactions between the metal halides and terminal diynes in amine solvents, or by displacement of labile ligands from the metal by lithiated diynes. For example, high yields of M(C≡CC≡CH)(CO)$_3$Cp [M = Mo, W (**2**)] have been obtained from reactions of buta-1,3-diyne with the corresponding metal halides in NHEt$_2$ the presence of CuI (Scheme 2)[104] and similar reactions were used to prepare W(C≡CC≡CR)(CO)$_3$Cp* (R = H,[105] Ph[106]). However, reactions with the more electron-rich substrate MoBr(CO)(dppe)Cp and HC≡CC≡CH were rather less successful, and the terminal diynyl complex was isolated in only moderate yield.[107] Deprotonation of the latter was achieved with LiBu^s or LiNPr$^i{}_2$, the resulting lithio derivative being trapped with SiClMe$_3$.

Several diynyl complexes of Group 6 metals have been prepared directly from LiC≡CC≡CSiMe$_3$. In the cases of WCl(CO)$_3$Cp,[104] MoCl(CO)(dppe)Cp,[107] and MoBr(CO)$_2$(η-C$_7$H$_7$)[108] the silyl–diynyl complexes could be isolated in low to moderate yields. In all cases, proto-desilylation was achieved using KF in MeOH/thf. The terminal diynyl complexes were also obtained from MoBr(CO)(L) (η-C$_7$H$_7$) (L = CO, PMe$_3$) and Li$_2$C$_4$ in low yield.

The chemistry of W(C≡CC≡CH)(CO)$_3$Cp (**2**) has been explored in some detail and in general the diynyl ligand behaves as a somewhat electron-rich 1-alkyne (Scheme 2). Not surprisingly, the reactivity of the molybdenum analogue is similar to that of the tungsten complex in the cases studied to date. The deprotonated complexes formed following treatment of W(C≡CC≡CH)(CO)$_3$Cp with LDA can be quenched with SiClMe$_3$ or PClPh$_2$ to give W(C≡CC≡CX)(CO)$_3$Cp [X = SiMe$_3$ (**3**) or P(O)Ph$_2$ (**4**), respectively; the latter was formed by oxidation during work-up].[104]

Coupling of the diynyl ligand in **2** with 4-IC$_6$H$_4$R (R = H, OMe, Me, CO$_2$Me) is catalyzed by Pd(0)/Cu(I) in amine solvents and affords W(C≡CC≡CC$_6$H$_4$R-4) (CO)$_3$Cp (**5**).[109] The formation of Mo(C≡CC≡CPh)(CO)$_3$Cp in a similar manner was also reported. Iodo-alkynes can also be coupled, the triynyl complexes W(C≡CC≡CC≡CAr)(CO)$_3$Cp (**6;** Ar = Ph, Fc) being isolated in good yield from the reaction of ArC≡CI with W(C≡CC≡CH)(CO)$_3$Cp.[110] Oxidative homo-coupling reactions of M(C≡CC≡CH)(CO)$_3$Cp(M = Mo, W) afforded bis(metallated) tetraynes, e.g., **7**[104,109]; the hetero-coupling reaction with HC≡CC≡CFc gave W{(C≡C)$_4$Fc}(CO)$_3$Cp (**8**) together with both homo-coupled products.[111]

Substitution of CO by tertiary phosphines in W(C≡CC≡CPh)(CO)$_3$Cp is induced by trimethylamine-*N*-oxide and gives *cis*-W(C≡CC≡CPh)(CO)$_2$(PPh$_3$)Cp (**9**) in modest yields.[104] Similarly, mono-substitution of CO by dppm occurs in the series W{(C≡C)$_n$Fc}(CO)$_3$Cp (n = 1–3).[111]

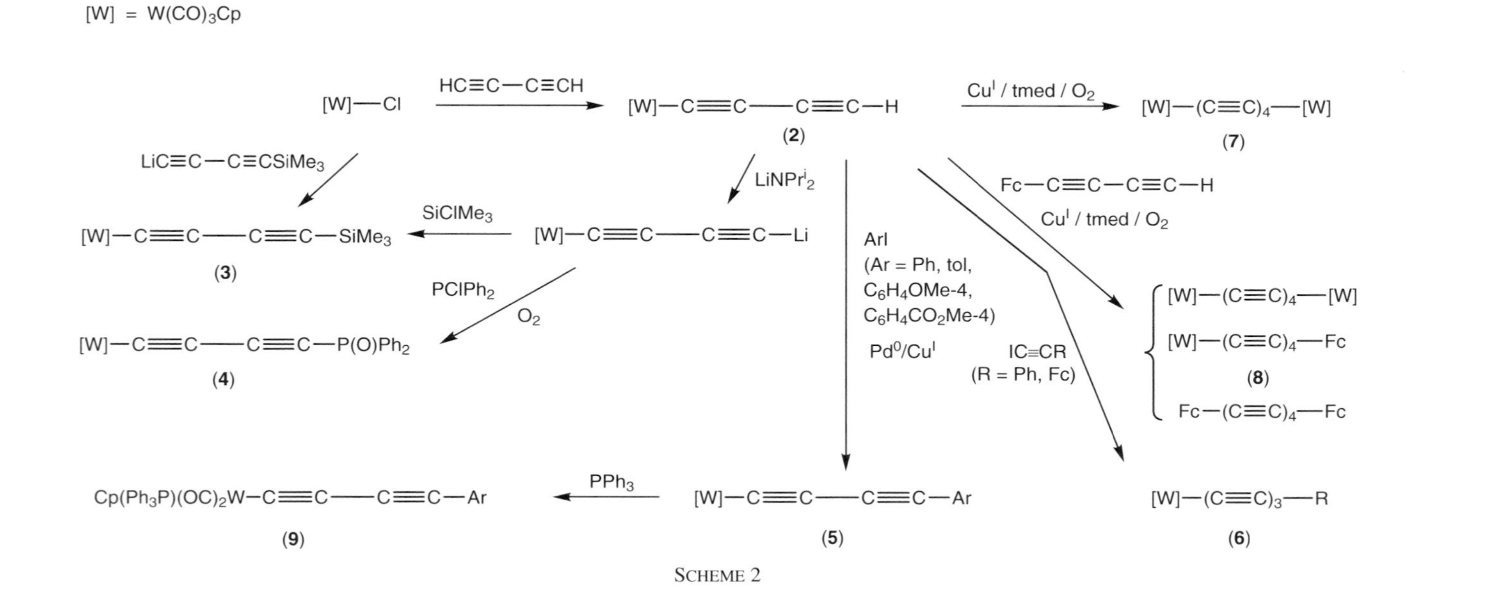
[W] = W(CO)3Cp
[W]—Cl
HC≡C—C≡CH
[W]—C≡C—C≡C—H
(2)
Cu^I / tmed / O2
[W]—(C≡C)4—[W]
(7)
LiC≡C—C≡CSiMe3
LiNPr^i2
SiClMe3
[W]—C≡C—C≡C—SiMe3
(3)
[W]—C≡C—C≡C—Li
PClPh2
O2
[W]—C≡C—C≡C—P(O)Ph2
(4)
Fc—C≡C—C≡C—H
Cu^I / tmed / O2
ArI
(Ar = Ph, tol,
C6H4OMe-4,
C6H4CO2Me-4)
Pd^0/Cu^I
IC≡CR
(R = Ph, Fc)
[W]—(C≡C)4—[W]
[W]—(C≡C)4—Fc
(8)
Fc—(C≡C)4—Fc
Cp(Ph3P)(OC)2W—C≡C—C≡C—Ar
(9)
PPh3
[W]—C≡C—C≡C—Ar
(5)
[W]—(C≡C)3—R
(6)

SCHEME 2

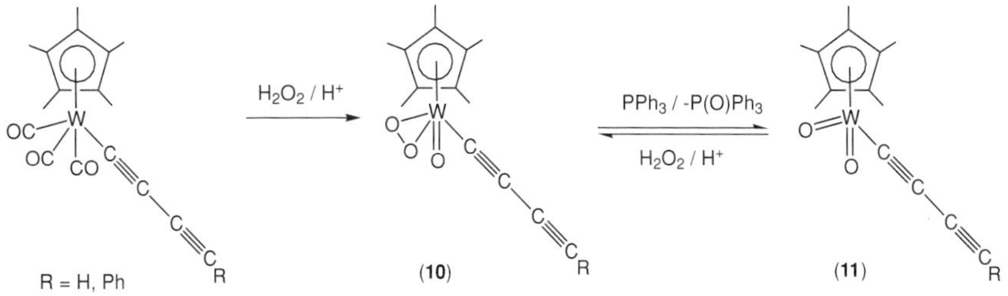

H_2O_2 / H^+ PPh_3 / -P(O)Ph_3 H_2O_2 / H^+

R = H, Ph   (10)   (11)

SCHEME 3

The W(VI) oxo–peroxo complex W(C≡CC≡CPh)(O_2)(O)Cp* (**10;** Scheme 3) is obtained by oxidation of W(C≡CC≡CPh)$(CO)_3$Cp* with hydrogen peroxide in strongly acidic media. Subsequent treatment with PPh_3 gave the dioxo-diynyl W(C≡CC≡CPh)$(O)_2$Cp* (**11**).[112] The terminal diynyl di-oxo complex can be prepared similar fashion.[113]

3. *Rhenium*

While no simple diynyl complexes of manganese or technetium have been reported to date the Gladysz group has conducted an extensive study of poly-ynyl and poly-yndiyl complexes containing the chiral rhenium fragment Re(NO)(PPh_3)Cp*. The key complex Re(C≡CC≡CH)(NO)(PPh_3)Cp* (**12;** Scheme 4) has been prepared as a mixture of stereo-isomers by reaction of the labile complex [Re(ClC_6H_4)(NO)(PPh_3)Cp*][BF_4] with HC≡CC≡$CSiMe_3$ to give both isomers of Re(η^2-HC≡CC≡$CSiMe_3$)(NO)(PPh_3)Cp* (**13**). Treatment of this mixture with BuLi afforded the silyl–diynyl complex Re(C≡CC≡$CSiMe_3$)(NO)(PPh_3)Cp*, which was proto desilylated by reaction with K_2CO_3 in methanol to give **12.**[87,114] Lithiation of **12** affords Re(C≡CC≡CLi)(NO)(PPh_3)Cp* which reacts with MeI to give Re(C≡CC≡CMe)(NO)(PPh_3)Cp* (**14**).[87]

Luminescent diynyl complexes Re(C≡CC≡CR)$(CO)_3$(Bu^t_2bpy) (R = H, Ph) have been prepared from reactions of ReCl$(CO)_3$(Bu^t_2bpy) with Me_3SiC≡CC≡$CSiMe_3$ or HC≡CC≡CPh, respectively, in the presence of AgOTf as a halide abstracting agent and a suitable base such as KF (R = H) or NEt_3 (R = Ph). In the case of the terminal diynyl complex, the KF also serves to protodesilylate the protected 1,3-diyne *in situ*.[80]

4. *Iron, Ruthenium, and Osmium*

Conversion of FeX$(CO)_2$(η-C_5R_5) (X = halide, R = H, Me, Ph) to the corresponding trimethylsilylbutadiynyl complexes has been achieved by reaction with LiC≡CC≡$CSiMe_3$ at low temperatures.[24,26,27,115] The carbonyl halides

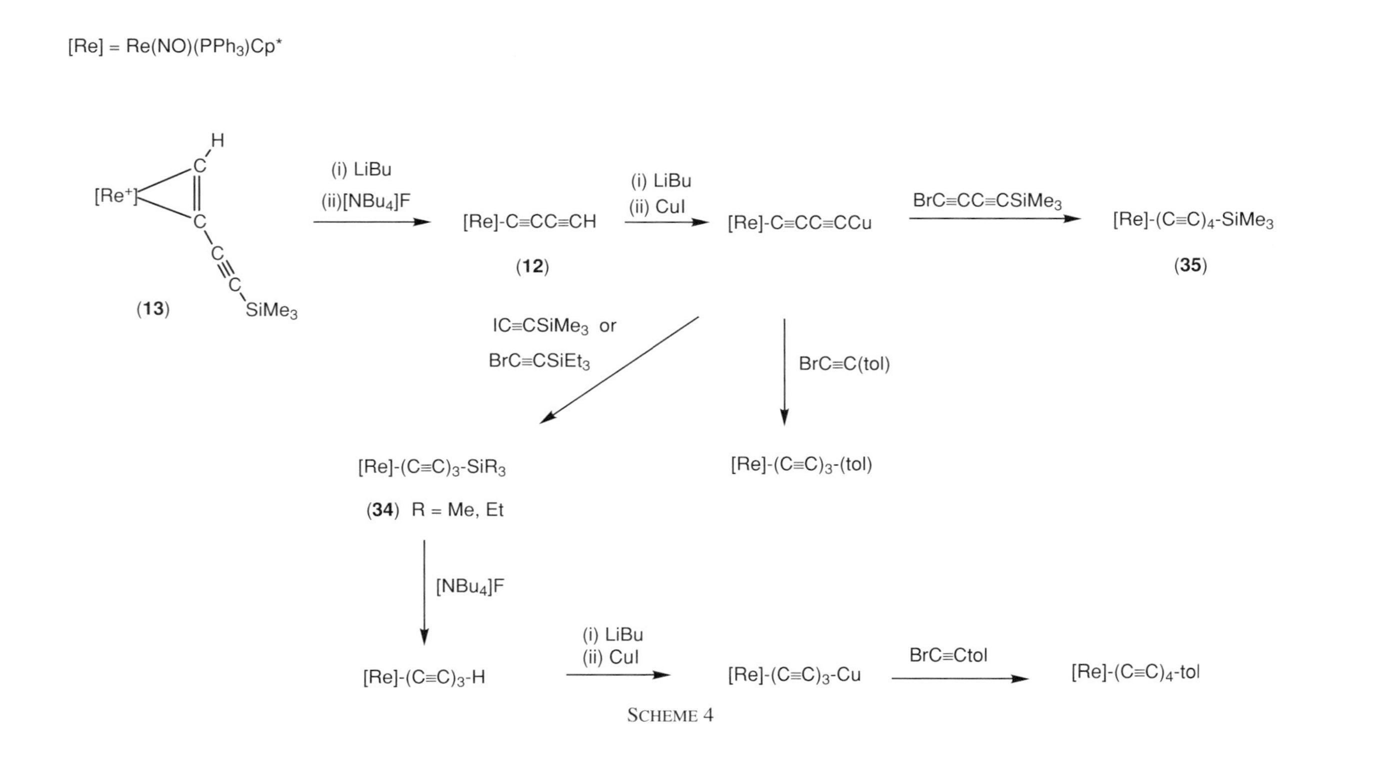
[Re] = Re(NO)(PPh3)Cp*
[Re+]
H
C
C
C
C
SiMe3
(13)
(i) LiBu
(ii)[NBu4]F
[Re]-C≡CC≡CH
(12)
(i) LiBu
(ii) CuI
[Re]-C≡CC≡CCu
BrC≡CC≡CSiMe3
[Re]-(C≡C)4-SiMe3
(35)
IC≡CSiMe3 or
BrC≡CSiEt3
BrC≡C(tol)
[Re]-(C≡C)3-SiR3
(34) R = Me, Et
[Re]-(C≡C)3-(tol)
[NBu4]F
[Re]-(C≡C)3-H
(i) LiBu
(ii) CuI
[Re]-(C≡C)3-Cu
BrC≡Ctol
[Re]-(C≡C)4-tol

SCHEME 4

$MX(CO)_2Cp$ (M = Fe, X = I, Cl; M = Ru, X = Cl) also couple readily with stannyl-diynes in the presence of Pd(0) catalysts to give metal–diynyl complexes[66,83,116] in reactions reminiscent of the Stille-coupling protocol. Carbonyl substitution reactions with these diynyl complexes have been performed both thermally and photochemically to give complexes such as $Fe(C{\equiv}CC{\equiv}CSiMe_3)(CO)(PPh_3)Cp$[24] and $Fe(C{\equiv}CC{\equiv}CSiMe_3)(dppe)Cp^*$[26] in excellent yield. In all cases, protodesilylation gives the the corresponding buta-1,3-diynyl complexes.[24,26,27,115]

Other diynyl anions may be used to construct M–C(diynyl) bonds with similar success. For example, Fischer and colleagues have employed the anion derived from the diynyl-carbene $W\{=C(NMe_2)C{\equiv}CC{\equiv}CH\}(CO)_5$ and LiBu in reactions with $FeI(CO)_2Cp$ to form $Fe\{C{\equiv}CC{\equiv}CC(NMe_2){=}[W(CO)_5]\}(CO)_2Cp$.[66] In constrast, the ruthenium analogue was prepared via the Pd(0)-catalyzed coupling of $W\{=C(NMe_2)C{\equiv}CC{\equiv}CSnBu_3\}(CO)_5$ with $RuCl(CO)_2Cp$.[66,116]

Reactions between $LiC{\equiv}CC{\equiv}CR$ (R = H, $SiMe_3$) and $[Ru(thf)(PPh_3)_2Cp][PF_6]$ or *trans*-$RuCl_2(CO)_2(PEt_3)_2$ afford $Ru(C{\equiv}CC{\equiv}CH)(PPh_3)_2Cp$ and *trans*-$Ru(C{\equiv}CC{\equiv}CSiMe_3)_2(CO)_2(PEt_3)_2$, respectively.[62,117,118] The latter was converted into *trans*-$Ru(C{\equiv}CC{\equiv}CH)_2(CO)_2(PEt_3)_2$ with $[NBu_4]F$, both complexes being quite air-, light-, and moisture sensitive and susceptible to oligomerization.

In addition to their role as a source of lithiated diynes or terminal diynes silyl-protected diynes have found direct application in the preparation of Group 8 diynyl complexes. Reactions of $Me_3SiC{\equiv}CC{\equiv}CR$ (R = Ph, C≡CPh, C≡CC≡CPh) with $RuCl(PPh_3)_2Cp$ in the presence of KF afford the corresponding poly-ynyl complexes $Ru(C{\equiv}CC{\equiv}CR)(PPh_3)_2Cp$.[119]

Reactions of $HC{\equiv}CC{\equiv}CCPh_2(OSiMe_3)$ with $RuCl_2(PR_3)(\eta\text{-}C_6Me_6)$ (PR_3 = PMe_3, PMe_2Ph, $PMePh_2$)[90,120] in the presence of strong, non-nucleophilic bases such as NEt_3 or $NHPr^i_2$ give $RuCl\{C{\equiv}CC{\equiv}CCPh_2(OSiMe_3)\}(PR_3)(\eta\text{-}C_6Me_6)$ (**15**) via silyl-vinylidene intermediates which are desilylated *in situ* (Scheme 5).[89] In contrast, $Me_3SiC{\equiv}CC{\equiv}CC(C_6H_4NMe_2\text{-}4)_2(OSiMe_3)$ reacts with $RuCl_2(PMe_3)(\eta\text{-}C_6Me_6)$ in the presence of $NaPF_6$ but without base to give the stable pentatetraenylidene complex $[RuCl\{C{=}C{=}C{=}C{=}C(C_6H_4NMe_2\text{-}4)_2\}(PMe_3)(\eta\text{-}C_6Me_6)]^+$ (**16**). On the basis of ^{13}C and IR data, a significant contribution to the structure of this complex from the cationic diynyl form has been proposed.[89,121]

Products obtained from *cis*-$RuCl_2(dppm)_2$ (**17**) and $LiC{\equiv}CC{\equiv}CSiMe_3$ depend on solvent and reagent stoichiometry (Scheme 6). For example, *cis*-$Ru(C{\equiv}CC{\equiv}CSiMe_3)_2(dppm)_2$ (**18**) was formed at −78°C in Et_2O, while the *trans* isomer **19** was obtained from reactions conducted in thf. With a 1/1 ratio of reactants in thf at −78°C, *trans*-$RuCl(C{\equiv}CC{\equiv}CSiMe_3)(dppm)_2$ was isolated. An excess of diynyllithium reagent $LiC{\equiv}CC{\equiv}CSiMe_3$ displaces both Cl and C≡CH ligands from *trans*-$RuCl(C{\equiv}CH)(dppm)_2$ to give the *trans*-bis(diynyl) complex.[122] The stannyl-diyne $Bu_3SnC{\equiv}CC{\equiv}CCPh_2(OSiMe_3)$ and *cis*-$RuCl_2(dppm)_2$ react without a catalyst but in the presence of $NaPF_6$ to give *trans*-$RuCl\{C{\equiv}CC{\equiv}CCPh_2(OSiMe_3)\}(dppm)_2$ (**20**) by elimination of $SnClBu_3$.[85,86]

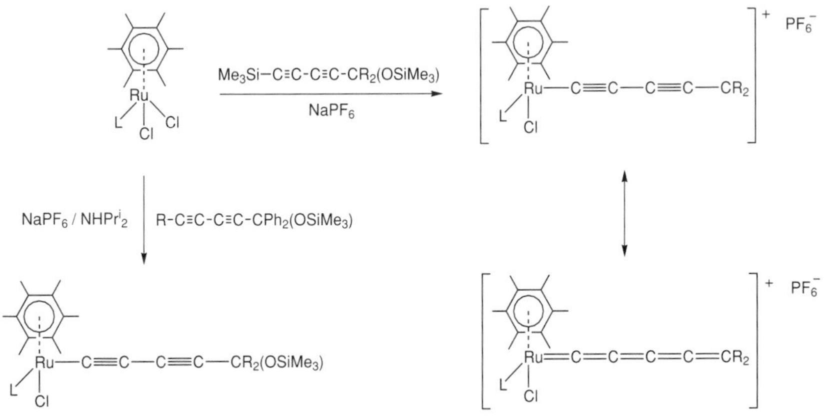

$Me_3Si{-}C{\equiv}C{-}C{\equiv}C{-}CR_2(OSiMe_3)$ / $NaPF_6$

$NaPF_6$ / $NHPr^i_2$ | $R{-}C{\equiv}C{-}C{\equiv}C{-}CPh_2(OSiMe_3)$

(**15**) R = H, $SiMe_3$; L = PMe_3, PMe_2Ph (**16**) R = $C_6H_4NMe_2$-4; L = PMe_3

SCHEME 5

In the presence of NEt_3, HC≡CC≡CPh reacts with *cis*-$RuCl_2(dppe)_2$ (**21**) and $NaPF_6$ (effectively a source of $[RuCl(dppe)_2]^+$) to give *trans*-RuCl(C≡CC≡CPh) $(dppe)_2$ (**22**),[95] while analogous reactions with HC≡CC≡CFc gave *trans*-$RuCl_{2\text{-}n}$ $(C{\equiv}CC{\equiv}CFc)_n(dppe)_2$ ($n = 1$, 2), according to reagent stoichiometry. Mixed alkynyl–diynyl complexes were obtained similarly from *trans*-RuCl(C≡CR) $(dppe)_2$ (**23**; R = Fc, $C_6H_4NO_2$-4) and HC≡CC≡CFc.[123] The bis(diynyl) complex *trans*-$Ru\{C{\equiv}CC{\equiv}CCPh_2(OSiMe_3)\}_2(dppm)_2$ was obtained from $RuCl_2(dppm)_2$ and an excess of $HC{\equiv}CC{\equiv}CCPh_2(OSiMe_3)$ in the presence of $NaPF_6$ and $NHPr^i_2$. The acetylide complexes *trans*-$RuCl(C{\equiv}CR)(dppm)_2$ (R = Bu, Ph) behave similarly with $HC{\equiv}CC{\equiv}CCPh_2(OSiMe_3)$ in CH_2Cl_2 and in the presence of $NaPF_6$ and NEt_3 to give *trans*-$Ru(C{\equiv}CR)\{C{\equiv}CC{\equiv}CCPh_2(OSiMe_3)\}(dppm)_2$ (**24**) in reactions that are both solvent (thf is unsuitable) and counterion (Na^+ rather than NH_4^+) specific.[85,124,125] These complexes are assumed to form via intermediate vinylidenes, which are deprotonated *in situ*.

5. *Rhodium and Iridium*

The chloro complex $\{RhCl(PPr^i_3)_2\}_2$ reacts readily with HC≡CC≡CPh, giving a mixture of $RhCl(\eta^2\text{-}HC_2C{\equiv}CPh)(PPr^i_3)_2$ (**25**) and $RhHCl(C{\equiv}CC{\equiv}CPh)(PPr^i_3)_2$ (**26**) (Scheme 7).[71] The acetato complex $Rh(\eta^2\text{-}O_2CCH_3)(PPr^i_3)_2$ reacts with 2 equiv of 1-phenylbuta-1,3-diyne in the presence of Na_2CO_3 to afford five-coordinate $RhH(C{\equiv}CC{\equiv}CPh)_2(PPr^i_3)_2$ (**27**) which readily adds Lewis bases to give $RhH(C{\equiv}CC{\equiv}CPh)_2(L)(PPr^i_3)_2$ (**28**; L = py, CO). For L = CO, the complex undergoes reductive elimination of HC≡CC≡CPh on warming in pentane to yield square-planar $Rh(C{\equiv}CC{\equiv}CPh)(CO)(PPr^i_3)_2$ (**29**), also obtained directly by sequential treatment of $Rh(CH_2Ph)(PPr^i_3)_2$ with CO and HC≡CC≡CPh. The

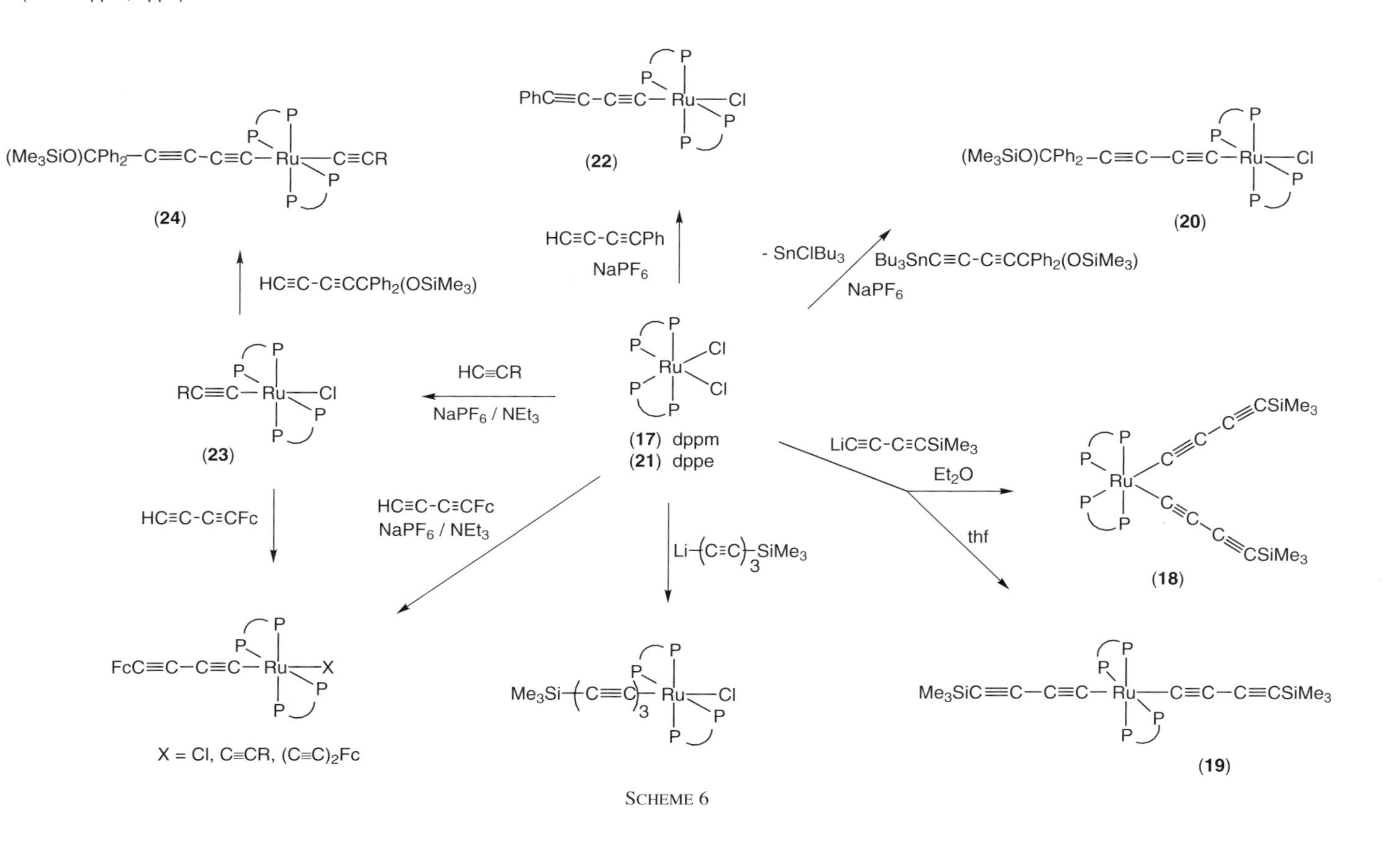

(P-P = dppm, dppe)
(24)
HC≡C-C≡CCPh2(OSiMe3)
(22)
HC≡C-C≡CPh
NaPF6
- SnClBu3
Bu3SnC≡C-C≡CCPh2(OSiMe3)
NaPF6
(20)
(23)
HC≡CR
NaPF6 / NEt3
(17) dppm
(21) dppe
LiC≡C-C≡CSiMe3
Et2O
thf
(18)
HC≡C-C≡CFc
HC≡C-C≡CFc
NaPF6 / NEt3
Li-(C≡C)3-SiMe3
X = Cl, C≡CR, (C≡C)2Fc
(19)

SCHEME 6

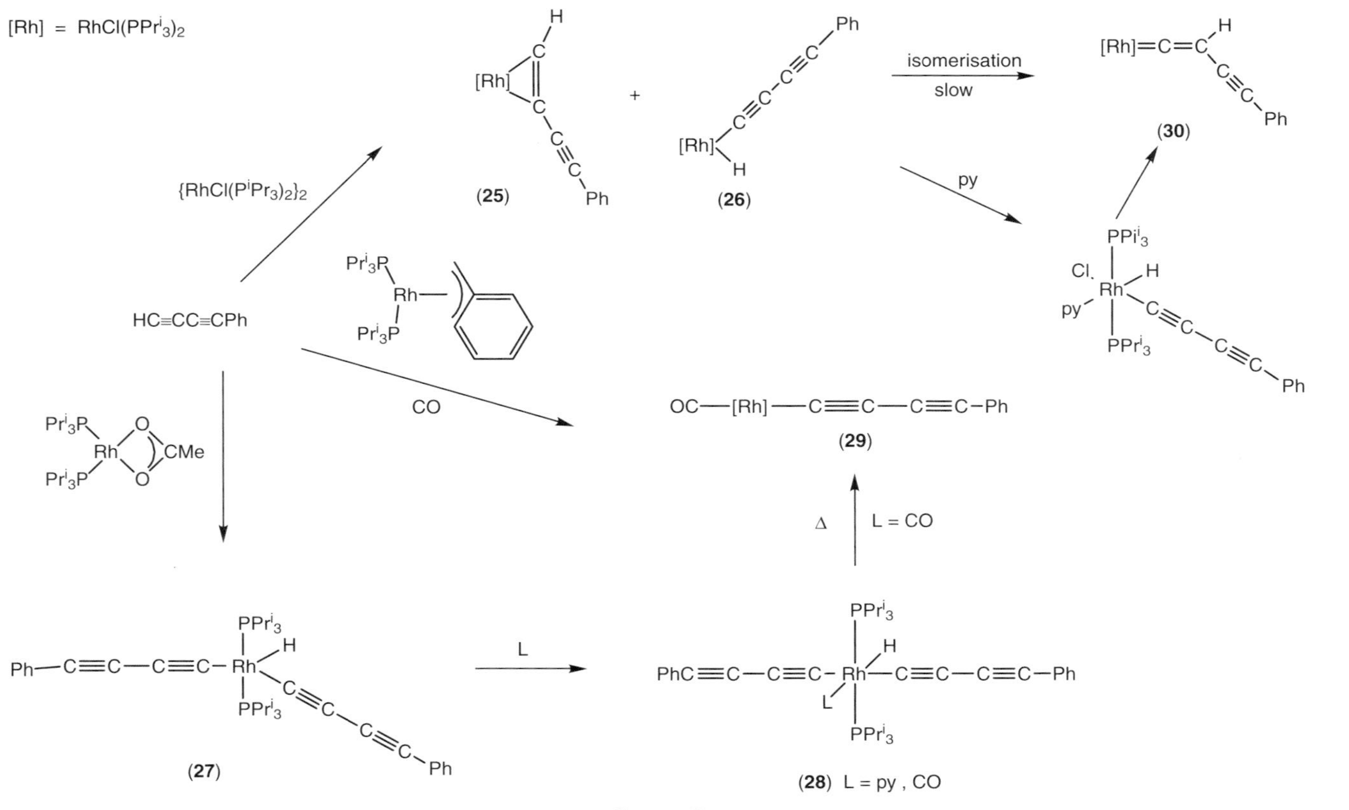

[Rh] = RhCl(PPr$^i_3$)$_2$
{RhCl(P^{i}Pr$_3$)$_2$}$_2$
HC≡CC≡CPh
(25)
(26)
isomerisation
slow
py
(30)
CO
(29)
Δ
L = CO
L
(27)
(28) L = py , CO

SCHEME 7

complex *trans*-$Rh(C{\equiv}CC{\equiv}CSiMe_3)(CO)(PPr^i_3)_2$ is formed from *trans*-Rh(OH) $(CO)(PPr^i_3)_2$ and $Me_3SiC{\equiv}CC{\equiv}CSnPh_3$ by preferential cleavage of the C–Sn bond.[126] The reaction is presumably driven by the elimination of $Sn(OH)Ph_3$ in preference to $Si(OH)Me_3$, and proceeds without cleavage of the C–Si bond.

Slow isomerization of *trans*-$RhHCl(C{\equiv}CC{\equiv}CPh)(PPr^i_3)_2$ and RhHCl(C≡C C≡CPh)(py)$(PPr^i_3)_2$ occurs to give RhCl{=C=CH(C≡CPh)}$(PPr^i_3)_2$ (**30**).[71] Similar reactions have been found to occur with the products from $RhCl(PPr^i_3)_2$ and $HC{\equiv}CC{\equiv}CCPh_2(OSiMe_3)$ to give eventually $RhCl({=}C{=}C{=}C{=}C{=}CPh_2)$ $(PPr^i_3)_2$, via intermediate η^2-alkyne, hydrido-diynyl and vinylidene complexes.[127] Closely related iridium compounds are obtained from reactions of $IrH_2Cl(PPr^i_3)_2$ with HC≡CC≡CR[R = $CPh_2(OH)$,[72] $SiMe_3$[79]]. Photochemical [R = $CPh_2(OH)$] or thermal (R = $SiMe_3$) rearrangements of IrHCl(C≡CC≡CR)$(PPr^i_3)_2$ give the ethynylvinylidene species IrCl{=C=CH(C≡CR)}$(PPr^i_3)_2$. Five-coordinate MHX (C≡CC≡CR)$(PPr^i_3)_2$ (M = Rh, R = Ph, X = Cl, C≡CC≡CPh; M = Ir, R = CPh_2OH, $SiMe_3$, X = Cl) readily add pyridine to give six-coordinate adducts MHX (C≡CC≡CR)(py)$(PPr^i_3)_2$.[71,72,79]

6. *Nickel, Palladium, and Platinum*

Diynyl and bis(diynyl) complexes of all three metals (Ni, Pd, Pt) are prepared by similar reactions. In early work, these materials were prepared from *cis*- or *trans*-$MX_2(PR_3)_2$ via halide displacement by diynyl anions. For example, the complexes Ni{$(C{\equiv}C)_n$H}(PPh_3)Cp (n = 1–3) were prepared from $NiCl(PPh_3)Cp$ and $H(C{\equiv}C)_nMgBr$[128] while a fourfold excess of HC≡CC≡CH reacts with $NiBr(PPh_3)$ Cp in NEt_3 in the presence of CuI to give Ni(C≡CC≡CH)(PPh_3)Cp.[129] Additionally, *trans*-$Pd(C{\equiv}CC{\equiv}CH)_2(PBu_3)_2$ was prepared from LiC≡CC≡CH and $PdCl_2(PBu_3)_2$ in Et_2O in 90% yield,[130] while *trans*-$Pt(C{\equiv}CC{\equiv}CSiMe_3)_2(PR_3)_2$ (R = Et, Bu) have been obtained from $LiC{\equiv}CC{\equiv}CSiMe_3$ and *trans*-$PtCl_2(PR_3)_2$. Protodesilylation ($[NBu_4]F$) gave *trans*-$Pt(C{\equiv}CC{\equiv}CH)_2(PR_3)_2$.[131]

However, the preparative method of choice for these complexes is the CuI-catalyzed reaction of a terminal diyne with the halide precursors. The reaction is generally applicable and numerous compounds, including the complete series of $M(C{\equiv}CC{\equiv}CH)_2(PBu_3)_2$ (M = Ni, Pd, Pt) complexes, have been prepared in this manner. In the case of nickel, the sensitivity of the halide species $NiCl_2(PBu_3)_2$ to amine solvents employed necessitated the use of *trans*-$Ni(C{\equiv}CH)_2(PBu_3)_2$, in which the ethynyl ligands are considered to be pseudo-halides and readily replaced in reactions with buta-1,3-diyne in $NHEt_2$ and a CuI catalyst.[132] Addition of HC≡CC≡CH to a mixture of $PtCl_2(PBu_3)_2$ and CuI in $NHEt_2$ at or below r.t. gave *cis*- or *trans*-$Pt(C{\equiv}CC{\equiv}CH)_2(PBu_3)_2$ according to the geometry of the starting material in high yield.[70] Subsequently, complexes $Pt(C{\equiv}CC{\equiv}CH)_2(L)_2$ (L = PEt_3; L_2 = dppe, dppp) have been prepared and structurally characterized.[133] Coupling of 4-HC≡CC≡C-terpy with *trans*-$PtCl_2(PBu_3)_2$ (CuI/$NHPr^i_2$) gives the *trans*-bis(diynyl) complex.[134]

Functionalization of the diynyl ligand in $Pt(C\equiv CC\equiv CH)_2(dppe)$ has been achieved with LiBu, the resulting lithio derivative reacting with MeI, $SiClMe_3$, or $AuCl(PPh_3)$ to give mono- or di-substituted derivatives according to stoichiometry. Coupling with iodobenzene gave $Pt(C\equiv CC\equiv CPh)_2(dppe)$.[133]

7. *Copper and Gold*

While diynylcopper complexes are implicated as intermediates in the various Cu(I)-catalyzed coupling reactions described earlier, and alkynyl–Au(I) complexes are well known, there are few reports of simple diynyl complexes of this Group. A copper(I) derivative has been made from $HC\equiv CC\equiv CMe$ and a large excess of ammoniacal CuCl in the presence of NaOAc[135] while synthesis of $CuC\equiv CC\equiv CSiMe_3$ has been achieved by successive reactions of $Me_3SiC\equiv CC\equiv CSiMe_3$ with LiMe.LiBr and CuI.[136] However, to our knowledge, none of these derivatives have been structurally characterized nor have they been used to any great extent in synthesis of organometallic diynyl complexes.

Reactions between $[ppn][Au(acac)_2]$[137] and buta-1,3-diyne give $[ppn][Au(C\equiv CC\equiv CH)_2]$.[134] Similarly, $(AuCl)_2(\mu\text{-dppm})$ affords $\{Au(C\equiv CC\equiv CH)_2\}(\mu\text{-dppm})$ while serendipitous use of a stoichiometric amount of CuI catalyst resulted in migration of dppm from gold to copper and formation of the curious bis(diynyl) species $Cu_3(\mu_3\text{-I})(\mu_3\text{-}C\equiv CC\equiv CAuC\equiv CC\equiv CH)(\mu\text{-dppm})_3$ (**31**).[133] Direct reaction between $[ppn][Au(C\equiv CC\equiv CH)_2]$ and $[Cu_2(\mu\text{-dppm})_2(NCMe)_2]^+$ afforded the "dumbbell" complex **32.**

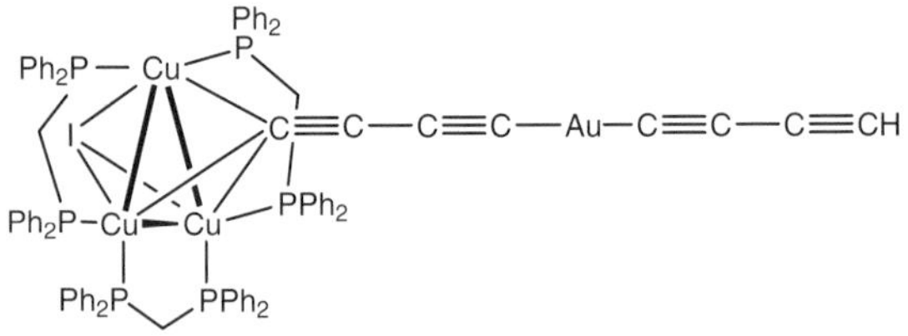

(**31**)

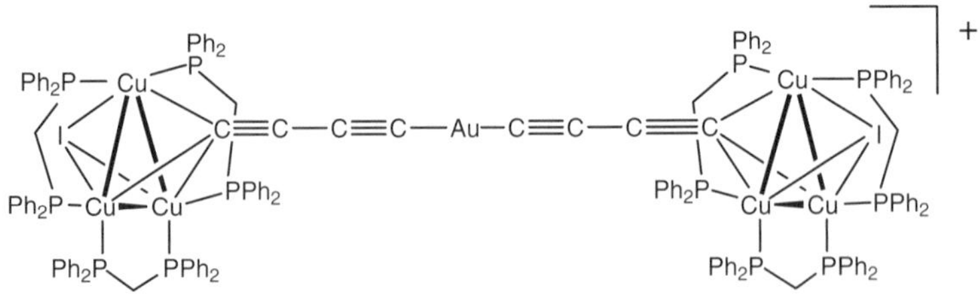

(**32**)

8. *Zinc and Mercury*

A brief account of the preparation of $ZnCl(C{\equiv}CC{\equiv}CSiMe_3)$ from $ZnCl_2$ and $LiC{\equiv}CC{\equiv}CSiMe_3$ is available.[138] In a recent report, Dehnicke and co-workers have described the preparation of the heterocubane zinc diynyl complexes $\{Zn(\mu_3\text{-}NPR_3)(C{\equiv}CC{\equiv}CSiMe_3)\}_4$ (**33**; R = Me, Et) from the reaction of $\{Zn(\mu_3\text{-}NPR_3)Br\}_4$ with $LiC{\equiv}CC{\equiv}CSiMe_3$. The Zn_4N_4 cores are only slightly distorted from the ideal cubic arrangement.[139]

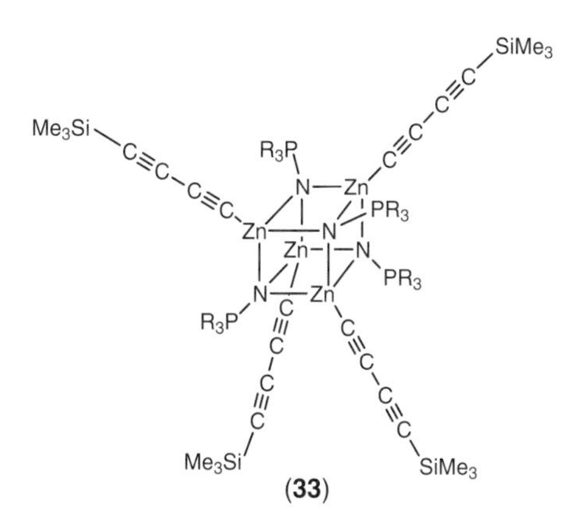

(**33**)

A few mercury complexes featuring diynyl ligands are known. Deprotonation of $W\{{=}C(NMe_2)C{\equiv}CC{\equiv}CH\}(CO)_5$ with BuLi, followed by treatment with $HgCl_2$ has afforded $Hg\{C{\equiv}CC{\equiv}CC(NMe_2){=}W(CO)_5\}_2$[66] and mercury(I) derivatives of more common terminal diynes $HC{\equiv}CC{\equiv}CR$ (R = Me, Ph) have also been described.[135]

C. *Poly-ynyl Complexes*

The poly-ynes $Me_3Si(C{\equiv}C)_nH$ ($n > 2$) are generally less thermally stable than the corresponding diynes, and the bis(silylated) compounds $Me_3Si(C{\equiv}C)_nSiMe_3$ ($n > 2$) are more robust.[140] By using of an excess of the doubly protected triyne $Me_3Si(C{\equiv}C)_3SiMe_3$ over the alkyllithium reagent (4.1/3.5) and a limiting amount of $FeI(CO)_2Cp^*$, Akita and colleagues successfully prepared the triynyl complex $Fe\{(C{\equiv}C)_3SiMe_3\}(CO)_2Cp^*$,[141] and the related *trans*-$RuCl\{(C{\equiv}C)_3SiMe_3)\}(dppe)_2$ has also been prepared.[142] However, attempts to desilylate η^2-alkyne rhenium complexes of 1,8-bis(trimethylsilyl)octa-1,3,5,7-tetrayne did not proceed cleanly.[143]

More often, the preparation of mono-metallic complexes featuring long poly-ynyl ligands $\{ML_n\}(C{\equiv}C)_nR$ has been accomplished by a sequence of cross-coupling reactions between terminal poly-ynyl complexes and 1-halo alkynes and poly-ynes involving intermediate ynyl-Cu(I) species (Scheme 4). Lithiation of **12** in the presence of CuI, or direct reaction of Re(C≡CC≡CH)(NO)(PPh_3)Cp* with Cu(OBu^t) gave Re(C≡CC≡CCu)(NO)(PPh_3)Cp*. This heterometallic diyndiyl derivative reacts with $X(C{\equiv}C)_nR$ (**34;** $n = 1$, X = I, R = $SiMe_3$; X = Br, R = $SiEt_3$, tol; **35;** $n = 2$, X = Br, R = $SiMe_3$). Other synthetic approaches for Re{$(C{\equiv}C)_3$ $SiMe_3$}(NO)(PPh_3)Cp* include reaction of **12** with IC≡C$SiMe_3$ in pyrrolidine with a CuI catalyst, or addition of [PhIC≡C$SiMe_3$]OTf to lithiated **12.** Similar combinations of reactions have given Re{$(C{\equiv}C)_mR$}(NO)(PPh_3)Cp* ($m = 5$, R = H, $SiMe_3$, $SiEt_3$; $m = 6$, R = $SiMe_3$).[29,143,144] Analogous coupling reactions have proved to be useful for the preparation of tungsten poly-ynyl complexes (see earlier).

D. *Reactions of Diynyl and Poly-ynyl Complexes*

1. *Reactions at the Metal Center*

The presence of the η^1-diynyl ligand seems to have little effect on the normal reactivity associated with the metal–ligand fragment. Ligand substitution and addition reactions at the metal center have also been widely employed to give other examples of diynyl complexes and, for example, carbonyl substitution by phosphine ligands has been accomplished using the standard techniques of thermolysis, photolysis, and CO oxidation (Me_3NO).[24,26,104] Diynyl complexes of coordinatively and electronically unsaturated metal centers participate in ligand addition reactions, and have been extensively investigated, particularly by Werner.[71,72,79]

2. *Reactions of the Diynyl Ligand*

The diynyl ligand behaves chemically as a rather electron-rich organic alkyne. The complexes {ML_n}C≡CC≡C$SiMe_3$ (and those with longer chains) are readily protodesilylated under standard conditions, such as treatment with fluoride in protic solvents or carbonate in methanol, to give the corresponding terminal diynyl complexes in high yield.[24,26,27,87,107,118]

Complexes featuring C≡CC≡CH ligands are highly versatile precursors of diynyl ligand complexes. The terminal diynyl ligand may be deprotonated by suitable bases, such as LiBu or $LiNPr^i_2$, to give nucleophilic conjugate bases that are useful reagents for the preparation of novel diynyl and diyndiyl complexes. The choice of base appears to be critical. Thus, while the phosphine-containing complex Re(C≡CC≡CH)(NO)(PPh_3)Cp*[87] and Fe(C≡CC≡CH)(CO)(PPh_3)Cp[24] may be satisfactorily deprotonated with LiBu, more sterically hindered bases such as $LiBu^s$ or $LiNPr^i_2$ are required to deprotonate the carbonyl complexes Fe(C≡CC≡CH)$(CO)_2$Cp[24] and W(C≡CC≡CH)$(CO)_3$Cp.[104] Reactions of these carbonyl-rich complexes with LiBu are possibly complicated by nucleophilic attack on the CO ligands by the lithio derivatives.

The diyne may be functionalized by other standard methods appropriate for 1-alkynes including the Sonogashira cross-coupling reaction[67] with aryl and alkynyl halides carried out in the presence of a mixed Pd(0)/Cu(I) catalyst.[104] The Cadiot–Chodkiewicz procedure[145] for oxidative coupling of alkynes has been applied to the synthesis of rhenium complexes containing extended poly-yne chains using copper derivatives prepared *in situ*. Reactions using catalytic amounts of CuI and $Pd(PPh_3)_4$ are also successful.[143,144,146,147] Homo-coupling affords complexes containing chains end-capped by metal–ligand fragments.[26,104,146,147]

In addition, the π-system of diynyl and poly-yndiyl complexes is available for coordination to other metal centers. Numerous examples of these reactions have been reported, as summarized later.

3. *Addition of Electrophiles and Cycloaddition Reactions*

Addition of electrophiles to diynyl complexes is expected to occur at either C_β or C_δ, the latter being favored if sterically demanding ligands shielding C_α and C_β are present. The products are butatrienylidenes and the chemistry of these species is closely related to the chemistry of the related unsaturated carbene ligands (Section VIII.B).[91,121]

Protonation of $Re\{(C{\equiv}C)_3(tol)\}NO)(PPh_3)Cp^*$ occurred at C_β to give a 60/40 equilibrium mixture of the *ac*/*sc* geometrical isomers of $[Re\{{=}C{=}CHC{\equiv}C\,C{\equiv}C(tol)\}(NO)(PPh_3)Cp^*]^+$, the position of electrophilic attack being confirmed by a 2D INADEQUATE ^{13}C NMR spectrum which gave values of $^1J(CC)$ within the carbon chain.[144] However, protonation of $Ru(C{\equiv}CC{\equiv}CH)(PPh_3)_2Cp$ is thought to afford the butatrienylidene cation $[Ru({=}C{=}C{=}C{=}CH_2)(PPh_3)_2Cp]^+$ which has not yet been isolated.[62] The steric bulk of the metal fragment prevents attack at C_α and nucleophiles add exclusively to C_γ.[91]

Treatment of *trans*-$RuCl\{C{\equiv}CC{\equiv}CCPh_2(OSiMe_3)\}(dppm)_2$ with HBF_4 in MeOH gives *trans*-$[RuCl\{{=}C{=}C{=}C(OMe)CH{=}CPh_2\}(dppm)_2]^+$ (**36**), possibly via intermediate formation of the pentatetraenylidene *trans*-$[RuCl({=}C{=}C{=}C{=}C{=}CPh_2)(dppm)_2]^+$ which rapidly adds MeOH at C_γ (Scheme 8). In the absence of an external nucleophile electrophilic attack at an ortho carbon of a pendant Ph ring with proton transfer to C_δ occurs to give **37**. Similarly, formation of the bis-allenylidene *trans*-$[Ru\{{=}C{=}C{=}C(OMe)CH{=}CPh_2\}_2(dppm)_2]^{2+}$ occurred with *trans*-$Ru\{C{\equiv}CC{\equiv}CCPh_2(OSiMe_3)\}_2(dppm)_2$[85] while protonation of *trans*-$RuCl(C{\equiv}CC{\equiv}CPh)(dppe)_2$ with CF_3SO_3H in wet solvents affords the vinylidene cation **38**.[95] Much of this chemistry has been reviewed.[121]

Regioselective cycloaddition of the vinylidene ligand in $Cr({=}C{=}CMe_2)(CO)_5$ to the $C_\alpha{\equiv}C_\beta$ bond of butadiynyl complexes $Fe(C{\equiv}CC{\equiv}CR)(CO)(PPh_3)Cp$ (R = $SiMe_3$, Bu, Ph) has given binuclear cyclobutenylidene complexes $\{(OC)_5Cr\}\{\mu\text{-}C_4Me_2(C{\equiv}CR)\}\{Fe(CO)(PPh_3)Cp\}$ (**39**) (Scheme 9). Protodesilylation of the $SiMe_3$ derivative (TBAF/thf) afforded the corresponding ethynyl complex;

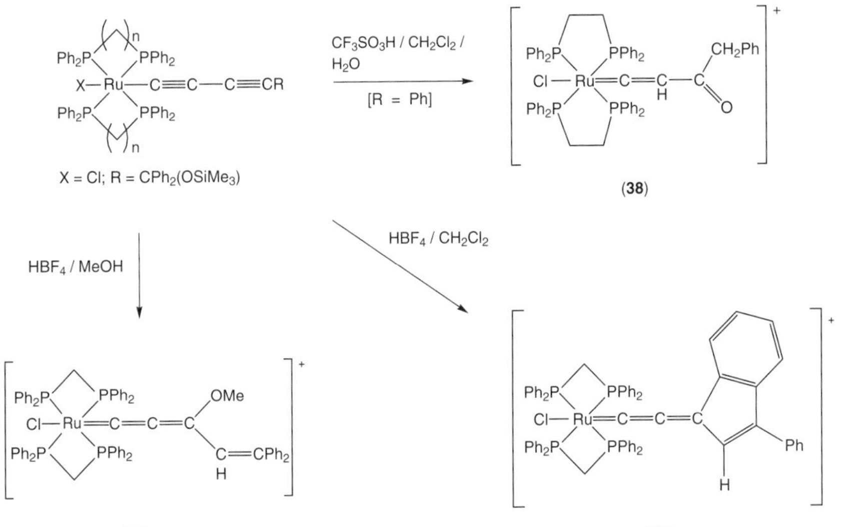

SCHEME 8

further chemistry, including coupling to a third ML_n fragment, has also been described.[148]

Reactions of $W(C{\equiv}CC{\equiv}CH)(CO)_3Cp$ with tetracyanoethene [$C_2(CN)_4$, tcne] result in addition to the C≡C triple bond further from the tungsten to give cyclobutenyl **40,** followed by ring opening to give $W\{C{\equiv}CC[{=}C(CN)_2]CH{=}C(CN)_2\}(CO)_3Cp$ (**41**) (Scheme 9).[104] Related complexes have been obtained from $W\{(C{\equiv}C)_nR\}(CO)_3Cp$ ($n=2$, 3, R = Ph, Fc) and $Ru\{(C{\equiv}C)_nPh\}(PPh_3)_2Cp$ ($n=1$–3).[81,111]

E. *Structural Studies*

There are few strictly comparable structurally characterized alkynyl/diynyl/poly-ynyl systems, which make detailed comparisons between the various unsaturated ligands difficult. For diynyl and poly-ynyl complexes, this is no doubt due to the poor packing qualities of the rod-like mono-metallic systems and resulting poor crystal quality. Structural features of poly-ynyl complexes are summarized in Table II.

In general, the lengths of the C≡C triple bonds in these complexes reportedly vary between 1.153(6) and 1.28(2) Å, with an average of ca. 1.21 Å. The conjugated ≡C–C≡ single bonds are shorter [1.33(2)–1.40(5) Å] than the C–C separation in ethane (1.54 Å)[149] as the bonds are between two C(sp) rather than C(sp^3) atoms. In buta-1,3-diyne itself, C–C and C≡C distances of 1.383(2) and 1.217(1) Å, respectively, were determined by electron diffraction.[150] In the case of the iron

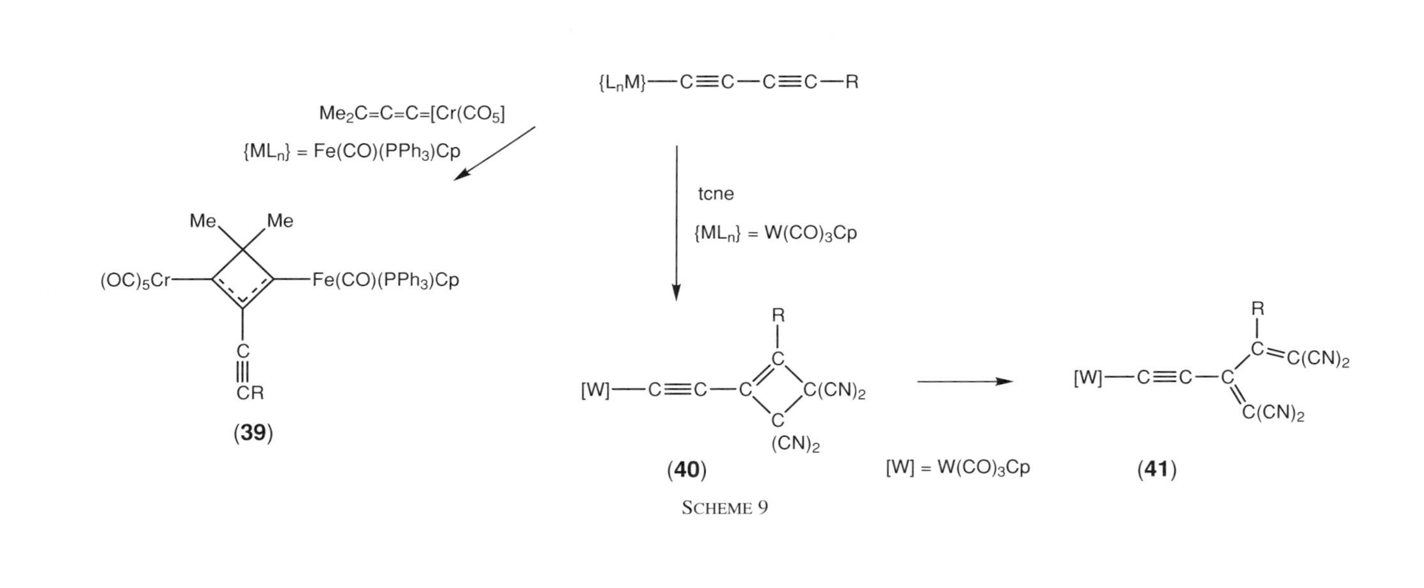
{L_nM}—C≡C—C≡C—R
Me2C=C=C=[Cr(CO5]
{MLn} = Fe(CO)(PPh3)Cp
tcne
{MLn} = W(CO)3Cp
Me
Me
(OC)5Cr
Fe(CO)(PPh3)Cp
CR
(39)
[W]—C≡C—C
R
C(CN)2
(CN)2
(40)
[W] = W(CO)3Cp
C(CN)2
C(CN)2
(41)

SCHEME 9

TABLE II

SOME STRUCTURAL DATA FOR DIYNDIYL AND POLY-YNDIYL COMPLEXES $\{L_nM\}$ $(C{\equiv}CC{\equiv}CR)_n$

ML_n	R	n	Bond Lengths (Å) M—C	C≡C	C—C	C—R	Angles (°) M—C—C	C≡C—C, C—C≡C	C—C—R	Reference
$TiCp^{Si}{}_2$	Fc	2	2.099(7)	1.224(10) 1.195(11)	1.377(10)	1.440	178.37	177.52 178.58	177.55	97
			2.090(7)	1.224(10) 1.210(11)	1.377(10)	1.415	174.0	177.96 175.29	178.62	
$Zr\{2\text{-}N(SiMe_3)\text{-}4\text{-}MeC_5H_4N\}_3$	$SiMe_3$	1	2.254(6)	1.228 1.228	1.390	1.804	169.54	174.5(7) 178.0(7)	175.4(6)	100
$Cr(CO)_5$	$C(NEt_2)C(Me){=}C(NMe_2)_2$	1	2.041(6)	1.219(8) 1.209(9)	1.379(8)	1.433(8)	175.4(6)	177.1(7) 178.8(7)	180.0(10)	102
$W(CO)_3Cp$	H	1	2.148(4)	1.191(5) 1.176(6)	1.376(6)		178.4(3)	178.3(4), 178.7(5)		133
$W(CO)_3Cp$	$SiMe_3$	1	2.124(8)	1.22(1) 1.22(1)	1.36(1)	1.82(1)	1.79(1)	179(1) 177(1)	177(1)	104
$W(CO)_5$	$C(NMe_2)_2$	1	2.144(6)	1.223(9) 1.192(9)	1.361(9)	1.399(9)	180.0(1)	180.0(1) 180.0(1)	180.0(1)	101
$Re(NO)(PPh_3)Cp^*$	$C{\equiv}CC_6H_4Me\text{-}4$	1	1.998(12)	1.28(2) 1.23(2) 1.22(2)	1.35(2) 1.33(2)	1.46(2)	169.1(10)	171.6(12) 176.7(12) 174.7(11) 178.8(13)	177.1(12)	144
$Re(NO)(PPh_3)Cp^*$	$C{\equiv}CC{\equiv}CSiMe_3$	1	2.032(7)	1.208(9) 1.21(1) 1.194(9) 1.20(1)	1.35(1) 1.36(1) 1.37(1)	1.848(9)	176.4(6)	177.4(8) 178.2(8) 176.4(8) 178.9(8) 175.9(8) 179(1)	178.0(9)	143
$Re(NO)(PPh_3)Cp^*$	$C{\equiv}CC{\equiv}C(tol)$	1	2.016(8)	1.241(11) 1.233(11) 1.242(12) 1.223(11)	1.380(11) 1.338(11) 1.337(12)	1.439(12)	174.5(7)	170.0(9) 176.9(9) 173.6(10) 178.1(10) 177.6(10) 178.8(10)	175.8(10)	144

$Re(CO)_3(Bu^t{}_2bpy)$	H	1	2.114(8)	1.199(10) 1.19(1)	1.369	0.948	175.22	178.3(8) 179.8(10)	179.89	80
$Re(CO)_3(Bu^t{}_2\text{-}py)$	Ph	1	2.126(5)	1.198(7) 1.189(7)	1.385	1.439	175.7	174.4(6) 178.6(6)	178.62	80
$Fe(CO)_2Cp^*$	H	1	1.907(4)	1.207(5) 1.153(6)	1.378(6)	0.87(4)	178.0(4)	179.0(5) 178.2(6)	175(3)	115
$Ru(PPh_3)_2Cp$	Ph	1	1.994(4)	1.206(5) 1.200(6)	1.389(6)	1.416(6)	178.3(3)	169.5(4) 175.9(4)	176.9(4)	119
$Ru(CO)_2(PEt_3)_2$	$SiMe_3$	2	2.057(2)	1.226(2) 1.209(2)	1.370(2)	1.831(2)	176.5(2)	178.9(2) 179.8(3)	177.2(2)	117, 118
$Ru(CO)_2(PEt_3)_2$	H	2	2.078(2)	1.194(2) 1.196(3)	1.386(3)		177.9(1)	176.9(2) 178.9(2)		117, 118
$RuCl(PMe_3)(C_6Me_6)$	CPh_2OSiMe_3	1	1.93(3)	1.26(4) 1.16(6)	1.40(5)	1.53(6)	174(3)	175(4) 176(4)	171(4)	90
$OsHCl(PPr^i{}_3)_2(NO)$	CPh_2OH	1	2.016	1.194(5) 1.187(5)	1.369(5)	1.480(5)	173.2(3)	168.7(4) 175.5(4)	176.3(4)	557
$Ni(PPh_3)Cp$	H	1	1.8383(15)	1.212(2) 1.187(3)	1.370(2)	0.96(3)	177.61(14)	177.94(14) 179.5(2)	173.81	129
Pt(dppe)	H	2	2.02(1)	1.17(2) 1.16(2)	1.36(2)		171.7(9)	176(1) 175(1)		133
$Pt(PMe_2Ph)_2$	Ph	2	2.009(5)	1.175(8) 1.188(8)	1.406(8)	1.442(8)	177.8(5)	177.1(6) 177.1(6)	174.9(6)	154
Pt(dcype)	H	2	2.026	1.153 1.173	1.387	0.950	176.61	175.19 176.63	179.96	464
			2.060	1.174 1.184	1.372	0.951	174.78	179.31 178.45	179.84	
trans-$Pt(PBu_3)_2$	H	2	1.987	1.212 1.176	1.370	0.950	178.22	177.58 179.07	179.93	464
			1.984	1.211 1.159	1.372	0.950	178.52	176.00 179.32	179.92	
$Au(PPh_3)$	H	1	2.001(5)	1.204(7) 1.190(8)	1.379(7)		170.6(4)	177.7(5) 178.0(6)		133
$[Zn(\mu_3\text{-}NPMe_3)]_4$	$SiMe_3$	1	1.93(1)	1.22(1) 1.18(1)	1.42(1)	1.839	175.4(9)	178(1) 177(1)	175(1)	139

complexes Fe(C≡CH)$(CO)_2$Cp* and Fe(C≡CC≡CH)$(CO)_2$Cp*, the contraction of the Fe(1)—C(1) distance (by 0.014Å) and lengthening of the C≡C distance (by 0.034 Å) in the diynyl complex have been interpreted in terms of a greater contribution from a cumulenic structure, together with the implication that the C_4H ligand is a better electron-withdrawing/accepting ligand than C_2H[115] (see also Section X.D).

Structural comparisons of the complexes Ru(C≡CC≡CR)$_2$$(CO)_2$$(PEt_3)_2$ (R = H, $SiMe_3$) indicate some delocalization along the Ru—C_4—Si chain, said to be facilitated by the $SiMe_3$ groups by hyperconjugation [σ(Si—C_{sp})-pπ(C_{sp})-dπ(M)] or by dπ(M)-pπ(C_{sp})-dπ(Si) conjugation.[117,118] These effects are also apparent from a comparison of the structures W(C≡CC≡CR)$(CO)_3$Cp (R = H, $SiMe_3$) (Table II).

A striking feature in many structures is the curvature of the C_n chain. Many distortions of C_n(sp) chains have been attributed to crystal packing forces rather than to any inherent feature of the ligand itself. Thus, the solid-state structures of Re{(C≡C)$_n$(tol)}(NO)(PPh_3)Cp* ($n = 3$, 4) showed a marked bending of the carbon chain, as evidenced either by the average ReCC, CCC, and CCC(tol) angles (between 174.7 and 175.7°) or the angles of particular carbons to the vector joining the end atoms of the chain (maximum values: 17.7°, $n = 3$; 17.08°, $n = 4$).[144] This feature probably results from low bending force constants as found for alkynes and, indeed, has previously been conjectured as a necessity for long polyalkynes (carbynes) as precursors of fullerenes.[143,151,152] Although significant curvature is not found in the Re—C_8SiMe_3 compound, elongated thermal ellipsoids are interpreted as indicating librational motion.[143] Similar librational motion has been found in other diynyl-platinum complexes.[133,153,154]

III

DI- AND POLY-YNE π COMPLEXES WITH ONE OR TWO METAL CENTERS

The first π complexes of 1,3-diynes were reported by Greenfield.[155] Shortly thereafter, Tilney-Bassett described the first heterometallic derivatives.[156] This area has grown steadily since these initial reports and many complexes of this type are now known. Diyne complexes are often simply alkyne-substituted analogues of conventional π-alkyne complexes. Indeed, transition metal compounds that form π-complexes with mono-alkynes can be expected to form complexes with diynes. However, the thermal sensitivity of terminal diynes, especially 1,3-butadiyne, may limit the application of routine reaction conditions in some cases. Further coordination of the ynyl ligand by additional metal fragments is usually determined by the reagent stoichiometry and by steric effects.

Complexes containing diyne—metal bonding of types **G–K** are known. In the following, we have used the following convention in formulas: C≡C represents a

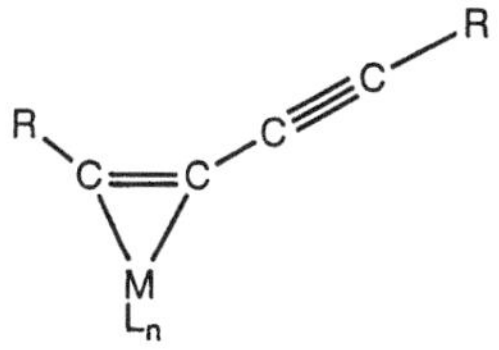

η^2

(G)

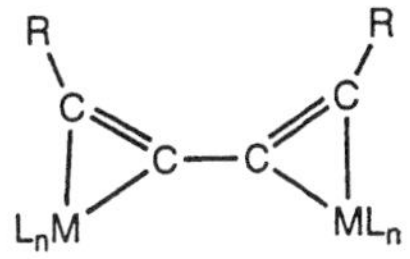

$\eta^2{:}\eta^2$

(H)

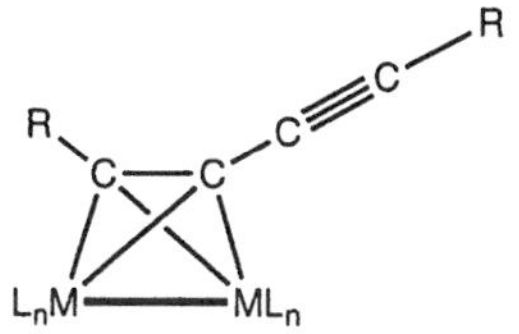

μ-η^2

(I)

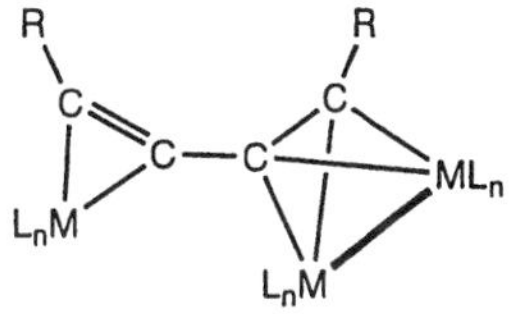

μ-$\eta^2{:}\eta^2$

(J)

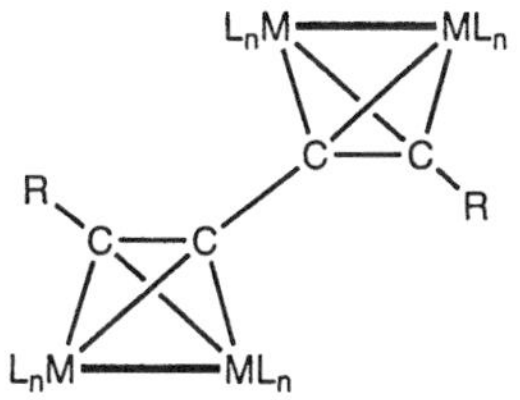

μ-$\eta^2{:}\mu$-η^2

(K)

carbon—carbon triple bond which is not π coordinated to a metal center, whereas C_2 represents a π-complexed carbon—carbon triple bond. For example, in $Co_2(\mu$-η^2-$PhC_2C{\equiv}CSiMe_3)(CO)_6$, the $PhC{\equiv}C$ unit bridges the Co—Co bond, whereas the $C{\equiv}CSiMe_3$ unit is free.

Generally, the preparation of these species is achieved by (a) coordination of preformed diynes to transition metal centers, and (b) coupling reactions of the alkynyl ligand in mono-alkynyl metal complex precursor.

A. *Coordination of Preformed Diynes to Transition Metal Centers*

Structural and spectroscopic evidence as well as computational work[157] indicate a considerable degree of delocalization throughout the four-carbon chain of a conjugated diyne. However, individual C≡C moieties retain sufficient electron density to sequester a wide variety of metal fragments, in processes which are often accompanied by the displacement of an equally diverse array of labile ligands. Thus, treatment of $Pt(cod)(PR_3)_2$ with buta-1,3-diyne results in displacement of the labile cyclooctadiene ligand and the formation of $Pt(\eta^2\text{-}HC_2C{\equiv}CH)(PR_3)_2$,[158] while CO and NCMe ligands are displaced from $WI_2(CO)_3(NCMe)_2$ by PhC≡CC≡CPh to give complexes such as $WI_2(CO)(NCMe)(\eta^2\text{-}PhC_2C{\equiv}CPh)_2$.[159] Similarly, $Co_2(CO)_8$ and $\{Ni(\mu\text{-}CO)Cp\}_2$ react with PhC≡CC≡CPh via the loss of two carbonyl ligands to give $Co_2(\mu\text{-}\eta^2\text{-}PhC_2C{\equiv}CPh)(CO)_6$[160] and $Ni_2(\mu\text{-}\eta^2\text{-}PhC_2C{\equiv}CPh)Cp_2$,[156] respectively.

B. *Coupling Reactions of the Alkynyl Ligand in Mono-Alkynyl Metal Complex Precursors*

As an alternative to the use of preformed diyne reagents, the diynyl ligand may be formed via the coupling of two alkynyl moieties within the metal coordination sphere. Thus, just as metal-catalyzed coupling reactions of terminal alkynes, typically using Cu(I/II) and/or Pd(0/II) redox couples, are the basis for many convenient methods of preparing free conjugated diynes, a combination of oxidative coupling and reductive elimination reactions on appropriate metal centres have been used in the formation of diyne complexes from alkynyl–metal precursors. In these reactions, coupling of alkynyl fragments occurs to give diynes or related ligands, either directly, or with concomitant condensation of metal fragments to form a binuclear or cluster core. In many cases, these reactions have not yet been fully investigated and the requirements for them to proceed are not known in detail. Examples of diyne complexes prepared in this manner include $Mo_2(\mu\text{-}\eta^2\text{-}PhC_2C{\equiv}CPh)(CO)_4Cp_2$ by thermolysis of $Mo(C{\equiv}CPh)(CO)_3Cp$[161] and $\{Co_2(CO)_6\}_2(\mu\text{-}\eta^2{:}\mu\text{-}\eta^2\text{-}Me_3SiC_2C_2SiMe_3)$ from the sequential reactions of $Co_2(\mu\text{-}\eta^2\text{-}HC_2SiMe_3)(CO)_6$ with $LiNPr^i_2$ and H_2O.[162]

C. *Survey of Complexes by Group*

1. *Vanadium, Niobium, and Tantalum*

The reaction of an excess of VCp_2 with RC≡CC≡CR (R = $SiMe_3$, PPh_2) results in addition of one VCp_2 moiety to each C≡C moiety and the formation of *cis*-$\{VCp_2\}_2(\mu\text{-}\eta^2{:}\mu\text{-}\eta^2\text{-}RC_2C_2R)$ (**42**).[163] Magnetic susceptibility measurements indicate a greater antiferromagnetic interaction between the V d^1 centers in the case of R = PPh_2 than $SiMe_3$. This has been attributed to the in-plane π-geometry and arrangement of the VCp_2 centers in the R = PPh_2 complex and the more favorable orbital interactions (mainly a linear combination of carbon p orbitals with d_{z^2} of the HOMO in a d^1 system) that result in this case.

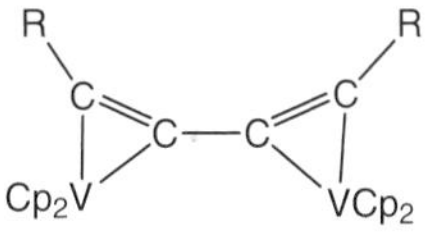

(**42**)

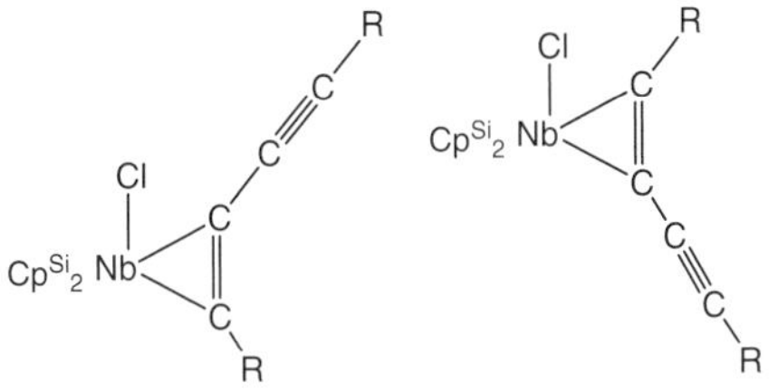

endo exo

(**43**)

Hexa-2,4-diyne reacts with $NbI(CO)_3(PEt_3)_3$ (from an *in situ* reaction between $[NEt_4][Nb(CO)_6]$, PEt_3 and I_2) to give *all-trans*-$NbI(CO)_2(PEt_3)_2(\eta^2\text{-}MeC_2C{\equiv}CMe)$ in which the complexed C≡C triple bond is *trans* to I.[164] Reactions of $NbCl_2Cp^{Si}{}_2$ with RC≡CC≡CR (R = Ph, $SiMe_3$) give $NbCl(\eta^2\text{-}RC_2C{\equiv}CR)Cp^{Si}{}_2$. Recrystallization of the R = $SiMe_3$ complex at low temperatures (−50°C, pentane) results in a first-order intramolecular isomerization, the two forms being *endo* and *exo* with respect to the Cl atom (**43**-*endo*, -*exo*), the *endo* isomer being thermodynamically favoured. With HCl, selective formation of (*Z*)-PhCH=CHC≡CPh is found.[165] The chloride ligand in **43** is relatively labile and treatment with $MgMe_2$ gives $NbMe(\eta^2\text{-}Me_3SiC_2C{\equiv}CSiMe_3)Cp^{Si}{}_2$. The phenyl analogue could not be isolated, but a similar reduction, followed by addition of allyl bromide, gave $NbBr(\eta^2\text{-}PhC_2C{\equiv}CPh)Cp^{Si}{}_2$. Reduction of **43** with sodium amalgam gives paramagnetic $Nb(\eta^2\text{-}Me_3SiC_2C{\equiv}CSiMe_3)Cp^{Si}{}_2$, reoxidation of which with $[FcH]^+$ in the presence of a Lewis base affords $[Nb(\eta^2\text{-}Me_3SiC_2C{\equiv}CSiMe_3)(L)Cp^{Si}{}_2]^+$ (L = MeCN, $CNBu^t$).

The tantalum(V) calix[4]arene complex **44** provides an interesting scaffold for the construction of an η^2-$PhC_2C{\equiv}CPh$ ligand upon reaction with an excess of phenylethynyllithium (Scheme 10). The coupling reaction is presumed to proceed via bis-alkynyl **45**, which is subsequently attacked at the α-carbon of one of the acetylide ligands to give anion **46**, isolated as its lithium salt. The addition of a further equivalent of LiC≡CPh is probably prohibited by orbital constraints.[166]

2. *Molybdenum and Tungsten*

The reactions of MCl_n (M = Mo, $n = 5$; M = W, $n = 6$) with RC≡CC≡CR′ (M = Mo, R = R′ = Ph; M = W, R = R′ = Ph, $SiMe_3$, I; R = Ph, R′ = $SiMe_3$) in

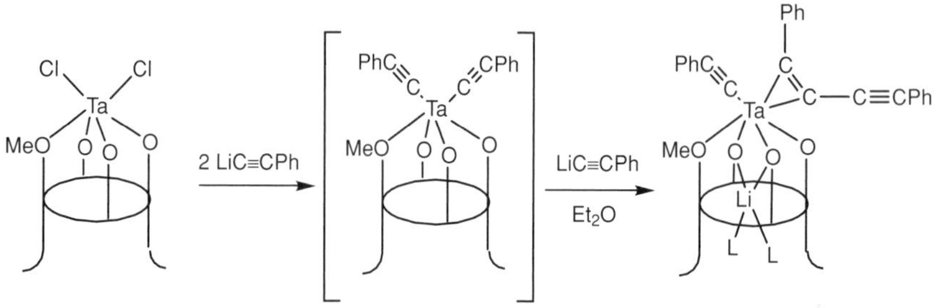

(**44**) (**45**) (**46**) L = Et_2O

SCHEME 10

CCl_4 afford compounds of the type $\{M(\mu\text{-Cl})Cl_3(\eta^2\text{-RC}_2C{\equiv}CR')\}_2$ which feature $M(\mu\text{-Cl})_2M$ bridged structures. The C≡C triple bonds act as 4-e donors in each of these complexes. In the case of reactions with M = W, tetrachloroethene C_2Cl_4 was required as an additional reducing agent. In many cases, concomitant chlorination of the diyne occurred, resulting in the formation of PhC_4Cl_2Ph and PhC_4Cl_4Ph as by-products.[167] The selective coordination of the C≡CPh moiety in the reaction of WCl_6 with $PhC{\equiv}CC{\equiv}CSiMe_3$ has been attributed to favorable orbital interactions between this fragment and the electron-deficient WCl_4 fragment. In the presence of an excess of WCl_6 both alkyne moieties are complexed resulting in the formation of polymeric $\{[WCl_4]_2(\mu\text{-}\eta^2\text{:}\mu\text{-}\eta^2\text{-PhC}_2C_2SiMe_3)\}_n$.

The chloride bridges are readily cleaved by reaction with various nucleophiles. Neutral species $\{WCl_4(L)\}_2(\mu\text{-}\eta^2\text{:}\eta^2\text{-RC}_2C_2R')$ (L = py, Et_2O; R = R′ = Ph, I, $SiMe_3$; R = Ph, R′ = $SiMe_3$) are formed from reactions of the dimeric precursors with pyridine or Et_2O. Treatment of $\{W(\mu\text{-Cl})Cl_3(\eta^2\text{-RC}_2C{\equiv}CR')\}_2$ with $[PPh_4]Cl$ or $[P(CH_2I)Ph_3]Cl$ results in the formation of the ionic derivatives $[PPh_4][WCl_5(\eta^2\text{-RC}_2C{\equiv}CR')]$.[168,169] Subsequent reaction of $[P(CH_2I)Ph_3]_2[\{WCl_5\}_2(\mu\text{-}\eta^2\text{:}\mu\text{-}\eta^2\text{-IC}_2C_2I)]$ with AgCl results in alkynyl iodine/chlorine exchange to give $[P(CH_2I)Ph_3]_2[\{WCl_5\}_2(\mu\text{-}\eta^2\text{:}\mu\text{-}\eta^2\text{-ClC}_2C_2Cl)]$.[169] Exchange of the metal–halide ligands has been observed following the reaction of $\{[WCl_4]_2(\mu\text{-}\eta^2\text{:}\mu\text{-}\eta^2\text{-PhC}_2C_2SiMe_3)\}_n$ with sodium fluoride in acetonitrile containing 15-crown-5 to give [Na(15-crown-5)][$WF_5(\eta^2\text{-PhC}_2C{\equiv}CSiMe_3)$] (**47**). In the crystal, the Na^+ cation is coordinated by two F atoms of the anion as well as by the crown ether O atoms.[170] The Mo—diyne bonds in $\{Mo(\mu\text{-Cl})Cl_3(\eta^2\text{-PhC}_2C{\equiv}CPh)\}_2$ are less robust and treatment of this complex with $[PPh_4]Cl$ affords the diyne and $[PPh_4]_2[Mo_2Cl_{10}]$. The free alkyne in $\{W(\mu\text{-Cl})Cl_3(\eta^2\text{-PhC}_2C{\equiv}CPh)\}_2$ reacts in a manner similar to that of a conventional alkyne, and forms $\{W(\mu\text{-Cl})Cl_3(\eta^2\text{-PhC}_2CBr{=}CBrPh)\}_2$ upon reaction with Br_2.[170]

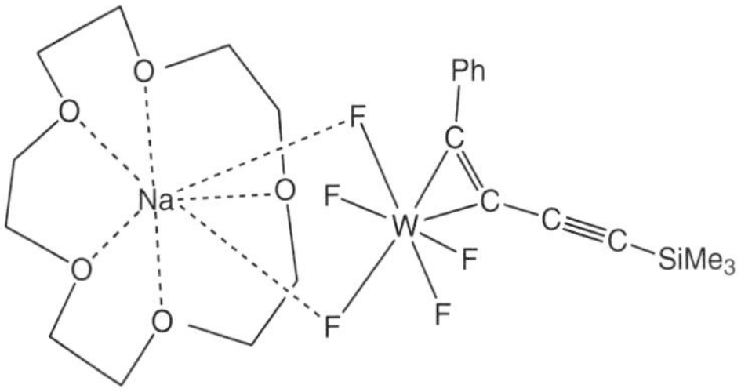

(47)

The reaction of equimolar quantities of $WI_2(CO)_3(NCMe)_2$ and $PhC{\equiv}CC{\equiv}CPh$ gave the iodo-bridged dimer $\{W(\mu\text{-}I)(I)(CO)(NCMe)(\eta^2\text{-}PhC_2C{\equiv}CPh)\}_2$ (**48;** Scheme 11), while the use of a twofold excess of the diyne yielded monomeric $WI_2(CO)(NCMe)(\eta^2\text{-}PhC_2C{\equiv}CPh)_2$ (**49**). Further reaction of the latter with $WI_2(CO)_3(NCMe)_2$ afforded the bimetallic complex $\{WI_2(CO)(NCMe)\}_2(\mu\text{-}\eta^2\text{:}\mu\text{-}\eta^2\text{-}PhC_2C_2Ph)_2$ (**50**). In the majority of cases, geometries for the various complexes have been assigned on the basis of IR evidence, and by comparison with well characterized mono-alkyne complexes. The analogy to mono-alkyne complexes is quite appropriate, as usually only one of the diyne alkyne moieties is coordinated to the metal center, acting as a four-electron donor.[159] An exception to this generalization is found in **50** in which the two diyne ligands donate a total of six electrons to each metal. A description of the various NMR parameters associated with alkyne ligands acting as two- or four-electron donors has previously been given.[171]

The dimeric complex **48** is a useful precursor of monomeric species $WI_2(CO)L_2(\eta^2\text{-}PhC_2C{\equiv}CPh)$ via cleavage of the halide bridge and ligand exchange reactions with $L = PPh_3$, PPh_2Cy, or the chelating ligands L_2 [2,2′-bipy, 1,10-phen or $Ph_2P(CH_2)_nPPh_2$ ($n = 1\text{–}6$)].[159] Similarly, iodide and NCMe ligands are displaced from **49** by treatment with equimolar amounts of bpy yielding $[WI(CO)(bpy)(\eta^2\text{-}PhC_2C{\equiv}CPh)_2]I$ (**51**), which exchanges anions in the presence of $Na[BPh_4]$.

Several complexes featuring metal—metal bonds have been described. The reaction of the unsaturated reagent $\{Mo(CO)_2Cp\}_2$ with $RC{\equiv}CC{\equiv}CR$ gave $Mo_2(\mu\text{-}\eta^2\text{-}RC_2C{\equiv}CR)(CO)_4Cp_2$ [$R = SiMe_3$[172], C_5H_4N-4[173]]. Similar reactions of $(CH_2{=}CHCH_2)Me_2SiC{\equiv}CC{\equiv}CSiMe_2(CH_2{=}CHCH_2)$ with an excess of $\{Mo(CO)_2Cp\}_2$ gave $\{Mo(CO)_2Cp\}_2\{\mu\text{-}\eta^2\text{:}\mu\text{-}\eta^2\text{-}(CH_2CH{=}CH_2)Me_2SiC_2C_2SiMe_2(CH_2CH{=}CH_2)\}$, which afforded $\{Mo(CO)_2Cp\}_2(\mu\text{-}\eta^2\text{:}\mu\text{-}\eta^2\text{-}Me_2FSiC_2C_2R)$ [$R = SiMe_2(CH_2CH{=}CH_2)$, $SiFMe_2$] following treatment with the stoichiometric amount of HBF_4.[174] Protodesilylation of the $SiMe_3$ complex with $[NBu_4]F$ in moist thf afforded $Mo_2(\mu\text{-}\eta^2\text{-}HC_2C{\equiv}CH)(CO)_4Cp_2$.[172]

In refluxing toluene, $\{W(CO)_2Cp\}_2$ reacts with $Me_3SiC{\equiv}CC{\equiv}CSiMe_3$ to give $\{W_2(CO)_4Cp_2\}_2(\mu\text{-}\eta^2\text{:}\mu\text{-}\eta^2\text{-}Me_3SiC_2C_2SiMe_3)$ in which each alkyne moiety

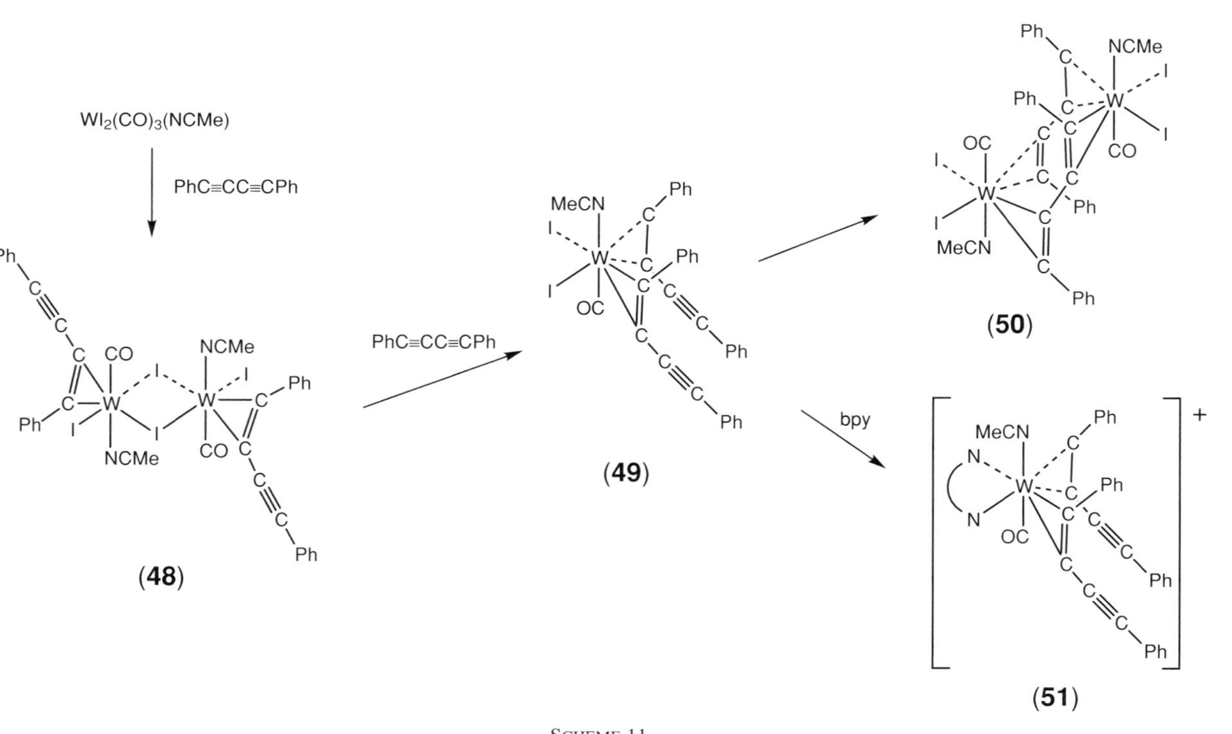

Scheme 11

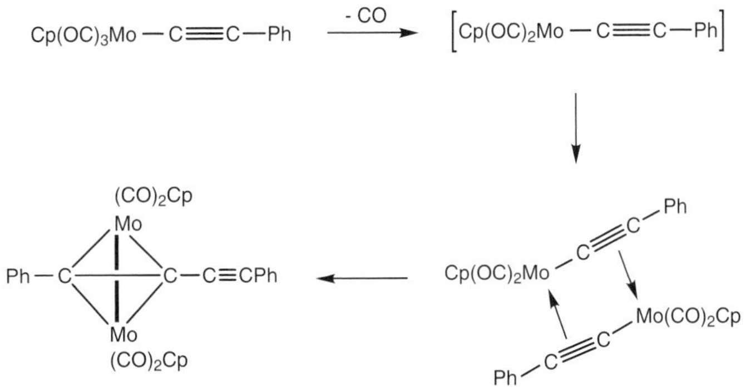

SCHEME 12

bridges a W_2 unit.[175] This complex is also formed in the reaction between $W(C{\equiv}CSiMe_3)(CO)_3Cp$ and $\{W(CO)_3Cp\}_2$. Thermolysis of the mononuclear complexes $Mo(C{\equiv}CR)(CO)_3Cp$ ($R = SiMe_3$, Ph, 4-FC_6H_4) gave $\{Mo(CO)_2Cp\}_2(\mu$-η^2-$RC_2C{\equiv}CR)$ in modest yield (Scheme 12).[161,175] A cross-coupling experiment using the Ph and C_6H_4F-4 complexes afforded a mixture of the two homo-coupled products as well as the two possible isomers of $\{Mo(CO)_2Cp\}_2(\mu$-η^2-4-$RC_6H_4C_2C{\equiv}CC_6H_4R'$-4) (R, R′ = H, F; F, H).[161] Trace amounts of the Cr analogue were obtained by pyrolysis of $Cr(C{\equiv}CPh)(CO)_3Cp$, but the product is not thermally stable.

3. *Rhenium*

Treatment of the labile chlorobenzene complex $[Re(ClPh)(NO)(PPh_3)Cp^*]BF_4$ with $HC{\equiv}CC{\equiv}CSiMe_3$ gave two inseparable rotamers of $[Re(\eta^2$-$HC_2C{\equiv}CSiMe_3)(NO)(PPh_3)Cp^*]BF_4$.[87]

4. *Iron and Ruthenium*

Reactions of the equilibrium mixture $Fe(CO)_2(PEt_3)_2(N_2)/\{Fe(CO)_2(PEt_3)_2\}_2N_2$ with $Me_3SiC{\equiv}CC{\equiv}CSiMe_3$ gave $Fe(\eta^2$-$Me_3SiC_2C{\equiv}CSiMe_3)(CO)_2(PEt_3)_2$ in almost quantitative yield. The analogous $P(OMe)_3$ complex forms in a *ca* 2:1 equilibrium with the vinylidene $Fe\{{=}C{=}C(SiMe_3)C{\equiv}CSiMe_3\}(CO)_2\{P(OMe)_3\}_2$ following the reaction of the diyne with $Na[FeI(CO)_2\{P(OMe)_3\}_2]$.[76]

Reactions of $PhC{\equiv}CC{\equiv}CPh$ with either $Ru(CO)_2(PPh_3)_3$ or $[RuH(NCMe)(CO)_2(PPh_3)_2]ClO_4$ in the presence of dbu afford $Ru(\eta^2$-$PhC_2C{\equiv}CPh)(CO)_2(PPh_3)_2$.[176] This complex reacts with $HClO_4$, HBF_4, and HPF_6 to give $Ru(\eta^3$-$PhC_3{=}CHPh)(CO)_2(PPh_3)_2$ and with 1 equiv of HCl to give $RuCl\{C(C{\equiv}CPh){=}CHPh\}(CO)(PPh_3)_2$. Excess HCl affords $RuCl_2(CO)_2(PPh_3)_2$.

5. *Cobalt*

Mononuclear $CoCl(\eta^2\text{-}PhC_2C{\equiv}CPh)(PMe_3)_3$ has been obtained from $CoCl(PMe_3)_3$ and $PhC{\equiv}CC{\equiv}CPh$.[177] Reactions of diynes with a mixture of $Co(L)(PMe_3)_3$ ($L = C_2H_4$, cyclopentene, PMe_3) and $CoCl_2$ gave air-sensitive bimetallic, mixed-valence compounds $Co_2(\mu\text{-}\eta^2\text{-}RC_2C{\equiv}CR)(Cl)(PMe_3)_4$ ($R = Bu^t$, $SiMe_3$). The formation of $GeMe_3$ analogues has also been noted. While thermally stable, these compounds fragment by reaction with CO to give the free diyne, $CoCl(CO)_2(PMe_3)_2$, and $Co_2(CO)_4(PMe_3)_4$.

The diyne chemistry of cobalt is dominated by the formation of the dicobaltatetrahedrane systems containing the C_2Co_2 moiety. A large number of complexes of general form $Co_2(\mu\text{-}RC_2C{\equiv}CR)(CO)_4L_2$ ($L = CO$, tertiary phosphine or phosphite) are now known, and these are most commonly prepared by reaction of the diyne with $Co_2(CO)_8$ or a substituted derivative in nonpolar solvents at moderate temperatures. The first diyne complex of a metal carbonyl to be described was the bis-$Co_2(CO)_6$ derivative of $Me_2C(OH)C{\equiv}CC{\equiv}CCMe_2(OH)$[155] and early work in this area has been reviewed.[178] Dicobalt carbonyl complexes have long been used to protect alkynes and have the advantage that the dicobalt unit may be easily removed by gentle oxidation, e.g., with iron(III) or cerium(IV), or Me_3NO.[10,179]

As expected, the extent of coordination the di-or poly-yne following treatment with $Co_2(CO)_6L_2$ is determined by the stoichiometry of the reaction and by steric factors. While 1:1 stoichiometric reactions invariably lead to mono-adducts, more highly coordinated compounds are observed from reactions with excess cobalt reagent or from further reaction of mono-complexed compounds with $Co_2(CO)_6L_2$ (Scheme 13) [$R = Ph$[160], $CH_2(OH)$[180], $SiMe_3$[181]]. The bis-adducts $\{Co_2(CO)_6\}_2(\mu\text{-}\eta^2\text{:}\mu\text{-}\eta^2\text{-}RC_2C_2R)$ have also been obtained from $Co_4(CO)_{12}$ and $RC{\equiv}CC{\equiv}CR$ ($R = Me$, Ph). It has been suggested that these reactions involve an undetected $Co_4(\mu_4\text{-}\eta^2\text{-}RC_2C{\equiv}CR)(CO)_{10}$ intermediate, which fragments upon coordination of the second C≡C triple bond.[182]

The C≡C moieties in the asymmetric diynes $RC{\equiv}CC{\equiv}CSiMe_3$ ($R = Ph$, tol) show no selectivity towards $Co_2(CO)_8$ and give 1:1 mixtures of the two possible isomers in which either triple bond is coordinated.[183] While two adjacent C≡C triple bonds may be coordinated to $Co_2(CO)_6$ or $Co_2(\mu\text{-dppm})(CO)_4$ groups, introduction of a third such group to the next adjacent C≡C triple bond is restricted by steric constraints (see following). However, less sterically encumbered derivatives of $RC{\equiv}CC_6H_4C{\equiv}CC{\equiv}CC_6H_4C{\equiv}CR$ ($R = H$, $SiMe_3$) in which all four C≡C triple bonds are complexed to $Co_2(CO)_6$ fragments have been made.[184]

Cobalt carbonyl adducts of buta-1,3-diyne are best obtained via indirect methods. Very early work showed that the reaction between $Hg\{Co(CO)_4\}_2$

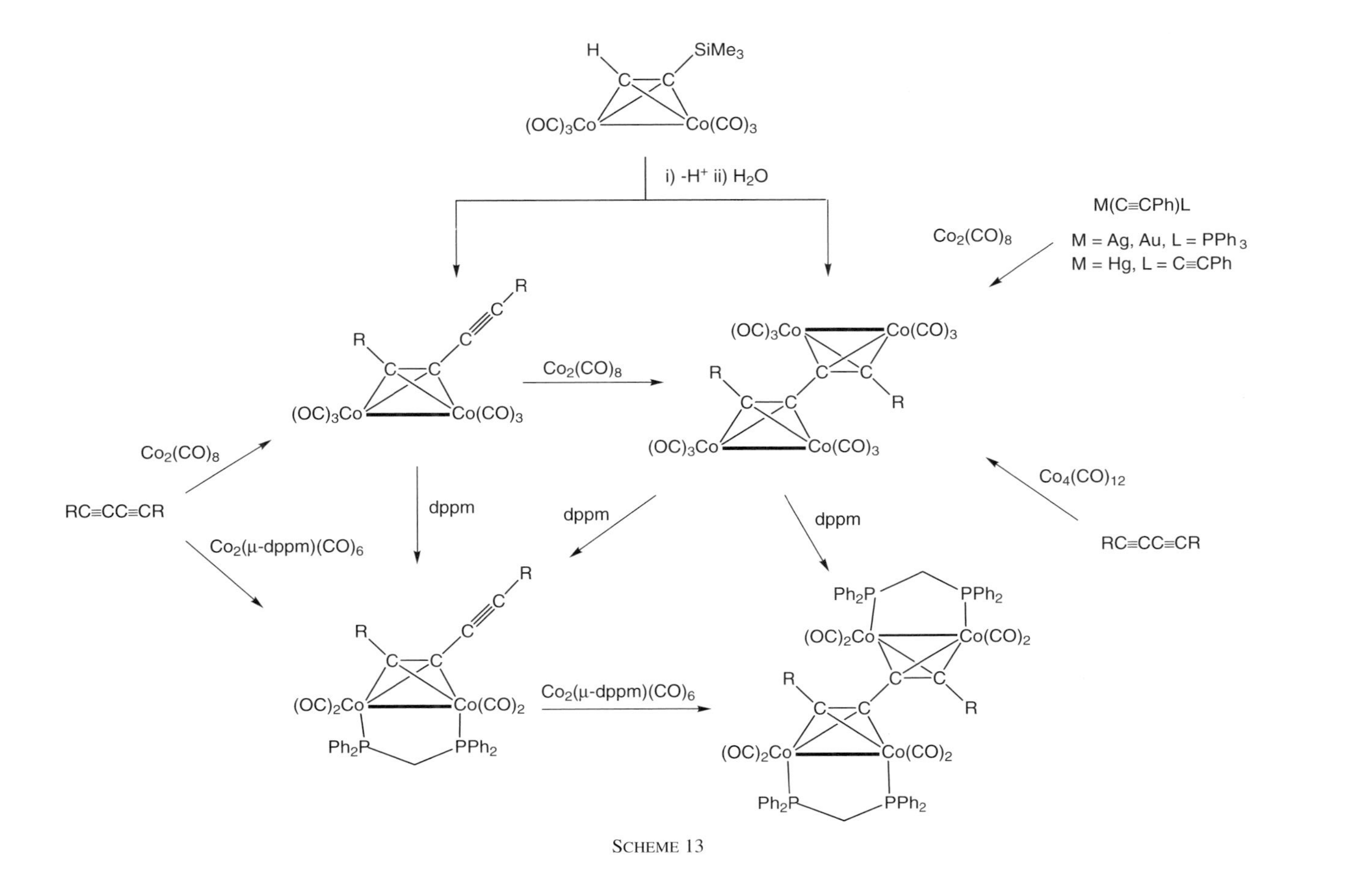

SiMe3
H
(OC)3Co
Co(CO)3
i) -H+ ii) H2O
M(C≡CPh)L
M = Ag, Au, L = PPh3
M = Hg, L = C≡CPh
Co2(CO)8
Co4(CO)12
RC≡CC≡CR
dppm
Co2(μ-dppm)(CO)6
(OC)2Co
Co(CO)2
Ph2P
PPh2
R

SCHEME 13

and $CH_2ClC{\equiv}CCH_2Cl$ afforded $\{Co_2(CO)_6\}_2(\mu\text{-}\eta^2:\mu\text{-}\eta^2\text{-}HC_2C_2H)$.[185] Cobalt complexes of terminal diynes $Co_2(\mu\text{-dppm})(\mu\text{-}\eta^2\text{-}RC_2C{\equiv}CH)(CO)_4$ have also been obtained as minor components following lithiation of $Co_2(\mu\text{-dppm})(\mu\text{-}\eta^2\text{-}RC_2CH{=}CHCl)(CO)_4$ (R = $SiMe_3$, tol, Ph, C_6H_4OMe-4, C_6H_4F-4) with $LiNPr^i_2$ and subsequent hydrolysis.[183] However, protodesilylation of complexes derived from $Me_3SiC{\equiv}CC{\equiv}CR$ (where R may also be $SiMe_3$) provides the simplest route to dicobalt derivatives of the terminal diynes. The uncomplexed $C{\equiv}CSiMe_3$ groups are desilylated more readily than silylalkynes coordinated by the Co_2 moiety (see following). Thus, the $SiMe_3$ group on the free C≡C group in $Co_2(\mu\text{-}\eta^2\text{-}RC_2C{\equiv}CSiMe_3)(\mu\text{-dppm})(CO)_4$ (R = $SiMe_3$, tol, Ph, C_6H_4OMe-4, C_6H_4F-4) is easily removed under standard conditions (KOH/MeOH or $[NBu_4]F$) to give $Co_2(\mu\text{-}\eta^2\text{-}RC_2C{\equiv}CH)(\mu\text{-dppm})(CO)_4$,[172,183,186] while the isomeric $Co_2(\mu\text{-dppm})(\mu\text{-}\eta^2\text{-}Me_3SiC_2C{\equiv}CR)(CO)_4$ complexes remain unchanged.

The complexes $M(C{\equiv}CPh)(PPh_3)$ (M = Au, Ag) or $Hg(C{\equiv}CPh)_2$ react with $Co_2(CO)_8$ in CH_2Cl_2 to form $\{Co_2(CO)_6\}_2(\mu\text{-}\eta^2:\mu\text{-}\eta^2\text{-}PhC_2C_2Ph)$ as the major product along with a small amount of a red compound tentatively formulated as $Co_2\{\mu\text{-}\eta^2\text{-}PhC_2[M(PPh_3)]\}(CO)_6$.[187] The reaction is thought to involve initial formation of the usual η^2 complex, which eliminates the bulky $M(PPh_3)$ group in a manner similar to the formation of $\{Co_2(CO)_6\}_2(\mu\text{-}\eta^2:\mu\text{-}\eta^2\text{-}Me_3SiC_2C_2SiMe_3)$ from $Co_2(\mu\text{-}\eta^2\text{-}Me_3SiC_2Li)(CO)_6$ (see following). This proposal is given credence by the formation of $\{Co_2(CO)_6\}_2(\mu\text{-}\eta^2:\mu\text{-}\eta^2\text{-}Me_3SiC_2C_2SiMe_3)$ and $\{Fe(CO)_2Cp\}_2$ by thermolysis of $Fe\{C_2SiMe_3[Co_2(CO)_6]\}(CO)_2Cp$.[188] The reaction of $FcC{\equiv}CI$ with $Co_2(CO)_8$ gives $\{Co_2(CO)_6\}_2(\mu\text{-}\eta^2:\mu\text{-}\eta^2\text{-}FcC_2C_2Fc)$ by deiodinative coupling of the iodoalkyne followed by complexation of the resulting diyne[189]; the same complex has also been obtained by more conventional methods.[190] Other examples of cobalt–diyne complexes derived from alkyne coupling reactions include the formation of a 1:1 mixture of $\{Co_2(CO)_6\}_2(\mu\text{-}\eta^2:\mu\text{-}\eta^2\text{-}Me_3SiC_2C_2SiMe_3)$ and $Co_2(\mu\text{-}\eta^2\text{-}Me_3SiC_2C{\equiv}CSiMe_3)(CO)_6$ by deprotonation of $Co_2(\mu\text{-}\eta^2\text{-}HC_2SiMe_3)(CO)_6$ with $LiNPr^i_2$ or $LiN(SiMe_3)_2$ [which gave a dark green-black solution, presumed to contain $Co_2(\mu\text{-}LiC_2SiMe_3)(CO)_6$], followed by quenching with H_2O.[162]

Reactions of $(CH_2{=}CHCH_2)Me_2SiC{\equiv}CC{\equiv}CSiMe_2(CH_2CH{=}CH_2)$ with dicobalt carbonyl have given $\{Co_2(CO)_6\}_n\{(\mu\text{-}\eta^2)_n\text{-}(CH_2{=}CHCH_2)Me_2SiC_2C_2SiMe_2(CH_2CH{=}CH_2)\}$ ($n = 1, 2$). Treatment of these derivatives with HBF_4 resulted in stepwise replacement of the allyl group by F to give bis adducts of the fluorosilanes $(CH_2{=}CHCH_2)Me_2SiC{\equiv}CC{\equiv}CSiFMe_2$ or $Me_2FSiC{\equiv}CC{\equiv}CSiFMe_2$, and the mono adduct $Co_2\{\mu\text{-}\eta^2\text{-}(CH_2{=}CHCH_2)SiMe_2C_2C{\equiv}CSiFMe_2\}(CO)_6$. Electrophilic attack at the silicon center results in elimination of propene and formation of the silyl fluoride; silyl cations are unlikely to be involved.[174,191]

Two CO ligands of $Co_2(\mu\text{-}\eta^2\text{-}RC_2R)(CO)_6$ are labile, and compounds of general type $Co_2(\mu,\eta^2\text{-}RC_2R)(CO)_4L_2$ are readily obtained via thermal substitution reactions. More highly substituted complexes require forcing conditions. In the case of compounds featuring two Co_2 fragments coordinated to adjacent C≡C moieties, the increased steric demands of these more highly substituted derivatives may lead to decomplexation of one of the Co_2 units (Scheme 13).[192]

The complexes $\{Co_2(CO)_6\}_2(\mu$-η^2:μ-η^2-$RC_2C_2R)$ (R = Ph, Fc) react with $P(OMe)_3$ at room temperature to give small amounts of the substituted compounds $\{Co_2(CO)_5[P(OMe)_3]\}\{Co_2(CO)_{6-n}[P(OMe)_3]_n\}(\mu$-$\eta^2$:$\mu$-$\eta^2$-$RC_2C_2R)$ ($n = 0$–2) in addition to greater quantities of $Co_2(\mu$-η^2-$RC_2C{\equiv}CR)(CO)_{6-n}\{P(OMe)_3\}_n$ ($n = 1$–3). Product complexes with 0:1, 1:1 and 1:2 substitution patterns at each metal have been identified. At higher temperatures dicobalt compounds were formed by dissociation of the $Co_2(CO)_6$ fragment aided by coordination of $P(OMe)_3$ to the adjacent centre. Reintroduction of a $Co_2(CO)_6$ group onto the substituted complexes $Co_2(\mu$-η^2-$RC_2C{\equiv}CR)(CO)_{6-n}\{P(OMe)_3\}_n$ was readily achieved by reaction with $Co_2(CO)_8$, indicating that steric hindrance between the $P(OMe)_3$ ligands on adjacent Co_2 centers rather than phosphite–carbonyl interactions result in the initial dissociation (Scheme 14). For $\{Co_2(CO)_5[P(OMe)_3]\}_2(\mu$-$\eta^2$: μ-η^2-$RC_2C_2R)$, two isomers (ratio 1/1, R = Ph; 2/1, Fc) result from occupation of the two sets of axial sites. The trisubstituted complex is labile (by steric interaction between phosphite ligands) and contains one equatorial and two axial $P(OMe)_3$ ligands.[192]

The complexes $Co_2(\mu$-η^2-$Me_3SiC_2C{\equiv}CSiMe_3)(CO)_4(L)_2$ [$L_2 = (PMePh_2)_2$ or $NH(PPh_2)_2$] have been prepared directly from the phosphines and the parent carbonyl complex or from $Co_2\{\mu$-$(PPh_2)_2NH\}(CO)_6$ and the diyne. A second $Co_2\{\mu$-$(PPh_2)_2NH\}(CO)_4$ moiety can be added to give $\{Co_2[\mu$-$(PPh_2)_2NH](CO)_4\}_2(\mu$-$\eta^2$:$\mu$-$\eta^2$-$Me_3SiC_2C_2SiMe_3)$, while protodesilylation by $[NBu_4]F$ gives $Co_2\{\mu$-$(PPh_2)_2NH\}(\mu$-η^2-$Me_3SiC_2C{\equiv}CH)(CO)_4$.[107]

The $Co_2(CO)_4(dppm)$ complexes of thermally sensitive terminal di- and poly-ynes have been found to be particularly useful as these derivatives are stable enough to be used in further reactions under rather harsh conditions (Scheme 14). Reactions of $\{Co_2(CO)_6\}_2(\mu$-η^2:μ-η^2-$PhC_2C_2Ph)$ with dppm gave $\{Co_2(\mu$-dppm$)(CO)_4\}_2(\mu$-η^2:μ-η^2-$PhC_2C_2Ph)$ as the major product, together with some $\{Co_2(\mu$-dppm$)(CO)_4\}(\mu,\eta^2$-$PhC_2C{\equiv}CPh)$, a large excess of dppm being required to form $\{Co_2(\mu$-dppm$)_2(CO)_2\}(\mu$-η^2-$PhC_2C{\equiv}CPh)$. Reactions of dppm with $\{Co_2(CO)_6\}_2(\mu$-η^2:μ-η^2-$FcC_2C_2Fc)$ result in loss of $Co_2(CO)_6$ and formation of $\{Co_2(\mu$-dppm$)(CO)_4\}(\mu$-η^2-$FcC_2C{\equiv}CFc)$ together with some $Co_4(\mu$-dppm$)_2(CO)_8$.[192] The diyne $Me_3SiC{\equiv}CC{\equiv}CSiMe_3$ affords $\{Co_2(\mu$-dppm$)(CO)_4\}(\mu$-η^2-$Me_3SiC_2C{\equiv}CSiMe_3)$.[193,194]

The dppm-subsituted complexes of the asymmetric diynes $RC{\equiv}CC{\equiv}CSiMe_3$ (R = Ph, tol) are readily obtained from thermal reactions of the hexacarbonyl complexes with dppm or by direct reactions of the diynes with $Co_2(\mu$-dppm$)(CO)_6$. In these cases there is some evidence for regioselectivity due to steric interaction

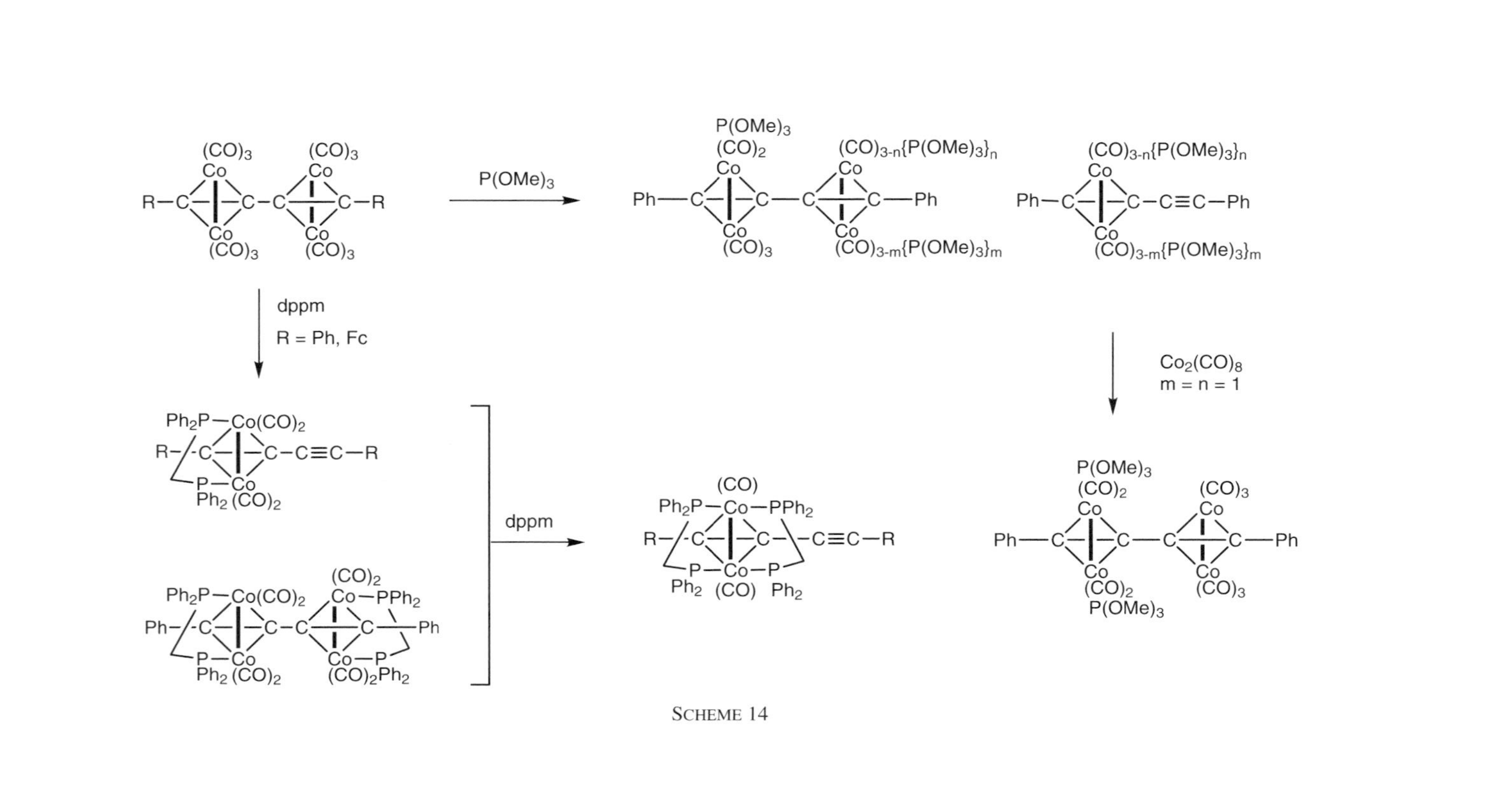

SCHEME 14

between the $SiMe_3$ and dppm, with the $Co_2(\mu\text{-dppm})(\mu\text{-}\eta^2\text{-}RC_2C{\equiv}CSiMe_3)(CO)_4$ isomer being formed preferentially.[183]

Addition of 2,3-bis(diphenylphosphino)maleic anhydride (bma) to $\{Co_2(CO)_6\}_2(\mu\text{-}\eta^2{:}\mu\text{-}\eta^2\text{-}PhC_2C_2Ph)$ under mild conditions (refluxing CH_2Cl_2 or addition of Me_3NO) gives thermally sensitive $\{Co_2(CO)_6\}(\mu\text{-}\eta^2{:}\mu\text{-}\eta^2\text{-}PhC_2C_2Ph)\{Co_2(CO)_4(\mu\text{-bma})\}$ (**52**) (Scheme 15).[195] Above 80°C, P—C bond cleavage and C—C bond formation occurred to give $Co_2\{\mu\text{-}\eta^2,P{:}\eta^2,P\text{-}(Z)\text{-}PPh_2CPh{=}C(C{\equiv}CPh)C{=}C(PPh_2)C(O)OC(O)\}(CO)_4$ (**53**) with competitive loss of the diyne ligand to give $Co_2(\mu\text{-bma})_2(CO)_2$ (**54**) which further reacted to afford $Co_2(\mu\text{-}PPh_2)(\mu\text{-bma})\{\mu\text{-}C{=}C(PPh_2)C(O)OC(O)\}(CO)_2$ (**55**). Complex **52** is the common precursor, by loss of a $Co_2(CO)_6$ group to give **53,** and by reaction with an excess of bma for **54** then **55.**

Metal complexes containing pendant alkynyl moieties coordinate readily to $Co_2(CO)_8$, as shown by an example derived from 2,2′:6′,2″-terpyridine-ruthenium centers linked by 1,3-diynyl groups, namely, $[(tpy)Ru(tpy\text{-}4'\text{-}OCH_2C_2\{Co_2(CO)_6\}C_2\{Co_2(CO)_6\}CH_2O\text{-}4'\text{-}tpy)Ru(tpy)][PF_6]_4$.[196] Although no M^+ ion was present in the electrospray mass spectrum, the IR and NMR spectra were consistent with the formation of a symmetrical complex. Addition of $Co_2(CO)_8$ to $M(C{\equiv}CC{\equiv}CH)(CO)_3Cp$ (M = Mo, W) or $Ru(C{\equiv}CC{\equiv}CPh)(PPh_3)_2Cp$ affords the usual adducts $Co_2\{\mu\text{-}\eta^2\text{-}RC_2C{\equiv}C[ML_n]\}(CO)_6$ [ML_n = Mo/W$(CO)_3Cp$, $Ru(PPh_3)_2Cp$] in which the $Co_2(CO)_6$ moiety is attached to the least hindered C≡C moiety.[81,197]

a. *Reactions of dicobalt-diyne complexes.* Both $SiMe_3$ groups were removed from $Co_2(\mu\text{-}\eta^2\text{-}Me_3SiC_2C{\equiv}CSiMe_3)(CO)_6$ simultaneously by treatment with KF/MeOH to afford bright red $Co_2(\mu\text{-}\eta^2\text{-}HC_2C{\equiv}CH)(CO)_6$ which is unstable even at low temperatures.[172] The bis(silyl) complex $Co_2(\mu\text{-dppm})(\mu\text{-}\eta^2\text{-}Me_3SiC_2C{\equiv}CSiMe_3)(CO)_4$ may be sequentially protodesilylated to afford the air-stable crystalline derivatives $Co_2(\mu\text{-dppm})(\mu\text{-}\eta^2\text{-}Me_3SiC_2C{\equiv}CH)(CO)_4$ and $Co_2(\mu\text{-dppm})(\mu\text{-}\eta^2\text{-}HC_2C{\equiv}CH)(CO)_4$.[172,186]

An example of the protection afforded the C≡C triple bond by complexation to a $Co_2(CO)_6$ fragment is afforded by the chemistry of **56** which in reactions with $BF_3(OEt_2)$ was converted to the fused 6/9 bicyclic ether **57** (Scheme 16).[198] The linking of two separated C≡C triple bonds in $(HC{\equiv}CCH_2)_2O$ in one of the products of its reaction with $Co_2(CO)_8$ affords **58,** formed by H migration and C—C bond formation to give a C_4O ring and a Co-spiked Co_3C cluster.[199]

Treatment of $\{Co_2(CO)_6\}_2\{\mu\text{-}\eta^2{:}\mu\text{-}\eta^2\text{-}(HO)CH_2C_2C_2CH_2(OH)\}$ with HBF_4 generates the dicarbonium ion. In the presence of dithiols or thioethers, a series of macrocycles of types **59, 60,** and **61** (Scheme 17) were obtained. Linking groups include $-S(CH_2)_2S-$, $-O(CH_2)_2S-$, $-O(CH_2)_2O-$, $-OGePh_2O-$, and $-CMe_2CH_2C({=}CH_2)-$.[200]

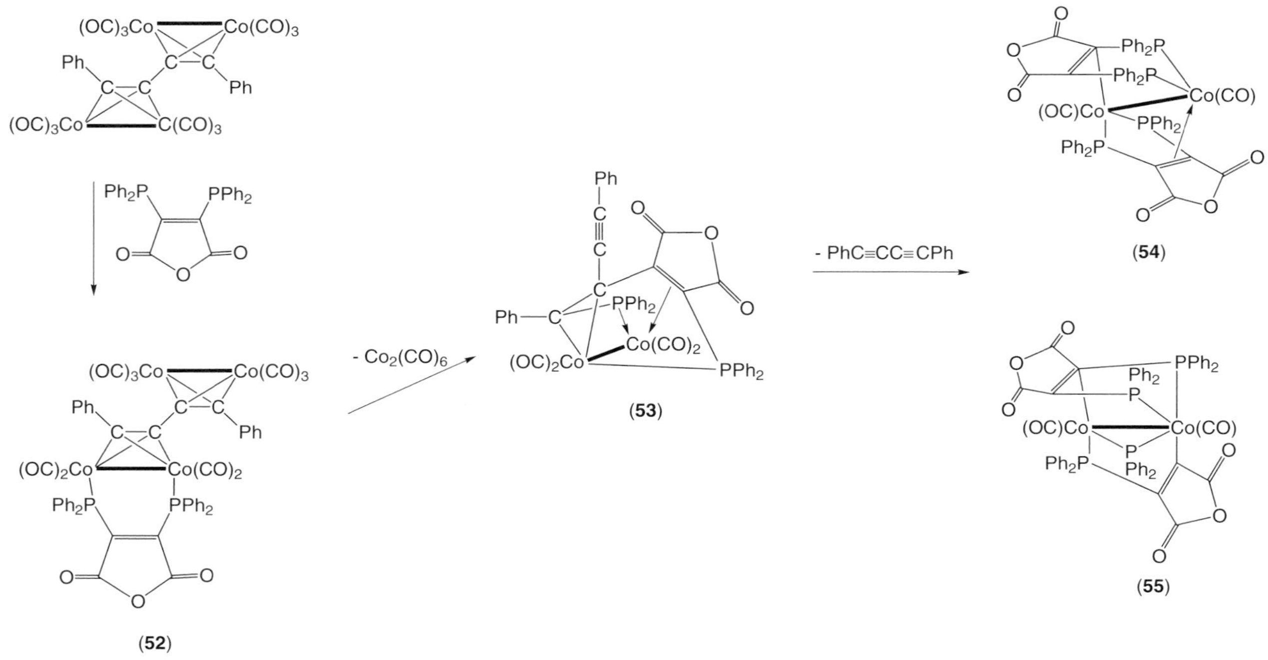

SCHEME 15

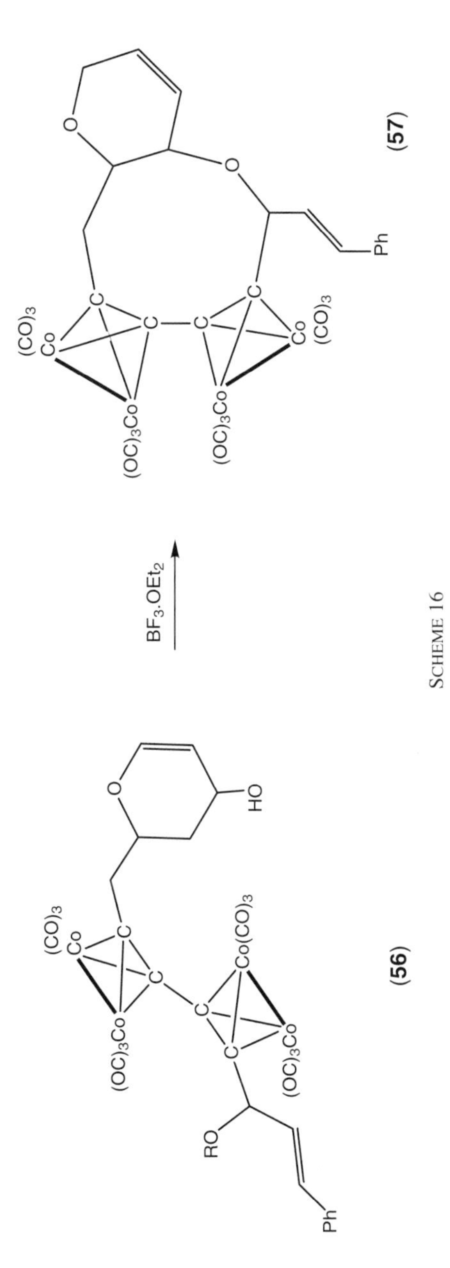

SCHEME 16

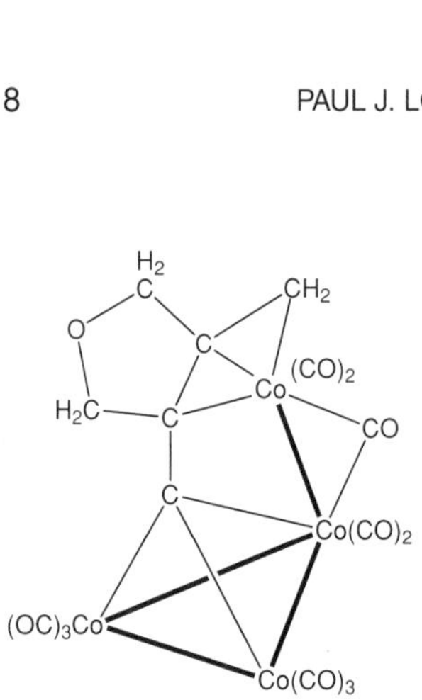

(58)

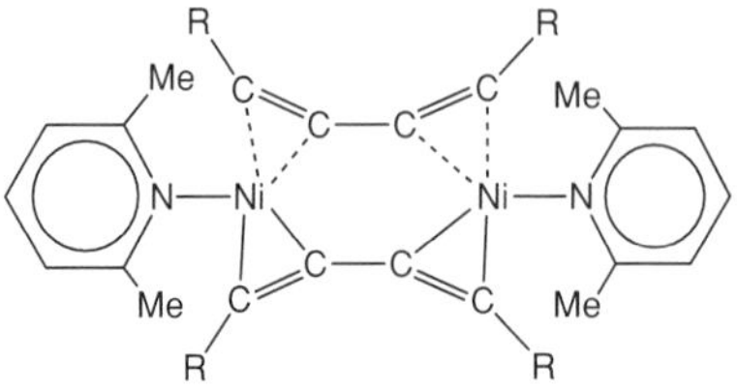

(67)

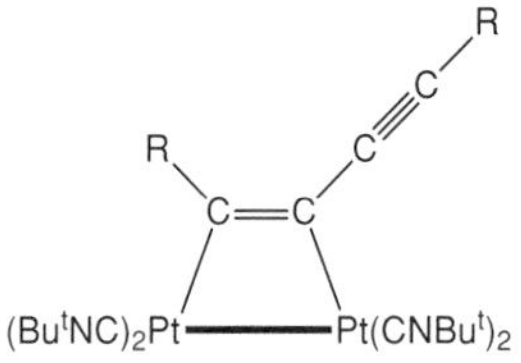

(70) R = Me, Ph

6. *Rhodium*

Treatment of $\{RhCl(PPr^i_3)_2\}_2$ with PhC≡CC≡CPh and NaI afforded *trans*-RhI(η^2-PhC$_2$C≡CPh)($PPr^i_3)_2$(**62;** Scheme 18). The iodo-complex was also obtained by coupling of the alkynyl ligands in Rh(C≡CPh)$_2$(SnPh$_3$)($PPr^i_3)_2$ following reaction of the bis(alkynyl) precursor **63** with I_2 when elimination of $SnIPh_3$ occurs.[201] The related complex *trans*-RhCl(η^2-Me$_3$SiC$_2$C≡CSiMe$_3$)($PPr^i_3)_2$, formed directly from $\{RhCl(PPr^i_3)_2\}_2$ and Me$_3$SiC≡CC≡CSiMe$_3$, isomerizes to *trans*-RhCl{=C=C(SiMe$_3$)C≡CSiMe$_3$}($PPr^i_3)_2$ upon UV photolysis and, in turn, can be hydrolysed to *trans*-RhCl(=C=CHC≡CSiMe$_3$)($PPr^i_3)_2$.[78] The complex RhCl($PPr^i_3)_2$\{η^2-HC$_2$C≡CCPh$_2$(OSiMe$_3$)\}, formed from \{Rh(μ-Cl)($PPr^i_3)_2\}_2$ and HC≡CC≡CCPh$_2$(OSiMe$_3$), rearranges on heating in toluene, first to RhHCl{C≡CC≡CCPh$_2$(OSiMe$_3$)}($PPr^i_3)_2$ and then to the vinylidene RhCl{=C=CHC≡CCPh$_2$(OSiMe$_3$)}($PPr^i_3)_2$ (**64**), which adds pyridine to give RhHCl{C≡CC≡CCPh$_2$(OSiMe$_3$)}(py)($PPr^i_3)_2$ (**65**). Both **64** and **65** react with triflic anhydride (Tf$_2$O) and NEt$_3$ to give the pentatetraenylidene RhCl(=C=C=C=C=CPh$_2$) ($PPr^i_3)_2$ (**66**).[128]

Complexes containing arsine ligands were prepared in nearly quantitative yields by displacement of ethene from *trans*-RhCl(η-C$_2$H$_4$)(L)$_2$ (L = $AsPr^i_3$, $AsPr^i_2$CH$_2$CH$_2$OMe) by RC≡CC≡CR (R = Me, SiMe$_3$) to give RhCl(η^2-RC$_2$C≡CR)(L)$_2$.[202]

7. *Nickel, Palladium, and Platinum*

Numerous complexes M(η^2-diyne)(L)$_2$ (diyne = HC≡CC≡CH, HC≡CC≡CR or RC≡CC≡CR; L = tertiary phosphine) have been obtained from reactions of

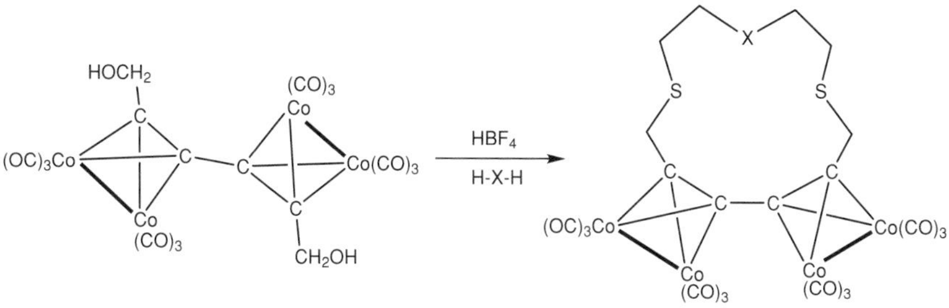

HBF$_4$

H-X-H

(**59**) X = O, S

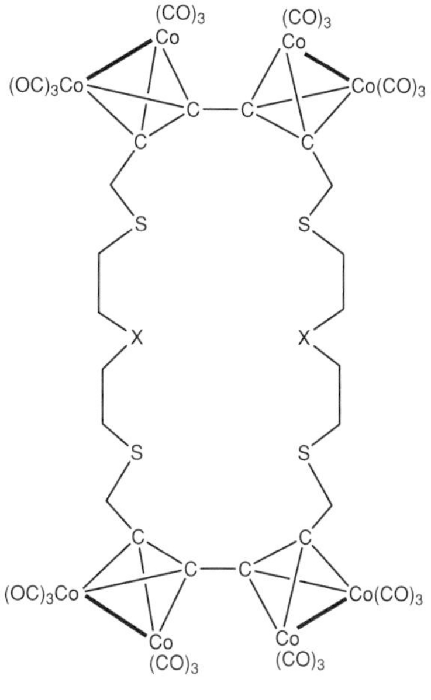

(**61**)

(**60**) X = O, S

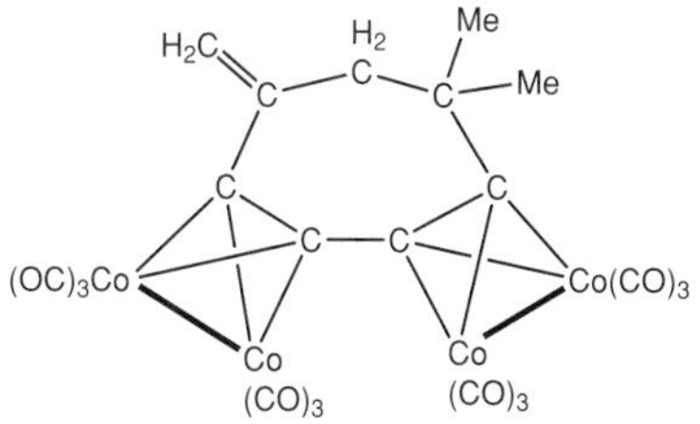

SCHEME 17

M(η-C_2H_4)L_2 [M = Pd, L_2 = dippe,[158] M = Pt, L=PPh_3, $PMePh_2$[203]], M(cod)L_2 (M = Ni, Pd, Pt; L_2 = cod, bipy, dippe, dippp),[158] and Pd(η^2-C_6H_{10})(dippe) [prepared *in situ* from Pd($CH_2CH{=}CH_2$)$_2$(dippe)].[158] The stoichiometry of the reaction is often the principal factor in determining the type of product formed. For example, the complexes $\{Ni(L_2)\}_n(\mu$-$\eta^2{:}\eta^2$-$HC_2C_2H)$ (n = 1, 2; L_2 = dippe, dippp) may be interconverted by addition of one equivalent of the nickel complex Ni(cod)L_2 or buta-1,3-diyne, as appropriate.[204] The nature of the diynyl R group may also influence the product distribution, probably due to steric effects. While reactions of RC≡CC≡CR′ (R = R′ = Ph, $SiMe_3$; R = Ph, R′ = $SiMe_3$) with

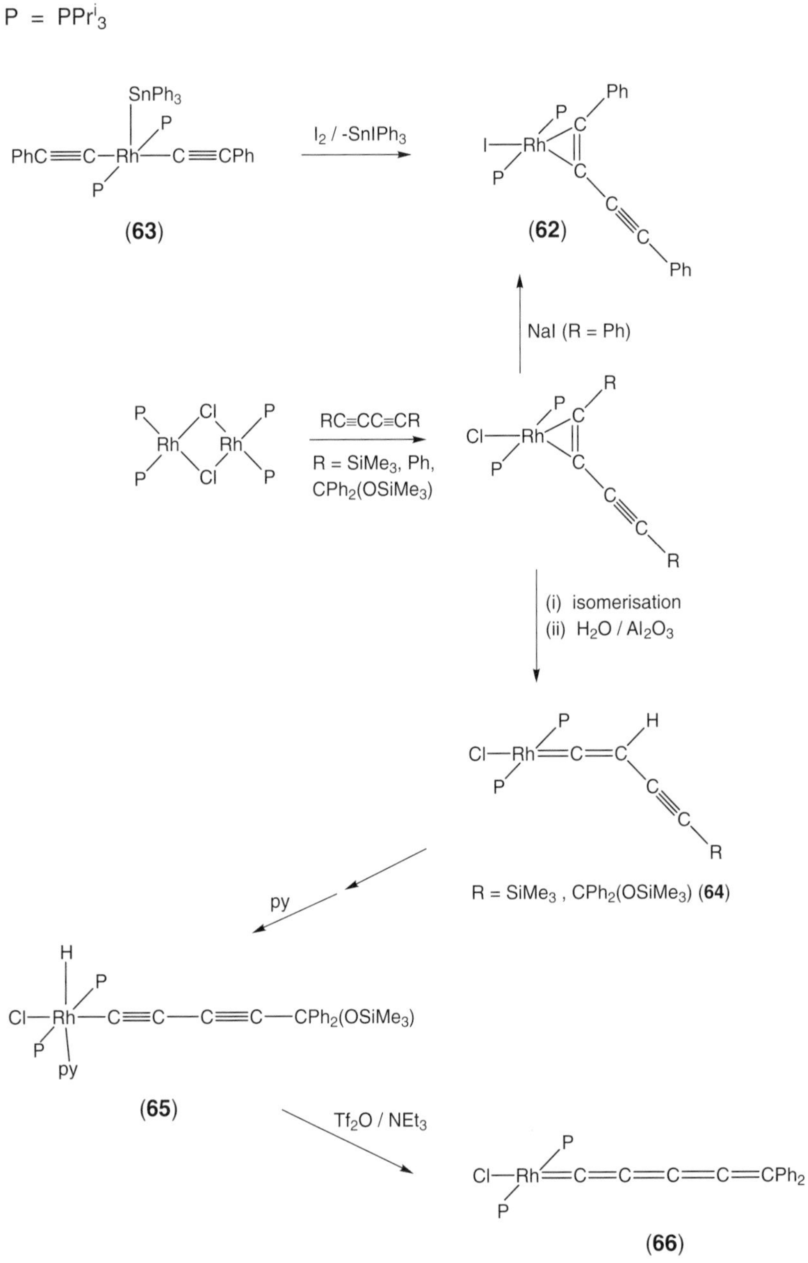

P = PPr^i_3

$PhC{\equiv}C{-}Rh(SnPh_3)(P)_2{-}C{\equiv}CPh$ (**63**) $\xrightarrow{I_2\ /\ -SnIPh_3}$ (**62**)

$[RhCl(P)_2]_2$ $\xrightarrow{RC{\equiv}CC{\equiv}CR}$ R = $SiMe_3$, Ph, $CPh_2(OSiMe_3)$

NaI (R = Ph)

(i) isomerisation
(ii) $H_2O\ /\ Al_2O_3$

R = $SiMe_3$, $CPh_2(OSiMe_3)$ (**64**)

py

$Cl{-}Rh(H)(P)_2(py){-}C{\equiv}C{-}C{\equiv}C{-}CPh_2(OSiMe_3)$

(**65**)

$Tf_2O\ /\ NEt_3$

$Cl{-}Rh(P)_2{=}C{=}C{=}C{=}C{=}CPh_2$

(**66**)

SCHEME 18

$Ni(cod)_2$ in the presence of PPh_3 gave mononuclear $Ni(\eta^2\text{-}RC_2C{\equiv}CR')(PPh_3)_2$ or binuclear $\{Ni(PPh_3)_2\}_2\{\mu\text{-}\eta^2{:}\eta^2\text{-}RC_2C_2R'\}$ according to reagent stoichiometry, only mononuclear compounds were isolated from reactions of $Bu^tC{\equiv}CC{\equiv}CR$ ($R = Bu^t$, $SiMe_3$).[205] A comparison of the structures of $NiL_2(\eta^2\text{-}Bu^tC_2C{\equiv}CBu^t)$ ($L_2 =$ bpy, $L = PPh_3$) suggests that changes in L have little effect on the geometry of the NiC_2 fragment.[206]

The unusual complex $\{Ni(NC_5H_3Me_2\text{-}2,6)\}_2(\mu\text{-}\eta^2{:}\eta^2\text{-}Me_3SiC_2C_2SiMe_3)_2$ (**67**) featuring twin $\eta^2{:}\eta^2$ diyne ligands has been obtained from the reaction of the lightly stabilized nickel complex $Ni(\eta^2,\eta^2\text{-}C_7H_{12})(NC_5H_3Me_2\text{-}2,6)$ and 1 equiv of the diyne. Attempts to isolate analogous Ph or Bu^t complexes were unsucessful.[207]

Reduction of $MCl_2(PR_3)_2$ in the presence of a suitable diyne has also been shown to afford complexes featuring η^2-diyne ligands. Treatment of $NiCl_2(PMe_3)_2$ with magnesium in the presence of $Me_3SiC{\equiv}CC{\equiv}CSiMe_3$ gave extremely air-sensitive mono- and bis-$Ni(PMe_3)_2$ complexes containing η^2- and $\eta^2{:}\eta^2$-diyne ligands, respectively.[208] The reduction of *cis*-$PtCl_2(PPh_3)_2$ with hydrazine in the presence of $MeC{\equiv}CC{\equiv}CMe$ gave only $Pt(\eta^2\text{-}MeC_2C{\equiv}CMe)(PPh_3)_2$, from which the diyne ligand is displaced by C_2Ph_2. A bis-platinum complex could not be prepared, probably for steric reasons.[209]

Pörschke and colleagues have observed a curious isomerization of the complex $\{Pd(dippe)\}_2\{\mu\text{-}\eta^2{:}\eta^2\text{-}HC_2C_2H\}$ in d_8-thf solution at $-80°C$. Over a period of about a week both Pd(dippe) moieties become coordinated to the same C≡C bond. Warming a solution of this lower symmetry isomer to 0°C results in only partial reversal of the isomerization. The isomers are apparently in slow equilibrium, with various subtle and unspecified factors determining which one is preferred and while the $\{Pd(dippe)\}_2\{\mu\text{-}\eta^2{:}\mu\text{-}\eta^2\text{-}HC_2C_2H\}$ (**68**) form is thermodynamically favored in solution, crystallization affords $\{Pd(dippe)\}_2\{\mu\text{-}\eta^2{:}\eta^2\text{-}HC_2C{\equiv}CH\}$ (**69**) (Scheme 19).[158]

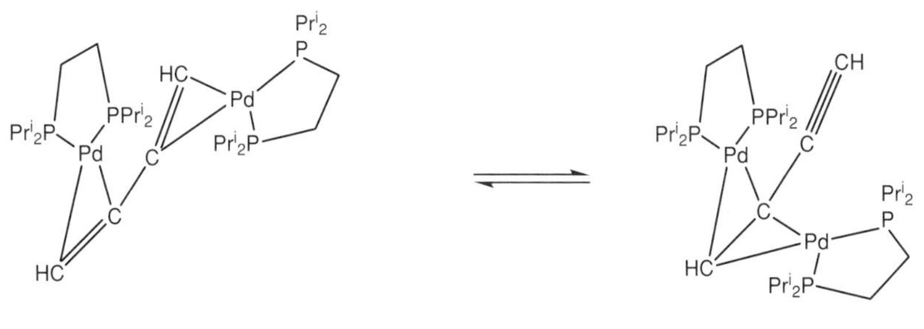

(**68**) (solid state) (**69**) (solution)

SCHEME 19

The reaction between aqueous K_2PdCl_4 and $CMe_2(OH)C{\equiv}CC{\equiv}CCMe_2(OH)$ (L) has been reported to give the palladium(I) complex PdCl(L), which exchanges L for X with salts MX (M = alkali metal, X = SCN, Br, I) and adds py to give PdCl(L)(py). However, it must be said that the dark brown to black solids so formed are not fully characterized by contemporary standards. It was assumed that hydropalladation of one C≡C triple bond has occurred.[210]

Reactions of $Pt(\eta\text{-}C_2H_4)L_2$ (L = PPh_3, $PMePh_2$) with equimolar amounts of RC≡CC≡CR [R = Me, Ph, $SiMe_3$, $CMe_2(OH)$, $CPh_2(OH)$, $Ph_2P\{M(CO)_n\}$ (n = 5, M = Mo, W; n = 4, M = Fe)][172,203,211] or of $Pt(PPh_3)_4$ with PhC≡CC≡CPh[212] proceed smoothly to give $Pt(\eta^2\text{-}RC_2C{\equiv}CR)L_2$. Successive formation of $\{PtL_2\}_n(\eta^2,\eta^2\text{-}RC_2C_2R)$ (n = 1, 2; L = PPh_3, $PMePh_2$; R = Me, Ph) occurs with some diynes.[203] The compound $Pt_3(CNBu^t)_6$ acts as a source of the reactive platinum species $Pt(CNBu^t)_2$ and reactions with RC≡CC≡CR (R = Me, Ph) gave the diplatinacyclobutene complexes $Pt_2(RC_2C{\equiv}CR)(CNBu^t)_4$ (**70;** R = Me, Ph).

Other examples of compounds containing metal—metal bonded fragments attached to either one or both C≡C triple bonds have been prepared either by coupling mononuclear precursors in situations where coordination of discrete metal fragments to adjacent C≡C moieties is sterically unfavorable, or by direct reactions of bimetallic reagents with diynes. Thus, while 1 equiv of Ni(cod)(bpy) reacts with the bulky diynes RC≡CC≡CR′ (R = R′ = Bu^t, Ph, $SiMe_3$; R = $SiMe_3$, R′ = Bu^t, Ph) to give $\{Ni(bpy)\}(\eta^2\text{-}R'C_2C{\equiv}CR)$, analogous reactions with an excess of the metal reagent gave the binuclear $\{Ni(bpy)\}_2(\mu\text{-}\eta^2\text{-}RC_2C{\equiv}CSiMe_3)$; in each, the $C{\equiv}CSiMe_3$ group remains uncoordinated.[206] The reaction of an excess of $Ni(cod)_2$ with RC≡CC≡CR (R = Ph, $SiMe_3$) afforded the tri- and tetranuclear complexes $\{Ni_2(cod)_2\}\{Ni(cod)\}(\mu\text{-}\eta^2{:}\eta^2\text{-}RC_2C_2R)$ (**71**) or $\{Ni_2(cod)_2\}_2(\mu\text{-}\eta^2{:}\mu\text{-}\eta^2\text{-}RC_2C_2R)$ (**72**), in which both alkynyl moieties are coordinated by Ni or Ni_2 fragments.[205,213] The diyne ligand in the latter (R = Ph) is cleaved in reactions with dppm to afford the mixed-valence bis-alkynyl complex $Ni_3(\mu\text{-dppm})_3(\mu\text{-}\eta^1\text{-}C{\equiv}CPh)_2$.[214] The reaction of $\{Ni(\mu\text{-}CO)Cp\}_2$ with an excess of PhC≡CC≡CPh gave a separable mixture of $Ni_2(\mu\text{-}\eta^2\text{-}PhC_2C{\equiv}CPh)Cp_2$ (**73**) and $\{Ni_2Cp_2\}_2(\mu\text{-}\eta^2{:}\mu\text{-}\eta^2\text{-}PhC_2C_2Ph)$ (**74**).[156]

In several cases the η^2-coordinated diyne ligand can be displaced by other ligands. The pentane-soluble, thermally stable compounds $\{Ni(L_2)\}_n(HC_2C_2H)$ (n = 1, 2; L_2 = dippe, dippp) react with $P(OPh)_3$ to give $Ni\{P(OPh)_3\}_4$ with liberation of diyne and the bis-phosphine. The reaction of $\{Ni(dippp)\}_2(\mu\text{-}\eta^2{:}\mu\text{-}\eta^2\text{-}HC_2C_2H)$ with four equivalents of CO resulted in polymerization of the butadiyne released from the metal center and the formation of $Ni(CO)_2(dippp)$.[204]

Low-temperature protonation ($HBF_4 \cdot OEt_2$) of $Pt(\eta^2\text{-}RC_2C{\equiv}CR)\{P(tol)_3\}_2$ (R = Me, $SiMe_3$), which was obtained from $Pt(\eta\text{-}C_2H_4)\{P(tol)_3\}_2$ and the diyne, gave hydrido complexes *trans*-$[PtH(\eta^2\text{-}RC_2C{\equiv}CR)\{P(tol)_3\}_2]^+$ (**75**) which rearranged to $[Pt(\eta^1,\eta^2\text{-}RC_2C{=}CHR)\{P(tol)_3\}_2]^+$ (**76**) on warming to −30°C (Scheme 20). In contrast, treatment of $Pt(\eta^2\text{-}MeC_2C{\equiv}CMe)(PPh_3)_2$ with CF_3CO_2H

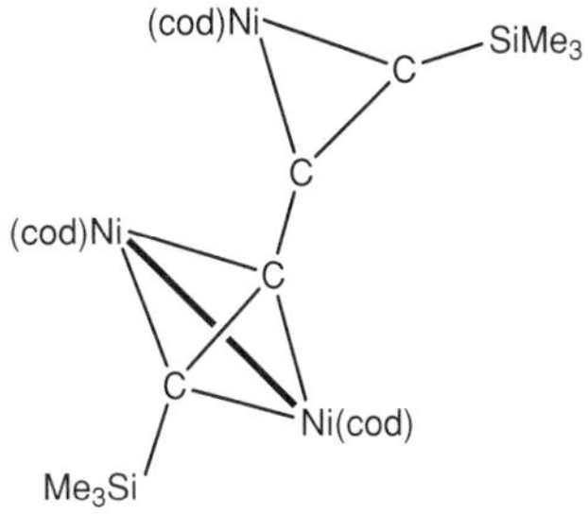

(**71**) 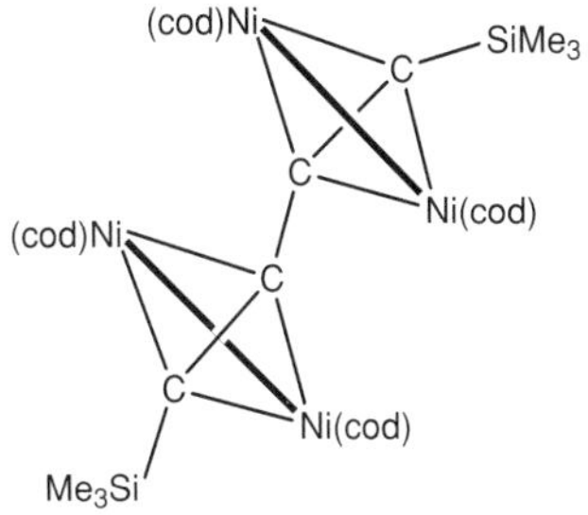 (**72**)

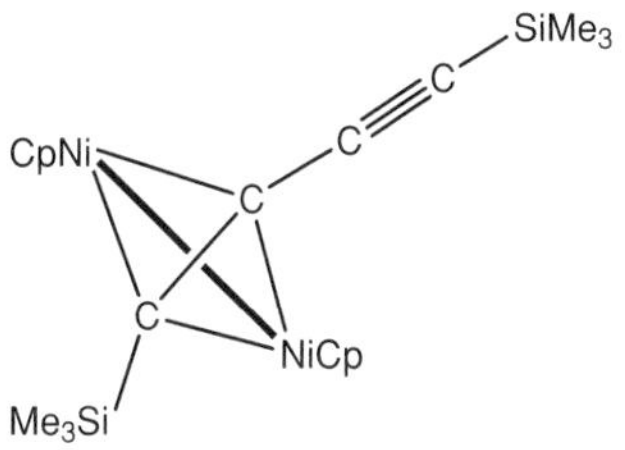

(**73**) 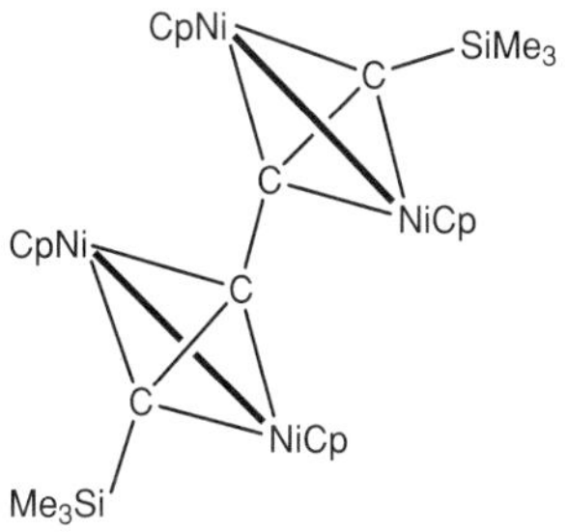 (**74**)

gave the vinyl *trans*-Pt{η^1-(*E*)-MeCH=CC≡CMe}(O_2CCF_3)(PPh_3)$_2$ (**77**), which was too unstable to isolate. However, subsequent addition of LiCl afforded *trans*-PtCl(η^1-MeCH=CCCl=CHMe)(PPh_3)$_2$ (**78**), possibly via an intermediate π-propargyl cation such as **79.** In the presence of water, the PPh_3-substituted η^2-diyne complex reacted with CF_3CO_2H to give Pt{η^1-(*Z*)-MeCH=CC(O)Et}(O_2CCF_3)(PPh_3)$_2$ (**80**) which exchanged trifluoroacetate ligand with chloride to give Pt{η^1-(*Z*)-MeCH=CC(O)Et}(Cl)(PPh_3)$_2$ (**81**).[215]

8. *Heterometallic Derivatives*

Mixed-metal complexes can be obtained from sequential reactions of diynes with different ynophilic metal fragments. Reaction of **43** with $Co_2(CO)_8$ affords NbCl{η^2:μ-η^2-$PhC_2C_2Ph[Co_2(CO)_6]$}$Cp^{Si}{}_2$, obtained as a mixture of *exo* and *endo* isomers.[165] The pendant C≡CSiMe$_3$ moiety in Mo_2(μ-η^2-Me_3SiC_2C≡CSiMe$_3$) $(CO)_4Cp_2$ reacts readily with $Co_2(CO)_8$ giving the mixed-metal complex

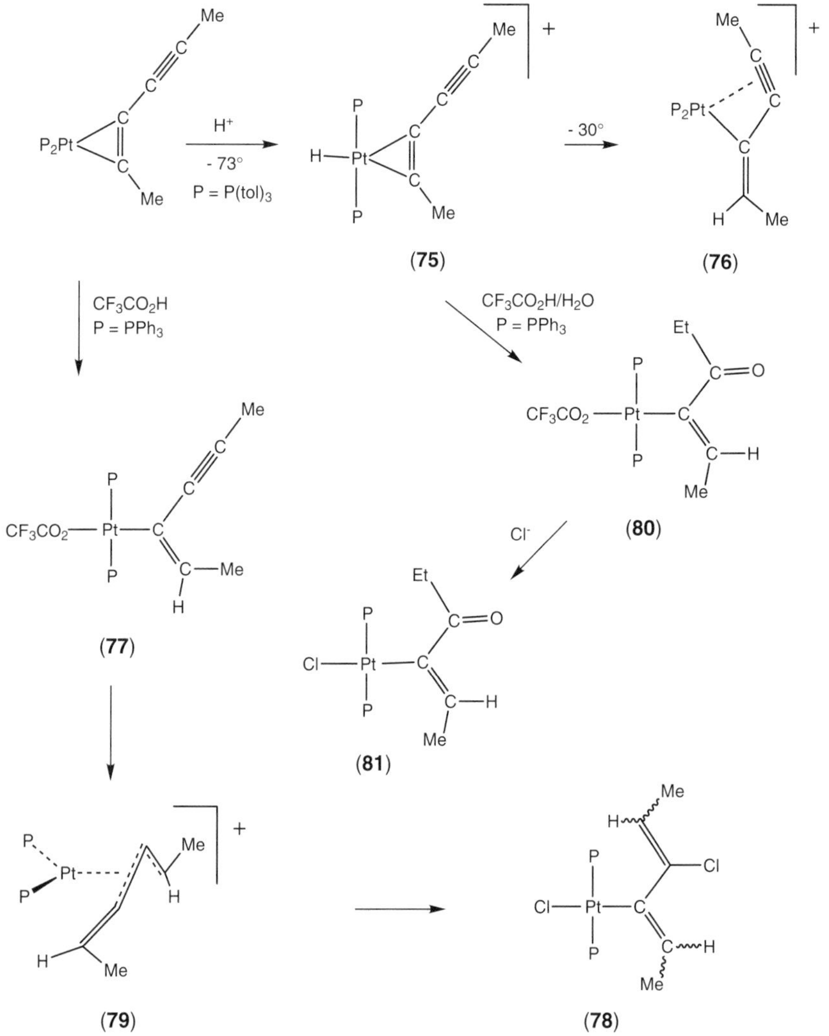

SCHEME 20

$\{Mo_2(CO)_4Cp_2\}\{Co_2(CO)_6\}(\mu\text{-}\eta^2\text{:}\mu\text{-}\eta^2\text{-}Me_3SiC_2C_2SiMe_3)$.[172,175] In the latter case, exposure to atmospheric oxygen on silica gel was sufficient to remove selectively the $Co_2(CO)_6$ group by oxidative decomposition.[172] Similarly, the free C≡C triple bond in $Ni_2(\mu\text{-}\eta^2\text{-}PhC_2C{\equiv}CPh)Cp_2$ reacts with $Co_2(CO)_8$ to give $\{Ni_2Cp_2\}\{Co_2(CO)_6\}(\mu\text{-}\eta^2\text{:}\mu\text{-}\eta^2\text{-}PhC_2C_2Ph)$, and with $Fe(CO)_5$ to give a mixture of dark blue-black and brown-black complexes, probably $FeNi_2(\mu_3\text{-}PhC_2C{\equiv}CPh)(CO)_3Cp_2$ (**82**) and $Fe_2Ni_2(\mu_4\text{-}PhC_2C{\equiv}CPh)(CO)_6Cp_2$ (**83**).[156,216]

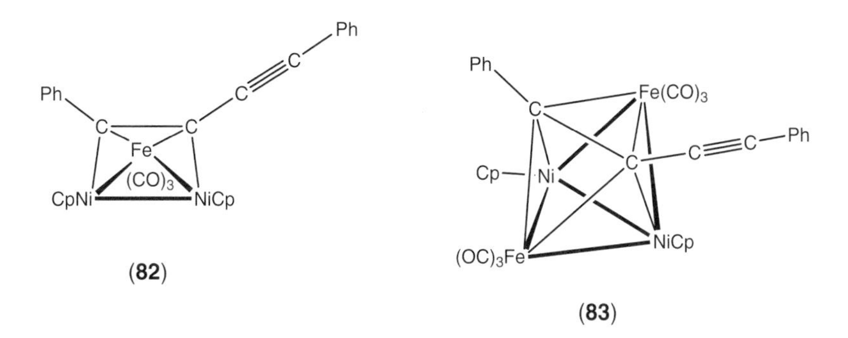

D. *Complexes Containing Poly-yne Ligands*

1. *Vanadium*

The vanadacyclopropene-like structure **84** has been suggested for the brown crystalline material (μ_{eff} 1.9 μ_B) obtained from the 1:1 reaction of VCp_2 with $Me_3Si(C{\equiv}C)_3SiMe_3$. In the presence of an excess of VCp_2 the *trans*-ene(diyne) complex $\{VCp_2\}_2(\mu^2\text{-}\eta^1{:}\eta^1\text{-}Me_3SiC{\equiv}CC_2C{\equiv}CSiMe_3)$ (**85**) was obtained as a black crystalline material (Scheme 21).[217] The triyne has a *trans* configuration, the two V atoms being coplanar with the SiC_6Si skeleton. The formation of this complex, featuring V(III) centers and V-C σ-type bonding, is in stark contrast to the reactions of VCp_2 with the diynes discussed previously and has been attributed to the greater electron density associated with the central C≡C moiety of the conjugated triyne.

2. *Rhenium*

The reaction of $[Re(ClPh)(NO)(PPh_3)Cp^*]BF_4$ with the 1,3,5,7-octatetrayne $Me_3Si(C{\equiv}C)_4SiMe_3$ yielded rotamers of $[Re(\eta^2\text{-}Me_3SiC{\equiv}CC_2C{\equiv}CC{\equiv}CSiMe_3)(NO)(PPh_3)Cp^*]BF_4$.[143]

Me3Si SiMe3 C C C C C═C V Cp2 (84) —VCp2→ VCp2 Me3Si—C≡C—C═C—C≡C—SiMe3 Cp2V (85)

SCHEME 21

3. *Cobalt*

Direct reactions of $Co_2(CO)_6L_2$ [$L_2 = (CO)_2$, dppm] with poly-ynes occur in a manner entirely consistent with the reactions of alkynes and diynes. Diederich has reported the formation of $Co_2(\mu\text{-}\eta^2\text{-}Pr^i_3SiC{\equiv}CC_2C{\equiv}CSiPr^i_3)(CO)_6$ from the reaction of $Co_2(CO)_8$ with $Pr^i_3Si(C{\equiv}C)_3SiPr^i_3$. At elevated temperatures (refluxing hexane) a dark blue oil tentatively formulated as $Co_4(\mu_4\text{-}\eta^2\text{-}Pr^i_3SiC{\equiv}CC_2C{\equiv}CSiPr^i_3)(CO)_{10}$ was also obtained.[186] Carbonyl substitution occurred readily upon treatment of $Co_2(\mu\text{-}\eta^2\text{-}Pr^i_3SiC{\equiv}CC_2C{\equiv}CSiPr^i_3)(CO)_6$ with dppm, yielding $Co_2(\mu\text{-}\eta^2\text{-}Pr^i_3SiC{\equiv}CC_2C{\equiv}CSiPr^i_3)(\mu\text{-dppm})(CO)_4$. Reaction of $Co_2(\mu\text{-dppm})(CO)_6$ with the less sterically hindered 1,3,5-hexatriyne $Me_3Si(C{\equiv}C)_3SiMe_3$ afforded the two possible isomers, $Co_2(\mu\text{-}Me_3SiC_2C{\equiv}CC{\equiv}CSiMe_3)(\mu\text{-dppm})(CO)_4$ (15%) and $Co_2(\mu\text{-}Me_3SiC{\equiv}CC_2C{\equiv}CSiMe_3)(\mu\text{-dppm})(CO)_4$ (42%).[194]

Cobalt carbonyl adducts of the unstable terminal poly-ynes $H(C{\equiv}C)_nR$ are best obtained indirectly. A suspension of $C_2(MgBr)_2$ reacts with $Hg\{Co(CO)_4\}_2$ to give $\{Co_2(CO)_6\}_3(\mu\text{-}\eta^2{:}\mu\text{-}\eta^2{:}\mu\text{-}\eta^2\text{-}HC_2C_2C_2H)$.[185] However, the simplest route is by protodesilylation of complexes derived from $R_3Si(C{\equiv}C)_nR'$ (R = alkyl; R′ may also be SiR_3). For example, deprotection of $Co_2(\mu\text{-dppm})\{\mu\text{-}\eta^2\text{-}Pr^i_3SiC{\equiv}CC_2C{\equiv}CSiPr^i_3)(CO)_4$, with $[NBu_4]F$ in moist thf gave stable $Co_2(\mu\text{-dppm})\{\mu\text{-}\eta^2\text{-}HC{\equiv}CC_2C{\equiv}CH)(CO)_4$, although the corresponding $Co_2(CO)_6$ complex decomposed under similar conditions.[186]

The tetrayne complexes 1,4-$\{Co_2(\mu\text{-dppm})(CO)_4\}_2\{\mu\text{-}\eta^2{:}\mu\text{-}\eta^2\text{-}RC_2C{\equiv}CC{\equiv}CC_2R\}$ have been obtained by oxidative coupling (Glaser) of the corresponding complexed terminal diynes.[183,186] The tetra-ynes are liberated by decomplexation of $\{Co_2(\mu\text{-dppm})(CO)_4\}_2(\mu\text{-}\eta^2{:}\mu\text{-}\eta^2\text{-}RC_2C{\equiv}CC{\equiv}CC_2R)$ with $Fe(NO_3)_3$ in MeOH at r.t.[218]

Other examples of cobalt carbonyl complexes containing poly-yne ligands involving coupling of various smaller fragments have been reported and usually proceed via sequential deprotonation and electron transfer to an electrophile. Thus, deprotonation of $\{Co_2(CO)_6\}_2(\mu\text{-}\eta^2{:}\mu\text{-}\eta^2\text{-}HC_2C_2SiMe_3)$ with $Li[N(SiMe_3)_2]$, followed by reaction with 4-$NO_2C_6H_4CH_2Cl$, gave 1,2,4-$\{Co_2(CO)_6\}_3(\mu\text{-}\eta^2{:}\mu\text{-}\eta^2{:}\mu\text{-}\eta^2\text{-}Me_3SiC_2C_2C{\equiv}CC_2SiMe_3)$ (**86;** Scheme 22).[162] A related reaction also presumed to proceed via C−C bond formation between $Co_2(\mu\text{-}\eta^2\text{-}RCC\bullet)(CO)_6$ fragments has been observed following exposure of a methanolic solution of $\{Co_2(CO)_6\}_2(\mu\text{-}\eta^2{:}\mu\text{-}\eta^2\text{-}Me_3SiC_2C_2SiMe_3)$ to air for 24 h. The black insoluble polymeric material that formed is thought to be a poly-yne (**87,** $n \sim 10$) with each (or most) C≡C triple bond attached to a $Co_2(CO)_6$ unit. Dehydrochlorination of $Co_2(\mu\text{-dppm})(\mu\text{-}\eta^2\text{-}RC_2CH{=}CHCl)(CO)_4$ (R = $SiMe_3$, tol, Ph, C_6H_4OMe-4, C_6H_4F-4) with $LiNPr^i_2$ and subsequent hydrolysis affords $\{Co_2(\mu\text{-dppm})(CO)_4\}_2(\mu\text{-}\eta^2{:}\mu\text{-}\eta^2\text{-}RC_2C{\equiv}CC{\equiv}CC_2R)$ as the major product.[183]

These studies have indicated that while two adjacent C≡C triple bonds may be coordinated to $Co_2(CO)_6$ or $Co_2(\mu\text{-dppm})(CO)_4$ groups, introduction of a third such group to the next adjacent C≡C triple bond is restricted by steric constraints. Thus, the reaction of $Me_3Si(C{\equiv}C)_4SiMe_3$ with a large excess of $Co_2(CO)_8$ gave **86**

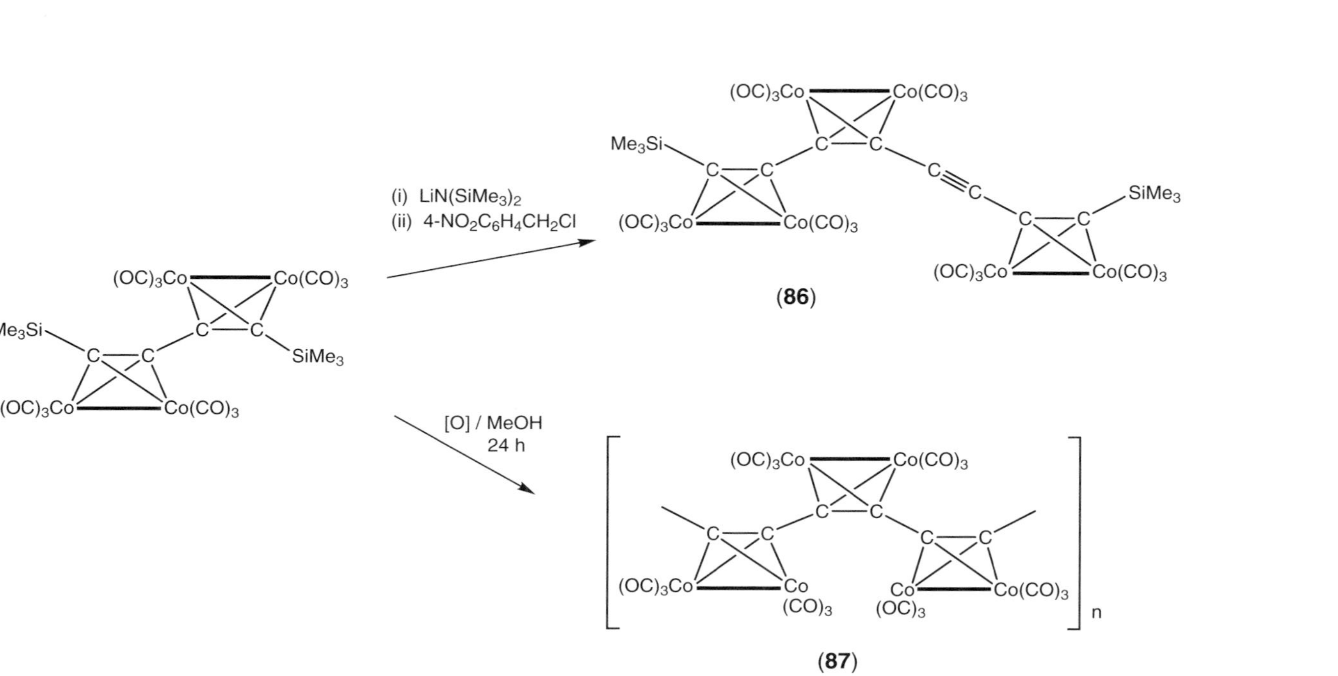

SCHEME 22

in low yield,[162] coordination of the remaining C≡C moiety being restricted by the carbonyl ligand of the adjacent dicobalt groups. Steric effects have also been cited as a likely cause for the coordination of the diagonally opposite C≡C triple bonds in **88** formed from the reaction of the cyclic tetrayne $C_6H_4(C{\equiv}CC{\equiv}C)_2C_6H_4$ with $Co_2(CO)_8$.[219] The silacyclyne ligand in **89** is distinctly nonplanar as a result of the bending of the coordinated alkyne (rather than diyne) group.[220] Structural characterization of **90** has also been reported.[221] Oxidative coupling reactions with the complex $Co_2(\mu\text{-dppm})(\mu\text{-}\eta^2\text{-HC}{\equiv}\text{CC}_2\text{C}{\equiv}\text{CH})(CO)_4$ afforded a mixture of the stable crystalline complexes 1,4,7-$\{Co_2(\mu\text{-dppm})(CO)_4\}_3$-cyclo[18]carbon (**91**) and 1,4,7,10-$\{Co_2(\mu\text{-dppm})(CO)_4\}_4$-cyclo[24]-carbon (33 and 5.4%, respectively).[218] The structural study of **91** shows the expected differences in bond lengths and

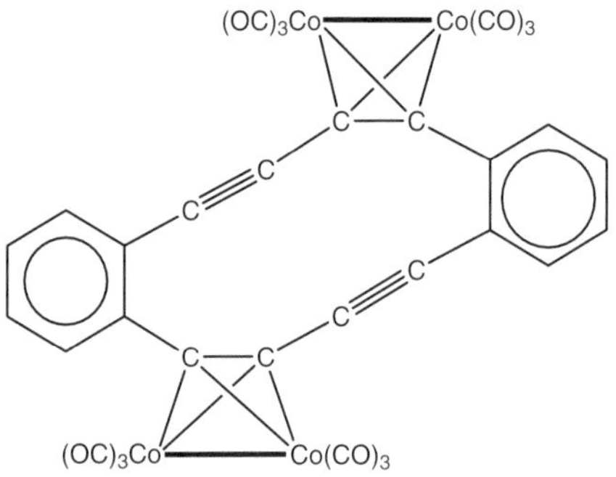

(**88**)

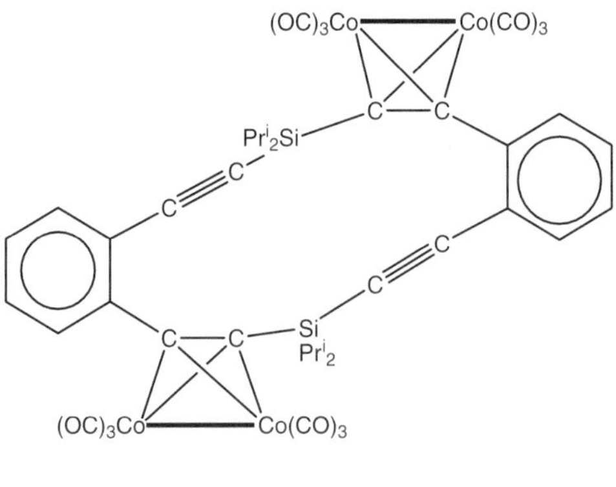

(**89**)

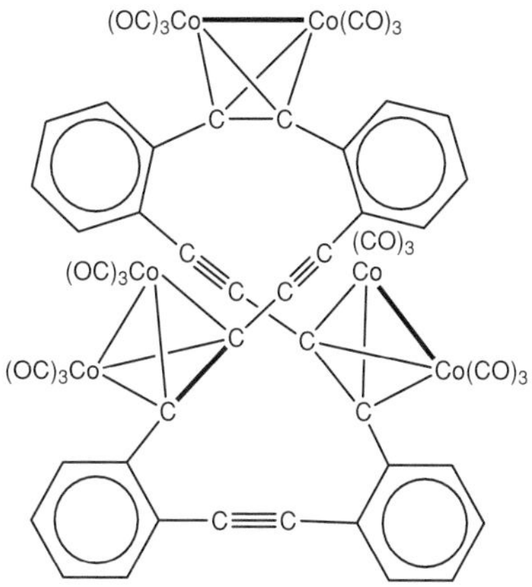
(OC)3Co
Co(CO)3
C
C
(CO)3
(OC)3Co
Co
C
C
C
C
(OC)3Co
Co(CO)3
C
C
C≡C

(**90**)

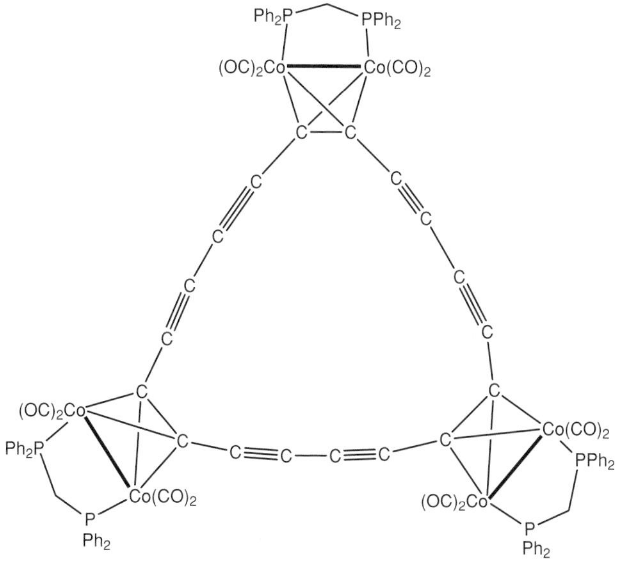
Ph2P
PPh2
(OC)2Co
Co(CO)2
C
C
(OC)2Co
Ph2P
Co(CO)2
P
Ph2
C≡C—C≡C
Co(CO)2
PPh2
(OC)2Co
P
Ph2

(**91**)

angles from those anticipated for free cyclo[18]carbon. Partial conjugation through the macrocyclic ring is suggested by the UV-vis spectra, which contain intense bands at 370 and 381 nm, respectively.

4. *Platinum*

While reactions of Group 10 reagents with diyne reagents are well documented, reports describing the analogous reactions with triynes $R(C{\equiv}C)_3R$ are rare. The reaction of $Me(C{\equiv}C)_3Me$ with $Pt(\eta\text{-}C_2H_4)(PPh_3)_2$ affords two complexes which have been identified from their ^{31}P NMR spectra as the symmetrical and asymmetrical adducts, together with a minor product formed by oxidative addition of the triyne, possibly *cis*-$Pt(C{\equiv}CMe)(C{\equiv}CC{\equiv}CMe)(PPh_3)_2$.[222]

E. *Structures*

In general, coordination of a single metal to the π-system of coordinated diynes and poly-ynes results in a minor elongation of the C≡C bond length. Direct comparison between coordinated and free C≡C triple bonds is possible, from which it is found that coordination of a C≡C moiety results in elongation of C≡C bond by about 0.06 Å. In monometallic complexes (type **G**), the substituents are bent back from the metal by up to 45°. Similar features are found in bimetallic complexes (type **H**), with further elongation to about 1.35 Å as a result of both π-bonds being used in bonding. There is little interaction with uncoordinated C≡C triple bonds, which retain their linearity and usual structural parameters.

As mentioned previously, the characteristic bend-back of substituents has enabled derivatives containing cyclo[n]carbons to be obtained, in which the carbon ring is relatively free of strain. Thus, for cyclo[18]carbon, the bend-back resulting from coordination of the Co_2 moieties in **91** amounts to between 131 and 134°, with the largest bending at noncoordinated carbons of about 19°.[218] Earlier calculations suggested that free cyclo[18]carbon might be stable enough to be isolated, the required 20° bending at each carbon atom requiring relatively little excess energy.[223]

IV

METAL CLUSTER COMPLEXES DERIVED FROM 1,3-DIYNES OR POLY-YNES

A. *Syntheses*

Many examples of metal cluster complexes featuring diyne or higher poly-yne ligands are known, and have generally been obtained by three main routes.

1. *Reactions of Cluster Carbonyls with Diynes or Poly-ynes*

In general, sources of metal cluster fragments, such as $Ru_3(CO)_{10}(NCMe)_2$ or $Ru_3(\mu\text{-dppm})(CO)_{10}$ react readily with 1,3-diynes or poly-ynes initially to give alkyne complexes, which readily undergo further reactions as a result of activation by the cluster core.

2. *Coupling of σ-Alkynyl Groups with Concomitant Aggregation of the Metal Fragments*

While many examples of alkyne coupling reactions on metal clusters are known, here we are only concerned with those reactions that result in the formation of diyne or diynyl ligands, or the poly-yne/poly-ynyl analogues, by the combination of alkynyl fragments on a cluster core.

3. *From σ-Diynyl–metal Complexes and Other Metal Fragments*

Several diynyl complexes react with other metal substrates, the proximity of the σ-bonded metal to a multimetal system often resulting in further cluster condensation.

B. *Survey of Complexes Formed in Reactions of Cluster Carbonyls with Diynes or Poly-ynes*

1. *Iron*

Most reactions of iron carbonyls with diynes have given mono- or binuclear products (see Section VII). Cluster build-up occurs in the reactions of $Fe_2(CO)_9$ with $Fe_2\{\mu\text{-}2\eta^1\text{:}\eta^4\text{-CPhC(NEt}_2\text{)CPhC(C}\equiv\text{CPh)}\}(CO)_6$ [from $PhC\equiv CC\equiv CPh$ and $Fe_2\{\mu\text{-}\eta^1\text{:}\eta^2\text{-PhCC(NEt}_2\text{)}\}(CO)_7$] to give the tri- and tetranuclear clusters $Fe_3\{\mu_3\text{-C(C}\equiv\text{CPh)CPhC(NEt}_2\text{)CPh}\}(\mu\text{-CO})_2(CO)_6$ (**92**) and $Fe_4\{\mu_4\text{-CPhCCCPhC(NEt}_2\text{)CPh}\}(CO)_{11}$ (**93**).[224]

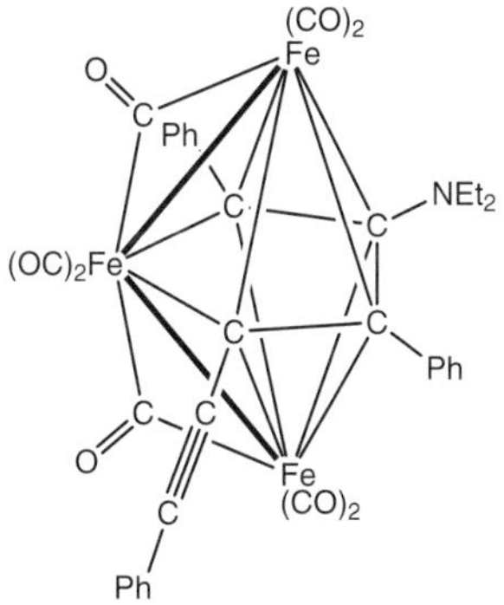

(**92**)

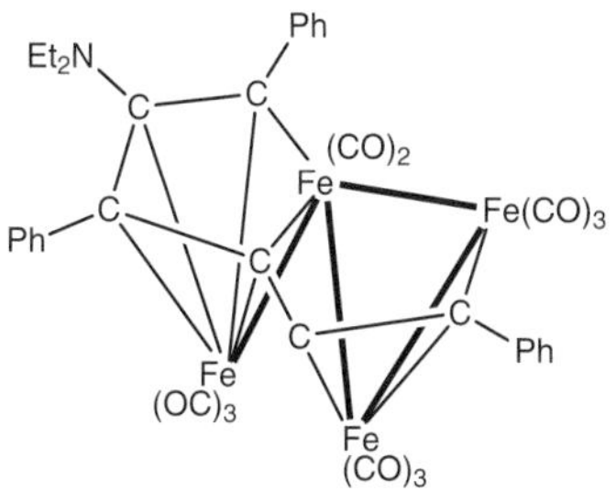

(**93**)

2. *Ruthenium*

As with the mono- and binuclear complexes described previously, reactions of 1,3-diynes with cluster carbonyls such as $Ru_3(CO)_{10}(NCMe)_2$ or $Ru_3(\mu\text{-dppm})(CO)_{10}$ have in general given complexes in which only one of the C≡C triple bonds is coordinated to the metal core. The structural types are similar to those obtained from monoalkynes.[225–227] Fragmentation of $Ru_3(CO)_{12}$ occurs in the reaction with PhC≡CC≡CPh (refluxing hexane, 3 h), from which the symmetrical metallacyclopentadiene complex $Ru_2\{\mu\text{-}2\eta^1\text{:}\eta^4\text{-CPh=C(C≡CPh)C(C≡CPh)=CPh}\}(\mu\text{-CO})(CO)_5$ (**94a;** R = Ph) was isolated (Scheme 23).[228] Fragmentation of the trinuclear core also occurs in reactions between FcC≡CC≡CFc and $Ru_3(CO)_{12}$ (refluxing hexane, 3 h), from which all three isomers of the metallacyclopentadiene $Ru_2\{\mu\text{-}2\eta^1\text{:}\eta^4\text{-}C_4Fc_2(C≡CFc)_2\}(CO)_6$ (**94;** R = Fc) and two isomers of dimetallacycloheptadienone $Ru_2\{\mu\text{-}\eta^1,\eta^2\text{:}\eta^1,\eta^2\text{-}C_4Fc_2(C≡CFc)_2CO\}(CO)_6$ (**95;** R = Fc) were isolated.[229] Reactions of PhC≡CC≡CPh with $Ru_3(CO)_{12}$ activated by Me_3NO carried out in thf afforded $Ru_2\{\mu\text{-C(C≡CPh)=CPhC(C≡CPh)=CPh}\}(CO)_6$ (**94b;** R = Ph) and $Ru_2\{\mu\text{-[C(C≡CPh)=CPh]}_2CO\}(CO)_6$ (**95;** R = Ph). The use of preformed $Ru_3(CO)_{10}(NCMe)_2$ gave mononuclear $Ru\{C(C≡CPh)=CPhC(C≡CPh)=CPh\}(CO)_3(NMe_3)$ (**96**), trinuclear $Ru_3(\mu_3\text{-}\eta^2\text{-}PhC_2C≡CPh)(\mu\text{-CO})(CO)_9$ (**97;** R = Ph) and tetranuclear $Ru_4(\mu_4\text{-}\eta^2\text{-}PhC_2C≡CPh)(CO)_{12}$ (**98**) in addition to **94b** and **95.**[230] Complexes **94** and **96** are formed by the coupling of two diyne molecules in head-to-tail and head-to-head fashion, respectively.

Thermolysis of $Ru_3(\mu\text{-}\eta^2\text{-}PhC_2C≡CPh)(\mu\text{-CO})(CO)_9$ gives two tetranuclear clusters, **98** and **99.** The latter contains the diyne as a 2,5-diphenylruthenacyclopentadiene, the 3,4-substituents of which are supplied by an $Ru_2(CO)_8$ fragment. The diyne is converted into a metallated 1,3-diene which chelates one Ru atom by virtue of the "rehybridization" of the carbons.[231] The Ru_6 cluster **100** bearing two methyleneindyne ligands attached in η^2 and η^4 modes to the cluster is also formed at high temperatures (refluxing xylene, 30 min). The latter has a novel geometry in which two Ru atoms are attached to one edge of a tetrahedron to give a puckered rhombus. The organic ligand is formed by attack of an acetylenic carbon on a phenyl group, resulting in cyclization to give the bicyclic system with concomitant migration of H from the phenyl ring to the end of the C_4 chain.[232]

The simple alkyne clusters $Ru_3(\mu_3\text{-}\eta^2\text{-}RC_2C≡CR)(\mu\text{-CO})(CO)_9$ [**97;** R = Ph,[230,233] $SiMe_3$,[172] $C_5H_8(OH)$,[234] $CH_2(OH)$[228]] are formed in low to moderate yields from $Ru_3(CO)_{10}(NCMe)_2$ and PhC≡CC≡CPh, Me_3SiC≡CC≡C$SiMe_3$ or 1,4-bis(1-hydroxycyclopentyl)buta-1,3-diyne, or as the sole product from (HO)CH_2C≡CC≡CCH_2(OH). In these complexes, the diyne is attached to the M_3 cluster by a conventional $\mu_3\text{-}\eta^1\text{:}\eta^1\text{:}\eta^2$ interaction and there are no obvious significant differences in the geometries of the clusters. In the crystal of the hydroxymethyl complex, an extensive hydrogen-bonding network involves the OH groups of both ligand and EtOH solvate molecule.[228]

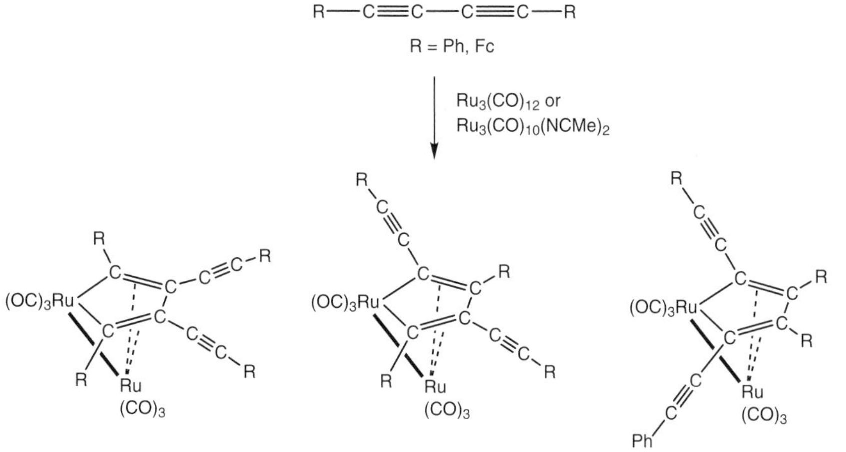

R—C≡C—C≡C—R

R = Ph, Fc

$Ru_3(CO)_{12}$ or

$Ru_3(CO)_{10}(NCMe)_2$

R C C—C≡C—R (OC)$_3$Ru C C C≡C R R Ru (CO)$_3$

(**94a**)

R C≡C C C R (OC)$_3$Ru C C C≡C R R Ru (CO)$_3$

(**94b**)

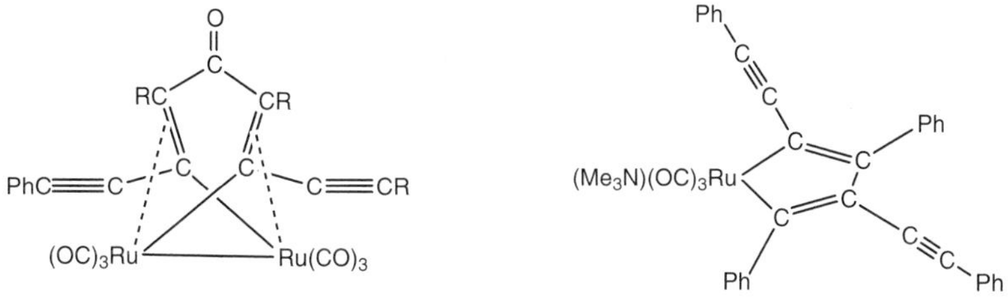

R C≡C C C R (OC)$_3$Ru C C R C≡C Ru (CO)$_3$ Ph

(**94c**)

O C RC CR PhC≡C C C C≡CR (OC)$_3$Ru Ru(CO)$_3$

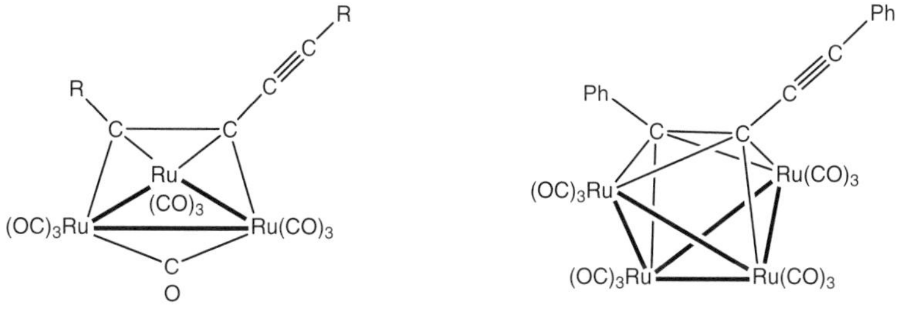

(**95**)

Ph C≡C C C Ph (Me$_3$N)(OC)$_3$Ru C C C≡C Ph Ph

(**96**)

R R C≡C C C Ru (CO)$_3$ (OC)$_3$Ru Ru(CO)$_3$ C O

(**97**)

Ph Ph C≡C C C (OC)$_3$Ru Ru(CO)$_3$ (OC)$_3$Ru Ru(CO)$_3$

(**98**)

SCHEME 23

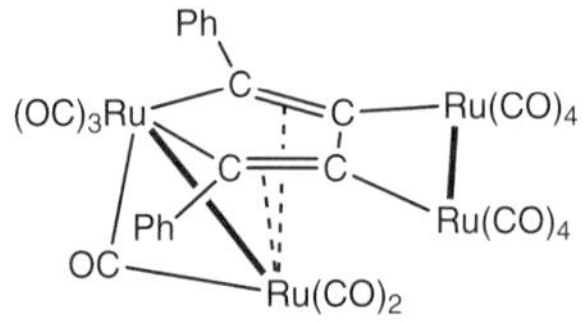

(**99**)

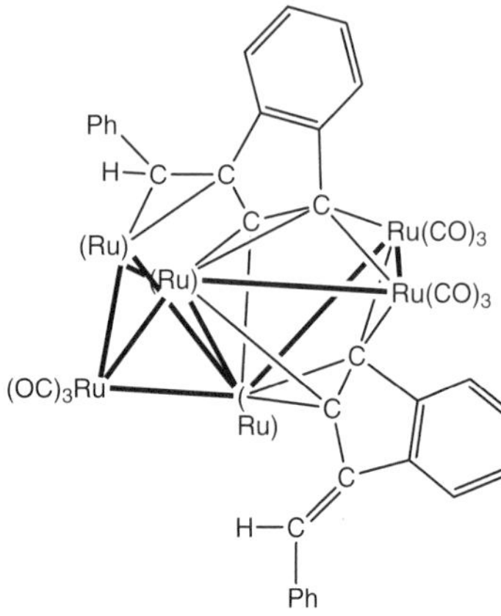

(**100**) (Ru) = $Ru(CO)_2$

Complexes **97** [$R = C_5H_8(OH)$], **101** (two isomers) and **102** were obtained from $Ru_3(CO)_{12}$ and 1,4-bis(1-hydroxycyclopentyl)buta-1,3-diyne ($CHCl_3$, 68°C) (Scheme 24).[234] Further reaction of **101** with the diyne resulted in formation of a second ligand analogous to that already present, together with intramolecular attack of OH on coordinated CO to give a carboxylate group in **103.** The organic ligand in **104** [from **102** and $Ru_3(CO)_{12}$] is derived by activation of C=C and C—H bonds forming an η^3-allyl group and bridging alkyl function. A by-product in this reaction is $Ru_6(\mu\text{-}H)(CO)_{15}Cp$, possibly originating from a hydroxycyclopentyl group by dehydration and dehydrogenation. Further transformations of the diyne, involving C≡C triple bond activation, intramolecular cyclization and coupling, occur on heating individual complexes with more $Ru_3(CO)_{12}$, resulting in formation of tetranuclear **104.** The bent Ru_3 chain is attached to the organic ligand formed by fragmentation to two alkyne units and coupling to the second diyne, together with formation of the furyl ring as found in **102.** Formal electron counting leads to a zwitterionic formulation of **104** with negative charges on the terminal Ru atoms counteracting formal positively charged oxygens. Its dark green color is attributed to strong MLCT transitions.

Reactions between $Ru_3(\mu\text{-dppm})(CO)_{10}$ and PhC≡CC≡CPh carried out in the presence of Me_3NO give $Ru_3(\mu_3\text{-}\eta^2\text{-}PhC_2C{\equiv}CPh)(\mu\text{-dppm})(\mu\text{-}CO)(CO)_7$ (**105**), also obtained from dppm and **97** (R = Ph), and $Ru_3(\mu\text{-dppm})\{\mu\text{-}C_4Ph_2(C{\equiv}CPh)_2\}(CO)_6$ (**106**) (Scheme 25). The former is a conventional μ_3-alkyne complex, whereas in **106**, two molecules of the diyne have combined to give a ruthenole, which is attached to the other two Ru atoms by an η^4 interaction from the ring and by η^2 coordination of one C≡C bond, respectively.[231] Thermolysis of **105** (refluxing xylene, 30 min) gave two further complexes identified crystallographically as $Ru_3\{\mu_3\text{-}CPhCHCC(C_6H_4)\}(\mu\text{-dppm})(CO)_8$ (**107**) and $Ru_3(\mu_3\text{-}C_4H_2Ph_2\text{-}1,4)(\mu\text{-}CO)(CO)_5(dppm)$ (**108**). In each of these, the "free" C≡C triple bond of

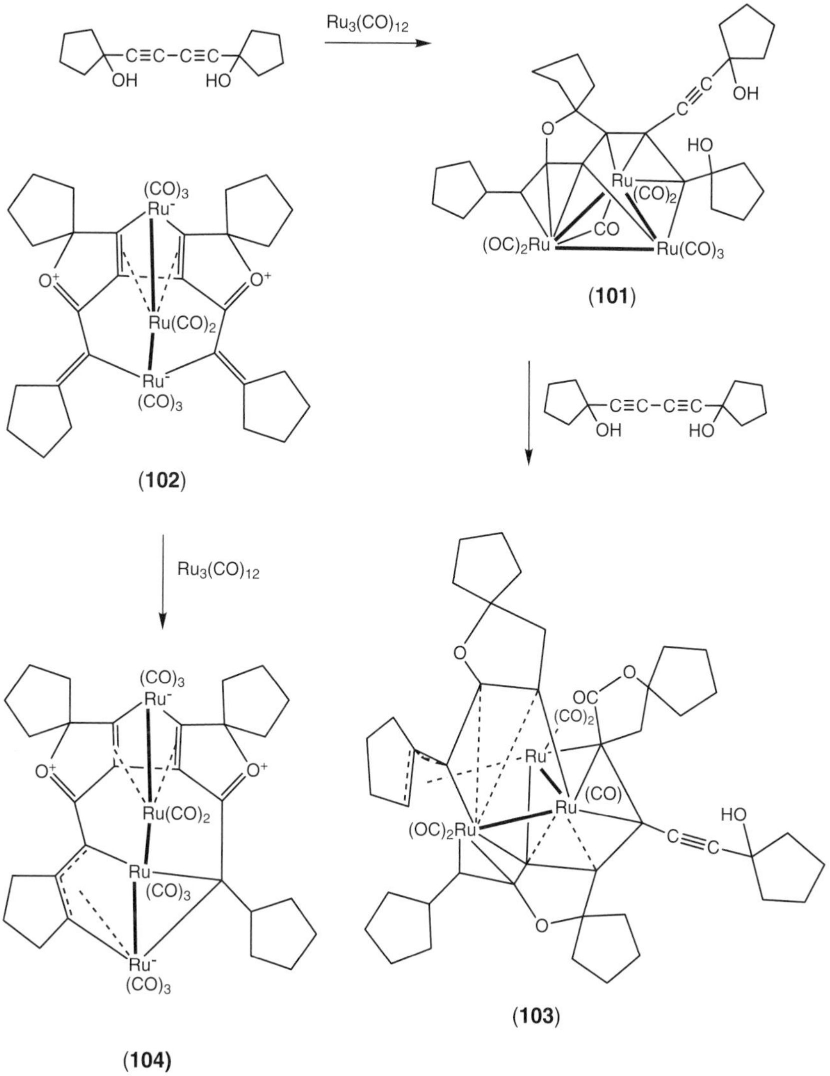

SCHEME 24

the diyne ligand in **105** has become involved in further bonding to the cluster by dint of forming a ruthenacyclopentadiene ligand. In **107,** further cyclization with the Ph group and the third Ru atom occurs to give an unusual 6/5/5 tricyclic system with transfer of the aromatic H atom to the ring, while in **108,** the ruthenole is part of a conventional thermodynamic isomer of a (substituted) $Ru_3(\mu_3\text{-}C_4R_4)(CO)_8$ cluster. In neither case is the dppm ligand degraded.

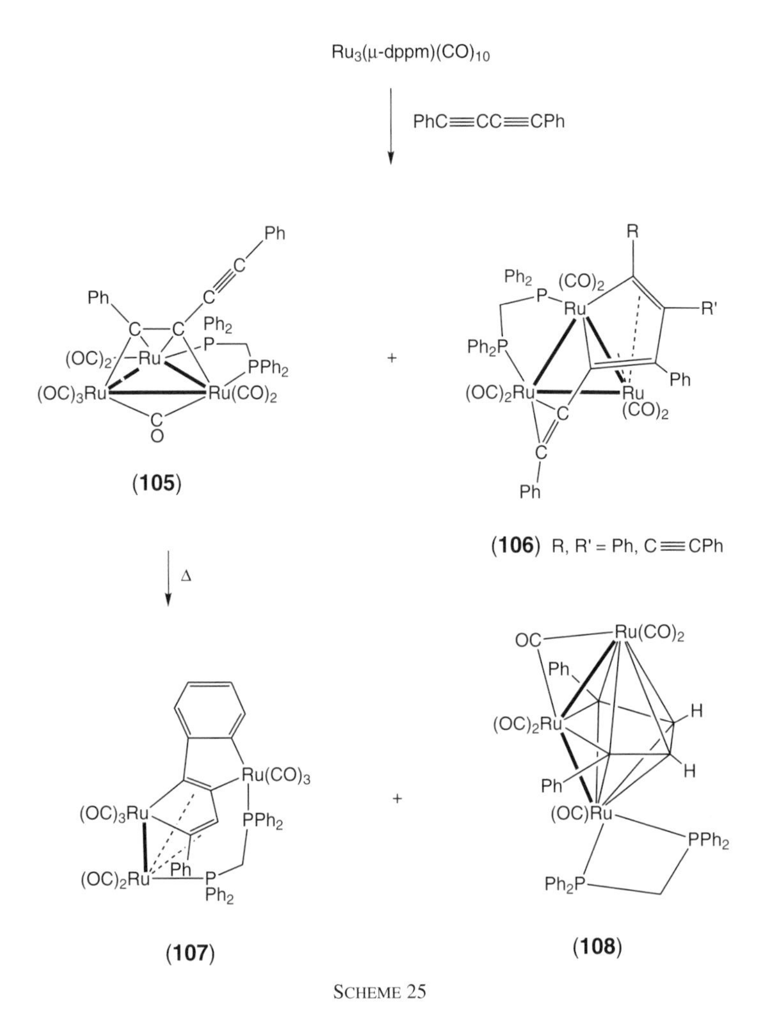

SCHEME 25

Among several products obtained from the reaction between **105** and $Me_3SiC{\equiv}CC{\equiv}CSiMe_3$ (Scheme 26) were **107**, $Ru_2(\mu\text{-dppm})\{\mu\text{-C(C}{\equiv}\text{CPh)}{=}\text{CPhC(SiMe}_3){=}\text{C(C}{\equiv}\text{CSiMe}_3)\}(CO)_4$ (**109;** head-to-head coupling of two diynes), $Ru_3\{\mu_3\text{-CPhCC(O)C(SiMe}_3)\text{C(C}{\equiv}\text{CSiMe}_3)\text{CCPh}\}(\mu\text{-dppm})(\mu\text{-CO})(CO)_6$ (**110;** formed by coupling of the two diynes and CO on the Ru_3 cluster to give a metallaindenone) and tetranuclear $Ru_4(\mu_4\text{-}\eta^2\text{-PhC}_2\text{C}{\equiv}\text{CPh})(\mu_4\text{-}\eta^2\text{-SiMe}_3\text{C}_2\text{C}{\equiv}\text{CSiMe}_3)(\mu\text{-dppm})(\mu\text{-CO})(CO)_8$ (**111**), in which the two diynes are on opposite sides of the Ru_4 puckered rhomboid. Again it is interesting that in these reactions, the often-found dephenylation of the dppm ligand does not occur.[235,236]

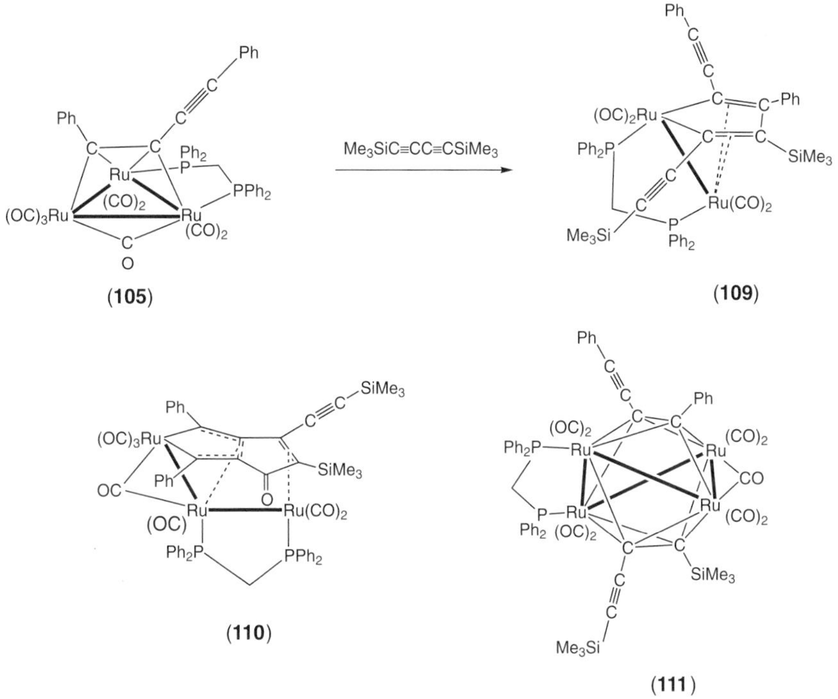

$Me_3SiC{\equiv}CC{\equiv}CSiMe_3$

(105) (109)

(110) (111)

SCHEME 26

The reaction of $HC{\equiv}CC{\equiv}CSiMe_3$ with $Ru_3(\mu\text{-dppm})_2(CO)_8$ in thf afforded $Ru_3(\mu\text{-H})(\mu\text{-dppm})_2(\mu\text{-}\eta^1{:}\eta^2\text{-}C_2C{\equiv}CSiMe_3)(CO)_5$ (**112**) via oxidative addition of the terminal C≡CH portion of the diynyl ligand across one of the dppm-bridged Ru—Ru bonds.[172] An unusual $\mu\text{-}\eta^3$-PhCCC=CHPh ligand is found in **113,** formed in the reaction of PhC≡CC≡CPh with $Ru_3(\mu\text{-H})(\mu_3\text{-PhNC}_5H_4N\text{-2})(CO)_9$.[237]

An extensive series of complexes has been obtained from reactions of $Ru_4(\mu_3\text{-PPh})(CO)_{13}$ (**114**) with RC≡CC≡CR (R = Ph, Me, $SiMe_3$) (Scheme 27).[238] The first products to be formed are the 62-e clusters $Ru_4\{\mu_4\text{-}\eta^1{:}\eta^1{:}\eta^2\text{-PPhC(C}{\equiv}\text{CR)CR}\}(\mu\text{-CO})_2(CO)_{10}$ (**115**), formed by an easy P—C bond formation and preserving a free C≡C triple bond. The structure contains one C≡C triple bond coordinated in the usual $2\sigma,\pi$ fashion, although one σ bond is to P and is consistent with the P atom behaving as part of the cluster framework. On heating (hexane, reflux, 4 h), complexes **115** are decarbonylated to form 62-e $Ru_4(\mu_4\text{-PPh})(\mu_4\text{-}\eta^1{:}\eta^1{:}\eta^2{:}\eta^2\text{-}RC_2C{\equiv}CR)(\mu\text{-CO})(CO)_{10}$ (**116;** R = Ph, Me). The rearrangements of R = Ph and R = Me feature opposite regiochemistry with respect to the diyne ligand and the Ru_4 face, which is reflected in different distributions of Ru—Ru bond lengths and only one symmetrical μ-CO in the latter. In solution, all CO groups exchange readily, even at -90°C. A compound with spectroscopic properties consistent with the analogous $SiMe_3$ compound was formed as a by-product in the initial reaction of **114** with $Me_3SiC{\equiv}CC{\equiv}CSiMe_3$.

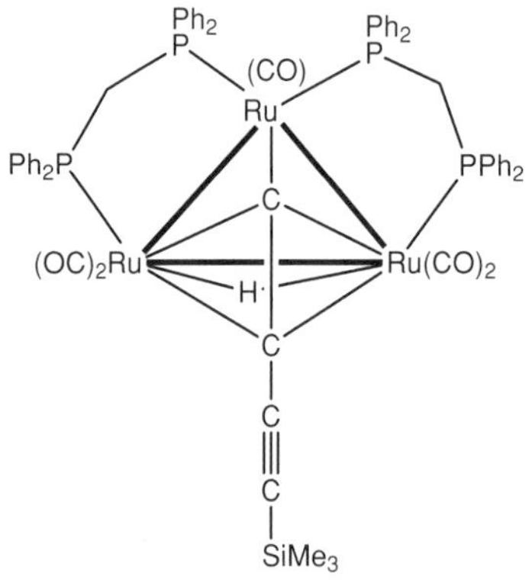

(**112**)

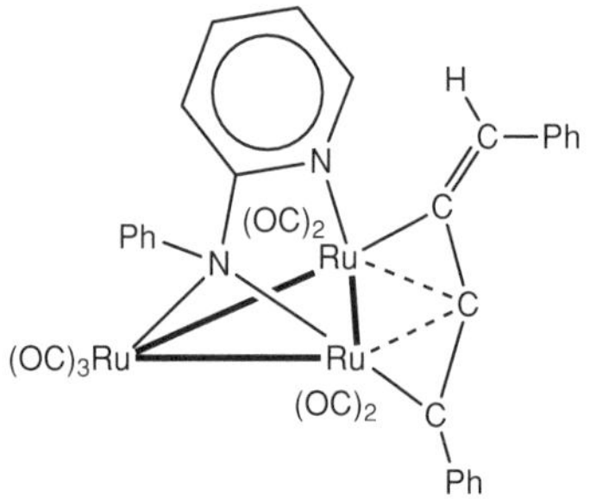

(**113**)

Further heating of **116-Ph** or **115-SiMe$_3$** affords $Ru_4(\mu_4\text{-}PPh)(\mu_4\text{-}\eta^1{:}\eta^1{:}\eta^3{:}\eta^3\text{-}RC_4R)(CO)_{10}$ (**117;** R = Ph, $SiMe_3$) in high yield.[238] In the case of the $SiMe_3$ derivative the molecule has a mirror plane containing the C_4 chain and substituent atoms (*ipso* C or Si). The Ru_4 square is highly distorted and is coordinated to both C≡C triple bonds of the diyne. The R = Me derivative could not be obtained under similar conditions. Pyrolysis of **115-Ph** (heptane, 80–90°C) with a H_2 purge results in hydrogenation of the diyne to *trans*-1,4-diphenylbut-1-ene, the cluster being recovered as $Ru_4(\mu\text{-}H)_2(\mu_3\text{-}PPh)(CO)_{12}$. Solid-state ^{31}P NMR investigations of *nido*-$Ru_4(\mu_3\text{-}\eta^1{:}\eta^1{:}\eta^2\text{-}PPhCMeCC{\equiv}CMe)(\mu\text{-}CO)_2(CO)_{10}$ and *closo*-$Ru_4(\mu_4\text{-}PPh)(\mu_4\text{-}\eta^1{:}\eta^1{:}\eta^3{:}\eta^3\text{-}SiMe_3CC_3SiMe_3)(CO)_{10}$ were undertaken to determine the phosphorus chemical shift tensors of the PPh groups, in the former a phosphido and in the latter, a phosphinidene group.[239]

The 62-e diyne complexes **115** and **116** are effective scaffolds for the trimerization of the diynes and codimerization of diyne and alkynes, resulting in extended carbon chains coordinated to the cluster.[240] Thus, reactions with $R'C{\equiv}CC{\equiv}CR'$ (R′ = Ph, Me) afforded $Ru_4(\mu_4\text{-}PPh)\{\mu_4\text{-}RCC(C{\equiv}CR')CPhC\text{-}\eta^4\text{-}CCPhCRC(C{\equiv}CR')\}(CO)_8$ (**118;** R = Ph, Me). The former was also found as a minor product from $Ru_4(\mu_4\text{-}PPh)(CO)_{13}$ and $PhC{\equiv}CC{\equiv}CPh$.[242] These are 64-e clusters containing a μ_4-PPh ligand capping an Ru_4 face, together with a C_{12} ligand formed from

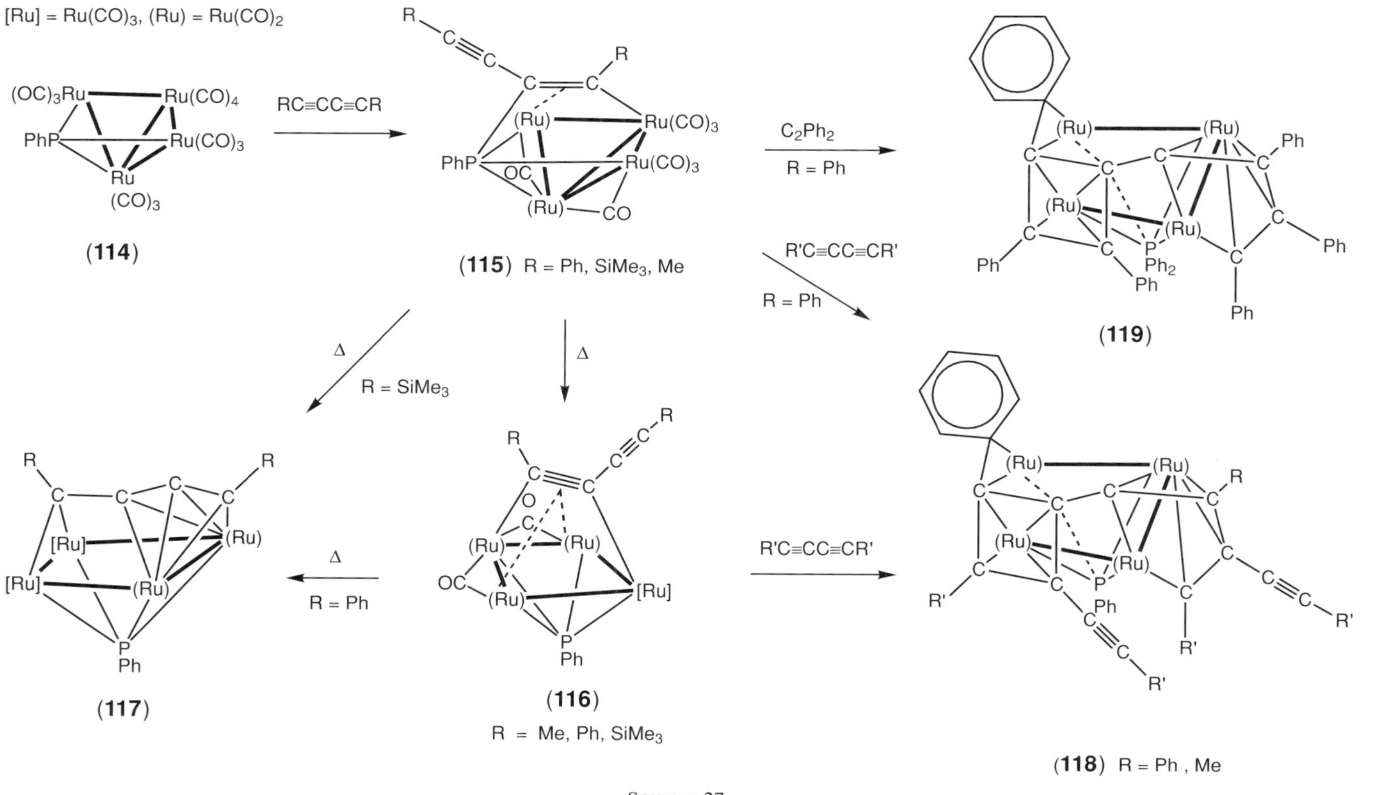
[Ru] = Ru(CO)3, (Ru) = Ru(CO)2
RC≡CC≡CR
(114)
(115) R = Ph, SiMe3, Me
C2Ph2
R = Ph
R'C≡CC≡CR'
R = Ph
(119)
Δ
R = SiMe3
Δ
Δ
R = Ph
R'C≡CC≡CR'
(117)
(116)
R = Me, Ph, SiMe3
(118) R = Ph , Me

SCHEME 27

three molecules of the diyne. Eight of the carbons are bonded to the four Ru atoms of the square-planar array on the opposite side to the PPh group. A central diyne is coordinated by all four carbons, the other two diyne molecules forming a metallacyclopentadiene with one Ru atom and an η^4-cyclobutadiene coordinated to another Ru. One of the Ph groups has an "agostic" C—C interaction with one Ru atom. The regiospecific formation of **118** (R = Me) demonstrates that the uncoordinated C≡C moieties in this complex arise from the added diyne. Reaction of **115-Ph** with PhC≡CPh affords **119,** which is an analogue of **118** with no uncoordinated C≡C triple bonds.

Reactions of $Ru_4(\mu\text{-H})_2(\mu_3\text{-PPh})(CO)_{12}$ (**120**) with PhC≡CC≡CPh afford non-cyclic *trans*-butatriene and *trans*-but-3-en-1-yne ligands in the complexes $Ru_4\{\mu_4\text{-}\eta^1\text{:}\eta^1\text{:}\eta^1\text{:}\eta^3\text{-PPhC(CHPh)CCHPh}\}(CO)_{12}$ (**121**), $Ru_4(\mu_4\text{-PPh})(\mu_4\text{-}\eta^1\text{:}\eta^1\text{:}\eta^2\text{:}\eta^2\text{-PhCHCCCHPh})(\mu\text{-CO})(CO)_{10}$ (**122**) and $Ru_4(\mu_4\text{-PPh})(\mu_4\text{-}\eta^1\text{:}\eta^1\text{:}\eta^1\text{:}\eta^3\text{-PhCHCHCCPh})(CO)_{11}$ (**123;** R = Ph) in reactions which involve easy P—C bond formation, hydrometalation, and skeletal isomerization (Scheme 28).[241] The related cluster $Ru_4(\mu_4\text{-PCF}_3)(\mu_4\text{-PhCHCHCCPh})(CO)_{11}$ (**123;** R = CF_3) has also been obtained.[242] The ligand in **121** lies on one side of the approximately planar Ru_3P face, which is capped on the opposite side by the fourth Ru atom.[241] In heptane (80°C, 4 hr) decarbonylation of **121** gives **122,** and migration and bond redistribution results in the organic ligand becoming located on the opposite side of the Ru_4 face to the PPh group. Both **121** and **122** originate from 1,4-addition of 2H to the diyne. In contrast, the other major product (**123**) results from 1,2-addition to give a μ_4-phenylethynyl group linked to an η^2-styryl unit, the entire ligand acting as a 6-e donor in the 64-e cluster. Further reaction of **123** (R = Ph) with PhC≡CC≡CPh (heptane, 80°C) results in displacement of the olefinic ligand and coupling to the diyne to give **124,** containing a C_8 ligand retaining a free C≡C triple bond.

In addition to $Ru_2(\mu\text{-}\eta^1,\eta^2\text{-}C_2C{\equiv}CR)(\mu\text{-PPh}_2)(CO)_6$ (**125;** R = Bu^t, Ph, $SiMe_3$),[243] a series of cluster complexes **126–131** containing ligands formed by Ru—Ru, C—P bond cleavage and fragment recombination reactions was obtained by heating $Ru_3(CO)_{11}\{PPh_2(C{\equiv}CC{\equiv}CR)\}$ in refluxing thf (Scheme 29). The most extensive range is formed by the Bu^t derivative. For R = Ph, only **126-Ph** and **130-Ph** were isolated, while for R = $SiMe_3$, only **126-Si** and **131-Si** were formed.[244] Of the three tetranuclear clusters, **126** (62 c.v.e.) contains a flattened butterfly cluster with two adjacent outer edges bridged by PPh_2 groups and supporting a six-electron donor $\mu_4\text{-}C_2C{=}CBu^tC{\equiv}CC{\equiv}CBu^t$ ligand formed by head-to-tail coupling of two diynyl groups. Small amounts of **126** were among several products formed by thermolysis of $Ru_2(\mu\text{-}\eta^1\text{:}\eta^2\text{-}C_2C{\equiv}CBu^t)(\mu\text{-PPh}_2)(CO)_6$. Although containing a distorted square-planar Ru_4 array, **127** is best regarded as containing a substituted pentagonal–bipyramidal C_2PRu_4 core, the carbon atoms bearing $C{\equiv}CBu^t$ and Ph substituents, the latter originating from the PPh_2 group. The structure is similar to that of **116** and also has 62 c.v.e.[238] Cluster **128** (64 c.v.e.), also obtained in

(Ru) = $Ru(CO)_2$, [Ru] = $Ru(CO)_3$

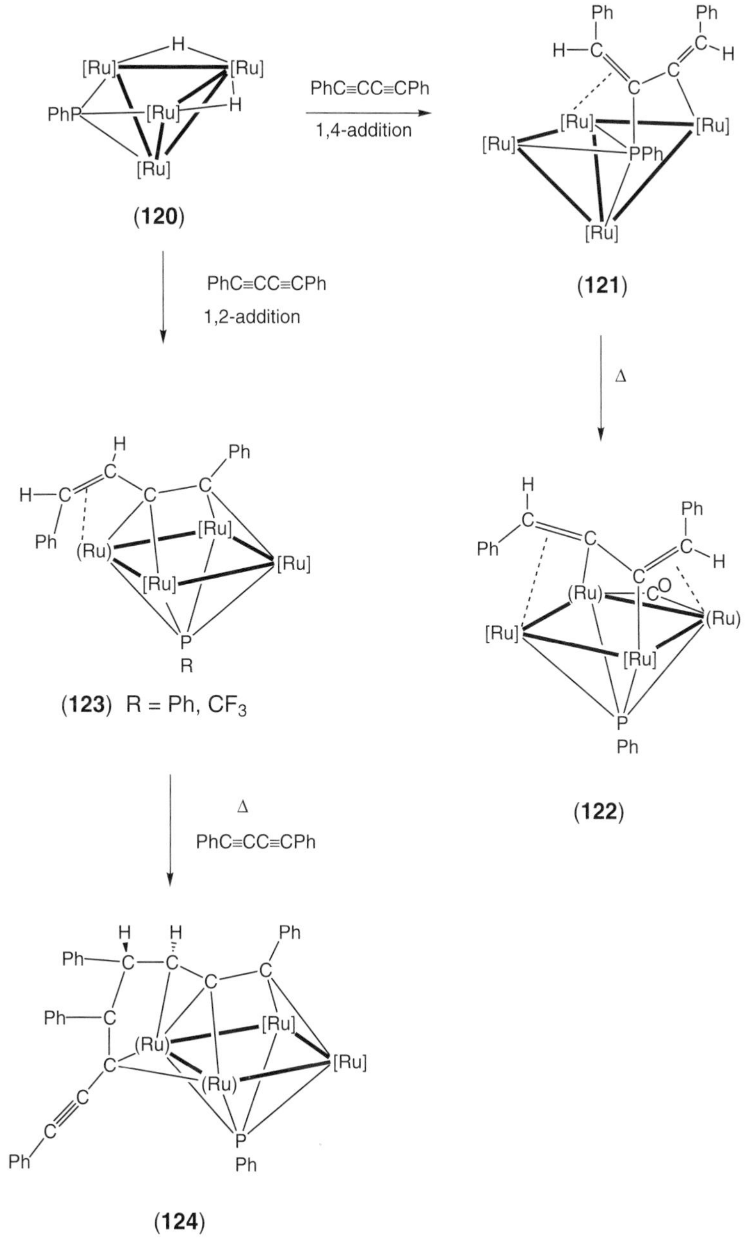

PhC≡CC≡CPh, 1,4-addition: (**120**) → (**121**)

PhC≡CC≡CPh, 1,2-addition: (**120**) → (**123**)

Δ: (**121**) → (**122**)

Δ, PhC≡CC≡CPh: (**123**) → (**124**)

(**123**) R = Ph, CF_3

SCHEME 28

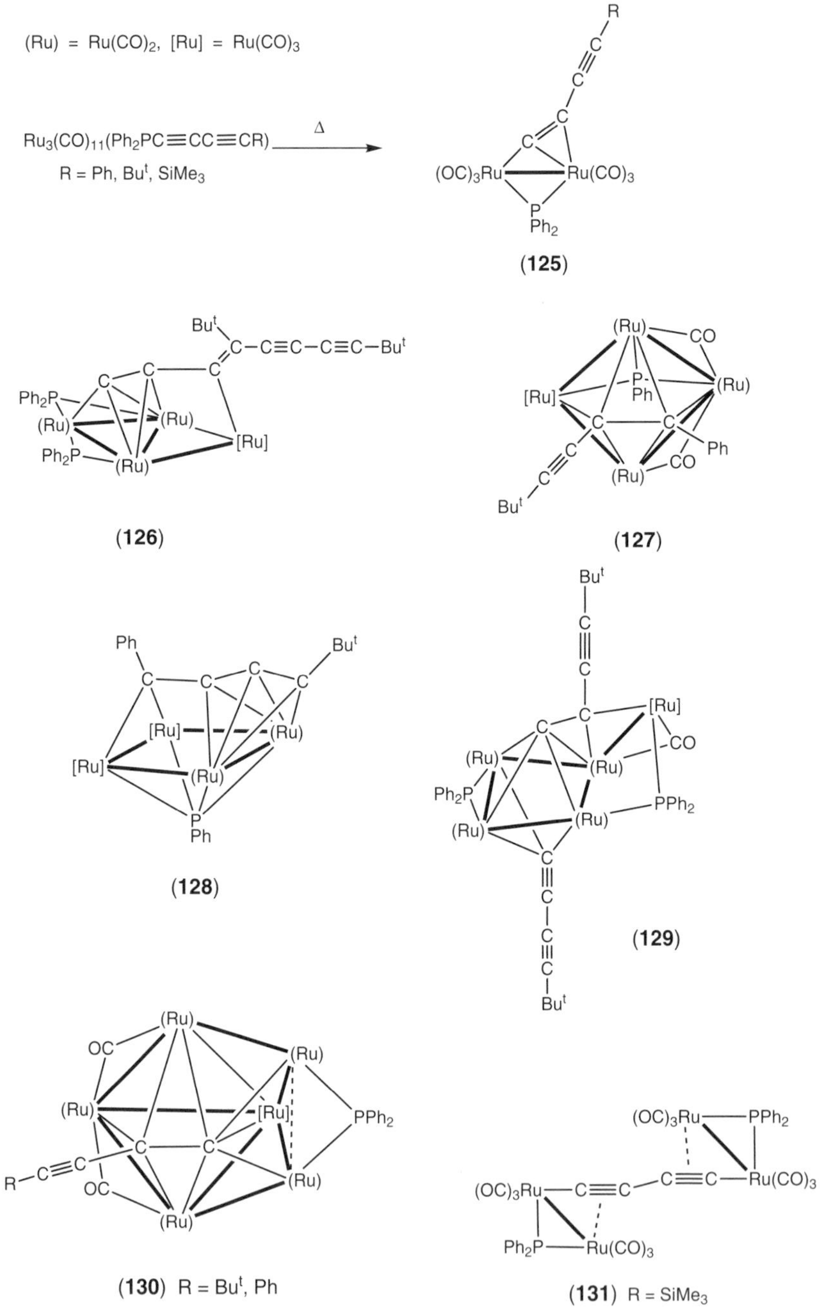

(Ru) = $Ru(CO)_2$, [Ru] = $Ru(CO)_3$

$Ru_3(CO)_{11}(Ph_2PC{\equiv}CC{\equiv}CR)$ $\xrightarrow{\Delta}$

R = Ph, Bu^t, $SiMe_3$

R
C
C
C
C
$(OC)_3Ru$ $Ru(CO)_3$
P
Ph_2

(**125**)

Bu^t
C−C≡C−C≡C−Bu^t
C C C
Ph_2P
(Ru) (Ru) [Ru]
Ph_2P (Ru)

(**126**)

(Ru) CO
[Ru] P Ph (Ru)
C C
C Ph
C (Ru) CO
Bu^t

(**127**)

Ph Bu^t
C C C C
[Ru] (Ru)
[Ru] (Ru)
P
Ph

(**128**)

Bu^t
C
C
C [Ru]
C CO
(Ru) (Ru)
Ph_2P PPh_2
(Ru) (Ru)
C
C
C
C
Bu^t

(**129**)

(Ru)
OC (Ru)
(Ru) [Ru] PPh_2
R−C≡C C C (Ru)
OC
(Ru)

(**130**) R = Bu^t, Ph

$(OC)_3Ru$ PPh_2
$(OC)_3Ru$−C≡C−C≡C−$Ru(CO)_3$
Ph_2P $Ru(CO)_3$

(**131**) R = $SiMe_3$

SCHEME 29

quantitative yield by thermolysis of **127,** contains a square pyramidal PRu_4 core. Further coordination of the C_4 ligand to the square face occurs via the $C\equiv CBu^t$ group pendant in **127.** Again, related complexes were obtained from $RC\equiv CC\equiv CR$ ($R = Ph, Bu^t, SiMe_3$) and $Ru_4(\mu_4\text{-}PPh)(CO)_{13}$[238] and by tail-to-tail coupling of two alkynyl moieties in $Ru_3\{\mu\text{-}P(C\equiv CBu^t)_2\}(\mu\text{-}\eta^1,\eta^2\text{-}C_2Bu^t)(CO)_9$ (see following).[245]

The pentanuclear cluster **129** contains a distorted spiked-square framework on which two PPh_2 groups bridge bonded and nonbonded Ru—Ru vectors. Two diynyl fragments are attached differently, one bonding via a terminal carbon to three Ru atoms, the other using two carbons to bridge all five Ru atoms. The latter ligand is formulated as an alkylidyne carbide, with the terminal C atom resonating at δ 294.5 ppm. This cluster is formally electron deficient with 76 c.v.e. rather than the expected 80 c.v.e., with the two C_4Bu^t ligands contributing one and five electrons, respectively. The 90-c.v.e. cluster **130** contains a bicapped octahedral C_2Ru_6 framework carrying a $\mu_6\text{-}\eta^2\text{-}C_2C\equiv CBu^t$ ligand, of which the terminal carbon is considered to be carbidic, interacting with five Ru atoms and therefore once again the ligand is considered as an alkylidyne-carbide.[246] An EHMO study analyzes the corresponding parent cluster $Ru_6(\mu_6\text{-}CCCCH)(\mu\text{-}PH_2)(\mu\text{-}CO)_2(CO)_{13}$ in terms of octahedral $[Ru_4(C_2CCH)(CO)_{11}]^{3-}$ and $[Ru_2(\mu\text{-}PH_2)(CO)_4]^{3+}$ fragments and reveals limitations to applications of the cluster condensation principle, resulting from the sharing of an edge by the two capping Ru atoms to give a CRu_3 rhombus containing only five frontier orbitals.[244]

In contrast to the reaction with $Co_2(CO)_8$ (q.v.), that between $Co_2(\mu\text{-}dppm)(CO)_6$ and $Ru_3(\mu_3\text{-}PhC_2C\equiv CPh)(\mu\text{-}CO)(CO)_9$ results in degradative fragmentation of the ruthenium cluster, coupling of two diyne ligands giving $Ru_2\{\mu_2\text{:}\eta^1,\eta^4\text{-}C_4Ph_2(C\equiv CPh)_2\}(CO)_6$ (**94**) and coordination of a dicobalt fragment to one of the free $C\equiv C$ triple bonds to give $Ru_2\{\mu\text{:}\mu\text{-}PhCC(C\equiv CPh)C[C_2Ph\{Co_2(\mu\text{-}dppm)(CO)_4\}]CPh\}(CO)_6$ (**132**).[247] The reaction of $Co_2(\mu\text{-}dppm)(\mu\text{-}\eta^2\text{-}SiMe_3C_2C\equiv CH)(CO)_4$ with $Ru_3(CO)_{12}$ gives an almost quantitative yield of $Ru_3(\mu\text{-}H)\{\mu_3\text{-}\eta^1,\eta^2\text{;}\mu\text{-}\eta^2\text{-}C_2C_2SiMe_3[Co_2(\mu\text{-}dppm)(CO)_4]\}(CO)_9$ (**133**).[172]

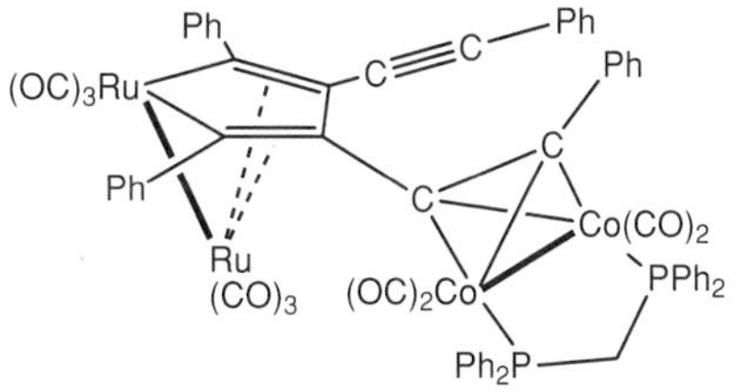

(**132**)

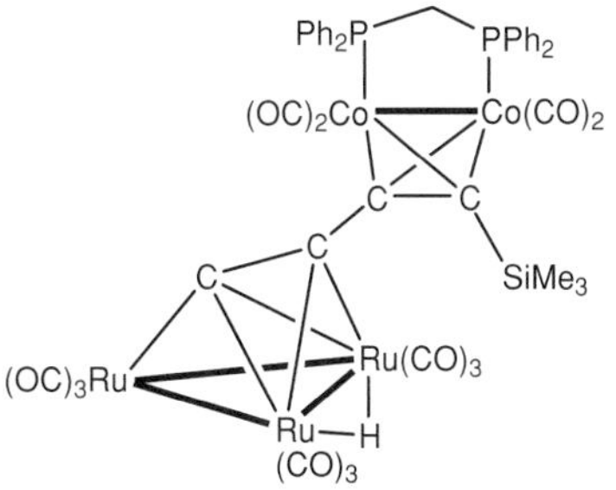

(**133**)

Facile oxidative addition of W(C≡CC≡CH)$(CO)_3$Cp to $Ru_3(CO)_{10}(L)_2$ (L = MeCN, L_2 = dppm) has given hydrido clusters containing the μ_3-η^1,η^2-alkynyl ligand, e.g., $Ru_3(\mu$-H){μ_3-η^1,η^2-C_2C≡C[W$(CO)_3$Cp]}(μ-dppm)$(CO)_7$ (**134**), which exhibits restricted fluxional behavior in solution (Scheme 30).[109,248] The μ_3-alkyne cluster Ru_3{μ_3-η^2-HC_2C≡C[W$(CO)_3$Cp]}(μ-CO)$(CO)_9$ (**135**) decarbonylates in refluxing benzene to give the hydrido-alkynyl $Ru_3(\mu$-H){μ_3-η^2-C_2C≡C[W$(CO)_3$Cp]}$(CO)_9$ (**136**).[249] Further reactions of **136** with metal carbonyls afford heterometallic systems such as **137,** obtained from $Fe_2(CO)_9$ or $Ru_3(CO)_{12}$, and **138,** formed with $Co_2(CO)_8$. In the iron-containing clusters, three of the Ru sites are partially occupied by up to two iron atoms. In **138,** the hydride ligand has migrated to the C_4 ligand, this time forming a vinylidene.[248]

The pentametallic cluster $Ru_5(\mu_5$-η^2-C_2Ph)(μ-PPh_2)(μ-CO)$(CO)_{13}$ reacts with PhC≡CC≡CPh to give $Ru_5(\mu_5$-η^2-C_2Ph)(μ_3-η^2-PhC_2C≡CPh)(μ-PPh_2)$(CO)_{12}$ (**139**) in which the diyne ligand acts as a four-electron donor.[250] Several complexes were obtained from reactions of $Ru_5(\mu_5$-C_2)(μ-SMe$)_2$(μ-$PPh_2)_2(CO)_{11}$ (**140**) with PhC≡CC≡CPh (Scheme 31).[251] The major products were formed by attack of the diyne on one of the carbons of the C_2 ligand, giving μ_5-CCCRCR′ ligands (R, R′ = Ph, C≡CPh). Minor products were also characterized, including two isomers of **141,** which contain only one cluster-bonded SMe group, the second migrating to the organic ligand to form a thioether. Double addition of the diyne to the same carbon atom afforded **142,** containing a multibranched C_{10} chain attached to the square face of an "open-envelope" Ru_5 cluster. Also notable are the electron counts for **142** and **143** (both 80 c.v.e.), which are two in excess of the number required for an electron-precise M_5 cluster with six M—M bonds. The extra electron density is accommodated by lengthening of two Ru—Ru bonds in each cluster.

3. *Osmium*

But-3-yn-1-ol and $Os_3(CO)_{12}$ (at 130°C) or $Os_3(CO)_{10}(NCMe)_2$ (at 90°C) give a 2,3-dihydrofuran-4,5-diyl ligand in $Os_3(\mu$-H$)_2(\mu_3$-η^2-$C_4H_4O)(CO)_9$ (**144**).[252] A related reaction between $Os_3(\mu$-H$)_2(CO)_{10}$ and CH_2(OH)C≡CC≡CCH_2(OH) gives $Os_3(\mu$-H)(μ-MeCC_4H_2O)$(CO)_{10}$ (**145**), in which the diyne is rearranged to a substituted furan bearing a methylcarbene substituent.[180] A possible route for its formation (Scheme 32) involves coordination of one C≡C triple bond and isomerization to the allene (cumulene) and intramolecular attack of the distant OH group at C_α, similar to that proposed for the rearrangement of CH_2(OH)C≡CCH_2(OH) to an allene on reaction with $Os_3(\mu$-H$)_2(CO)_{10}$.[253] Surprisingly, the reaction of the $Co_2(CO)_6$-protected diyne with $Os_3(\mu$-H$)_2(CO)_{10}$ gave only $Os_3(\mu$-H)(μ-OH)$(CO)_{10}$.

Complex **146** (R, R′ = H, $SiMe_3$) is obtained from $Os_3(CO)_{10}(NCMe)_2$ and HC≡CC≡C$SiMe_3$ as a 2/1 mixture of the two possible isomers. Subsequent

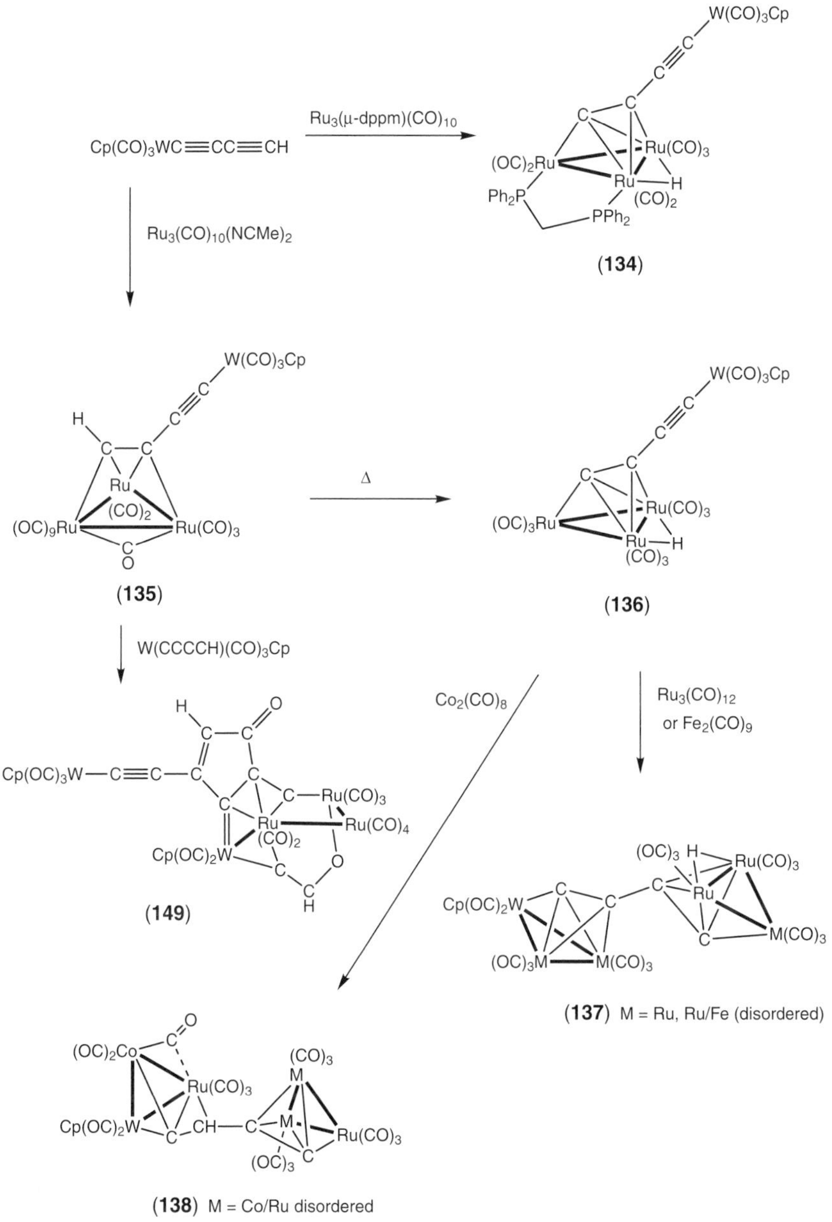

$Cp(CO)_3WC{\equiv}CC{\equiv}CH$

$Ru_3(\mu\text{-dppm})(CO)_{10}$

W(CO)$_3$Cp; C; C; C; C; (OC)$_2$Ru; Ru(CO)$_3$; Ru; H; (CO)$_2$; Ph$_2$P; PPh$_2$

(**134**)

$Ru_3(CO)_{10}(NCMe)_2$

W(CO)$_3$Cp; H; C; C; C; C; Ru; (CO)$_2$; (OC)$_9$Ru; Ru(CO)$_3$; C; O

(**135**)

Δ

W(CO)$_3$Cp; C; C; C; C; (OC)$_3$Ru; Ru(CO)$_3$; Ru; H; (CO)$_3$

(**136**)

W(CCCCH)(CO)$_3$Cp

H; C; C; O; Cp(OC)$_3$W—C≡C—C; C; C; C—Ru(CO)$_3$; Ru; Ru(CO)$_4$; (CO)$_2$; Cp(OC)$_2$W; C; O; C; H

(**149**)

$Co_2(CO)_8$

$Ru_3(CO)_{12}$ or $Fe_2(CO)_9$

(OC)$_3$ H; Ru(CO)$_3$; Cp(OC)$_2$W; C; C; C; Ru; C; M(CO)$_3$; (OC)$_3$M; M(CO)$_3$

(**137**) M = Ru, Ru/Fe (disordered)

O; (OC)$_2$Co; C; Ru(CO)$_3$; (CO)$_3$; M; Cp(OC)$_2$W; C; CH—C; M; Ru(CO)$_3$; (OC)$_3$; C

(**138**) M = Co/Ru disordered

SCHEME 30

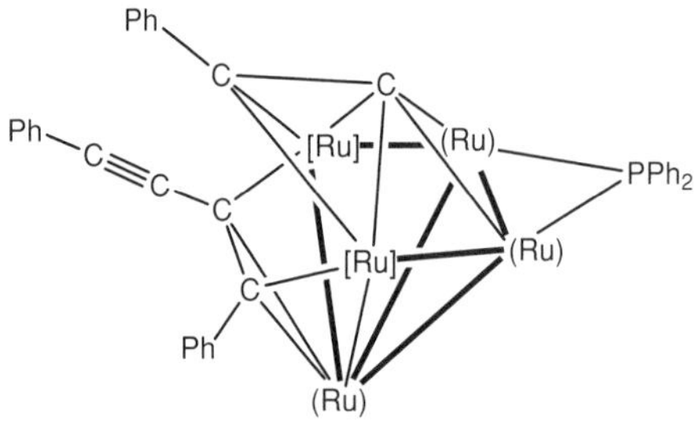

[Ru] = $Ru(CO)_3$, (Ru) = $Ru(CO)_2$

(**139**)

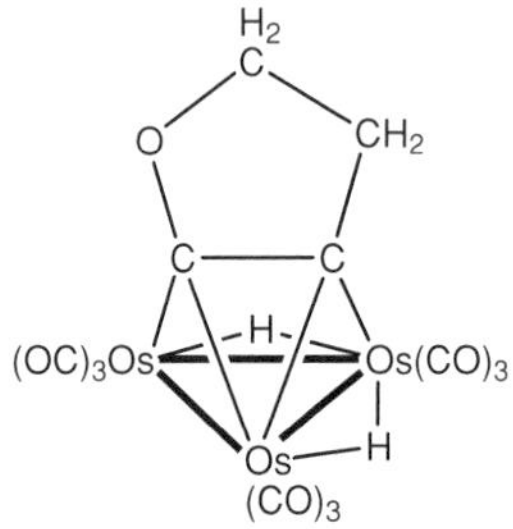

(**144**)

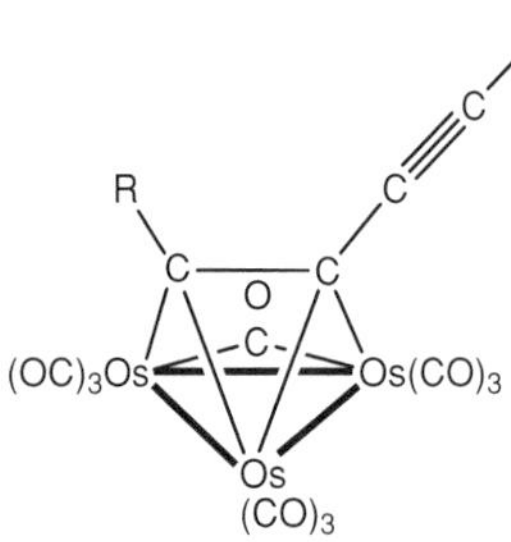

(**146**)

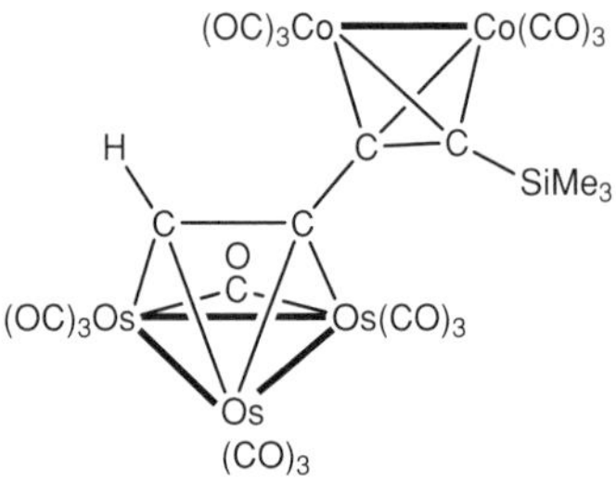

(**147**)

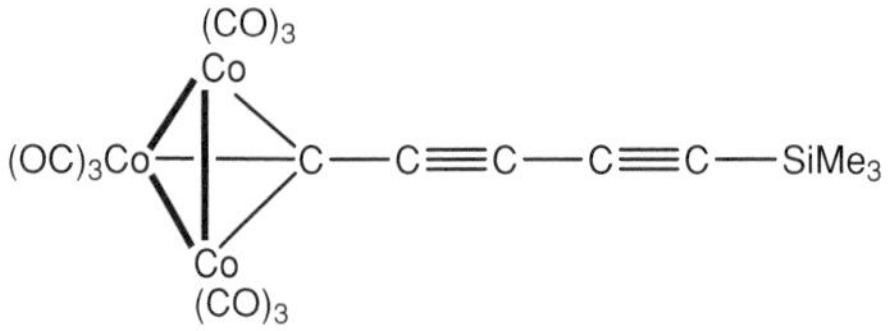

(**148**)

reaction of **146** with $Co_2(CO)_6$ gave $Os_3\{\mu_3\text{-}\eta^2\text{:}\mu\text{-}\eta^2\text{-}HC_2C_2SiMe_3[Co_2(CO)_6]\}(\mu\text{-}CO)(CO)_9$ (**147**), in which addition of the Co_2 fragment has only occurred to the C≡C triple bond adjacent to the $SiMe_3$ group, there being no evidence for formation of the other isomer.[172] Symmetrically disubstituted 1,3-diynes react with $Os_3(CO)_{10}(NCMe)_2$ to give $Os_3(\mu_3\text{-}\eta^2\text{-}RC_2C{\equiv}CR)(\mu\text{-}CO)(CO)_9$ (**146;** $R = R' =$ Me, Et, Bu^t, Ph, $SiMe_3$); the two non-interconvertible isomers of the analogous complex from $Os_3(CO)_{10}(NCMe)_2$ and $PhC{\equiv}CC{\equiv}CSiMe_3$ differ only by

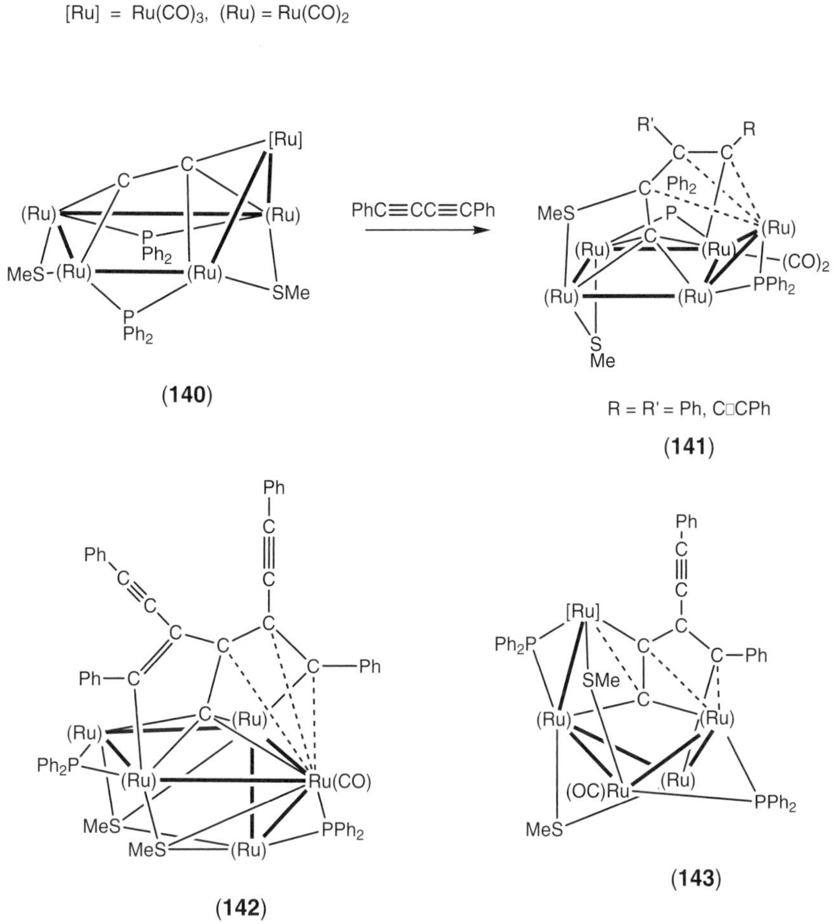

[Ru] = $Ru(CO)_3$, (Ru) = $Ru(CO)_2$

[Ru]
C
C
(Ru)
(Ru)
P
Ph_2
MeS-(Ru)
(Ru)
SMe
P
Ph_2

(**140**)

$PhC{\equiv}CC{\equiv}CPh$

R'
R
C
C
C
Ph_2
MeS
P
C
(Ru)
(Ru)
(Ru)
$-(CO)_2$
(Ru)
(Ru)
PPh_2
S
Me

R = R' = Ph, C□CPh

(**141**)

Ph
C
C
Ph
C
C
C
C
C
C
Ph—C
C—Ph
C
(Ru)
(Ru)
Ph_2P
(Ru)
Ru(CO)
MeS
MeS
(Ru)
PPh_2

(**142**)

Ph
C
C
[Ru]
C
Ph_2P
C
C—Ph
SMe
C
(Ru)
(Ru)
(OC)Ru
(Ru)
PPh_2
MeS

(**143**)

SCHEME 31

which of the C≡C triple bonds is coordinated.[254,255] The $PhC{\equiv}CC{\equiv}CPh$ derivative is hydrogenated (octane, 1 atm, 125°C) to give $Os_3H_3(\mu_3\text{-}CCH_2Ph)(CO)_9$.[254]

4. *Cobalt*

The tricobaltcarbon cluster $Co_3(\mu_3\text{-}CBr)(CO)_9$ reacts with $Me_3SiC{\equiv}CC{\equiv}CSiMe_3$ in the presence of $AlCl_3$ to afford $Co_3(\mu_3\text{-}CC{\equiv}CC{\equiv}CSiMe_3)(CO)_9$ (**148**). It is thought that this reaction proceeds via the abstraction of the halo ligand by the Lewis acid to give the carbo-cation $[Co_3C(CO)_9]^+$, which then participates in electrophilic attack on the diyne. Similar reactions with $Me_3SiC{\equiv}CSiMe_3$ and

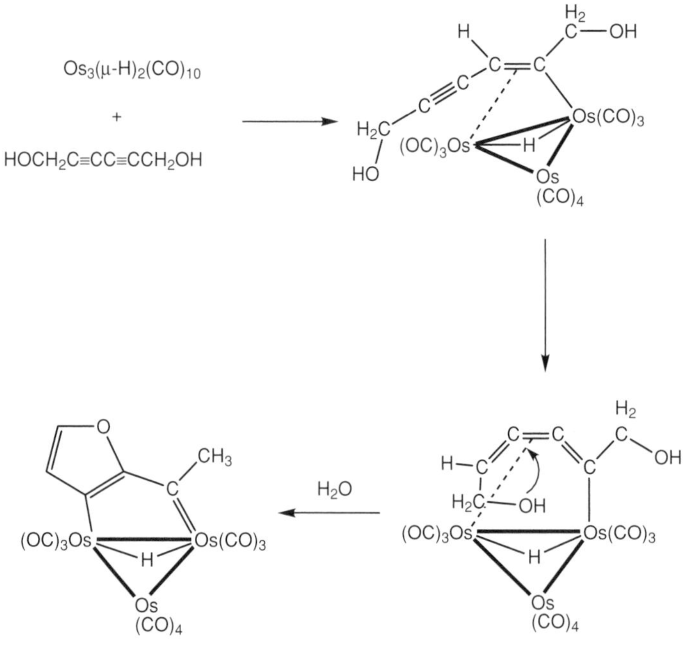

(145)

SCHEME 32

$Me_3Si(C{\equiv}C)_4SiMe_3$ failed. It was suggested that that the steric demands of the cluster carbonium ion prevented reaction with the bis(silyl)alkyne, while attempts to isolate the higher congeners $Co_3\{\mu_3\text{-}C(C{\equiv}C)_nSiMe_3\}(CO)_9$ were hampered by their instability.[256]

5. *Heterometallic Clusters*

The reaction of two equivalents of $W(C{\equiv}CC{\equiv}CH)(CO)_3Cp$ with $Ru_3(CO)_{10}(NCMe)_2$ gives the Ru_3W cluster **149** (Scheme 30), which is also obtained from **135** and $W(C{\equiv}CC{\equiv}CH)(CO)_3Cp$. The extended organic ligand is formed by coupling of two molecules of the diynyl complex with two of CO, to form a cyclopentadienone attached by a carbenic interaction to the cluster W atom, and featuring formylethynyl and $C{\equiv}CW(CO)_3Cp$ substituents.[249] One of the elementary steps in the reaction mechanism may involve formal rearrangement of the diyne to a dicarbyne.

Reaction of $W(C{\equiv}CC{\equiv}CPh)(O)_2Cp^*$ with $Os_3(\mu\text{-}H)_2(CO)_{10}$ gives $Os_3(\mu\text{-}H)\{\mu\text{-}\eta^1{:}\eta^2\text{-}C({=}CHPh)C_2[W(O)_2Cp^*]\}(CO)_{10}$ (**150**) (Scheme 33), in which the C_4 ligand has added one H atom from the cluster hydride precursor.[257] Cluster **150**

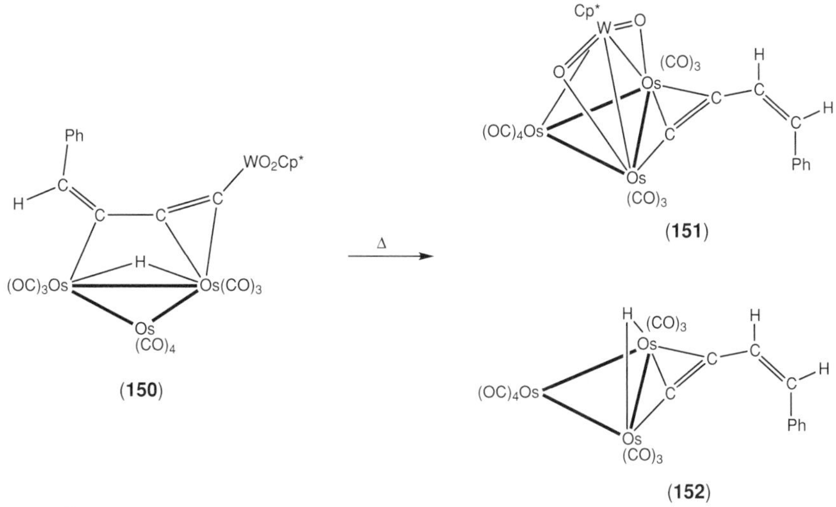

(150) $\xrightarrow{\Delta}$ (151) + (152)

SCHEME 33

exists as at least two structural isomers, as shown by structural determinations of two polymorphs. The structural differences arise from migration of the μ-H ligand between Os—Os bonds, the three possible isomers each being sufficiently long-lived to be observed by their $\nu(CO)$ and ^1H NMR spectra. Thermolysis of **150** (extended reflux in CH_2Cl_2) gives $Os_3W(\mu\text{-}O)_2(\mu\text{-}\eta^1{:}\eta^2\text{-}C_2CH{=}CHPh)(CO)_9Cp^*$ (**151**) and $Os_3(\mu\text{-}H)(\mu\text{-}\eta^1{:}\eta^2\text{-}C_2CH{=}CHPh)(CO)_{10}$ (**152**); the latter may be formed by reaction with traces of water during chromatography.

As with 1-alkynes, $Ir_2W_2(CO)_{10}Cp_2$ reacts with $W(C{\equiv}CC{\equiv}CH)(CO)_3Cp$ by formal insertion into a W—W bond to give butterfly cluster $Ir_2W_2\{\mu_4\text{-}\eta^2\text{-}HC_2C{\equiv}C[W(CO)_3Cp]\}(\mu\text{-}CO)_4(CO)_4$ (**153**) in which the diyne is attached by the terminal $C{\equiv}CH$ group parallel to the Ir–Ir hinge; the free $C{\equiv}C$ bond proved to be resistant to attempts to incorporate it into the cluster.[258] Similarly, $Ir_3W_2(\mu_4\text{-}\eta^2\text{-}C_2C{\equiv}CPh)(\mu\text{-}CO)(CO)_9Cp_2$ (**154**) is obtained from $W(C{\equiv}CC{\equiv}CPh)(CO)_3Cp$ and $Ir_3W(CO)_{11}Cp_2$.

A heterometallic cluster was obtained from the reaction of $Mo_2\{\mu\text{-}\eta^2\text{-}HC_2C{\equiv}C[Fe(CO)_2Cp^*]\}(CO)_4Cp_2$ with $Co_2(CO)_8$, which gave $Co_2Fe\{\mu_3\text{-}\eta^1,\eta^2{:}\mu\text{-}\eta^2\text{-}HC_2C_2[Mo_2(CO)_4Cp_2]\}(\mu\text{-}CO)_2(CO)_5$ (**155**).[259] The relief of steric congestion upon metal—metal bond formation apparently drives the reaction. No reaction occurred between $Mo_2\{\mu\text{-}\eta^2\text{-}HC_2C{\equiv}C[Fe(CO)_2Cp^*]\}(CO)_4Cp_2$ and an excess of $\{Mo(CO)_2Cp\}_2$.

Initial addition of $Co_2(CO)_8$ to $Fe(C{\equiv}CC{\equiv}CH)(CO)_2Cp^*$ occurs at the $C{\equiv}CH$ triple bond to give **156,** which adds a second Co_2 unit to give **157** (Scheme 34).[259] Reactions of these products with $Fe_2(CO)_9$ give mixed-metal clusters. Thus, **156**

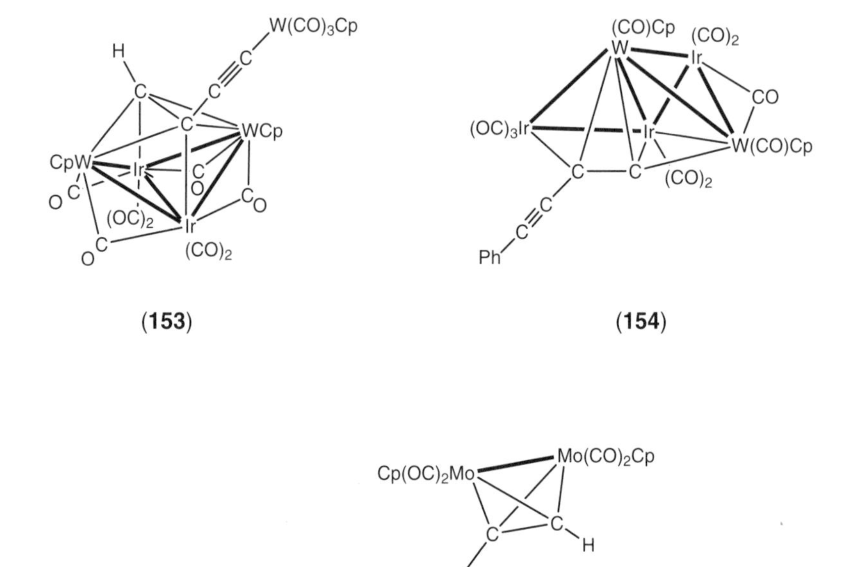

(**153**)

(**154**)

(**155**)

reacts to form $CoFe_2\{\mu_3\text{-}\eta^2\text{-}C_2C{\equiv}C[Fe(CO)_2Cp^*]\}(CO)_9$ (**158**), while both **159,** which features a $Fe(CO)Cp^*$-spiked $FeCo_2$ core, and the purple allenylidene cluster **160,** are obtained from **157.** Upon heating, **157** is converted into **161,** while **159** is transformed into the alkynylvinylidene $Co_2Fe\{\mu_3\text{-}CCHC_2[Co_2(CO)_6]\}(CO)_9$ (**162**). Formation of **159** and **160** (which are not interconvertible) is considered to occur by addition of a $Fe(CO)_n(thf)$ fragment to **157** to give a common intermediate which transforms by either Fe—C bond cleavage or Co transfer. Both $\mu_4\text{-}C_2$, $\mu_3\text{-}\eta^1{:}\eta^1{:}\eta^3\text{-}C_3$ and μ-vinylidene ligands are present in **161.**

Complexation of one C≡C triple bond in $\{M_n(CO)_{11}\}_2(\mu\text{-bdpp})$ [$M_n = Ru_3$, Os_3, $Re_3(\mu\text{-}H)_3$, $Ru_4(\mu\text{-}H)_4$] occurs in reactions with $Co_2(CO)_8$.[260] The related complex from $\{Re_3(\mu\text{-}H)_3(CO)_{11}\}(\mu\text{-bdpp})$ $\{Os_3(CO)_{11}\}$ was obtained as two separable isomers containing the $Co_2(CO)_6$ moiety attached to either C≡C triple bond. Thermolysis of $\{Os_3(CO)_{11}\}_2\{\mu\text{-}PPh_2C_2[Co_2(CO)_6]C{\equiv}CPPh_2\}$ afforded $Co_2Os_3\{\mu_5\text{-}C_2C{\equiv}CPPh_2[Os_3(CO)_{11}]\}(\mu\text{-}PPh_2)(CO)_{13}$ (**163**), while similar treatment of $Re_3(\mu\text{-}H)_3(CO)_{11}\{PPh_2C{\equiv}CC_2[Co_2(CO)_6]PPh_2[Os_3(CO)_{11}]\}$ (**164**) gave $Co_2Os_3Re\{\mu_6\text{-}C_2C{=}CH(PPh_2)\}(\mu\text{-}PPh_2)(\mu\text{-}CO)_2(CO)_{14}$ (**165**) (Scheme 35).

The reaction between $Ru_3(\mu_3\text{-}\eta^2\text{-}PhC_2C{\equiv}CPh)(\mu\text{-}CO)(CO)_9$ and $Co_2(CO)_8$ gives the bow-tie cluster $Co_2Ru_3(\mu_5\text{-}PhC_2C_2Ph)(\mu\text{-}CO)_3(CO)_{11}$ (**166**), in which

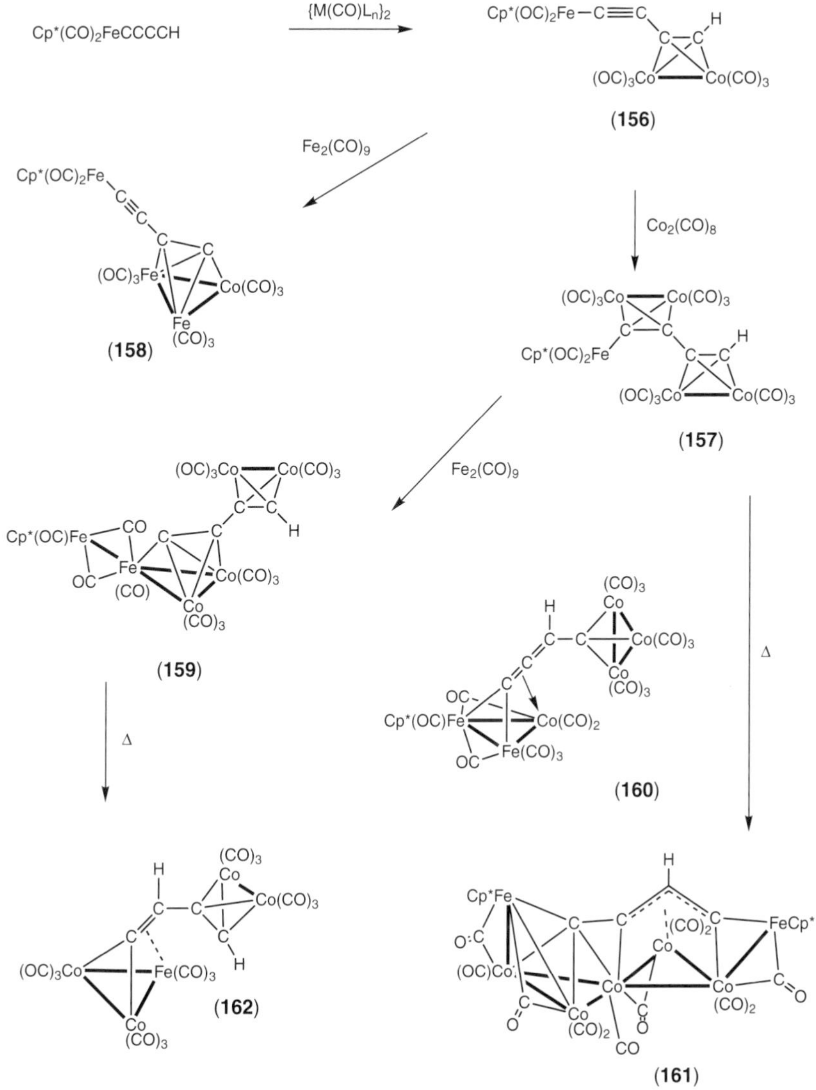

$Cp^*(CO)_2FeCCCCH$
$\{M(CO)L_n\}_2$
$Cp^*(OC)_2Fe{-}C{\equiv}C$
H
C—C
$(OC)_3Co$
$Co(CO)_3$
(**156**)
$Fe_2(CO)_9$
$Cp^*(OC)_2Fe$
C≡C
C
C
$(OC)_3Fe$
$Co(CO)_3$
Fe
$(CO)_3$
(**158**)
$Co_2(CO)_8$
$(OC)_3Co$
$Co(CO)_3$
C—C
H
$Cp^*(OC)_2Fe$
C—C
$(OC)_3Co$
$Co(CO)_3$
(**157**)
$(OC)_3Co$
$Co(CO)_3$
C—C
H
$Cp^*(OC)Fe$
CO
C
C
Fe
$Co(CO)_3$
OC
(CO)
Co
$(CO)_3$
(**159**)
$Fe_2(CO)_9$
$(CO)_3$
Co
H
C—C
$Co(CO)_3$
C
C
Co
$(CO)_3$
OC
$Cp^*(OC)Fe$
$Co(CO)_2$
$Fe(CO)_3$
OC
(**160**)
Δ
Δ
$(CO)_3$
H
Co
C—C
$Co(CO)_3$
C
$(OC)_3Co$
$Fe(CO)_3$
C
H
(**162**)
Co
$(CO)_3$
H
Cp^*Fe
C
C
$(CO)_2$
C
FeCp*
O:C
Co
$(OC)Co$
Co
Co
C=O
C
O
Co
$(CO)_2$
C
O
$(CO)_2$
$(CO)_2$
CO
(**161**)

SCHEME 34

the Ru_3 cluster has been opened and the Co—Co bond also cleaved, as the only product.[261] Its formation is thought to proceed by initial insertion of cobalt into the Ru_3 cluster, followed by Co—Co bond cleavage and coordination of the second C≡C triple bond to the enlarged cluster. The reaction contrasts with that of the analogous osmium complex **146** (*vide supra*). Attempts to form the Co_2Ru_3 cluster from $Ru_3(CO)_{10}(NCMe)_2$ and $Co_2(\mu\text{-}\eta^2\text{-}PhC_2C{\equiv}CPh)(CO)_6$ afforded only

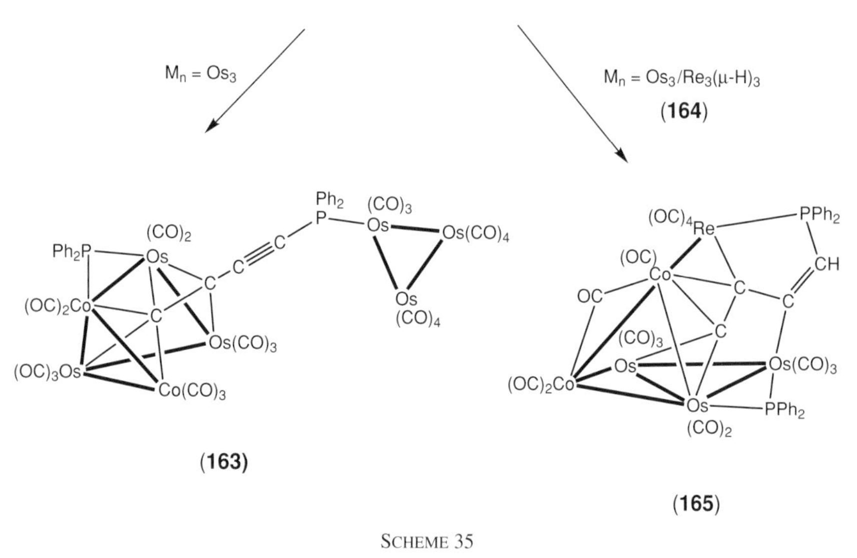

SCHEME 35

the disproportionation products $Ru_3(\mu_3\text{-}\eta^2\text{-}PhC_2C{\equiv}CPh)(\mu\text{-}CO)(CO)_9$ and $\{Co_2(CO)_6\}_2(\mu\text{-}\eta^2{:}\mu\text{-}\eta^2\text{-}PhC_2C_2Ph)$.

Reactions of $Pt(\eta^2\text{-}PhC_2C{\equiv}CPh)(PPh_3)_2$ with $Fe(CO)_5$ or $Ru_3(CO)_{12}$ give $MPt_2(\mu_3\text{-}\eta^1{:}\eta^1{:}\eta^2\text{-}PhC_2C{\equiv}CPh)(CO)_5(PPh_3)_2$ (**167;** M = Fe, Ru).[212] An isolobal interpretation of the structure is as an $M(CO)_3$ complex of the diplatinacyclobutadiene

$Pt_2(\mu\text{-}\eta^2\text{-}PhC_2C{\equiv}CPh)(CO)_2(PPh_3)_2$. Two structural polymorphs of the $FePt_2$ complex were obtained as benzene and $CHCl_3$ solvates, differing in parameters including the M—M distances, conformations of the PPh_3 ligands, and orientation of the $Fe(CO)_3$ groups. It would appear that there is either a weak or no Pt—Pt bond.

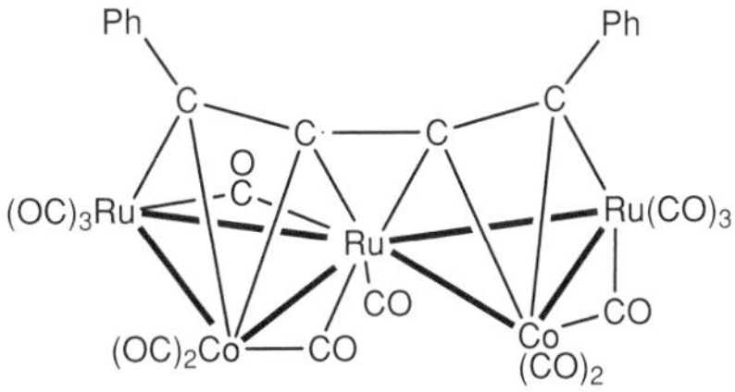

(**166**)

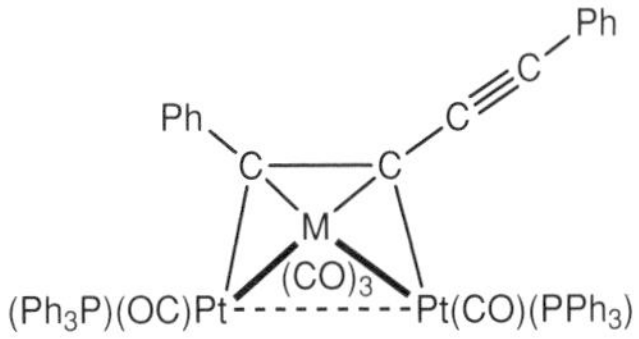

(**167**) M = Fe, Ru

C. *Formation of Diyne Complexes by Coupling of Alkynyl Moieties*

Independent investigations by Carty[262] and Mays[263] showed that two molecules of $Fe_2(\mu\text{-}PPh_2)(\mu\text{-}\eta^1{:}\eta^2\text{-}C_2Ph)(CO)_6$ couple upon heating in toluene (140°C, 2 hr) to give $Fe_4(\mu_4\text{-}\eta^1{:}\eta^1{:}\eta^2{:}\eta^2\text{-}PhC_4Ph)(\mu\text{-}PPh_2)_2(CO)_8$ (**168**) (Scheme 36). The two alkynyl C_α carbons are separated through the Fe_4 face by a long C—C bond (ca 1.6 Å). Formally, the two C_2R ligands (5-e each) or PhC_4Ph ligand (8-e) result in a 64-e or 62-e cluster. EH calculations suggest that the HOMO for a 64-e complex is 1.99 eV below the LUMO, favoring a structure with no through-cluster C ··· C bonding. However, the C ··· C overlap population (+0.66 e) indicates an attractive C ··· C interaction.[262] An alternative view considers that the distorted octahedral C_2Fe_4 cluster with face-capping CPh groups has 68 c.v.e.[263] The reaction of CO with **168** (PhMe, 100°C) gives $Fe_3\{\mu\text{-}Ph_2PC(CPh){=}C(CPh)PPh_2\}(CO)_8$ (**169**), formed by coupling of the two alkynyl groups with the two PPh_2 ligands and through the C_α atoms. The net process, given the source of the phosphido-alkynyl complex precursor, is the coupling of two phosphino-alkynes mediated by the iron cluster (Scheme 36).[262]

Thermolysis of $Ru_2(\mu\text{-}PPh_2)(\mu\text{-}\eta^1{:}\eta^2\text{-}C_2Bu^t)(CO)_6$ in refluxing toluene also results in coupling to give an inseparable 1/1 mixture of tetranuclear $Ru_4(\mu\text{-}PPh_2)_2(C_2Bu^t)_2(CO)_9$ (**170**) and $Ru_4(\mu_4\text{-}Bu^tCCCCBu^t)(\mu\text{-}PPh_2)_2(CO)_8$ (**171**) in which the μ_4-diyne ligand is derived from alkynyl coupling on the same cluster face (Scheme 37).[264] Extended heating times give only **171,** while pure **170** was obtained from the mixture by treatment with CO (PhMe, 80°C, 5 min). Cluster **171** eliminates a $Ru(CO)_n$ fragment upon reaction with CO to afford $Ru_3(\mu\text{-}PPh_2)_2(\mu_3\text{-}\eta^1{:}\eta^1{:}\eta^2\text{-}Bu^tC_2C{\equiv}CBu^t)(CO)_7$ (**172**). Complex **171** contains a flattened butterfly core upon which the new butadiyne ligand formed by head-to-tail coupling of the two C_2Bu^t moieties is attached by a series of η^2 interactions with all four Ru

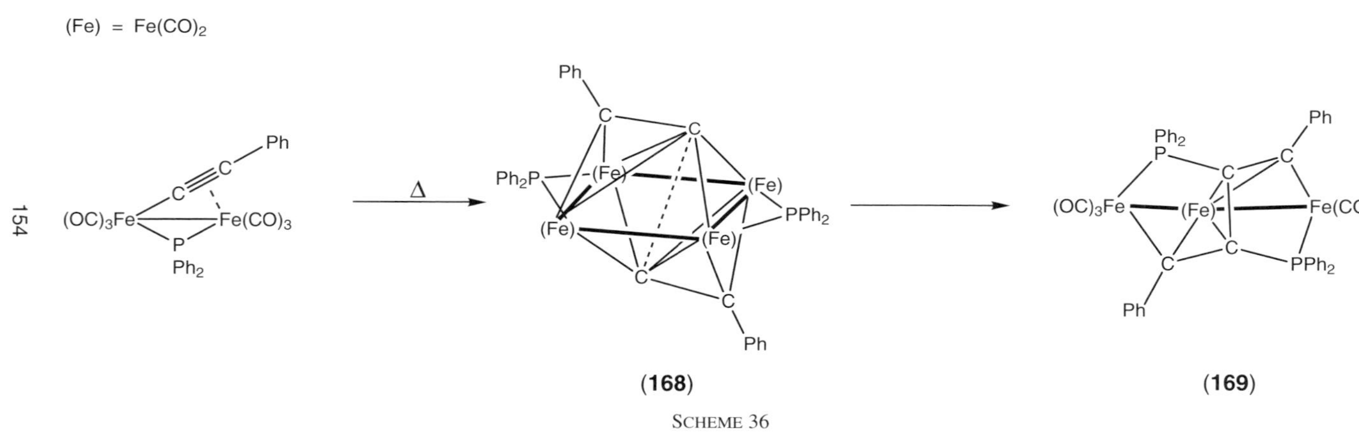

(Fe) = Fe(CO)2
(OC)3Fe
Fe(CO)3
C
C
Ph
P
Ph2
Δ
Ph
C
Ph2P
(Fe)
(Fe)
(Fe)
(Fe)
PPh2
C
C
C
Ph
(168)
Ph2
P
Ph
C
C
(OC)3Fe
(Fe)
Fe(CO)3
C
C
PPh2
Ph
(169)

SCHEME 36

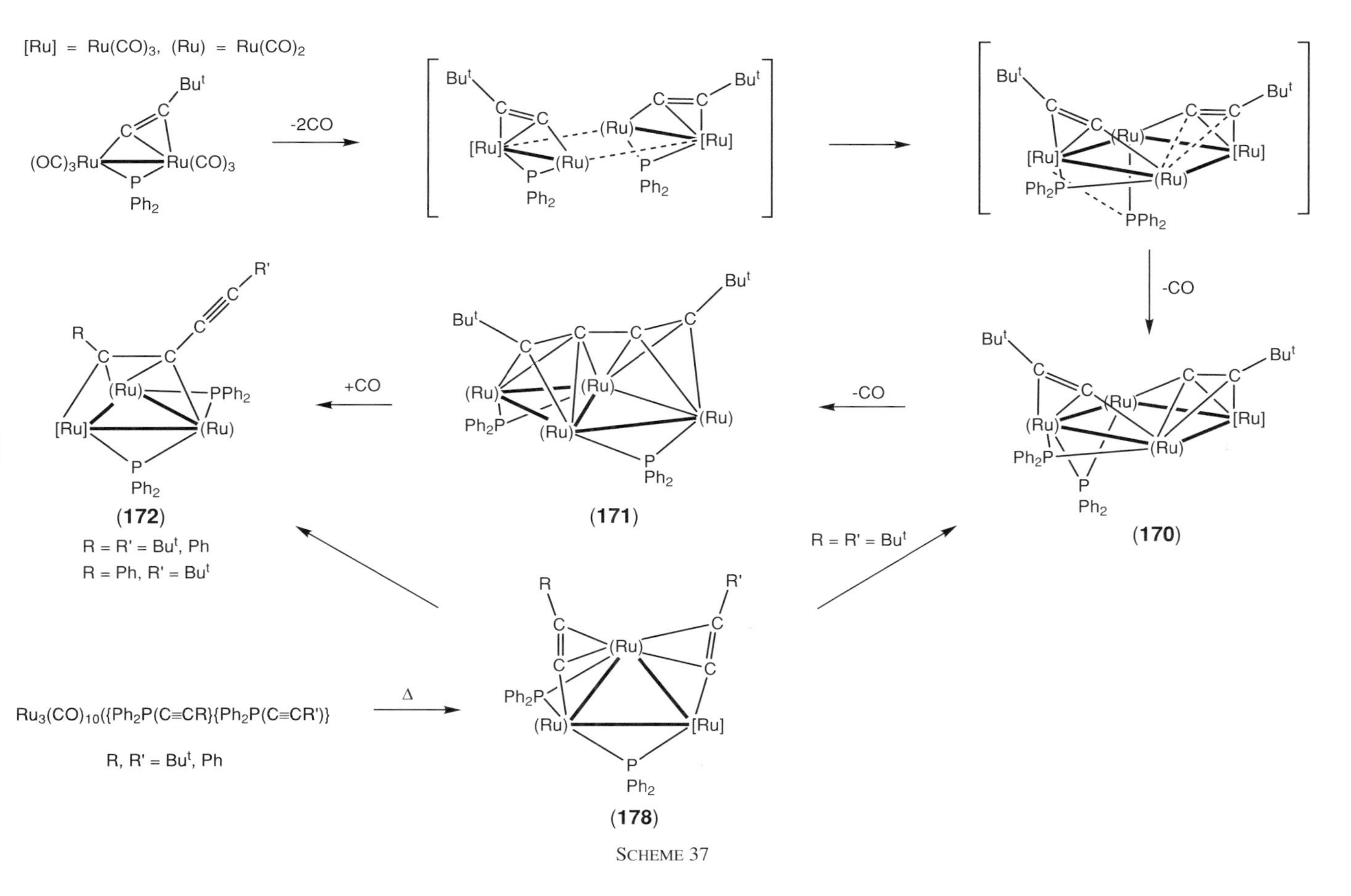

[Ru] = Ru(CO)3, (Ru) = Ru(CO)2
-2CO
-CO
-CO
+CO
(172)
R = R' = Bu^t, Ph
R = Ph, R' = Bu^t
(171)
(170)
R = R' = Bu^t
Δ
Ru3(CO)10({Ph2P(C≡CR}{Ph2P(C≡CR')}
R, R' = Bu^t, Ph
(178)

SCHEME 37

atoms. Although the solid-state structure shows one long Ru—C bond, in solution only single Bu^t and P resonances are found, indicating dynamic C_2 symmetry. Possible reaction sequences involve dimerization of the Ru_2 complex, aligning of the two C_2Bu^t ligands in **170,** which is followed by C—C bond formation to give **171.** Degradation of **171** could proceed by formation of a 64-e spiked triangular core by addition of CO and loss of "$Ru(CO)_4$" as $Ru_3(CO)_{12}$. Complex **172** shows the usual "windscreen-wiper" dynamic process which is frozen out at 203 K. In contrast, thermolysis (refluxing toluene) of $Ru_2(\mu\text{-}PPh_2)(\mu\text{-}\eta^1{:}\eta^2\text{-}C_2Ph)(CO)_6$ produced unsaturated clusters $Ru_4(\mu_4\text{-}PPh_2CCPhCCPh)(\mu\text{-}PPh_2)(CO)_9$ (**173;** 62-e) and $Ru_4(\mu_4\text{-}PPh_2CPhCCCPh)(\mu\text{-}PPh_2)(CO)_{10}$ (**174;** 64-e) (Scheme 38).[265] Carbonylation of **173** results in an unusual reversible addition of three CO molecules to give **175.**

Thermolysis (thf, 60°C, 1 hr) of $Ru_3(CO)_{11}\{P(C{\equiv}CBu^t)_3\}$ affords $Ru_3(\mu_3\text{-}\eta^2\text{-}C_2Bu^t)\{\mu\text{-}P(C{\equiv}CBu^t)_2\}(CO)_9$ (**176**) which on further heating (refluxing xylene, 18 hr) gives square-planar $Ru_4\{\mu_4\text{-}P(C{\equiv}CBu^t)\}(\mu_4\text{-}Bu^tC_2C_2Bu^t)(CO)_{10}$ (**177**), the diyne ligand being attached to all four Ru atoms and being formed by tail-to-tail coupling of two $C{\equiv}CBu^t$ groups (Scheme 39); this complex is structurally

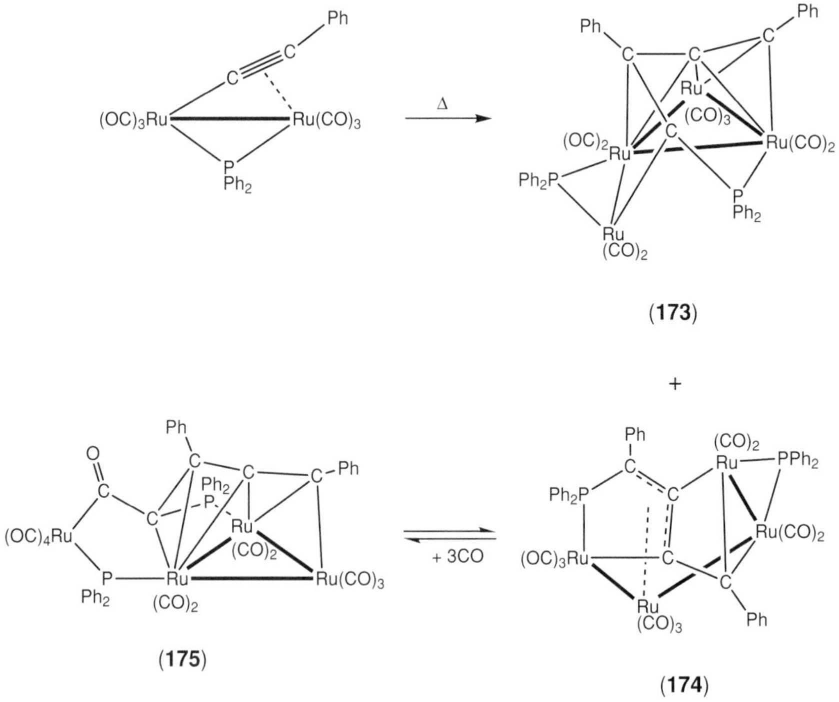

SCHEME 38

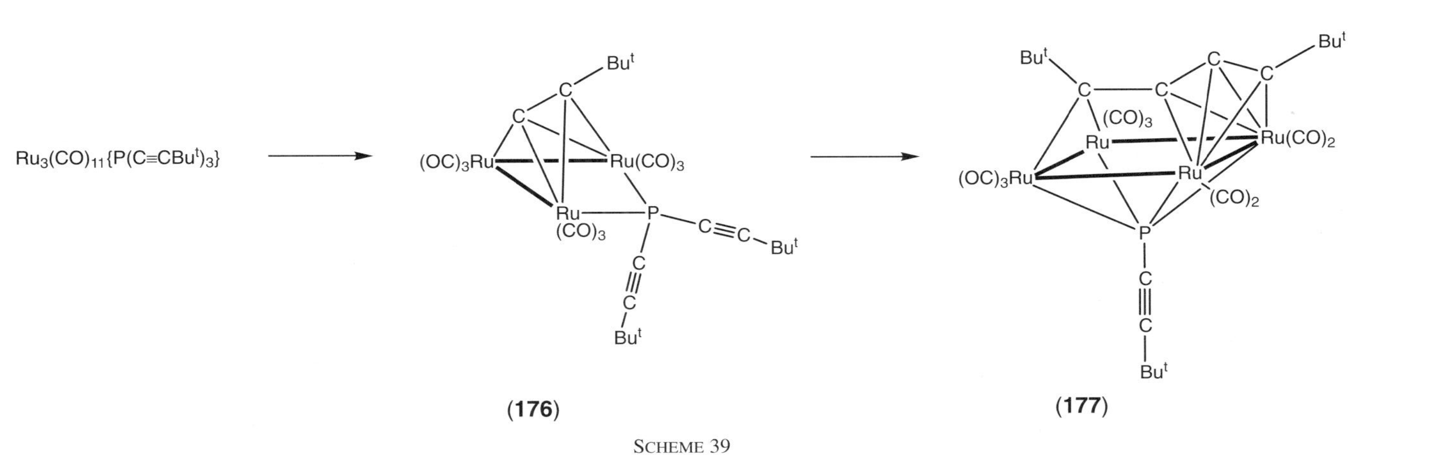

Ru3(CO)11{P(C≡CBut)3}
(OC)3Ru
Ru(CO)3
Ru
(CO)3
P
C≡C
But
C
C
(176)
(177)
(CO)2
Ru(CO)2

SCHEME 39

related to **128.**[245] The diyne acts as an 8-e donor, and has been described as a bis(alkylidyne)dicarbide. Coupling of the adjacent alkynyl ligands may be a result of their bent geometry. The related complexes $Ru_3(CO)_{10}(Ph_2PC{\equiv}CPh)(Ph_2PC{\equiv}CR)$ (R = Ph, Bu^t) transform under similar conditions to $Ru_3(\mu\text{-}PPh_2)_2(\mu\text{-}\eta^1,\eta^2\text{-}C_2Ph)(\mu\text{-}\eta^1,\eta^2\text{-}C_2R)(CO)_7$ (**178**) which convert smoothly at elevated temperatures (toluene, 110°C) to **172** (R = Ph, Bu^t) (Scheme 37).[266]

Several of the complexes obtained from reactions of $Ru_3(CO)_{12}$ with RC≡CSEt (R = Me, Ph) contain C_4 ligands formed by coupling of two alkynyl units after cleavage of the C—S bond, in addition to SEt groups which bridge Ru—Ru bonds. Coupling may occur head-to-tail, as in $Ru_3\{\mu_3\text{-SEtCCPhC(SEt)CPhCCPh}\}(\mu\text{-SEt})(CO)_7$ or $Ru_5(\mu_5\text{-CPhCCPhC})(\mu\text{-SEt})_2(CO)_{13}$, or head-to-head, as in the 90-e Ru_6 cluster **179,** in which a MeCCCCMe ligand spans all six Ru atoms of a rhombic Ru_6 raft.[267]

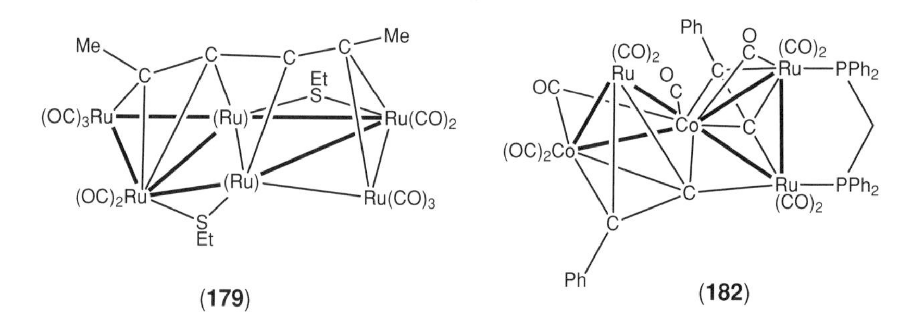

(179) **(182)**

Coupling of phenylethynyl groups is also found in reactions of $Fe_2(CO)_9$ with $Ir(C{\equiv}CPh)(CO)(PPh_3)_2$ to give $FeIr_2(\mu_3\text{-}\eta^2\text{-}PhC_2C{\equiv}CPh)(CO)_7(PPh_3)_2$[268] and with $Ni(C{\equiv}CPh)(PPh_3)Cp$, which gives $FeNi_2(\mu_3\text{-}\eta^2\text{-}PhC_2C{\equiv}CPh)(CO)_2Cp_2$.[269]

D. *C—C Bond Cleavage Reactions on Clusters*

The complexes $Os_3(\mu_3\text{-}\eta^2\text{-}RC_2C{\equiv}CR')(\mu\text{-}CO)(CO)_9$ (R = R′ = $SiMe_3$, Ph, Bu^t; R, R′ = $SiMe_3$, Ph both isomers) undergo thermal decarbonylation reactions with associated cleavage of the diyne ligand C—C bond yielding $Os_3(\mu\text{-}\eta^1\text{-}C_2R^1)(\mu_3\text{-}\eta^2\text{-}C_2R^2)(CO)_9$ (**180;** $R^1 = R^2 = SiMe_3$, Ph, Bu^t; R^1 = Ph, $R^2 = SiMe_3$) (Scheme 40).[254,255] One of the resulting acetylide ligands is a conventional $\mu_3\text{-}\eta^2$ system, while the other bridges the nonbonded Os · · · Os vector asymmetrically using only one carbon [the Os—C_β distance is 2.771(17) Å]. Nevertheless,

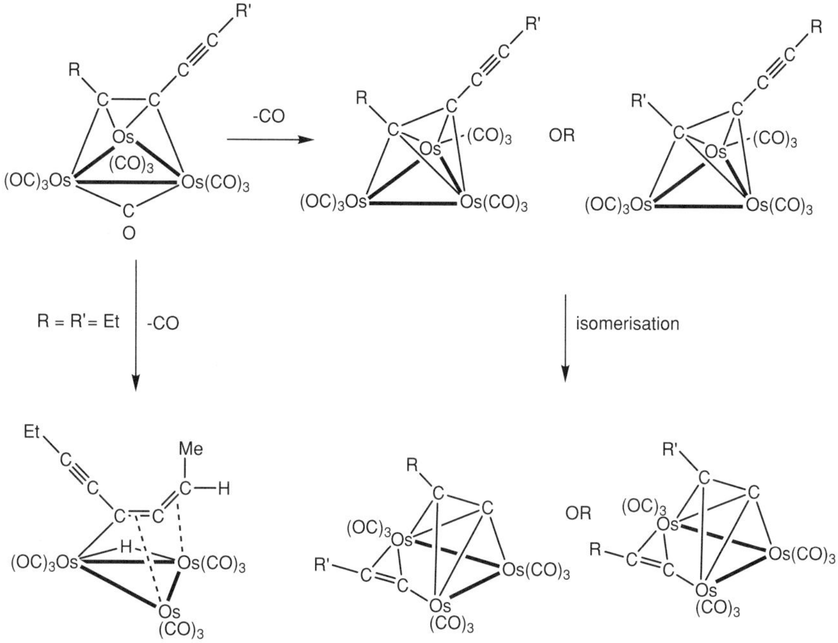

(**181**) (**180**)

SCHEME 40

bending of the CCR group toward one of the Os atoms suggests that this group should be considered as a 3-e donor. The nature of the unusual μ-η^1-C_2R group can be clarified by comparison with similar groups present in $AuOs_3(\mu$-η^2-$C_2Ph)(CO)_{10}$ (PMe_2Ph)[270] and $Ru_3(\mu$-$PPh_2)_2(\mu$-η^2-$C_2Bu^t)(\mu$-η^1-$C_2Bu^t)(CO)_5$ (PBu^tPh_2)[271] and it is considered to interact only via the π orbital on C_α. Both isomers derived from PhC≡CC≡CSiMe$_3$ give only $Os_3(\mu_3$-η^2-$C_2SiMe_3)(\mu$-η^1-$C_2Ph)(CO)_9$. Decarbonylation of the R = R′ = Et complex did not result in C—C bond cleavage, but rather in rupture of a methylene C—H bond and formation of $Os_3(\mu$-H$)(\mu_3$-EtC_2C=C=CHMe$)(CO)_9$ (**181**).[255] For R = Me, no hydride is present in the product which is formulated as $Os_3(C_4Me_2)(CO)_6$, and is possibly $Os_3(\mu_3$-CMe$)(\mu_3$-CC≡CMe$)(CO)_9$.

The reaction of PhC≡CC≡CPh with a mixture of $Ni(cod)_2$ and dppm afforded $Ni_3(\mu_3$-η^1-C≡CPh$)_2(\mu$-dppm$)_3$, which is unstable toward both air and moisture. In this complex the alkynyl fragment is bonded via only one carbon, with the CCPh fragment retaining its linearity and the ν(C≡C) frequency being found at 1933 cm^{-1}.[214]

Similar cleavage of the central C—C bond occurs in the reaction of $Ru_3(\mu_3$-η^2-$PhC_2C{\equiv}CPh)(\mu$-dppm$)(CO)_8$ with $Co_2(CO)_8$, which gives $Co_2Ru_3(\mu_4$-$C_2Ph)(\mu_3$-$C_2Ph)(\mu$-dppm$)(\mu$-$CO)(CO)_9$ (**182**).[272] Here, the C_2Ph fragments are attached to the two sides of the severely twisted Co_2Ru_3 bow-tie cluster. One of these spans the $CoRu_2$ face in the usual μ_3-η^1:η^2:η^2 mode, the C—C bond being perpendicular to a Co—Ru edge. The other C_2Ph group is similarly attached to the Co_2Ru face, which itself is twisted from the $CoRu_2$ face, so that the σ-bonded carbon bridges a Co—Ru vector. This reaction contrasts with that found for the dppm-free cluster **166.** Comparison of the two complexes shows that while bond lengths of analogous moieties are similar, the cluster configuration differs, the central atoms being Ru in **166** and Co in **182.** In **166,** an open Ru_3 array has each Ru—Ru bond bridged by Co, whereas in **182,** the Ru_3 unit is no longer preserved.

The reaction of $Os_3(CO)_{10}(NCMe)_2$ with $W(C{\equiv}CC{\equiv}CPh)(CO)_3Cp^*$ gives $Os_3W(\mu_4$-$C_2)(\mu_3$-$C_2Ph)(CO)_9Cp^*$ (**183;** Scheme 41) by cleavage of the diynyl C—C single bond. Subsequent reaction with O_2 gives $Os_3W(\mu_4$-$C_2)(\mu_3$-OCCPh)

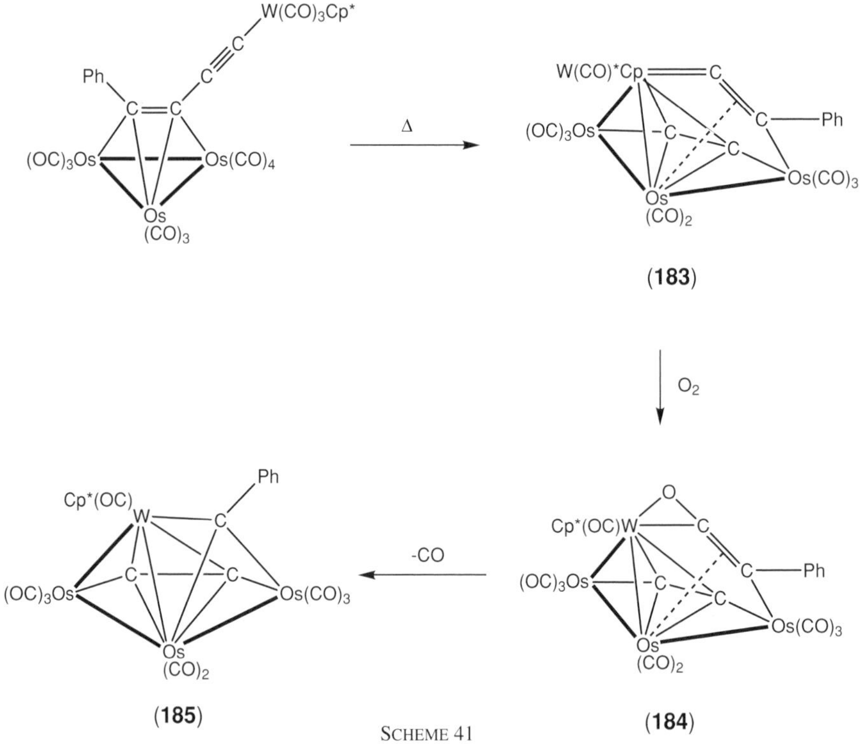

SCHEME 41

$(CO)_9Cp^*$ (**184**) which decarbonylates slowly in refluxing toluene to $Os_3W(\mu_4\text{-}C_2)(\mu_3\text{-}CPh)(CO)_9Cp^*$ (**185**). This sequence corresponds to oxidative decarbonation of alkynyl → ketenyl → alkylidyne.[106]

E. *Complexes Derived from Poly-ynes*

The reaction of the air-stable crystalline 1,6-bis(trimethylsilyl)hexa-1,3,5-triyne with $Ru(CO)_5$ afforded $Ru_3(\mu_3\text{-}\eta^2\text{-}Me_3SiC{\equiv}CC_2C{\equiv}CSiMe_3)(\mu\text{-}CO)(CO)_9$ (**186,** M = Ru; Scheme 42) (6%) and $Ru_4(\mu_4\text{-}\eta^2\text{-}Me_3SiC{\equiv}CC_2C{\equiv}CSiMe_3)(CO)_{12}$ (**187**) (36%).[57] The former was obtained in greatly improved yield (60%) from the reaction of the triyne with $Ru_3(CO)_{10}(NCMe)_2$, subsequent treatment with $Ru(CO)_5$ giving **187** in 43% yield. With $Ru_3(CO)_{12}$ the triyne gives **187** and $Ru_2\{\mu\text{-}2\eta^1{:}\eta^4\text{-}C_4(C{\equiv}CSiMe_3)_4\}(CO)_6$ (**188**).[55,56] In all cases the products are formed by exclusive reaction of the central C≡C triple bond. This may result from hyperconjugation of the Si d orbitals with the π-system, leading to deactivation of the "outer" C≡C triple bonds. However, reaction of **187** with $Co_2(CO)_8$ gives $Ru_4\{\mu_4\text{-}Me_3SiC_2C{\equiv}CC_2[Co_2(CO)_6]\}(CO)_{12}$ (**189**), resulting from displacement of the Ru_4 cluster to a terminal C≡C triple bond, while the $Co_2(CO)_6$ group is attached to the other end; NMR data suggest that there is some degree of electronic communication between the Co_2 and Ru_4 centers.[56] Further reactions of **187** with $Me_3SiC{\equiv}CC{\equiv}CR$ (R = $SiMe_3$, $C{\equiv}CSiMe_3$) result in insertion of an alkyne C≡C triple bond into the Ru—Ru-hinge yielding $Ru_4(\mu_4\text{-}SiMe_3C{\equiv}CC_2C{\equiv}CSiMe_3)(\mu_4\text{-}RC_2C{\equiv}CSiMe_3)(\mu\text{-}CO)_3(CO)_8$ (**190**).[273] With $Os_3(CO)_{10}(NCMe)_2$, $Me_3SiC{\equiv}CC{\equiv}CC{\equiv}CSiMe_3$ gives **186** (M = Os), which with $Ru_3(CO)_{12}$ gives $Os_3Ru(\mu_4\text{-}Me_3SiC{\equiv}CC_2C{\equiv}CSiMe_3)(CO)_{12}$ by formal substitution of the μ-CO ligand by an isolobal $Ru(CO)_3$ group.[56]

The reaction between $Me_3Si(C{\equiv}C)_3SiMe_3$ and **114** proceeds in a manner closely related to the reactions of the same cluster reagent with 1,3-diynes (*vide supra*) (Scheme 43).[55] Allowing an equimolar amount of each reagent to react in refluxing pentane afforded $Ru_4\{\mu\text{-}PhPC(C{\equiv}CSiMe_3)C(C{\equiv}CSiMe_3)\}(\mu\text{-}CO)_2(CO)_{10}$ (**191**) which rearranged smoothly to $Ru_4\{\mu_4\text{-}\eta^1{:}\eta^1{:}\eta^2{:}\eta^2\text{-}Me_3SiC{\equiv}CC_2C{\equiv}CSiMe_3)(\mu\text{-}CO)(CO)_{10}$ (**192**). Attempts to coordinate one of the pendant $C{\equiv}CSiMe_3$ moieties to the Ru_4 cluster face by pyrolysis in toluene gave instead **193.**[273] Reaction of either **114** or **192** with an excess of $Me_3Si(C{\equiv}C)_3SiMe_3$ gave $Ru_4\{\mu_4\text{-}2\eta^1{:}\eta^2{:}\eta^4{:}\eta^4\text{-}C_4(C{\equiv}CSiMe_3)_2(C_2SiMe_3)C_4(C{\equiv}CSiMe_3)_3\}(\mu_3\text{-}PPh)(CO)_9$ (**194**), containing an unusual C_{18} hydrocarbon ligand incorporated into the open cluster with η^2-alkyne, ruthenacyclopentadiene and cyclobutadiene fragments.[55]

The major product (57%) from the reaction of $Co_2(CO)_8$ and $Pr^i_3Si(C{\equiv}C)_3SiPr^i_3$ in refluxing hexane is $Co_4(\mu_4\text{-}\eta^2\text{-}Pr^i_3SiC{\equiv}CC_2C{\equiv}CSiPr^i_3)(\mu\text{-}CO)_2(CO)_8$.[186]

Pentanuclear $NiRu_4(\mu_4\text{-}\eta^1{:}\eta^1{:}\eta^2{:}\eta^4\text{-}Bu^tC{\equiv}CC_4C{\equiv}CBu^t)(\mu\text{-}PPh_2)_2(CO)_{12}$ (**195**) is obtained from **125** (R = Bu^t) and $Ni(cod)_2$ or $Ni(CO)_4$ by formal head-to-head

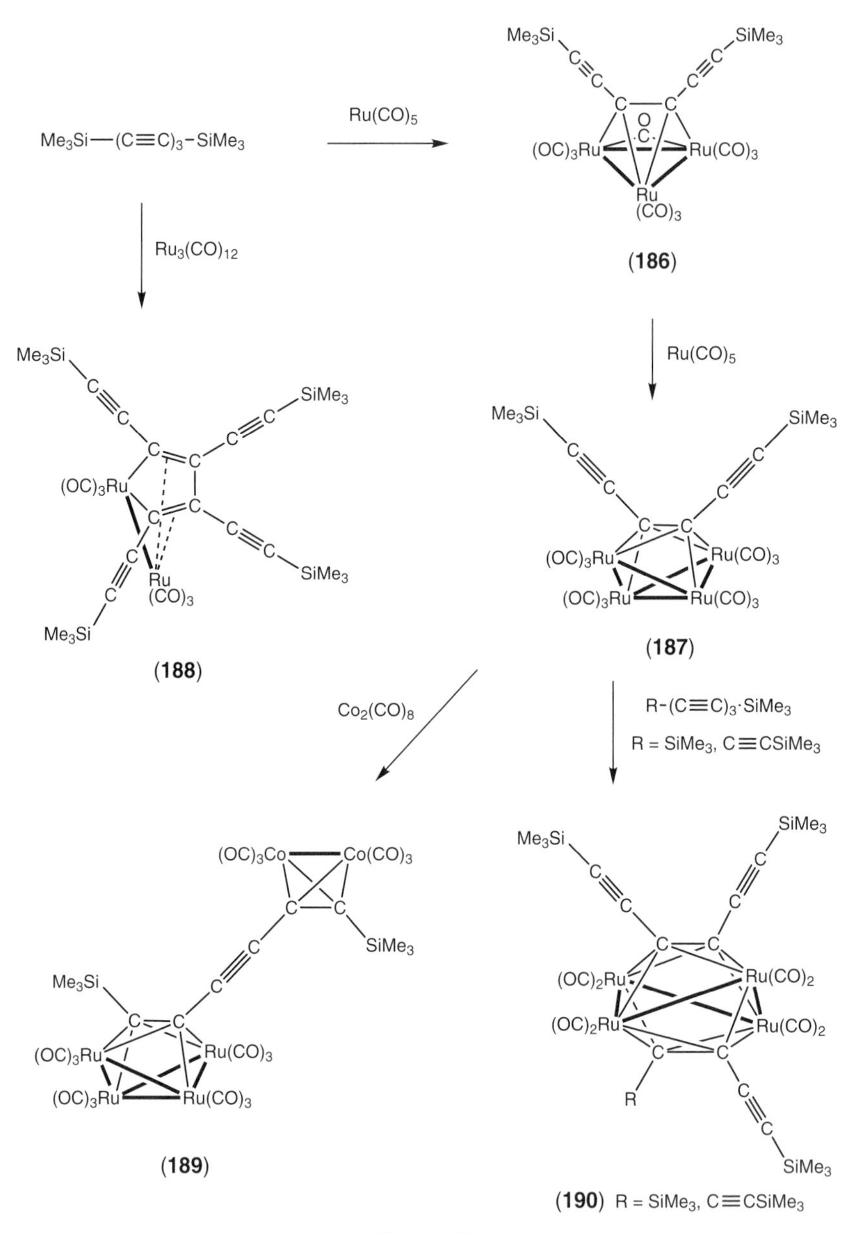
Me3Si—(C≡C)3–SiMe3
Ru(CO)5
Ru3(CO)12
(186)
Ru(CO)5
(188)
(187)
Co2(CO)8
R-(C≡C)3·SiMe3
R = SiMe3, C≡CSiMe3
(189)
(190) R = SiMe3, C≡CSiMe3

SCHEME 42

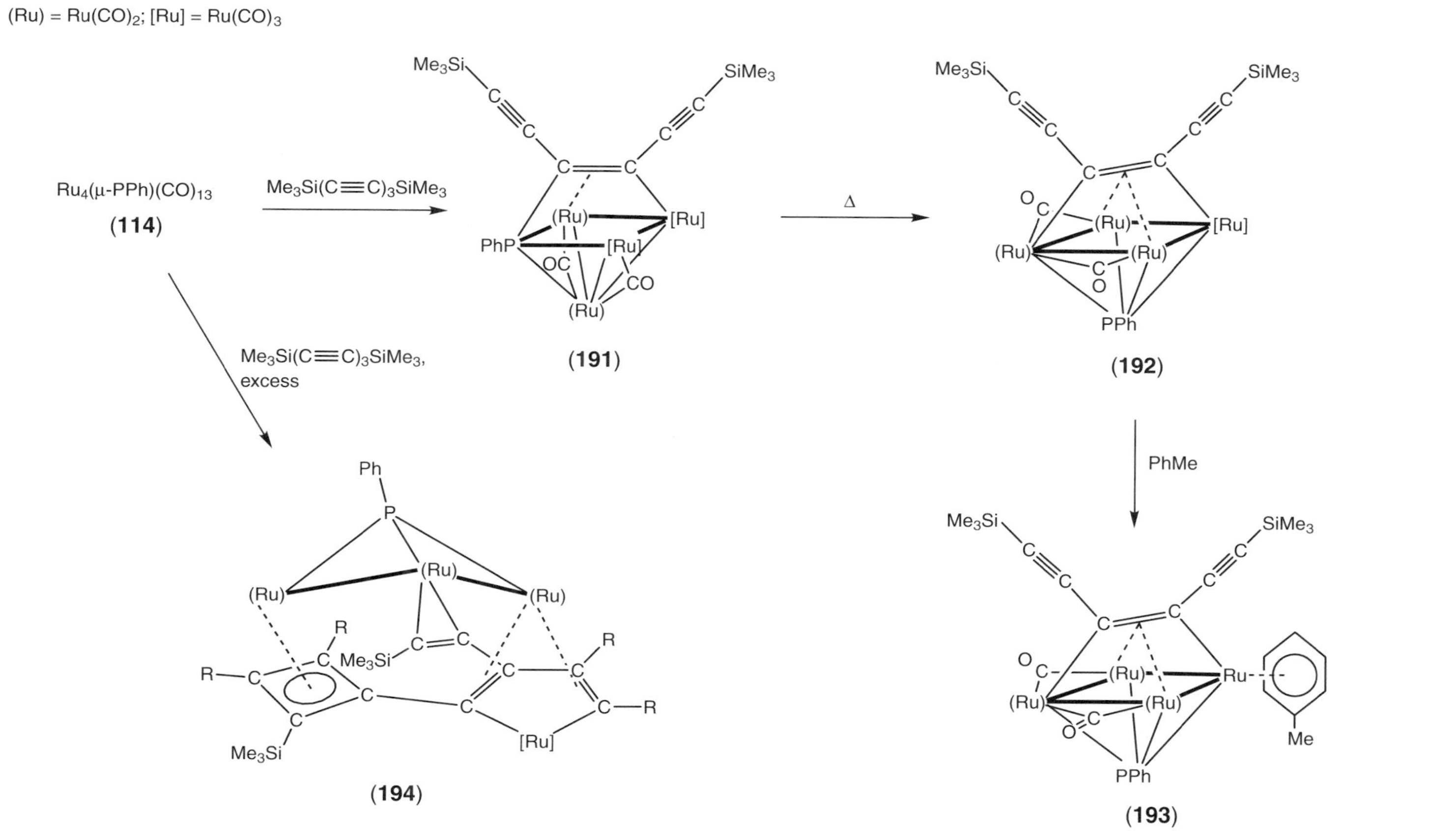

(Ru) = $Ru(CO)_2$; [Ru] = $Ru(CO)_3$

$Ru_4(\mu\text{-}PPh)(CO)_{13}$ (**114**)

$Me_3Si(C{\equiv}C)_3SiMe_3$ → (**191**)

Δ → (**192**)

PhMe → (**193**)

$Me_3Si(C{\equiv}C)_3SiMe_3$, excess → (**194**)

SCHEME 43

coupling of two molecules of **125** about the Ni atom. In this reaction, head-to-tail coupling of two C_4Bu^t groups has occurred to give a C_8 chain, the central four carbons of which are attached to all four Ru atoms.[54]

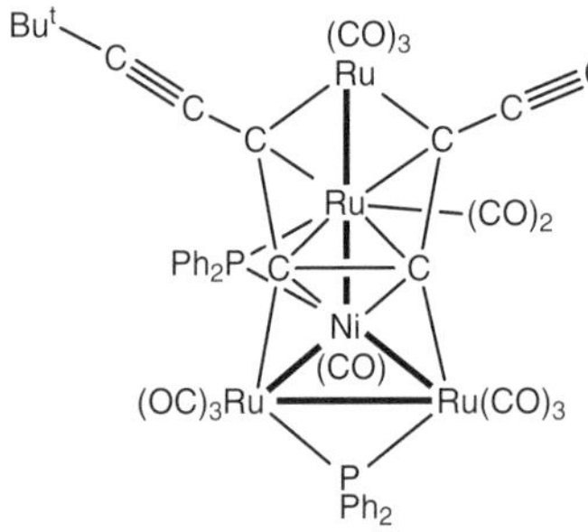

(**195**)

The heterometallic clusters $FeM_2(\mu_3\text{-}\eta^2\text{-}Bu^tC{\equiv}CC_2C{\equiv}CBu^t)(CO)_7Cp_2$ and $MoMRh(\mu_3\text{-}\eta^2\text{-}Bu^tC{\equiv}CC_2C{\equiv}CBu^t)(CO)_4Cp_2(\eta^5\text{-}C_9H_7)$ (M = Mo, W) were obtained from reactions of $M({\equiv}CC{\equiv}CBu^t)(CO)_2Cp$ (M = Mo, W) with $Fe_2(CO)_9$ or $Rh(CO)_2(\eta^5\text{-}C_9H_7)$.[274] The triyne ligand is formed by coupling of two alkynylmethylidyne groups. In the case of the molybdenum-containing complexes, the triyne ligand is fluxional, and oscillates between two of the three possible M—M vectors.[274]

V

σ,π-DIYNYL COMPLEXES

The C≡C triple bond of σ-diynyl complexes can act as a good two- or four-electron π-ligand to one or more metal centers. In these complexes, juxtaposition of σ- and π-bonded metal atoms often results in subsequent aggregation of the metal centers to form clusters in which the alkynyl groups interact with several or all of the metal atoms. However, in the case of the longer chain poly-ynyl systems the propensity for metal centers (particularly bi- or trinuclear) to coordinate in the close proximity required for cluster formation has to be balanced against the possibility of relieving steric congestion by coordination to more remote C≡C fragments. Aspects of this chemistry including a discussion of the range of known bridging modes, the synthesis and properties of bis-alkynyl tweezer complexes, and their relationship to the coupling reactions found with Group 4 complexes of this type (see Section VI) have been reviewed, although there are few references to diynyl systems.[275] Tables III and IV list known examples with some IR, NMR, and structural data.

A. *Synthetic Methods*

1. *Reactions of σ-Diynyl-Metal Complexes with Sources of ML_n Fragments*

The reactions of titanocene derivatives $TiX(C{\equiv}CC{\equiv}CR)Cp'_2$ (X = Cl, $C{\equiv}CC{\equiv}CR$; R = $SiMe_3$, Et) or of *cis*-$Pt(C{\equiv}CC{\equiv}CR)_2(PR'_3)_2$ with mononuclear metal complexes have given numerous products in which the diynyl ligand(s) are chelated to a low-valent metal via the internal C≡C fragment(s). These *cis*–bis(diynyl) complexes are often referred to as molecular tweezers.

Binuclear complexes including $\{M(CO)_2Cp\}_2$ (M = Mo, W), $Co_2(CO)_6(L)_2$ (L = CO, $L_2 = \mu$-dppm) and $\{Ni(\mu\text{-}CO)Cp\}_2$ and trimetallic reagents, such as $M_3(CO)_{10}L_2$ [M = Ru, $L_2 = (CO)_2$, $(NCMe)_2$, dppm; M = Os, $L_2 = (NCMe)_2$] react readily with diynyl complexes. However, the proximity of a multimetal system to the σ-bonded metal may result in further condensation to afford metal clusters of moderate nuclearity (see Section IV).

2. *Metallation of Pendant C≡CR (R = H, $SiMe_3$) Groups*

The pendant C≡CR moiety of the cobalt complexes $Co_2\{\mu\text{-}\eta^2\text{-}RC_2(C{\equiv}C)_nR'\}(\mu\text{-dppm})(CO)_4$ ($n = 1,2$; R, R′ = $SiMe_3$, H) can be metallated in much the same way as more conventional organic alkynes to afford the σ/π complexes $Co_2\{\mu\text{-}\eta^2\text{-}RC_2(C{\equiv}C)_n[ML_n]\}(\mu\text{-dppm})(CO)_4$.

3. *P—C Bond Cleavage Reactions*

The P—C bond in $PPh_2(C{\equiv}CC{\equiv}CR)$ is prone to oxidative addition across metal—metal bonds, particularly in the case of Group 8 metal systems. Complexes featuring $\mu_2\text{-}\eta^1\text{:}\eta^2$ and $\mu_3\text{-}\eta^1\text{:}\eta^2$-diynyl ligands have been isolated in this fashion.

B. *Survey of Complexes*

1. *Complexes Derived from $Ti(C{\equiv}CC{\equiv}CR)_2Cp^R_2$*

Reaction of $TiCl_2(\eta\text{-}C_5HMe_4)_2$ with $LiC{\equiv}CC{\equiv}CSiMe_3$ gave the paramagnetic adduct $[Li(thf)_2][Ti(C_2C{\equiv}CSiMe_3)_2(\eta\text{-}C_5HMe_4)_2]$ (**196**), in which the lithium ion is associated with the inner C≡C triple bonds. The two thf ligands complete pseudo-tetrahedral coordination for the lithium. The lithium ion has little effect on the geometry of the bis(diynyl)titanium fragment and the interaction is essentially ionic.

TABLE III

SOME σ/π-DIYNYL COMPLEXES $\{[L_nM](C{\equiv}CC{\equiv}CR)_n\}$ $\{M'L'_m\}$

Complex		^{13}C NMR	Reference
$\{L_nM\}(C{\equiv}CC{\equiv}CR)_n$($\sigma$-bonded)	$\{M'L'_m\}$ (π-bonded)	(Assignments given where possible)	
$[Ti(C_2C{\equiv}CSiMe_3)_2(\eta\text{-}C_5HMe_4)_2]^-$	$[Li(thf)_2]^+$	N/R	276
$Ti(C_2C{\equiv}CEt)_2Cp^{Si}{}_2$	Ni(CO)	69.5 (≡*C*Et), 91.3 (*C*≡CEt), 102.8 (TiC≡*C*), 178.0 (Ti*C*≡)	99
$Ti(C_2C{\equiv}CSiMe_3)_2Cp^{Si}{}_2$	Ni(CO)	93.6 (≡*C*SiMe$_3$), 95.2 (*C*≡CSiMe$_3$), 102.4 (TiC≡*C*), 181.2 (Ti*C*≡)	99
$Ti(C_2C{\equiv}CEt)_2Cp^{Si}{}_2$	$Pd(PPh_3)$	69.0 (≡*C*Et), 92.6 (*C*≡CEt), 102.0 (TiC≡*C*)	99
$Ti(C_2C{\equiv}CEt)_2Cp^{Si}{}_2$	$Pt(PPh_3)$	n.d.	99
$TiCl(C_2C{\equiv}CEt)Cp^{Si}{}_2$	CuBr	65.3 (≡*C*Et), 108.8 (*C*≡CEt), 122.9 (TiC≡*C*), 129.8 (Ti*C*≡)	98
$Ti(C_2C{\equiv}CEt)_2Cp^{Si}{}_2$	CuBr	65.4 (≡*C*Et), 101.3 (*C*≡CEt), 126.4 (TiC≡*C*), 134.9 (Ti*C*≡)	98
$Ti(C_2C{\equiv}CSiMe_3)_2Cp^{Si}{}_2$	CuBr	88.2 (≡*C*SiMe$_3$), 105.7 (*C*≡CSiMe$_3$), 125.2 (TiC≡*C*), 138.3 (Ti*C*≡)	98
$[Ti(C_2C{\equiv}CFc)_2Cp_2]_2$	Ag^+	n.d.[a]	277
$[Ti(C_2C{\equiv}CFc)_2Cp^{Si}{}_2]_2$	Ag^+	n.d.	277
trans-$Pt(C{\equiv}CC_2SiMe_3)_2(PEt_3)_2$	$Co_2(CO)_6$	128.8 (Pt*C*≡), 105.3 (PtC≡*C*), 91.4, 78.3 (*C*$_2$Co$_2$)	131
trans-$Pt(C{\equiv}CC_2SiMe_3)_2(PBu_3)_2$	$Co_2(CO)_6$	129.5 (Pt*C*≡), 104.2 (PtC≡*C*), 91.5, 78.5 (*C*$_2$Co$_2$)	131
trans-$Pt(C{\equiv}CC_2H)_2(PEt_3)_2$	$Co_2(CO)_6$	n.d.	131
trans-$Pt(C{\equiv}CC_2H)_2(PBu_3)_2$	$Co_2(CO)_6$	n.d.	131
$W(C{\equiv}CC_2H)(CO)_3Cp$	$Co_2(CO)_6$	123.13 (W*C*≡), 94.34 (WC≡*C*), 73.84, 71.56 (*C*$_2$Co$_2$)	109, 197
$Mo(C{\equiv}CC_2H)(CO)_3Cp$	$Co_2(CO)_6$	123.43 (Mo*C*≡), 109.52 (MoC≡*C*), 74.07, 71.35 (*C*$_2$Co$_2$)	197
$Fe(C{\equiv}CC_2H)(CO)_2Cp^*$	$Co_2(CO)_6$	132.2/106.7, 78.9 (J_{CH} 32), 70.2 (J_{CH} 224)	259
$Fe(C{\equiv}CC_2H)(CO)_2Cp^*$	$Mo_2(CO)_4Cp_2$	110.2/109.8, 68.3 (J_{CH} 5), 62.1 (J_{CH} 211)	259
$Fe(C{\equiv}CC_2H)(CO)_2Cp^*$	$CoMo(CO)_5Cp$	120.9/108.0, 77.9 (J_{CH} 9), 70.6 (J_{CH} 217)	259
$Fe(C_2C_2H)(CO)_2Cp^*$	$\{Co_2(CO)_6\}_2$	129.5/115.4/90.0, 81.3 (J_{CH} 219)	259
$Ru(C{\equiv}CC_2Ph)(PPh_3)_2Cp$	$Co_2(CO)_6$	n.d.	81
$Ru(C{\equiv}CC_2SiMe_3)(PPh_3)_2Cp$	$Co_2(\mu\text{-dppm})(CO)_4$	n.d.	278
$Ru(C{\equiv}CC_2H)(PPh_3)_2Cp$	$Co_2(\mu\text{-dppm})(CO)_4$	n.d.	81, 278
$W(C{\equiv}CC_2SiMe_3)(CO)_3Cp$	$Co_2(\mu\text{-dppm})(CO)_4$	108.97, 95.77, 86.80, 84.43	172, 278

TABLE III (*continued*)

Complex		^{13}C NMR	Reference
$V(C{\equiv}CC_2H)(CO)_3Cp$	$Co_2(\mu\text{-dppm})(CO)_4$	Not observed	172
$\text{Au}(C{\equiv}CC_2H)(PPh_3)$	$Co_2(\mu\text{-dppm})(CO)_4$	78.74, 70.63; Other resonances n.d.	172
$\text{Ni}(C{\equiv}CC_2SiMe_3)Cp$	$Co_2(\mu\text{-dppm})(CO)_4$	118.16, 93.14, 87.81, 84.54	278
$\text{Re}_2(\mu\text{-H})\{\mu\text{-}\eta^1,\eta^2\text{:}\mu\text{-}\eta^2\text{-}C_2C_2SiMe_3\}(CO)_8$	$Co_2(\mu\text{-dppm})(CO)_4$	96.86, 94.37, 86.77, 80.01	172
$\text{Ru}_3(\mu\text{-H})\{\mu\text{-}\eta^1,\eta^2\text{:}\mu\text{-}\eta^2\text{-}C_2C_2SiMe_3\}(CO)_9$	$Co_2(\mu\text{-dppm})(CO)_4$	165.10, 92.78; Other resonances n.d.	172
rans-$\{Rh(C{\equiv}CC_2Ph)(CO)(PPr^i_3)_2$	$RhCl(PPr^i_2)_2$	139.32, 106.29, 82.28, 65.51	71
$\text{Ru}_2(\mu\text{-}\eta^1\text{:}\eta^2\text{-}C_2C{\equiv}CBu^t)(\mu\text{-}PPh_2)(CO)_6$		100.2 (C_α), 88.1 (C_δ), 74.5 (C_β), 66.8 (C_γ)	243
$\text{Ru}_2(\mu\text{-}\eta^1\text{:}\eta^2\text{-}C_2C{\equiv}CPh)(\mu\text{-}PPh_2)(CO)_6$		109.0 (C_α), 78.2 (C_δ), 77.4 (C_γ), 74.0 (C_β)	243
$\text{Ru}_2(\mu\text{-}\eta^1\text{:}\eta^2\text{-}C_2C{\equiv}CSiMe_3)(\mu\text{-}PPh_2)(CO)_6$		106.4 (C_α), 91.0 (C_β), 85.7 (C_δ), 74.0 (C_γ)	243
$\text{Ru}_2(\mu\text{-}\eta^1\text{:}\eta^2\text{-}C_2C_2Bu^t)[Co_2(CO)_6])(\mu\text{-}PPh_2)(CO)_6$		115.7 (C_δ), 113.1 (C_α), 93.6 (C_β), 74.6 (C_γ)	243
$\text{Ru}_2Pt(\mu_3\text{-}\eta^1,\eta^1,\eta^1\text{-}C_2C{\equiv}CBu^t)(\mu\text{-}PPh_2)(PPh_3)(CO)_7$		n.d.	54
$\text{Ru}_2Pt(\mu_3\text{-}\eta^1,\eta^1,\eta^1\text{-}C{=}CC{\equiv}CPh)(\mu\text{-}PPh_2)(PPh_3)(CO)_7$		n.d.	54
$\text{Ru}_2Pt(\mu_3\text{-}\eta^1,\eta^1,\eta^1\text{-}C_2C{\equiv}CBu^t)(\mu\text{-}PPh_2)(CO)_6(dppb)$		n.d.	54
$\text{Ru}_2(\mu\text{-}\eta^1\text{:}\eta^2\text{-}C_2C{\equiv}CBu^t)(\mu\text{-}PPh_2)(CO)_5(PPh_3)$		n.d.	54
$\text{Ru}_2(\mu\text{-}\eta^1\text{:}\eta^2\text{-}C_2C{\equiv}CBu^t)(\mu\text{-}PPh_2)(CO)_4(PPh_3)_2$		n.d.	54
$\text{Ru}_2\{\mu\text{-}\eta^1,\eta^2\text{-}C_2C_2PPh_2[Ru_3(CO)_{11}]\}(\mu\text{-}PPh_2)(CO)_6$	$Co_2(CO)_6$	n.d.	260
Some σ/π-poly-ynyl complexes			
$\text{Ru}(C{\equiv}CC{\equiv}CC_2SiMe_3)(PPh_3)_2Cp$	$Co_2(\mu\text{-dppm})(CO)_4$	90.09 (Ru*C*≡C), 100.08 (RuC≡*C*), 89.82, 64.69 (C≡C), 84.68, 89.72 (C_2Co_2)	194
$\text{Ru}(C{\equiv}CC_2C{\equiv}CSiMe_3)(PPh_3)_2Cp$	$Co_2(\mu\text{-dppm})(CO)_4$	111.08 (Ru*C*≡C), 114.78 (RuC≡*C*), 95.78, 99.75 (C≡C); Other resonances n.d.	194

[a]n.d. = not reported or not determined.

In contrast to the bis(alkynyl) titanocene complex $[Mg(thf)Cl][Ti(C{\equiv}CSiMe_3)_2(\eta^5\text{-}C_5H_4Me)_2]$, which displays a single pair of high intensity absorption bands (380/400 nm), the UV/vis spectrum of the bis(diynyl) complex $[Li(thf)_2][Ti(C_2C{\equiv}CSiMe_3)_2(\eta\text{-}C_5HMe_4)_2]$ exhibits two pairs of bands at 363/392 and 380/400 nm. These transitions are tentatively attributed to d→π^* transitions and are taken as an indication of a lower energy LUMO in the case of the species containing conjugated diynyl ligands.[276]

TABLE IV

SOME STRUCTURAL DATA FOR σ/π-DIYNYL COMPLEXES

Complex	Bond Distances (Å)			Bond Angles (°)			Reference
	M—C (σ)	C≡C (free; complexed)	C—C	M—C—C	C—C—C	C—C—R	
$[Ti\{(C_2C{\equiv}CSiMe_3)_2[Li(thf)_2]\}(\eta\text{-}C_5HMe_4)_2]$	2.129(5)	1.213(8); 1.200(10)	1.395(9)	177.5(5)	178.8(6) 177.2(7)	177.2(6)	276
$Ti\{(C_2C{\equiv}CEt)_2[Pd(PPh_3)]Cp^{Si}{}_2$	2.057(6), 2.052(8)	1.238(9), 1.24(1); 1.15(1), 1.17(1)	1.38(1), 1.40(1)	163.2(7), 163.4(5)	159(1), 153.9(6) (C≡C—C) 176(1), 179.1(6) (C—C≡C)	164(1), 177.2(7)	99
$[Ti\{(C_2C{\equiv}CFc)_2[Ag]\}Cp^{Si}{}_2]PF_6$	2.10(4), 2.07(4); 2.12(5), 2.14(4)	n.d.[a]					277
$Co_2(\mu\text{-dppm})\{\mu\text{-}\eta^2\text{-}Me_3SiC_2C{\equiv}C[Ru(PPh_3)_2Cp]\}(CO)_4$	2.012(2)	1.223(3); 1.368(3)	1.410(3)	166.61(18)	177.6(2), 144.7(2)	131.45(17)	278
$Co_1(\mu\text{-dppm})(\mu\text{-}\eta^2\text{-}HC_2C{\equiv}C[Ru(PPh_3)_2Cp])(CO)_4$	2.01(2)	1.19(3); 1.32(3)	1.43(3)	168(2)	177(2), 145(2)		81
$Co_2\{\mu\text{-}\eta^2\text{-}PhC_2C{\equiv}C[Ru(PPh_3)_2Cp]\}(CO)_6$	1.985(4)	1.227(5); 1.354(5)	1.394(5)	171.3(2)	176.7(2), 140.0(3)	135.1(3)	81
$Co_2(\mu\text{-dppm})\{\mu\text{-}\eta^2\text{-}Me_3SiC_2C{\equiv}C[W(CO)_3Cp]\}(CO)_4$	2.134(2)	1.217(3); 1.362(3)	1.412(3)	169.1(2)	175.2(2), 140.8(2)	142.03(19)	172, 278
$Co_2(\mu\text{-dppm})\{\mu\text{-}\eta^2\text{-}Me_3SiC_2C{\equiv}C[Ni(PPh_3)Cp]\}(CO)_4$	1.8401(16)	1.220(2); 1.356(2)	1.405(2)	175.79(14)	175.07(16), 143.55(15)	145.84(12)	278
$Re_2(\mu\text{-H})\{\mu\text{-}\eta^1,\eta^2{:}\mu\text{-}\eta^2\text{-}C_2C_2SiMe_3[Co_2(\mu\text{-dppm})(CO)_4]\}(CO)_8$	2.35(1)	1.24(2); 1.36(2)	1.43(2)	168(1)	161(1) 144(1)		172
$Ru_3(\mu\text{-H})\{\mu\text{-}\eta^1,\eta^2{:}\mu\text{-}\eta^2\text{-}C_2C_2SiMe_3[Co_2(\mu\text{-dppm})(CO)_4]\}(CO)_9$	1.97(3), 1.98(2)[b]	1.26(4), 1.29(3); 1.32(3), 1.35(3)	1.45(3), 1.40(3)	159(2), 157(2)	148(2), 147(2); 147(2), 143(2)	151(2), 149(2)	172
$Ru_2(\mu\text{-}\eta^1{:}\eta^2\text{-}C_2C{\equiv}CBu^t)(\mu\text{-PPh}_2)(CO)_6$	2.031(3)	1.227(5); 1.172(6)	1.400(5)	159.7(3)	164.9(4) 179.2(4)	174.3(8)	243

$Ru_2\{\mu\text{-}\eta^1\text{:}\eta^2\text{-}C_2C_2Bu^t[Co_2(CO)_6]\}(\mu\text{-}PPh_2)(CO)_6$	2.024(2)	1.241(3); 1.365(4)	1.408(3)	163.1(2)	158.2(3) 136.8(2)	108.6(3)	243
$Ru_2Pt(\mu_3\text{-}\eta^1,\eta^1,\eta^1C_2C{\equiv}CBu^t)(\mu\text{-}PPh_2)(PPh_3)(CO)_7$	2.266 (M = Pt)	1.171; 1.198	1.399	91.29	136.34; 172.04	176.29	54
$Ru_2Pt(\mu_3\text{-}\eta^1,\eta^1,\eta^1C_2C{\equiv}CBu^t)(\mu\text{-}PPh_2)(dppb)(CO)_6$	2.224 (M = Pt)	1.293; 1.198	1.410	121.07	135.05; 173.99	176.84	54
$Ru_2\{\mu\text{-}\eta^1,\eta^2\text{-}C_2C_2[Co_2(CO)_6PPh_2[Ru_3(CO)_{11}]\}(\mu\text{-}PPh_2(CO)_6$	2.25(2)	1.26(3); 1.42(3)	1.39(3)	165(2)	160(2) 140(2)	141(2)	260
trans-$\{Rh(CO)(PPr^i{}_3)_2\}(\mu\text{-}\eta^1\text{:}\eta^2\text{-}C{\equiv}CC_2Ph)\{RhCl(PPr^i{}_2)_2\}$	2.061(4)	1.213(5); 1.271(5)	1.393(6)	167.9(3)	171.9(4) 149.5(4)	147.5(4)	71
$Co_2(\mu\text{-}dppm)\{\mu\text{-}\eta^2\text{-}Me_3SiC_2C{\equiv}CC{\equiv}C[Ru(PPh_3)_2Cp]\}(CO)_4$	1.981(4)	1.230(6), 1.219(6) (C≡C), 1.361(6) (C≡C/Co_2)	1.359(6), 1.382(6)	171.9(4)	178.7(5) 178.1(5) 176.2(5) 140.5(4)	141.6(3)	194
$Co(\mu\text{-}dppm)_t(\mu\text{-}\eta^2\text{-}Me_3SiC{\equiv}CC_2C{\equiv}C[Ru(PPh_3)_2Cp]\}(CO)_4$	1.997(2)	1.211(3) (C≡C); 1.370(3) (C≡C/Co_2); 1.217(3) (C≡C)	1.407(3), 1.395(3) (C—C)	178.0(2)	176.1(2) 143.9(2) 138.6(2) 176.7(2)	175.7(2)	194

[a] n.d., not determined or not reported.
[b] Two molecules.

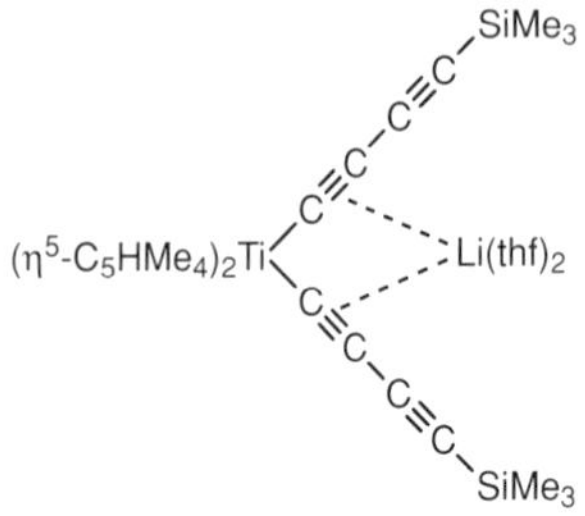

(**196**)

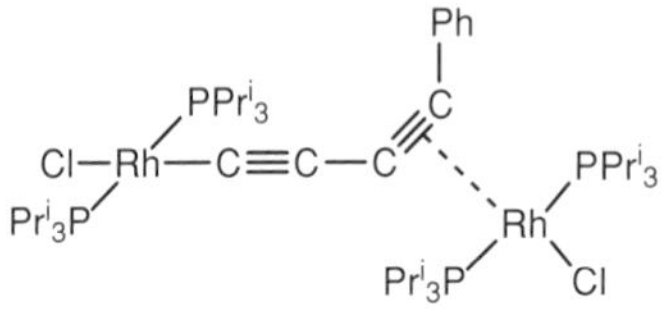

(**199**)

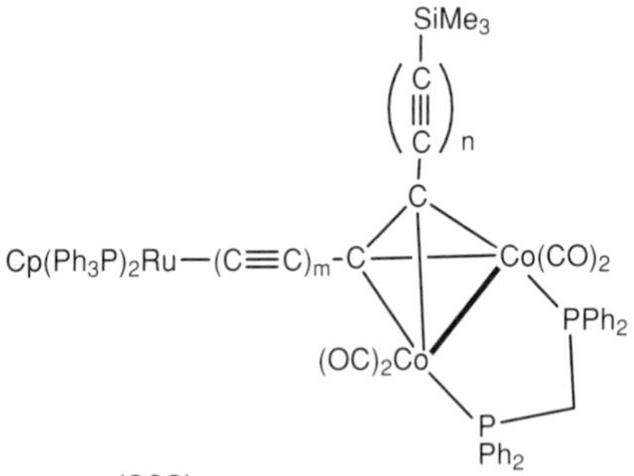

(**203**) m = 2, n = 0
(**204**) m = 1, n = 1

Reactions of $Ti(C{\equiv}CC{\equiv}CR)_2Cp^{Si}{}_2$ (R = Et, $SiMe_3$) with $Ni(CO)_4$, $Pd(PPh_3)_4$ or $Pt(\eta\text{-}C_2H_4)(PPh_3)_2$ give similar "tweezer" complexes $ML_n(\eta^2\text{-}RC{\equiv}CC_2)(X)TiCp_2^{Si}$ [X = Cl, ML_n = CuBr; X = $C_2C{\equiv}CR$, ML_n = CuBr, Ni(CO), $Pd(PPh_3)$, $Pt(PPh_3)$] in which the *cis*-diynyl groups chelate the new metal center.[98,99]

Addition of $AgPF_6$ to $Ti(C{\equiv}CC{\equiv}CFc)_2Cp'_2$ (Cp′ = Cp, Cp^{Si}) resulted in the formation of $Fc(C{\equiv}C)_4Fc$ via the cationic intermediate $[Ag\{(\eta^2\text{-}FcC{\equiv}CC_2)_2TiCp'_2\}2]^+$, in which the Ag^+ cation is tetrahedrally but asymmetrically coordinated [Ag−C(1) 2.29, 2.33; Ag−C(2) 2.41, 2.42 Å] to four inner C≡C triple bonds of two titanium complexes. Mössbauer studies show the iron nuclei in the ferrocene moieties remain Fe(II) and are electronically similar in both the precursor complex and the silver complex.[277]

2. *Complexes Derived from trans-*$Pt(C{\equiv}CC{\equiv}CR)_2(PR'_3)_2$

Reactions of *trans*-$Pt(C{\equiv}CC{\equiv}CR)_2(PR'_3)_2$ (R = H, $SiMe_3$; R′ = Et, Bu) with $Co_2(CO)_8$ give complexes in which both diynyl ligands are coordinated by

$Co_2(CO)_6$ fragments at the C≡C triple bonds further from the Pt center.[131] The Pauson–Khand reaction between *trans*-Pt{C≡CC$_2$H[$Co_2(CO)_6$]}$(PR_3)_2$ (R = Et, Bu) (**197**) and norbornene or cyclopentene gives the corresponding cyclopentenones **198** (Scheme 44).[131]

3. *Complexes Derived from M(C≡CC≡CR)(CO)$_n$Cp^R (M = Mo, W, Fe)*

Reactions of M(C≡CC≡CR)$(CO)_3$Cp (M = Mo, R = H; M = W, R = H, $Fe(CO)_2$Cp) with $Co_2(CO)_8$ have given M{C≡CC$_2$R[$Co_2(CO)_6$]}$(CO)_3$Cp, in which the dicobalt fragment has added to the C≡C triple bond furthest from the Group 6 metal center, i.e., the least sterically hindered site.[109,197] The iron diynyl complex Fe(C≡CC≡CH)$(CO)_2$Cp* also reacts smoothly with $Mo_2(CO)_4Cp_2$, MoCo$(CO)_7$Cp or $Co_2(CO)_8$, to afford the simple adducts M_2L_n{μ-η^2-HC$_2$C≡C [$Fe(CO)_2$Cp*]} [M_2L_n = $Mo_2(CO)_4Cp_2$, MoCo$(CO)_5$Cp, $Co_2(CO)_6$ (**156**)], in which the binuclear group is coordinated to the sterically less encumbered C≡CH moiety (see Scheme 34).[259] Further reaction of Co_2{μ-η^2-HC$_2$C≡C[$Fe(CO)_2$Cp*]} $(CO)_6$ with $Co_2(CO)_8$ resulted in coordination of the free C≡CFp* unit and the formation of {$Co_2(CO)_6$}$_2${μ-η^2 : μ-η^2-HC$_2$C$_2$[$Fe(CO)_2$Cp*]}.

4. *Complexes Derived from Ru{(C≡C)$_n$R}$(PPh_3)_2$Cp*

Addition of $Co_2(CO)_8$ to Ru(C≡CC≡CPh)$(PPh_3)_2$Cp afforded Co_2{μ,η^2-PhC$_2$C≡C[Ru$(PPh_3)_2$Cp]}$(CO)_6$ in which the $Co_2(CO)_6$ moiety is attached to the least hindered C≡C bond.[81] Similar complexes have been obtained from the reactions of Co_2(μ-dppm)(μ,η^2-RC$_2$C≡CR′)$(CO)_4$ with RuCl$(PPh_3)_2$Cp (see following).

5. *Complexes Derived from trans-Rh(C≡CC≡CPh)(CO)(PPr$^i_3$)$_2$*

The outer C≡C moiety of the diynyl ligand in *trans*-Rh(C≡CC≡CPh) (CO)(PPr$^i_3$)$_2$ reacts with {RhCl(PPr$^i_3$)$_2$}$_n$ to afford *trans*-{Rh(CO)(PPr$^i_3$)$_2$} (μ-η^1 : η^2-C≡CC≡CPh){RhCl(PPr$^i_3$)$_2$} (**199**).[71]

6. *Complexes Derived from Co_2(μ-η^2-RC$_2$C≡CH)(μ-dppm)$(CO)_4$*

The acetylenic C≡CH moiety in Co_2(μ-η^2-RC$_2$C≡CH)(μ-dppm)$(CO)_4$ (R = H, $SiMe_3$) is readily metallated to afford a range of diynyl complexes in which one C≡C moiety is σ-bound to a mononuclear metal center while the other remains π-bound to a Co_2(μ-dppm)$(CO)_4$ fragment (Scheme 45). For example, Cu(I)-catalyzed reactions between Co_2(μ-η^2-RC$_2$C≡CH)$(CO)_4$(dppm) and WCl$(CO)_3$Cp (R = H, $SiMe_3$) or NiBr(PPh_3)Cp (R = $SiMe_3$) in amine solvents afforded Co_2{(μ-η^2-RC$_2$C≡C[ML$_n$]}(μ-dppm)$(CO)_4$ (**200**) [ML$_n$ = W$(CO)_3$Cp,[171] Ni(PPh_3)Cp,[278] respectively]. The Co_2(μ-dppm)$(CO)_4$ adduct of Au(C≡CC≡CH)(PPh_3) has been obtained from Co_2(μ-dppm)(μ-HC$_2$C≡CH) $(CO)_4$ and

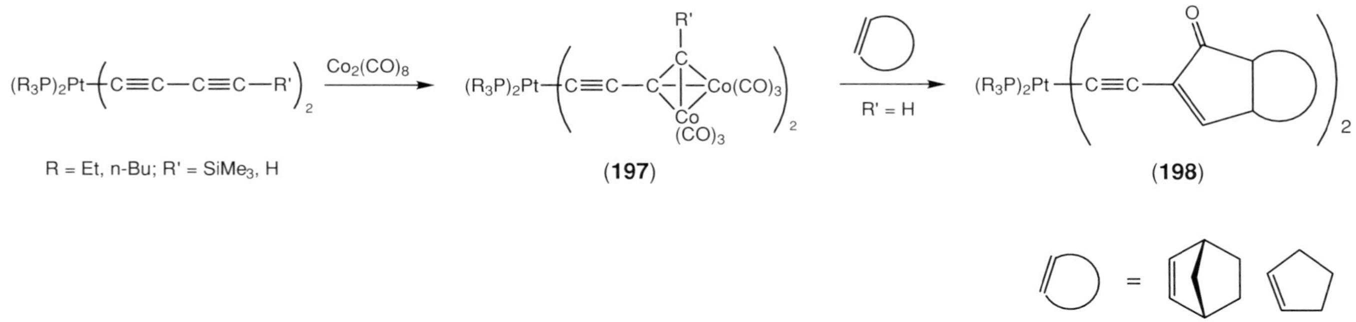

$(R_3P)_2Pt\text{-}(C\equiv C\text{—}C\equiv C\text{—}R')_2 \xrightarrow{Co_2(CO)_8} (R_3P)_2Pt\text{-}(C\equiv C\text{—}C(R')C[Co(CO)_3]_2)_2$ (197) $\xrightarrow[R' = H]{\text{alkene}}$ (198)

R = Et, n-Bu; R' = $SiMe_3$, H

SCHEME 44

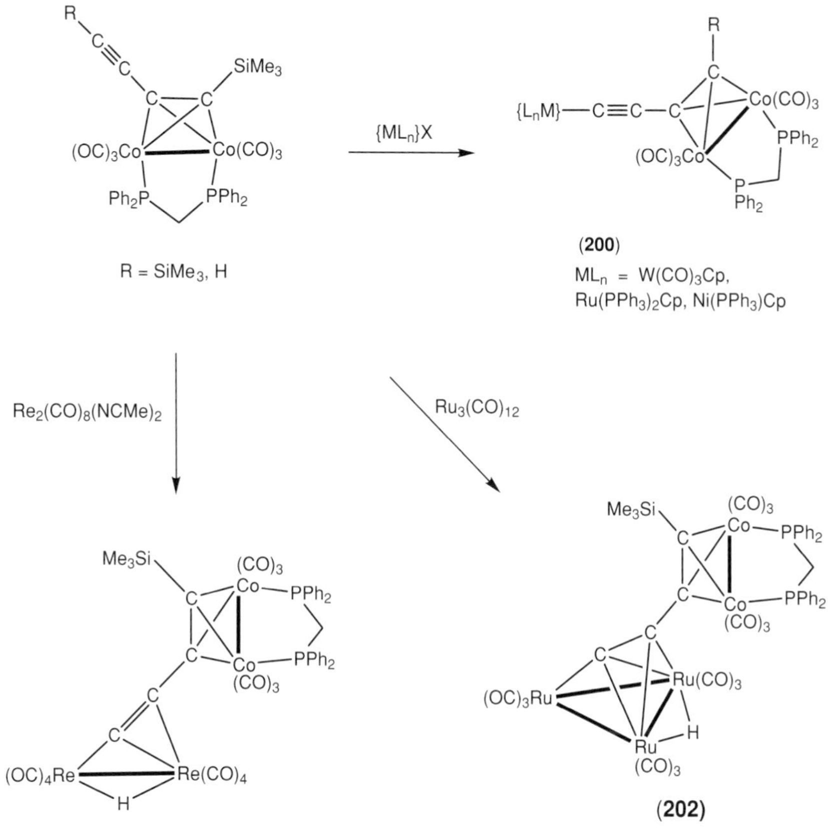

SCHEME 45

$AuCl(PPh_3)$ in the presence of 1,8-diazabicyclo[5.4.0]undec-7-ene (dbu).[171] Treatment of $Co_2(\mu\text{-}\eta^2\text{-}RC_2C{\equiv}CH)(\mu\text{-dppm})(CO)_4$ ($R = H$, $SiMe_3$) with $RuCl(PPh_3)_2Cp$ and NH_4PF_6 in MeOH gave the corresponding vinylidene complexes which were not isolated, but deprotonated with NaOMe or dbu *in situ* to give **200** [$ML_n = Ru(PPh_3)_2Cp$].[81,278]

The unsubstituted $C{\equiv}CH$ moiety in $Co_2(\mu\text{-dppm})(\mu\text{-}\eta^2\text{-}Me_3SiC_2C{\equiv}CH)(CO)_4$ oxidatively adds across the Re—Re bond in $Re_2(CO)_8(NCMe)_2$ to give $Re_2(\mu\text{-}H)\{\mu\text{-}\eta^1,\eta^2:\mu\text{-}\eta^2\text{-}C_2C_2SiMe_3[Co_2(\mu\text{-dppm})(CO)_4]\}(CO)_8$ (**201**) while a similar reaction with $Ru_3(CO)_{12}$ afforded $Ru_3(\mu\text{-}H)\{\mu_3\text{-}\eta^1:\eta^2;\mu\text{-}\eta^2\text{-}C_2C_2SiMe_3[Co_2(\mu\text{-dppm})(CO)_4]\}(CO)_9$ (**202**) (Scheme 45).[171]

KF-induced desilylation (Section II.B.4) of $Co_2(\mu\text{-}\eta^2\text{-}Me_3SiC_2C{\equiv}CC{\equiv}CSiMe_3)(\mu\text{-dppm})(CO)_4$ and $Co_2(\mu\text{-}\eta^2\text{-}Me_3SiC{\equiv}CC_2C{\equiv}CSiMe_3)(\mu\text{-dppm})(CO)_4$

in reactions with $RuCl(PPh_3)_2Cp$ afforded the two mono-desilylated complexes $Co_2\{\mu-\eta^2-Me_3SiC_2C{\equiv}CC{\equiv}C[Ru(PPh_3)_2Cp]\}(\mu\text{-dppm})(CO)_4$ (**203**) and $Co_2\{\mu-\eta^2-Me_3SiC{\equiv}CC_2C{\equiv}C[Ru(PPh_3)_2Cp]\}(\mu\text{-dppm})(CO)_4$ (**204**).[194]

7. *Complexes Formed by P—C Bond Cleavage Reactions*

Electron transfer-catalyzed reactions of $Ru_3(CO)_{12}$ with $PPh_2(C{\equiv}CC{\equiv}CR)$ (R = Bu^t, Ph, $SiMe_3$) afford excellent yields of $Ru_3(CO)_{11}\{PPh_2(C{\equiv}CC{\equiv}CR)\}$ which undergo thermal P—C bond cleavage to give **125,**[243,279] in addition to smaller amounts of higher nuclearity clusters (Section IV). Structurally, the dimetallodiynyl cores of these complexes are closely related to the mono alkynyl analogues $Ru_2(\mu\text{-PPh}_2)(\mu\text{-C}_2R)(CO)_6$.[280] The uncoordinated C≡C bond in $Ru_2(\mu\text{-PPh}_2)(\mu-\eta^1:\eta^2\text{-C}_2C{\equiv}CBu^t)(CO)_6$ is very short [1.172(6) Å] compared with 1.227(5) Å for the coordinated C≡C bond, and on the basis of structural trends and ^{13}C NMR data this moiety is considered to be an electronegative substituent on the $\mu-\eta^1:\eta^2$-ynyl fragment thereby enhancing Ru—C back-bonding, and elongating the π-coordinated ynyl moiety.[243] A contribution from a Ru=C=C=C=C^+Bu^t form is also consistent with the reactivity of these species (see following).[244] The binuclear complexes exchange the $\eta^1:\eta^2$ ligand between the two metals in a "windshield-wiper" process.[279] In the ^{13}C NMR spectra, long-range P—C coupling to C_δ is observed, suggesting enhanced electronic communication along the C_4 chain.[243]

C. *Reactions of σ,π-Diynyl Complexes*

1. *With Nucleophiles*

Treatment of **125** (R = Bu^t, Ph) with $NHEt_2$ resulted in exclusive nucleophilic attack at C_α and H migration from N to C_δ (Scheme 46). The resulting 1,4-addition products $Ru_2(\mu-\eta^1,\eta^2\text{-Et}_2NC{=}C{=}C{=}CHR)(\mu\text{-PPh}_2)(CO)_6$ (**205**) are best described as diethylaminobutatrienes. The Ru_2C_2 metallocycle in the R = Bu^t product is characterized by short C—C single [1.470(5) Å] and C=N bonds [1.315(4) Å] and asymmetric Ru—C bonds [Ru—C(N) 2.121(3) Å; Ru—C(C) 2.158(4) Å]. These structural data, together with a relatively high field shift for the Ru—C(N) carbon (δ_C 220.4) suggest that the ligands in these products are best represented by contributions from zwitterionic iminium and neutral amino-carbene forms.[279]

Reactions of **125** (R = Bu^t, Ph) with the carbene precursors R'_2CN_2 (R′ = H, Ph) also resulted in addition at C_α and afforded 1-alkynylallenyl complexes $Ru_2\{\mu-\eta^1,\eta^2\text{-C(C}{\equiv}\text{CR)}{=}C{=}CR'_2\}(\mu\text{-PPh}_2)(CO)_6$ (**206**). In the case of R′ = Ph, η^1-indenyl derivatives **207** and **208** resulting from attack at C_β were also isolated. In the former, the C≡CBu^t group is attached to the second Ru atom, while in **208,**

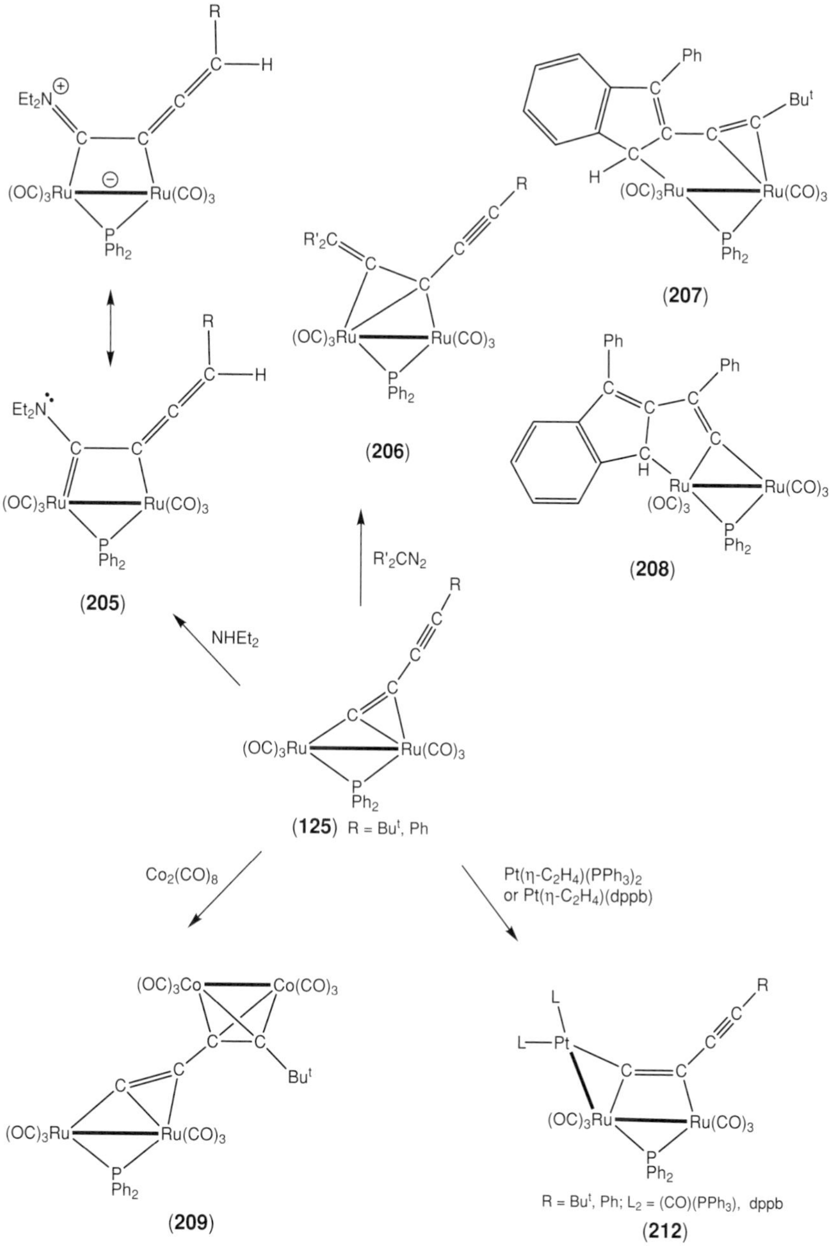

(**205**)

(**206**)

(**207**)

(**208**)

$NHEt_2$

R'_2CN_2

(**125**) R = Bu^t, Ph

$Co_2(CO)_8$

$Pt(\eta\text{-}C_2H_4)(PPh_3)_2$ or $Pt(\eta\text{-}C_2H_4)(dppb)$

(**209**)

R = Bu^t, Ph; L_2 = $(CO)(PPh_3)$, dppb

(**212**)

SCHEME 46

isomerization to a μ-vinylidene has occurred with coordination of C_γ to the second Ru atom.[281]

The preference for reactions of the σ,π-diynyl ligands in **125** (R = $SiMe_3$, Bu^t, Ph) with amines and carbenes at the coordinated, and hence activated, triple bond has been rationalized with assistance from EH MO calculations on the model complex $Ru_2(\mu\text{-}\eta^1,\eta^2\text{-}C_\alpha{\equiv}C_\beta C_\gamma{\equiv}C_\delta H)(\mu\text{-}PH_2)(CO)_6$.[281] In the ground state, attack at C_α is favored on the grounds of orbital control. Attack of the bulky carbene :CPh_2 at C_α is sterically disfavored in the ground state, but when the fluxional σ,π diynyl ligand passes through the transition state in which the diynyl ligand is perpendicular to the Ru—Ru vector, attack at C_β becomes favored by both orbital and charge factors. Generation of the indenyl group could follow attack of :CPh_2 at C_β, generating an electrophilic C_α center followed by attack at C(ortho) with C—C bond formation and migration of a proton to C_α.

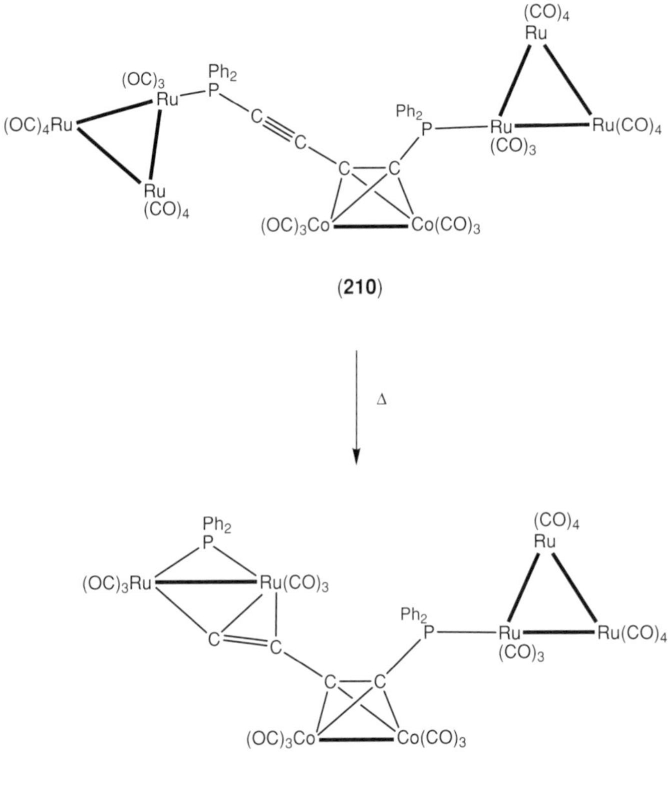

(**210**)

Δ

(**211**)

SCHEME 47

2. *With Metal Reagents*

The free $C{\equiv}CBu^t$ fragment in **125** ($R = Bu^t$) readily coordinates with $Co_2(CO)_8$ to give $Ru_2(\mu\text{-}PPh_2)\{\mu\text{-}\eta^1{:}\eta^2;\mu\text{-}\eta^2\text{-}C_2C_2Bu^t[Co_2(CO)_6]\}(CO)_6$ (**209**).[243] The reaction between $\{Ru_3(CO)_{11}\}_2(\mu\text{-bdpp})$ and $Co_2(CO)_8$ gives initially $\{Ru_3(CO)_{11}\}_2$ $\{\mu\text{-}\eta^2\text{-}PPh_2C_2[Co_2(CO)_6]C{\equiv}CPPh_2\}$ (**210**), which on heating transforms to $Ru_2\{\mu\text{-}\eta^1{:}\eta^2\text{-}C_2C_2[Co_2(CO)_6]PPh_2[Ru_3(CO)_{11}]\}(\mu\text{-}PPh_2)(CO)_6$ (**211**), which is closely related to **209** (Scheme 47).[260]

The complexed $C{\equiv}C$ moiety in **125** ($R = Bu^t$, Ph) is susceptible to attack by other metal reagents. Reactions with the cluster building blocks $Pt(\eta^2\text{-}C_2H_4)(PPh_3)_2$ and $Pt(\eta^2\text{-}C_2H_4)(dppb)$ have afforded $Ru_2Pt(\mu_3\text{-}\eta^1,\eta^1,\eta^1\text{—}C{=}C\text{—}C{\equiv}CR)(\mu_2\text{-}PPh_2)(CO)_6L_2$ [$L_2 = (CO)(PPh_3)$, dppb, respectively] (**212**) (Scheme 46).[54] The phosphine-substituted complexes $Ru_2(\mu\text{-}PPh_2)(\mu\text{-}\eta^1{:}\eta^2\text{-}C_2C{\equiv}CR)(CO)_{6-n}(PPh_3)_n$ ($n = 1, 2$) are formed as by-products in the reactions with $Pt(\eta^2\text{-}C_2H_4)(PPh_3)_2$.

VI

σ,π-DIYNE COMPLEXES OF GROUPS 3, 4, AND 5

The diyne chemistry of the elements of groups 3, 4, and 5 is intimately coupled with that of the alkynyl derivatives by virtue of the C—C bond coupling/cleavage reactions that are found. With few exceptions, the chemistry is confined to the metallocene derivatives, i.e., those containing the MCp_2 group. In one sense, these can be related to the tweezer complexes discussed earlier (Section V.B.1). The nature of the products obtained from particular reactions is strongly dependent on the diyne (or alkynyl) substituent(s) and the metal–ligand fragment. In the limit, catalytic C—C single bond metathesis can be achieved. In the following account, we shall consider the chemistry of each group separately.

A. *Group 3 (Sc, Y, Rare Earth Elements)*

The results of three independent groups converged in 1993 to allow an understanding of the reactions in which alkynyl groups are coupled at the metal centers to give compounds containing the lanthanides attached to 1,3-diynes. It is likely that an uncoupled bis-alkynyl is the immediate kinetic precursor to the coupled dimers.[282,283] The C—C coupling reactions are considered to be formally equivalent to C—C bond formation in transition metal-based reductive elimination.[283,284] For Group 3 metals, alkynyl coupling is promoted by a high degree of steric crowding, reactions not being found for sterically less demanding ligands, e.g., Cp. Coupling is also favored by alkyne substituents that are not too electron donating and is driven by the electrophilicity of the metal center. The tetrahedral geometry assumed by the four Cp* groups in binuclear lanthanide compounds is also important. Other studies have concluded that a redox-active lanthanide is unnecessary, both "concerted" and "insertion" type mechanisms being consistent with the data to date.[282]

Structural studies showed that a red complex, variously obtained from HC≡CPh and Sm{CH(SiMe$_3$)$_2$}Cp*$_2$,[285] SmCp*$_2$ or {Sm(μ-H)Cp*$_2$}$_2$, or by thermolysis of Sm(C≡CPh)(thf)Cp*$_2$ (120°C, 3 days; quantitatively, 145°C, 14 h),[284] and originally described as the alkyne-bridged dimer {Sm(μ-C≡CPh)Cp*$_2$}$_2$, actually contains the 1,4-dimetallated butatriene ligand, [PhC=C=C=CPh]$^{2-}$. Addition of Sm(thf)$_2$Cp*$_2$ to PhC≡CC≡CPh gives directly paramagnetic {SmCp*$_2$}$_2$ (μ-η^2:η^2-PhC$_2$C$_2$Ph) (**213-Sm/Ph**)* in which the Sm is formally Sm^{3+}.[283,286]

In general, lanthanide alkynyl derivatives {Ln(C≡CR)Cp*$_2$}$_2$ (**214;** Ln = Ce, La; R = Me, Bu^t) rearrange in solution (1 day at r.t.) to {LnCp*$_2$}$_2$(μ-η^3:η^3-RC$_2$C$_2$R) by a reversible C–C coupling reaction. The Me complexes couple much faster than the Bu^t derivatives. For the La/Me complex, ΔG 4.5(4) kJ mol^{-1}, k$_1$ 8.3(4) × 10^{-5} s^{-1}, k$_{-1}$ 1.1(7) × 10^{-5} s^{-1}, ΔG$^{\#}$ 96.3(1) kJ mol^{-1}; at 298 K, the equilibrium mixture contains 86/14 dimer/monomer, the coupled form being thermodynamically favored.[287] While La(C≡CBu^t)(thf)Cp*$_2$ does not couple after 48 h at 60°C, the reaction with HC≡CBu^t affords {LaCp*$_2$}$_2$(μ-η^2:η^2-Bu^{t}C$_2$C$_2$Bu^t) (**213-La/Bu^t**) quantitatively (60°C, 4 h). However, at 0°C, uncoupled {La(C≡CBu^t)Cp*$_2$}$_2$ (**214-La/Bu^t**) is isolated in 60% yield. Conversion of **214** to **213** occurs on heating, with first-order kinetics between 50–60°C, but with deviations at 70°C, suggesting dissociation of **214** occurs at the higher temperatures.

Coupling did not occur on heating Sm(C≡CR′)(thf)Cp*$_2$ (R′ = CH$_2$CH$_2$Ph, CH$_2$NEt$_2$, CH$_2$CH$_2$Pr^i, Pr^i, Bu^t) obtained from Sm{N(SiMe$_3$)$_2$}Cp*$_2$ and HC≡CR′; however, HC≡CR does react with SmCp*$_2$ to give coupled products for R = Pr^i, (CH$_2$)$_2$Pr^i, (CH$_2$)$_2$Ph. Of interest is the agostic interaction between one of the CH$_2$ groups [C(24)] and the Sm center [Sm–C(24) 3.748 Å] in {SmCp*$_2$}$_2$(μ-η^2:η^2-R′C$_2$C$_2$R′) (**215**). A weak dimer, via intermolecular interaction of a Cp*-methyl group with Sm in {Sm(C≡CBu^t)Cp*$_2$}$_2$, is formed from HC≡CBu^t, together with Sm(Bu^{t}CH=CCBu^t=CH$_2$)Cp*$_2$.

Addition of HC≡CPh to La{CH(SiMe$_3$)$_2$}Cp*$_2$ gives {LaCp*$_2$}$_2$(μ-η^2:η^2-PhC$_2$C$_2$Ph) (**213-La/Ph**) as the only product. In contrast, Ln{N(SiMe$_3$)$_2$}Cp*$_2$ (Ln = Ce, Nd, Sm) and HC≡CPh formed Ln(C≡CPh)(thf)Cp*$_2$ which in turn are converted to {LnCp*$_2$}$_2$(μ-η^2:η^2-PhC$_2$C$_2$Ph) on heating in toluene. The complexes {Cp*$_2$Ln(μ-C≡CPh)$_2$K}$_n$ (Ln = Ce, Nd, Sm) do not undergo coupling. Monomeric alkynyl complexes, e.g., Ce(C≡CBu^t)(thf)Cp*$_2$, are formed by addition of Lewis bases to solutions of the initial alkynyl complexes or their dimers.

*Reference to structures of complexes mentioned in this section has been simplified by giving the general structure and indicating the metal (M) and diyne substituent (R).

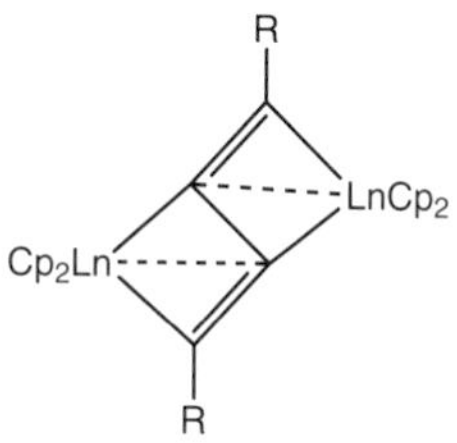

(**213-Ln/R**) 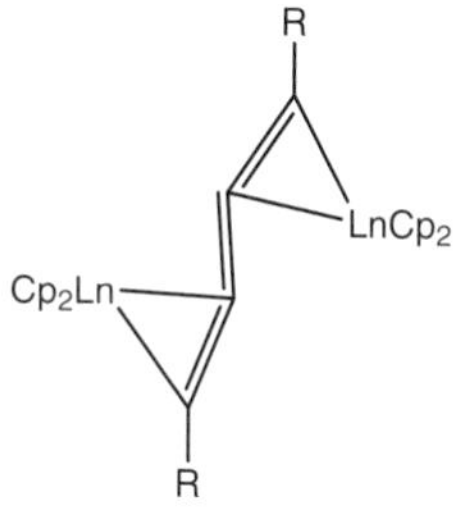 (**214-Ln/R**)

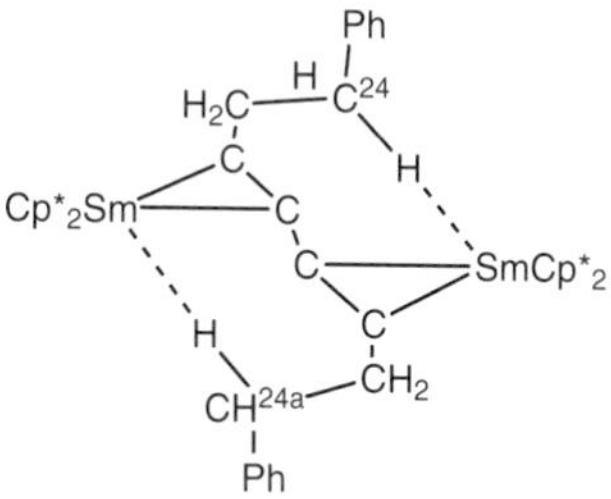

(**215**) 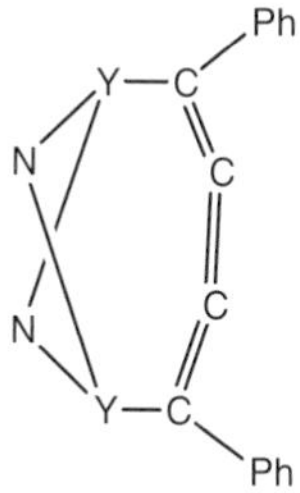 (**216**)

In these complexes, asymmetric attachment of the CC moiety is found, with Sm—C(1) [2.48(1) Å] comparable to the Sm—C(Ph) in SmPh(thf)Cp^*_2 but with Sm—C(2) longer at 2.76(1) Å; the C(1)—C(2) and C(2)—C(2′) separations are 1.33(2) and 1.29(2) Å, respectively.[283] Structural studies of **213-La/Ph** and **214-La/Bu^t** show that in the former, the C_4 chain is attached to La by three carbon atoms with separations 2.577 Å (consistent with an La—C σ-bond) and 2.823, 2.950 Å (π-bonds). For **214-La,** the three carbons are attached with La—C distances of 2.642, 2.761, and 2.912 Å ; C—C separations along the chain are 1.36 and 1.26 Å (R = Ph) and 1.310, 1.338 Å (R = Bu^t).[282]

The coupled product can be displaced by other alkynes in thf solution, e.g., **213-Sm/Ph** reacts with HC≡CPh to give (*E*)-PhCH=CHC≡CPh (tail-to-tail coupling). Reaction of **213-La/Ph** with D_2O gives $C_4D_2Ph_2$, while with H^+, a variety of products is formed. Only (*E*)-hex-2-en-4-yne was obtained from the Ln/Me compound and 2,6-Bu^t_2-4-MeC_6H_2OH (ArOH), together with $Ln(OAr)Cp^*_2$. In contrast, the Ce/Bu^t derivative reacts with ArOH to give three $C_4H_2Bu^t_2$ isomers: *cis*- and *trans*-$Bu^tCH{=}CHC{\equiv}CBu^t$ (60 and 20%) and $Bu^tCH{=}C{=}C{=}CHBu^t$ (20%).[287]

A strongly temperature-dependent equilibrium mixture [ΔH° −67.0(2.0) kJ mol^{-1}, ΔS° −228(8) J mol^{-1} K^{-1}, K_{eq} 0.68 (at 298 K)] of blue-purple {Y(4,13-diaza-18-crown-6)}$_2${μ-η^2:η^2-(*Z*)-PhC_2C_2Ph)} (**216**) and white {Y(μ-C≡CPh)(4,13-diaza-18-crown-6)}$_2$ is formed in the reaction of $Y(CH_2SiMe_3)$(4,13-diaza-18-crown-6) with HC≡CPh. A disordered X-ray structure is interpreted in terms of one of the N atoms of each cryptand bridging the two Y atoms. Although the system is still sterically congested, the (*Z*)-isomer is formed here, perhaps because the crown ether is more flexible.[288] Structural and low-temperature NMR data suggest that although the alkynyl is dimeric in the solid state, it probably dissociates in solution.

B. *Titanium, Zirconium, and Hafnium*

1. *General Features of Diyne and Bis-alkynyl Complexes*

The diyne chemistry of Group 4 metallocenes is largely derived from the "MCp_2" precursors Ti{η-$C_2(SiMe_3)_2$}Cp_2, Zr(L){η-$C_2(SiMe_3)_2$}Cp_2 (L = thf, py) and Zr{$OCMe_2C(SiMe_3){=}C(SiMe_3)$}$Cp_2$. The chemistry of these complexes has been reviewed.[289,290] Alternatively, metal reduction of MCl_2Cp_2 provides a source of "MCp_2." The reactions of di- and poly-ynes with "MCp_2" have been summarized recently.[291]

The first complexes to be described were obtained from reactions between MCl_2Cp_2 and metallated alkynes and initially formulated as the dimers {M(μ-C≡CR)Cp}$_2$. Further studies have clarified the nature of these complexes together with those formed from RC≡CC≡CR (R = Me, Bu^t, Ph, $SiMe_3$) and "MCp_2." The several structural types obtained contain 1/1, 1/2, 2/1, and 2/2 ratios of MCp_2 to diyne, as shown in Scheme 48. The nature of the complexes formed depends on diyne substituent and metal.

The course of these reactions is assumed to proceed via initial formation of the η^2-diyne complex **217** which, however, has not often been isolated in the MCp_2 series. In early work, the formation of enynes RCH=CHC≡CR in reactions of an excess of LiC≡CR with $ZrCl_2Cp_2$, followed by hydrolysis, was interpreted in terms of the formation of "ate" complexes $[Zr(C{\equiv}CR)_3Cp_2]^-$ and $[Zr(C{\equiv}CR)(\eta^2\text{-}C_2C{\equiv}CR)Cp_2]^-$ (**218;** R = Ph, n-C_6H_{13}, CMe=CH_2),[292] whereas the intermediate $[Zr(C{\equiv}CPh)(\eta^2\text{-}PhC_2C{\equiv}CPh)Cp_2]^-$ is obtained from an excess of LiC≡CPh with $ZrCl_2Cp_2$.[293] The η^2-diyne is stabilized in anionic or electron-rich species (see following for the Cp* analogue).

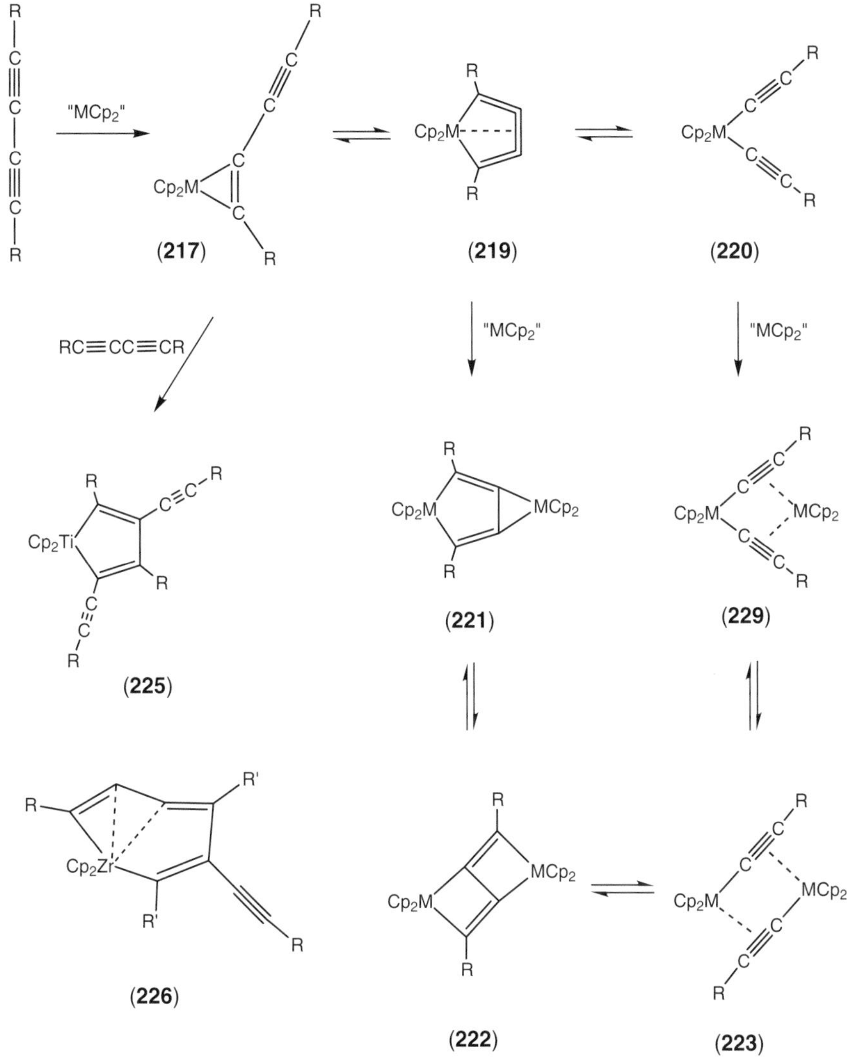

SCHEME 48

Rapid conversion to the cyclic cumulene (**219**) may be followed by C—C bond cleavage to give the bis-alkynyl (**220**); rapid interchange between these two forms has been demonstrated. Cleavage of the diyne is favored for "$ZrCp_2$" from $Zr(thf)\{\eta\text{-}C_2(SiMe_3)_2\}Cp_2$, which gives $Zr(C{\equiv}CR)_2Cp_2$ for all diynes except $Bu^tC{\equiv}CC{\equiv}CBu^t$.[294–296] Cleavage of the central C—C bond of the diyne is rationalized by formation of longer Zr—C bonds by the larger Zr atom. However,

$Zr(py)\{\eta\text{-}C_2(SiMe_3)_2\}Cp_2$ reacts with $Bu^tC{\equiv}CC{\equiv}CBu^t$ to give cyclocumulene $Zr(\eta^4\text{-}Bu^tC_4Bu^t)Cp_2$, in which the diyne remains intact and is symmetrically coordinated to only one Zr atom.[297]

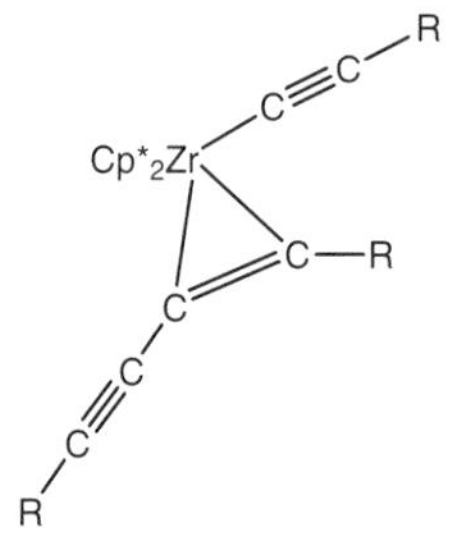

(**218**)

Form **219** provides a route to 2/1 complexes by coordination of the second "MCp_2" group to the central C=C bond to give postulated intermediate **221.** This has been demonstrated in the case of a titanium–nickel derivative (see following), but with $TiCp_2$, binuclear "zig-zag" diyne complexes **222** (or so-called tetradehydro-*trans,trans*-diene derivatives) are formed. Partial or complete conversion to the $\mu\text{-}\eta^1{:}\eta^2$-alkynyl **223** may occur. The stability of the five-membered titanacyclocumulenes **219** depends on the substituents on the diyne precursor; in one case, only $\{Ti(C{\equiv}CS_iMe_3)Cp_2\}_2$ is formed.[312b] It is interesting that similar derivatives of silicon have been obtained from reactions of the silylene $SiBu^t_2$ with $RC{\equiv}CC{\equiv}CR$ (R = Me, Bu^t, $SiMe_3$),[298,299] while mixed silicon/Ti or Zr derivatives **224** were obtained from $Si(C{\equiv}CR)_4$ (R = Ph, Bu^t, $SiMe_3$) (Scheme 49).[300]

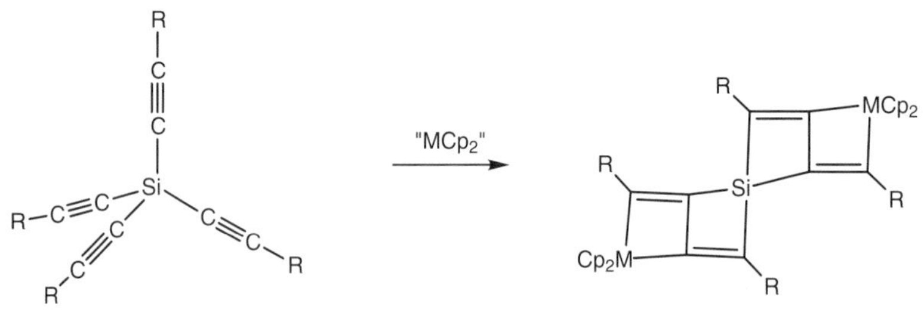

"MCp_2"

R = Ph, Bu^t, $SiMe_3$

(**224**)

M = Ti, Zr

SCHEME 49

Metallacyclic complexes containing two molecules of diyne per MCp_2 group have also been isolated, that with titanium containing 2,4-alkynyl substituents (**225**) while with zirconium, the unusual seven-membered metallacumulene structure **226** is adopted, which has only one alkynyl substituent. The bi- and tricyclic 2/2 complexes **227** and **228** have so far been obtained only from reactions of "$TiCp_2$" with PhC≡CC≡CPh.[301]

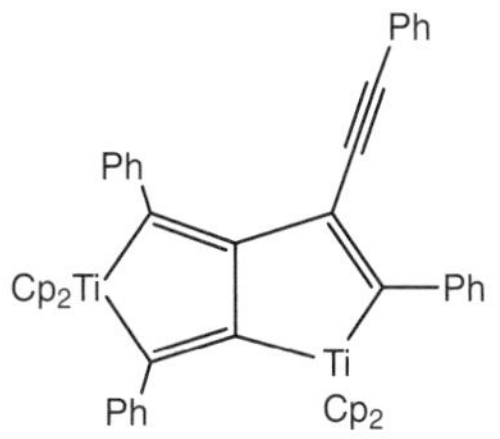

(**227**)

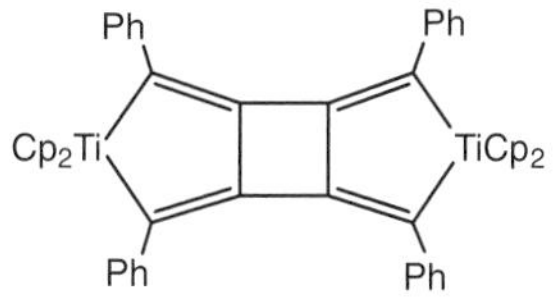

(**228**)

Extensive theoretical studies of the alkynyl coupling reactions have been reported. An early MO study of the relationship between $L_2M(\mu\text{-}C_2R)_2ML_2$ and $L_2M(\mu\text{-}RC_4R)ML_2$ used EH and MNDO techniques and encompassed a range of Main Group elements as well as Ti and Zr[302] and showed the transition between symmetrical, asymmetric, $\mu\text{-}\eta^1{:}\eta^2$ and linked C_4 ligands. The structural evidence suggests that the acetylenic MOs are not involved to a great extent in the bonding in the Main Group systems. Later more detailed calculations using *ab initio* and DFT methods comparing Ti and Zr complexes with H, Cl, or Cp ligands with variable alkynyl substituents (H, CN, or F) show that substituent changes can shift the equilibrium, particularly for the $SiMe_3$ complexes, which gives some insight into C—C bond activation by the bimetallic template. The relative stabilities are reversed if Ti is replaced by Zr. Differences between Ti and Zr can be traced to differences in ionic radii, larger Zr leading to longer Zr—C distances and in turn a longer central C · · · C bond. Calculations on likely transition states also indicate that C—C coupling is more likely for Ti than for Zr.[303,304] The metallacyclocumulene is thermodynamically more stable than the bis(alkynyl) as a result of additional π-coordination from the central C=C bond. Complexation of a second ML_2 fragment to the bis(alkynyl) **220** gives tweezer complex **229.** Electron-withdrawing substituents in the alkynyl group provide a method of stabilizing the coupled product on Zr (the coupled product with R = H is 13.5 kcal mol^{-1} less stable, while R = F is 17.6 kcal mol^{-1} more stable, than the bis-alkynyls).

2. *Reactions of Individual 1,3-Diynes, RC≡CC≡CR′*

a. *R = R′ = Me.* The only isolated products from Ti{η^2-$C_2(SiMe_3)_2$}Cp_2 and MeC≡CC≡CMe were {$TiCp_2$}$_2$(μ-η^2:η^2-MeC_2C_2Me) (**222-Ti/Me**) (containing the *trans, trans*-diyne) and Ti{CMe=C(C≡CMe)CMe=C(C≡CMe)}Cp_2 (**225-Ti/Me**).[301]

b. *R = R′ = Bu^t.* Reactions of "$TiCp_2$" or "$ZrCp_2$" with Bu^tC≡CC≡CBu^t afford metallacumulenes M(Bu^tC=C=C=CBu^t)Cp_2 (**219-M/Bu^t**) in which the ring is highly strained,[297,305] the bond from Ti to the central carbon (of the three) being 2.31 Å; the ZrC_4 system is coplanar, with equivalent C—C distances (1.28–1.31 Å) and angles at C of 147.2 and 150.0°.[297,306] The Ti complex has also been obtained by irradiation (390–450 nm) of Ti(C≡CBu^t)$_2$$Cp_2$ which then reacts further with "$TiCp_2$" to give **222-Ti/Bu^t**.[307] Addition of "$ZrCp_2$" gave the mixed Ti–Zr complex, also formed from Zr(Bu^tC=C=C=CBu^t)Cp_2 (**219-Zr/Bu^t**) and "$TiCp_2$." The Ti–Zr compound is fluxional by alkynyl group exchange.[305]

c. *R = R′ = Ph.* The reaction between {Ti(μ-Cl)Cp_2}$_2$ and NaC≡CPh results in coupling of the phenylethynyl groups to give the 1,4-diphenylbuta-1,3-dien-1,4-diyl ligand.[308,309] The anomalous ^{13}C NMR parameters found for M(C≡CR)$_2$$Cp_2$ (M = Ti, Zr) and the reaction of $TiCl_2Cp_2$ with LiC≡CBu or MgBr(C≡CBu) to give dark green paramagnetic solids had been noted earlier.[310] The unstable metallacumulene Ti(η^4-CPh=C=C=CPh)Cp_2 (**219-Ti/Ph**), formed from Ti{η^2-$C_2(SiMe_3)_2$}Cp_2 and PhC≡CC≡CPh, has similar spectroscopic properties to the Bu^t complex. Apparently, it is in equilibrium with the alkyne complex Ti(η^2-PhC_2C≡CPh)Cp_2 (**217-Ti/Ph**). On standing in toluene, green **227** is obtained, together with the symmetrical complex **228.**[301,309,311]

Reactions of TiCl(η-C_5H_4Me)$_2$ either with NaC≡CPh or with PhC≡CC≡CPh and sodium gave dark green {Ti(η-C_5H_4Me)$_2$}$_2$(μ-PhC_4Ph) containing a planar TiC_4Ti system in which there has been partial reduction of the diyne.[309]

d. *R = R′ = SiMe$_3$.* Reactions between {Ti(μ-Cl)Cp_2}$_2$ and NaC≡C$SiMe_3$ afforded dark burgundy {Ti(μ-η^1:η^2-C_2SiMe_3)Cp_2}$_2$ (**223-Ti/Si**)[310,312,313]; the same product was obtained from Me_3SiC≡CC≡C$SiMe_3$ and "$TiCp_2$."[312b] The X-ray structure confirmed that carbon–carbon coupling had not occurred, the central C ··· C distance being lengthened to 2.762(2) Å. The reaction with the diyne is interpreted as an oxidative addition to intermediate $TiCp_2$. Cleavage of the central

C—C bond results from the "β-effect" of $SiMe_3$ groups which renders these carbons electron deficient via $d\pi$(Si)–$p\pi$(C) interactions.[297,306,314] The reaction of $Me_3SiC{\equiv}CC{\equiv}CSiMe_3$ with Ti{η^2-$C_2(SiMe_3)_2$}Cp_2 also gives Ti{C($SiMe_3$)=C(C≡$CSiMe_3$)C($SiMe_3$)=C(C≡$CSiMe_3$)}Cp_2 (**225-Ti/Si**).[305]

Small cyclic cumulenes are formed from Zr(py){η-$C_2(SiMe_3)_2$}Cp_2 and diynes, e.g., $Me_3SiC{\equiv}CC{\equiv}CSiMe_3$ gives Zr{2η^1,η^2-$Me_3SiC{=}C_2{=}C(SiMe_3)$C(C≡$CSiMe_3$)=C($SiMe_3$)}$Cp_2$ (**226-Zr/Si**), in which two diyne molecules have coupled at the metal to give a seven-membered cyclic cumulene; some {Z-(C≡$CSiMe_3$)Cp_2}$_2$ is obtained as a by-product. Both complexes were also made from the diyne and $ZrBu_2Cp_2$ [a precursor of Zr(η^2-CH_2CHEt)Cp_2].[297,315]

The metallocyclic ring is essentially planar. The proposed mechanism of formation is via metallacycle Zr{C($SiMe_3$)=C(C≡$CSiMe_3$)C($SiMe_3$)=C(C≡$CSiMe_3$)}Cp_2 (**225-Zr/Si**).[316] The complex Zr(thf){η-$C_2(SiMe_3)_2$}Cp_2 reacts with $Me_3SiC{\equiv}CC{\equiv}CSiMe_3$ to give {Zr(μ-η^1:η^2-C_2SiMe_3)Cp_2}$_2$ which has also been obtained from $ZrBu_2Cp_2$ and the diyne.[296,310,316]

e. *R = SiMe*$_3$*, R′ = Bu*^{t}*, Ph.* The reactions of Ti{η^2-$C_2(SiMe_3)_2$}Cp_2 with $Me_3SiC{\equiv}CC{\equiv}CR$ (R = Bu^t, Ph) gave {TiCp_2}$_2$(μ-RC_4SiMe_3) (**222**) containing an intact zig-zag diyne; no C—C bond cleavage is found with an excess of the diyne. The ^{13}C NMR spectra are diagnostic while the central C—C bonds are 1.517(6) and 1.494(6) Å for R = Bu^t, R′ = $SiMe_3$, Bu^t, respectively. No symmetrization of the Bu^t/$SiMe_3$ complex occurs.[306,317]

Reactions between Zr(thf){η-$C_2(SiMe_3)_2$}Cp_2 and mixed 1,3-diynes $SiMe_3C{\equiv}CC{\equiv}CR$ lead to cleavage of the diyne, affording {ZrCp_2}$_2$(μ-η^1:η^2-C_2SiMe_3)(μ-η^1:η^2-C_2R) (**223**). For R = Ph, only one Cp signal (^1H, ^{13}C NMR) indicates that the complex is highly fluxional, with rapid intramolecular migration of alkynyl groups between the Zr centers. The complexes are diamagnetic, probably by electronic coupling via the alkynyl groups.[295,297] In contrast, reaction of Zr(η^2-$SiMe_3C_2SiMe_3$)(py)Cp_2 with $Bu^tC{\equiv}CC{\equiv}CSiMe_3$ gives the zirconacumulene Zr{η^t,η^3-$SiMe_3CC(C{\equiv}CBu^t)C(SiMe_3){=}CC{=}CBu^t$}$Cp_2$ (**226**) (38%).[318]

Whereas coupling of alkynyl groups occurs during the formation of {TiCp'_2}$_2$(μ-η^2:η^2-PhC_2C_2Ph) (Cp′ = Cp, Cp^{Me}) from {TiClCp'_2}$_2$, the reaction of Zr(C≡CPh)$_2Cp^{Me}{}_2$ with Zr(*trans*-C_4H_6)$Cp^{Me}{}_2$ gave {Zr(μ-η^1:η^2-C_2Ph)$Cp^{Me}{}_2$}$_2$ (**223-Zr/Ph**), possibly via tweezer complex **229**.[319] The related complex {Zr(μ-η^1:η^2-C_2Bu)Cp_2}$_2$ was isolated from the reaction between $ZrCl_2Cp_2$, Mg and BuC≡CTeBu.[320,321] The direct reaction between "ZrCp_2" (from $ZrCl_2Cp_2$ and LiBu) and HC≡$CSiMe_3$ gives {Zr(μ-η^1:η^2-C_2SiMe_3)Cp_2}$_2$ as the only characterized, but minor (1%), product.[296] All these complexes have rather long Zr—C π-bonds and short C—C multiple bonds, with low energy ν(CC) absorptions around 1750 cm^{-1}.

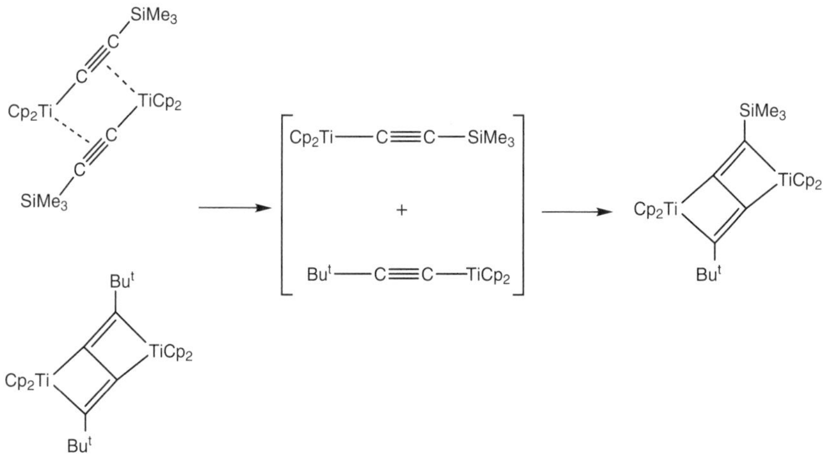

SCHEME 50

3. *C—C Bond Dismutation Reactions*

The challenge to combine cleavage of the central C—C bond with subsequent coupling of the alkynyl groups has been met by taking advantage of the acceleration of coupling of $Ti(C{\equiv}CBu^t)_2Cp_2$ which occurs upon irradiation.[322] Metathesis of disubstituted butadiynes is mediated by "$TiCp_2$." In practice, irradiation of equimolar amounts of $RC{\equiv}CC{\equiv}CR$ ($R = Bu^t$ and $SiMe_3$) in the presence of an excess of $Ti\{\eta^2\text{-}C_2(SiMe_3)_2\}Cp_2$ (100°C, toluene) gave $Bu^tC{\equiv}CC{\equiv}CSiMe_3$ (4.9% isolated) in addition to the symmetrical diynes; in solution, the three dimers $\{TiCp_2\}_2(\mu\text{-}\eta^1{:}\eta^2\text{-}C_2SiMe_3)_2$ (14%) and $\{TiCp_2\}_2(\mu\text{-}Bu^tC_2C_2R)$ [$R = Bu^t$ (36%), $SiMe_3$ (21%)] contained 71% of the diynes. The thermal reaction occurs at temperatures above 140°C. Irradiation of a mixture of the symmetrical complexes gives the mixed bis-alkynyl (and hence mixed diyne after oxidative work-up with AgOTf) (Scheme 50). The Zr analogue is inactive.[323]

4. *Other Diyne Systems*

Extension of these reactions of diynes to $1,3,5\text{-}(Bu^tC{\equiv}CC{\equiv}C)_3C_6H_3$ affords complexes containing three moieties (**230** and **231**) analogous to **221** (Ti) or **222** (Zr).[324] No similar products were obtained from $C_6(C{\equiv}CC{\equiv}CBu^t)_6$. Similar derivatives are formed in reactions with tetraynes $R(C{\equiv}C)_4R$ (Scheme 51).[325] For $R = Bu^t$, two "zig-zag" diyne fragments are linked in **232,** whereas for $R = SiMe_3$, only the central C≡C triple bonds are used in **233;** the latter appears to be identical with a complex obtained by reaction of $\{Ti(\mu\text{-}Cl)Cp_2\}_2$ with $LiC{\equiv}CC{\equiv}CSiMe_3$. The tetranuclear Zr complex **234** containing a $\mu\text{-}C_4$ ligand was obtained from $Me_3Si(C{\equiv}C)_4SiMe_3$.

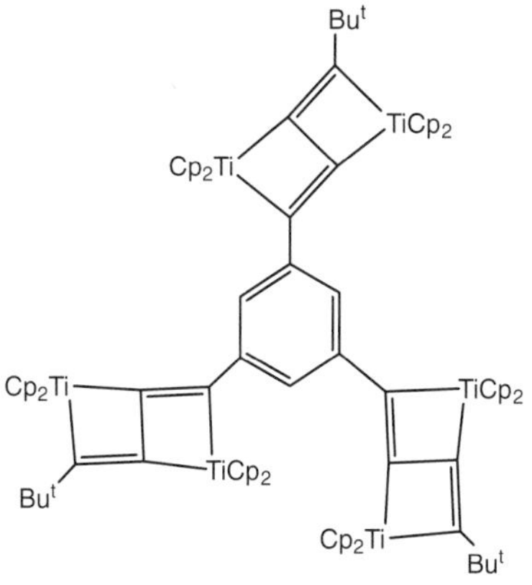

(**230**)

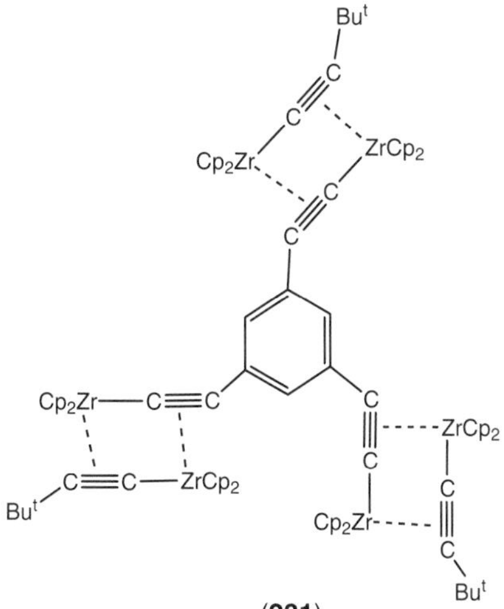

(**231**)

5. *Other Cp Ligands*

Two intermediate complexes are formed if $\{TiCl[(\eta\text{-}C_5H_4)_2SiMe_2]\}_2$ is used. Reactions with LiC≡CPh, carried out in thf at −10°C, afford first the red paramagnetic $\{Ti(\mu\text{-}\eta^1\text{-}C{\equiv}CPh)[(\eta\text{-}C_5H_4)_2SiMe_2]\}_2$ which after 12 h affords red-black diamagnetic (by antiferromagnetic coupling) $\{Ti(\mu\text{-}\eta^1{:}\eta^2\text{-}C{\equiv}CPh)[(\eta\text{-}C_5H_4)_2SiMe_2]\}_2$. On warming to r.t., conversion to dark green $\{Ti[(\eta\text{-}C_5H_4)_2SiMe_2]\}_2(\mu\text{-}\eta^2{:}\eta^2\text{-}C_4Ph_2)$ occurs. Treatment of each complex with HCl gave HC≡CPh for the first two, and 3/7 *cis*/*trans* PhCH=CHC≡CPh for the third complex.[326] The reaction between $ZrCl_2\{(\eta\text{-}C_5H_4)(\eta\text{-}C_5H_3SiMe_3\text{-}3)SiMe_2\}$ and $LiC{\equiv}CSiMe_3$ gives $\{Zr(\mu\text{-}\eta^1{:}\eta^2\text{-}C_2SiMe_3\text{-}3)[(\eta\text{-}C_5H_4)(\eta\text{-}C_5H_3SiMe_3\text{-}3)SiMe_2]\}_2$ as the only characterized complex, the structure being consistent with contributions from the resonance forms **L** and **M**.[327]

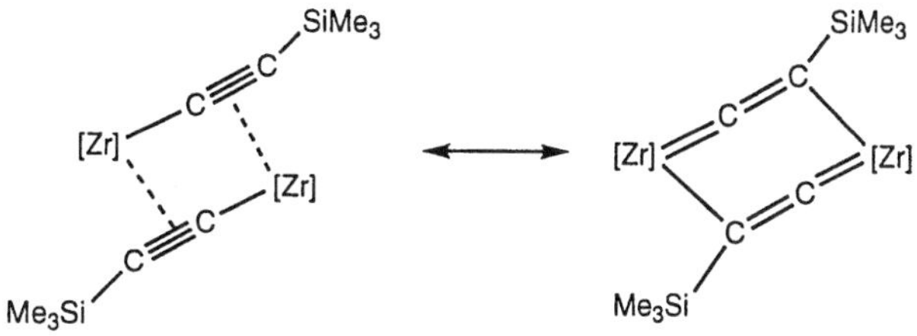

(**L**)

(**M**)

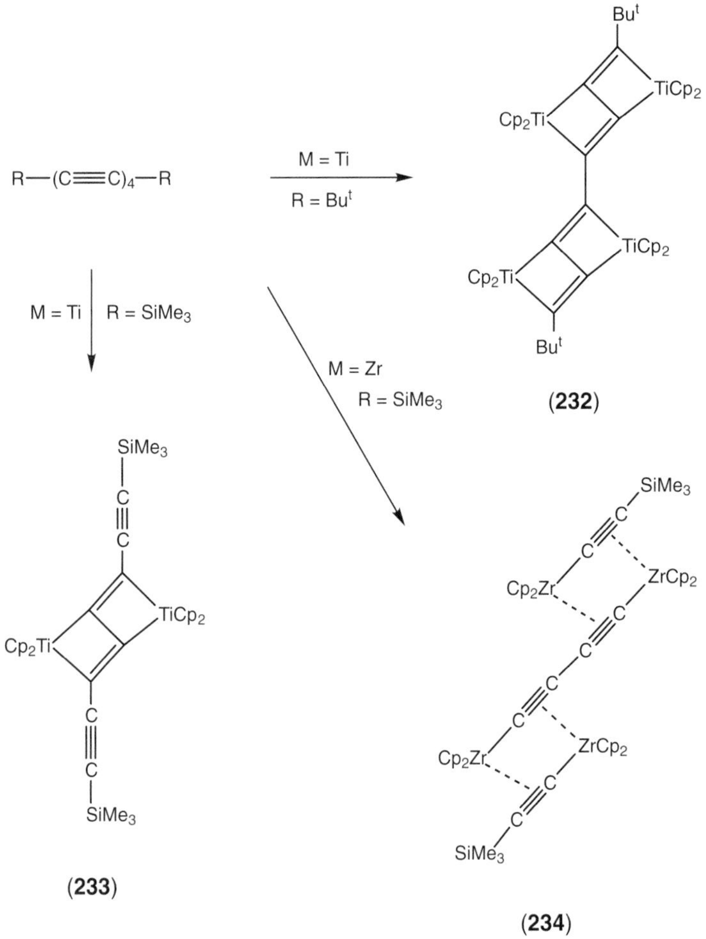

R—(C≡C)$_4$—R

M = Ti, R = Bu^t → (232)

M = Ti, R = SiMe$_3$ → (233)

M = Zr, R = SiMe$_3$ → (234)

SCHEME 51

Treatment of $\{Zr(\mu\text{-}Cl)Cp\}_2(\mu\text{-}Fv)$ with LiC≡CR (R = Ph, $SiMe_3$) gave orange $\{Zr(\mu\text{-}\eta^1\text{:}\eta^2\text{-}C_2R)Cp\}_2(\mu\text{-}Fv)$. The $SiMe_3$ complex is fluxional by exchange of the $\eta^1\text{:}\eta^2$ bonds on the Zr centers.[326,328] Comparison of kinetic parameters with the phenylethynyl complex showed no bond breaking occurs; the transition state for the latter is stabilized by conjugation with the Ph π-electrons.

Extension of this chemistry to complexes containing η^6-boratabenzene ligands confirmed the expectation that the derivatives would be analogous to the Cp complexes. Zirconacycles analogous to **219** were obtained from RC≡CC≡CR and $Zr(PMe_3)_2(\eta^6\text{-}C_5H_5BX)_2$ (R = Ph, X = Ph, NPr^i_2; R = Et, X = Ph). However, $Me_3SiC{\equiv}CC{\equiv}CSiMe_3$ did not react.[329]

6. *Derivatives Containing Cp* Ligands*

Replacement of the Cp group by Cp* on the Group 4 metals results in significant changes in the chemistry, largely as a result of participation of the ring-methyl groups in the chemistry. Other effects from increased steric bulk, stronger electron-donor properties, and even solubility also contribute. For Ti, η^2-alkyne complexes are formed, whereas for Zr, the metallacyclocumulenes $Zr(\eta^4\text{-}RC_4R)Cp^*_2$ predominate. As found in the Cp series, the nature of the product is dependent on metal and diyne substituent.[301,330]

The most useful synthetic route into these complexes is the reduction of $MCl_2Cp^*_2$ by magnesium in the presence of the diyne. With an excess of magnesium, the tweezer complex **235** is formed from $Me_3SiC{\equiv}CC{\equiv}CSiMe_3$,[331] while with a Ti/diyne/Mg ratio of 2/3/2, paramagnetic **236** is isolated, possibly being formed by coupling of ligands within an "ate" complex resembling **218** (Scheme 52).[332] With appropriate stoichiometry, the η^2-diyne complex **237** is formed.[333] In solution, a dynamic equilibrium involves shuttling of the $TiCp^*_2$ group between the two C≡C triple bonds, perhaps via the (unobserved) metallacumulene **238.** Reaction of **237** with CO_2 gives the titanafuranone **239.**

Similar reactions with RC≡CC≡CR [R = Me,[330] Ph[333]] result in activation of a Me group on each Cp* ring to give **240;** the phenyl derivative reacts with CO_2 to give **241,** perhaps via tautomer **240a** (Scheme 53). Two Me groups on the same Cp* ring are involved in the formation of complexes **242** in reactions of $Bu^tC{\equiv}CC{\equiv}CR$. Single isomers (with *exo*-Bu^t or *endo*-$SiMe_3$ groups) are formed (Scheme 54). The Bu^t derivative is converted to **243** on heating, which is also obtained from $Ti\{\eta^2\text{-}C_2(SiMe_3)_2\}Cp^*_2$ and the diyne at 140°.[334] Successive treatments of **242-Bu^t^** with HCl, Mg and HCl again result in hydrogenation of the eight-membered ring to give complexes **244–246.** Two chiral centers are present at the Bu^t-substituted carbons.[335]

The zirconium system reacts differently, zirconacumulenes **238-Zr** being isolated from reactions of $ZrCl_2Cp^*_2$ with RC≡CC≡CR (R = Me, Ph, $SiMe_3$) in the presence of magnesium, or by UV irradiation of $Zr(C{\equiv}CR)_2Cp^*_2$. The $SiMe_3$ derivative reacts with CO_2 to give **247.** With $Bu^tC{\equiv}CC{\equiv}CBu^t$, activation of Cp* ring methyl groups occurs to give **248** (Scheme 55).[330,333]

Detailed studies of these systems have enabled a rationalization of the observed chemistry to be made, involving as possible intermediates (a) the fulvene complexes $MHCp^*\{C_5Me_4(CH_2)\}$ or $MH_2Cp^*\{C_5Me_3(CH_2)_2\}$, (b) the η^2-diyne complexes $M(\eta^2\text{-}RC_2C{\equiv}CR)Cp^*_2$, and (c) the η^4-metallacumulenes, $M(\eta^4\text{-}RCCCCR)Cp^*_2$.[291,330]

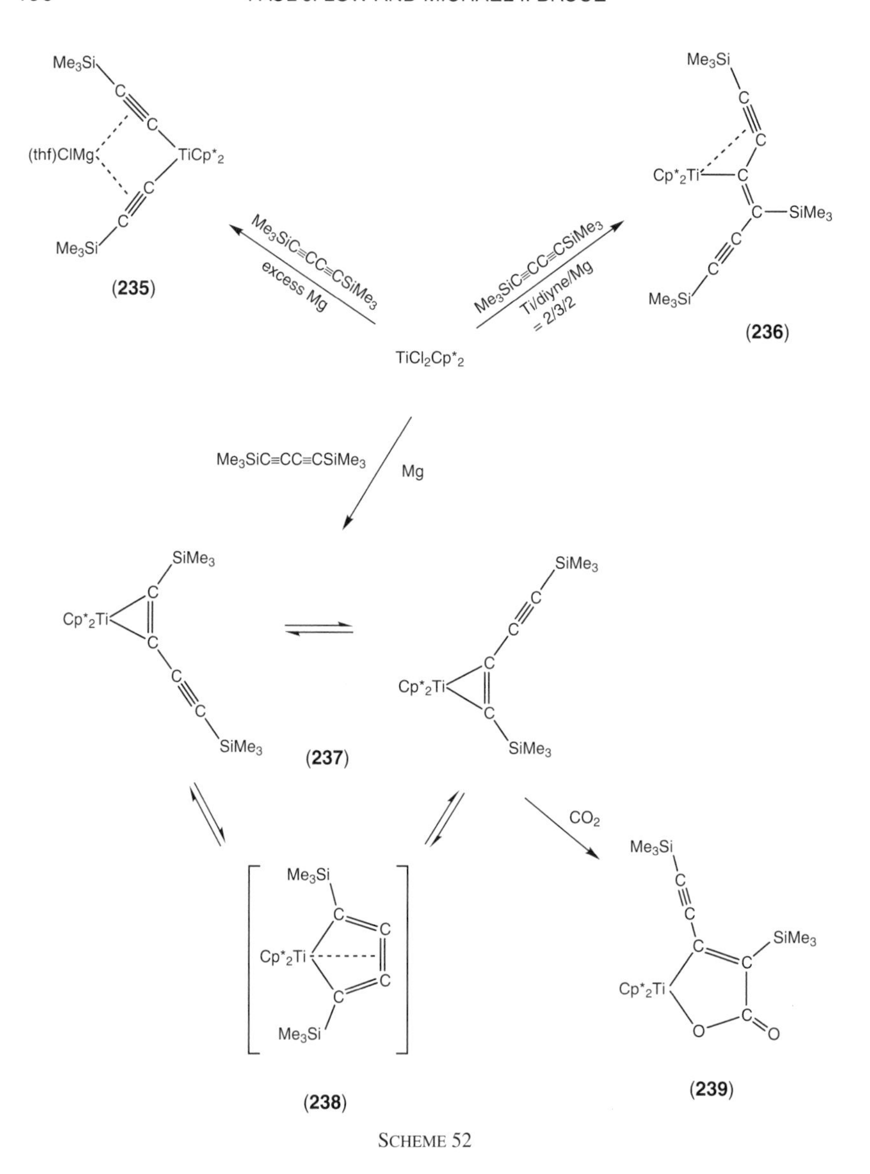

SCHEME 52

7. *Heterometallic Complexes*

The rare combination of two electron-poor metals in heterodimetallic complexes has been found in reactions of dialkynylmetallocenes with suitable metallocene precursors. Thus, reaction of $Zr(\eta^4\text{-}C_4H_6)Cp_2$ with $Hf(C{\equiv}CPh)_2Cp_2$ affords $\{Cp_2Zr\}(\mu\text{-}C_2Ph)_2\{HfCp_2\}$ (**222-Zr/Hf**).[336] Similar reactions of VCp_2 with

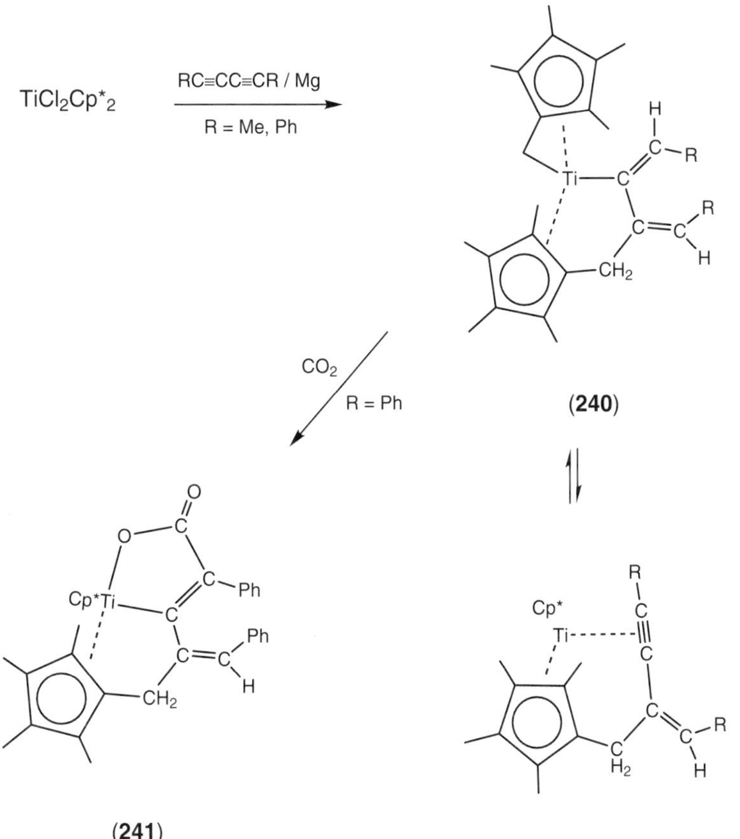

SCHEME 53

$M(C{\equiv}CPh)_2Cp'_2$ (M = Ti, Cp′ = Cp; M = Zr, Cp′ = Cp, Cp^{Me}, Cp^{Si}, Cp^{Bu^t}) give the paramagnetic (~1.73 μ_B) complexes of type **221,** in which the vanadium is considered to be V(IV), by coordination of a VCp_2 fragment to the zirconocumulene.[337] The trimethylsilyl diyne is, however, unreactive.

The reactions between η^2-diyne-nickel complexes and "MCp_2" have similarly given a range of products.[307,317,338] With the PhC≡CC≡CPh complex, formation of the cumulene system **249** is found, whereas with the $Me_3SiC{\equiv}CC{\equiv}CSiMe_3$ derivative, cleavage of the central C–C bond occurs to give **250;** rearrangement to **251** is found on treatment with CO (Scheme 56). The "mixed" diyne, $PhC{\equiv}CC{\equiv}CSiMe_3$, gives tweezer complex **252,** which with PPh_3 eliminates $PhC{\equiv}CSiMe_3$ to give **253.** It is concluded that while equilibria exist in solution

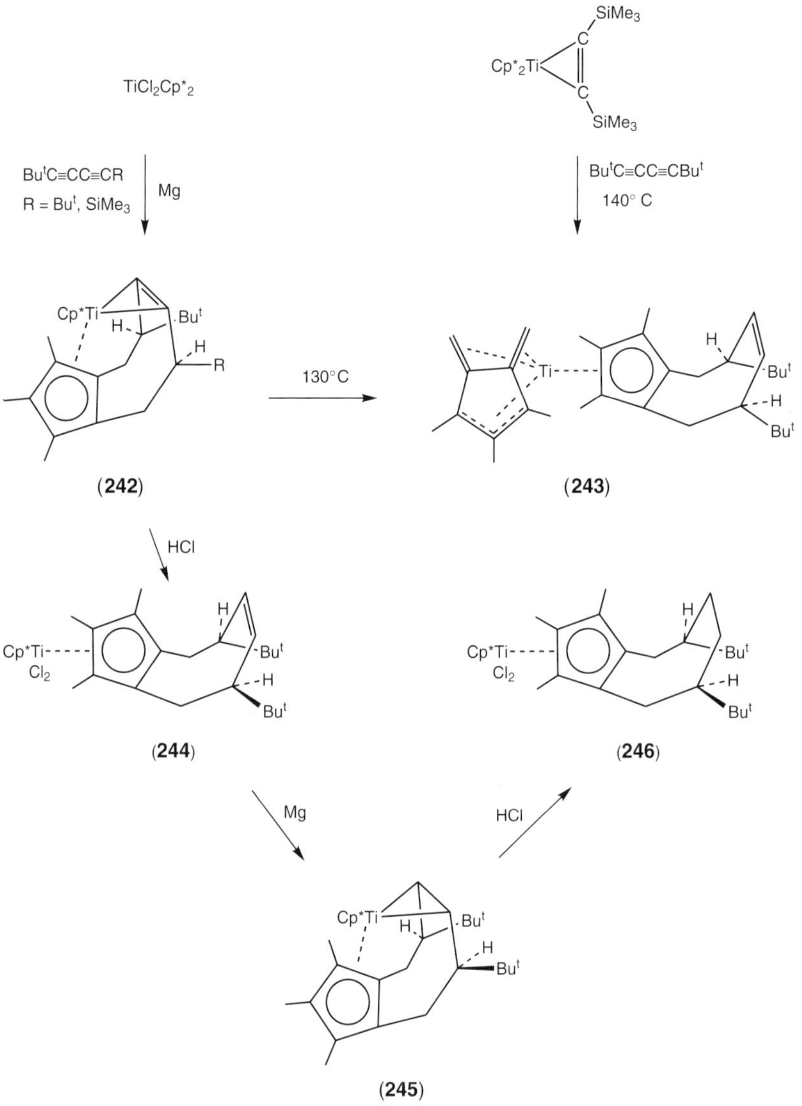

SCHEME 54

between the various structures, in the solid state, the energy minimum is determined by the metal, ligand, and diyne substituent. The tweezer complexes are favored by phenyl diynes, whereas more bulky substituents (or ligands) result in formation of the μ-η^1:η^2-alkynyl complexes such as **250.** The different products are rationalized by a common cyclocumulene complex which reacts either

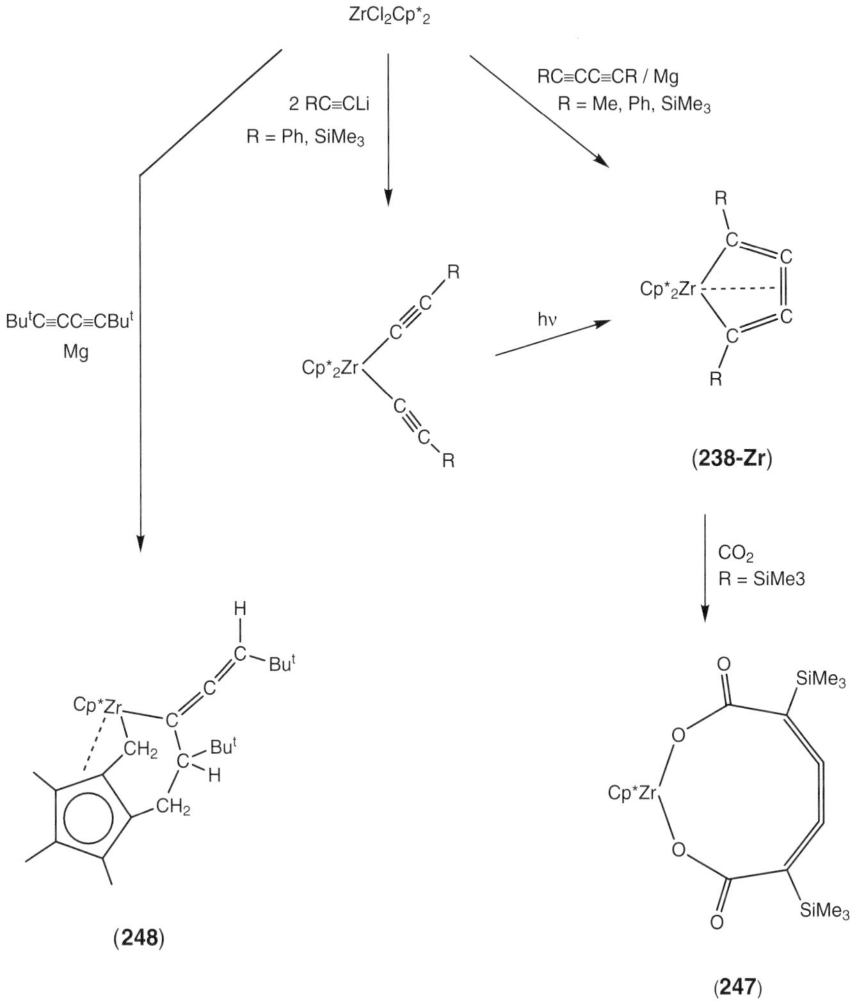

Scheme 55

intramolecularly via bis-alkynyl derivatives or intermolecularly, followed by C—C bond cleavage. The solid-state structures are energy minima, influenced by metal, substituents, and co-ligands, small ligands and substituents favoring the tweezer complexes, while bulky ligands result in unsymmetrical complexes. The formally M^{III}—Ni^{I} complexes are diamagnetic via an electronic coupling via the bridging groups, although a resonance contributor based on Ni^{0} and Ti^{IV} can also be considered.

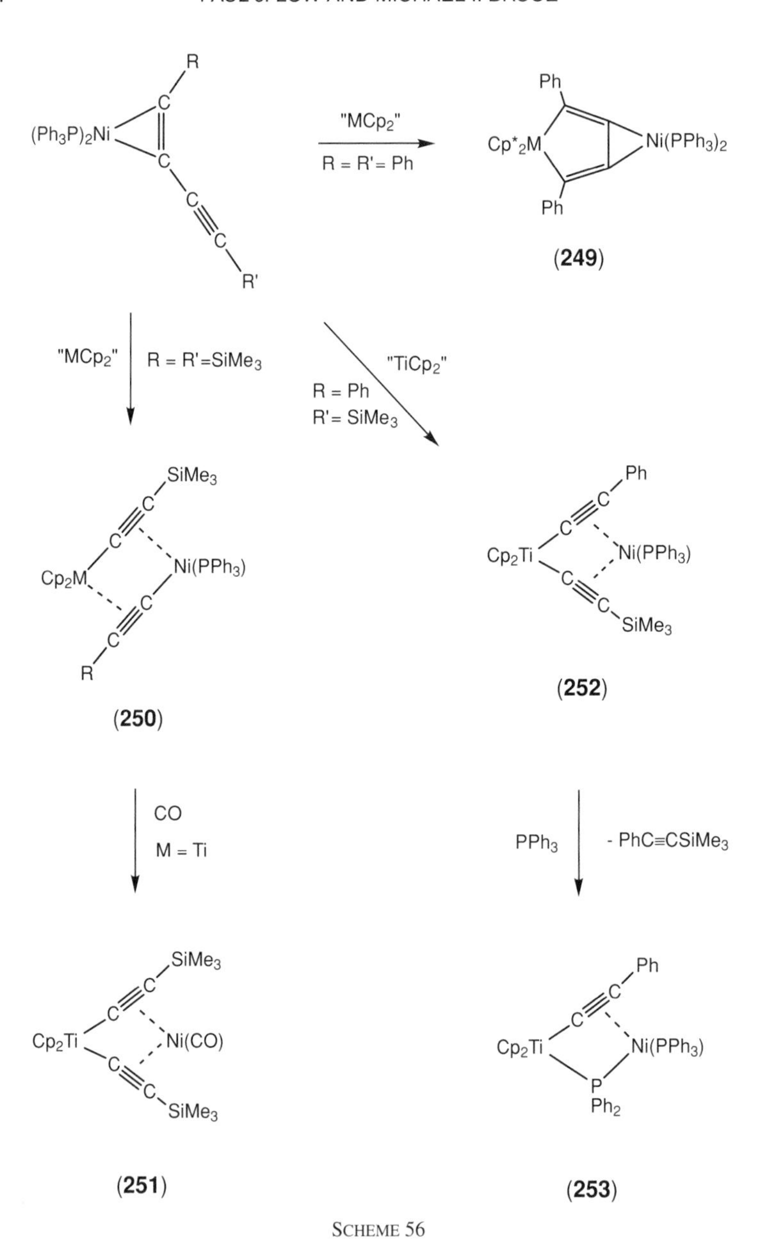

R
(Ph3P)2Ni
C
C
C
C
R'
"MCp2"
R = R'= Ph
Ph
Cp*2M
Ni(PPh3)2
Ph
(249)
"MCp2"
R = R'=SiMe3
"TiCp2"
R = Ph
R'= SiMe3
SiMe3
Cp2M
Ni(PPh3)
R
(250)
Ph
Cp2Ti
Ni(PPh3)
SiMe3
(252)
CO
M = Ti
PPh3
- PhC≡CSiMe3
SiMe3
Cp2Ti
Ni(CO)
SiMe3
(251)
Ph
Cp2Ti
Ni(PPh3)
P
Ph2
(253)

SCHEME 56

8. *Planar Tetracoordinate Carbon*

An interesting feature in some of these alkynyl/diyne complexes is the geometry of the bridging carbons, whereby one or two of these may be considered to have a planar tetracoordinate geometry. While many examples, particularly but not exclusively, have been found in the chemistry of the Group 4 metallocenes, these are resticted to having only one such atom.[293,339] Most approaches to the Group 4 examples involve the alkynyl-metallocene, which adds another species XMR_n to give the dimetallacycle **254,** in which C(2) interacts with four other atoms while retaining planar geometry.

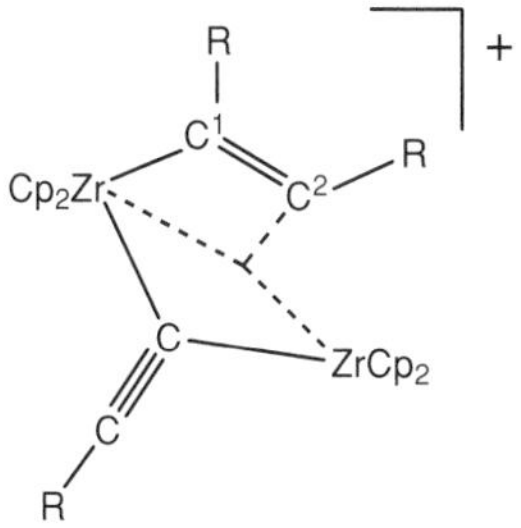

(**254**)

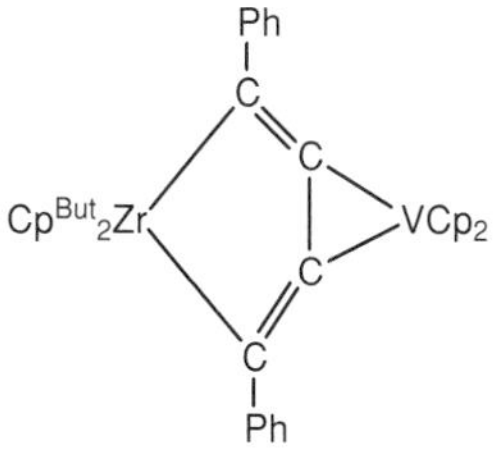

(**257**)

The β-carbon is planar tetracoordinate in the complexes $[\{ZrCp_2\}_2(\mu\text{-}C{\equiv}CMe)(\mu\text{-}MeC_2C{\equiv}CMe)]^+$ (**255**) and $[\{ZrCp_2\}_2(\mu\text{-}N{=}CHPh)(\mu\text{-}MeC_2C{\equiv}CMe)]^+$ (**256**) (Scheme 57). Complex **255** is obtained from $[Zr(C{\equiv}CMe)Cp_2]^+$ (from $Zr(C{\equiv}CMe)_2Cp_2$ and $[CPh_3]^+$) and $Zr(C{\equiv}CMe)_2Cp_2$,[340] while the same cation reacts with $Zr(C{\equiv}CMe)({=}N{=}CHPh)Cp_2$ to give **256.** The dynamic behavior of the latter is consistent with interconversion of the two diastereomers (*cis/trans* ratio 4:1) by rearrangement of the Zr_2-diyne framework with concomitant symmetrization and rotation of the μ-aldimino group.[341] Formation of **255** occurs by coupling of the two propynyl groups; the resulting hexa-2,4-diynyl group bridges two bent $ZrCp_2$ moieties in the $\eta^1{:}\eta^2$ mode. The asymmetric bridge results in atom C(2) becoming planar tetracoordinate as part of a three-center–two electron Zr—C—Zr fragment. The attachment of the remaining C≡C triple bond to Zr(2) is very unsymmetrical, apparently by C(3) only (a π-agostic interaction). The structural data obtained for the internal carbons of the C_4 ligand in $\{ZrCp^{But}{}_2\}(\mu\text{-}PhC_4Ph)\{VCp_2\}$ (**257**) confirm that the two carbons are planar tetracoordinate, a finding confirmed by detailed analysis of the electron localization function (ELF).[342]

In the alkyne chemistry described previously, combination of bis-alkynyl complexes with a second metal complex gives the "tweezer" complex, which may

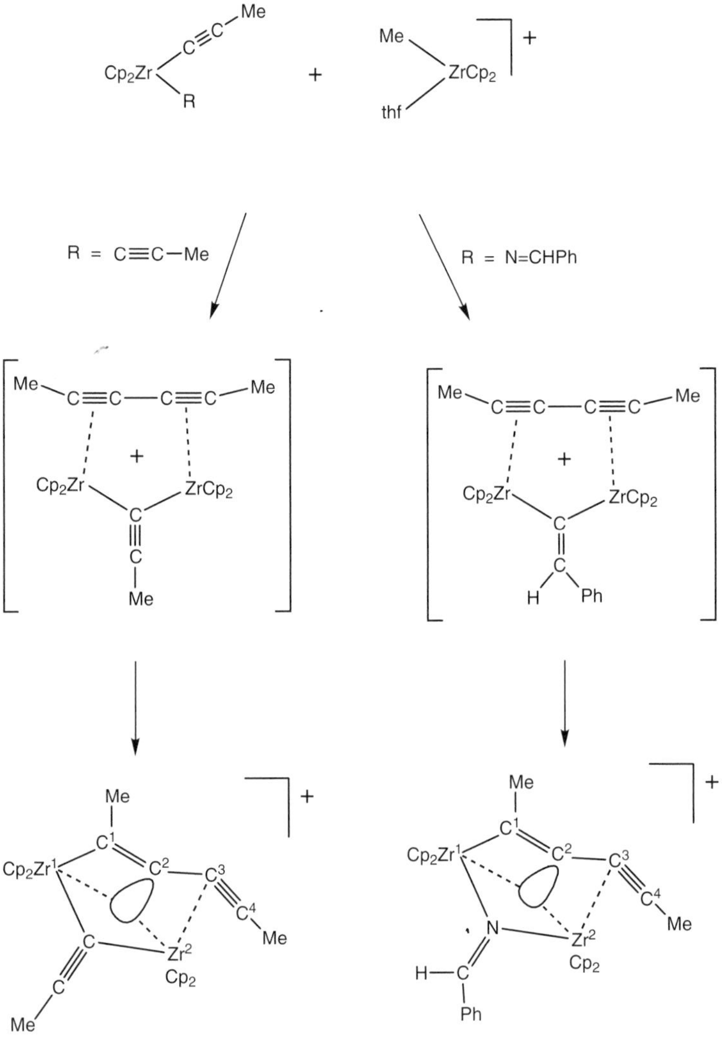

SCHEME 57

rearrange via a symmetrical intermediate μ-η^1-alkynyl to the μ-η^1:η^2 complex.[318] Coupling of alkynyl groups, particularly at a zirconium center in the presence of Lewis acids, has given a variety of products in which the Lewis acid is usually attached to the resulting C_4 ligand and therefore lies outside the scope of this article. However, while $Zr(C{\equiv}CMe)_2Cp_2$ reacts in this way with $B(C_6F_5)_3$ to give

$Zr\{\eta^1,\eta^2\text{-}CMe{=}C(BAr_3)C_2Me\}Cp_2$, with a deficiency of $B(C_6F_5)_3$ (only 1% is necessary) or with $[CPh_3]^+$, catalytic coupling of propynyl groups occurs to give **255**.[343] Dynamic behavior in CD_2Cl_2 indicates rapid exchange of Zr centers via a $\mu\text{-}\eta^2{:}\eta^2$-diyne intermediate, the unsymmetrically bridged structure being favored over the dimetallacyclopentene by ca 10–12 kcal mol^{-1}.[340,344]

9. *Other Related Chemistry*

trans-3,4-Dibenzylidene-1,6-diphenylhex-1-en-5-yne was the sole product from **228** and HCl, formed by H-shift and ring-opening of the radialene; complex **227** gives the *cis* isomer.[301] The intermediate $[Zr(C{\equiv}CPh)(\eta^2\text{-}PhC_2C{\equiv}CPh)Cp_2]^-$ reacts with HCl to give *E*-PhC≡CCH=CHPh and with I_2 to give PhC≡CC≡CPh.[292]

Trace amounts of $Ti\{OC(O)CBu^tC(C{\equiv}CBu^t)\}Cp_2$ (**258**) were obtained during attempts to recrystallize $Ti(C{\equiv}CBu^t)_2Cp_2$ from Et_2O cooled by dry ice; this

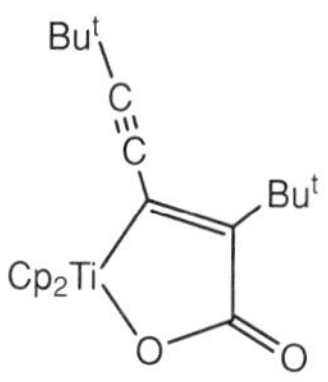

(**258**)

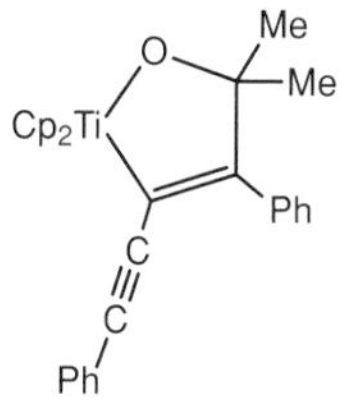

(**259**)

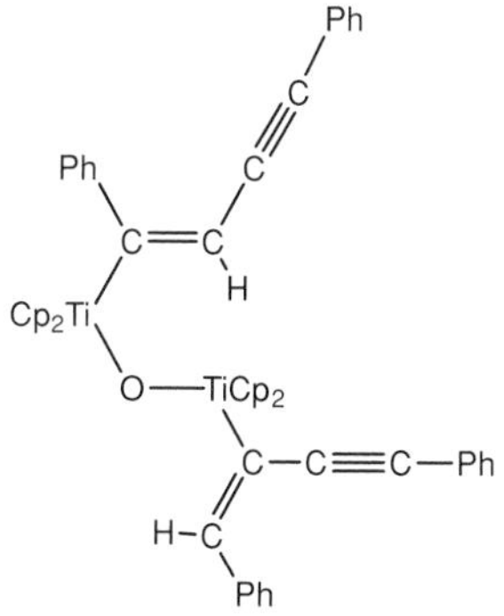

(**260**)

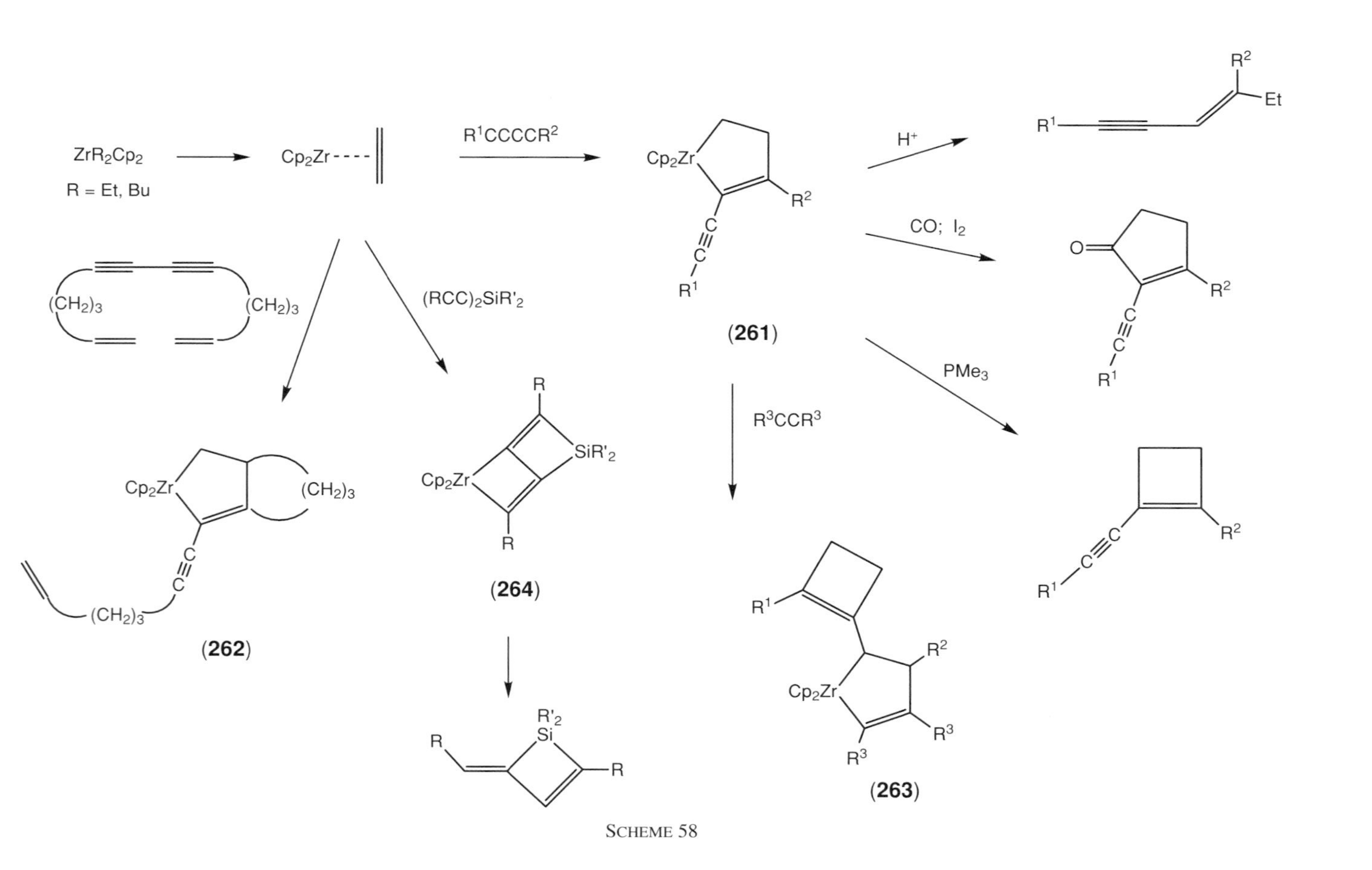
ZrR2Cp2
R = Et, Bu
Cp2Zr
R1CCCCR2
H+
CO; I2
PMe3
(RCC)2SiR'2
R3CCR3
(261)
(262)
(263)
(264)

SCHEME 58

interesting complex could not be obtained from CO_2 and $Ti(C{\equiv}CBu^t)_2Cp_2$, $[Li(thf)_2][Ti(C{\equiv}CBu^t)_2Cp_2]$ or $Ti(\eta^2\text{-}Bu^tC_2C{\equiv}CBu^t)Cp_2$.[345]. Other reactions of the diphenyl analogue noted briefly are those with acetone and water, which afford the titanafuran $Ti\{OCMe_2CPh{=}C(C{\equiv}CPh)\}Cp_2$ (**259**) and $\{Ti[C(C{\equiv}CPh){=}CHPh]Cp_2\}_2O$ **(260),** respectively.[311]

Zirconacyclopentenes are intermediates in the catalytic reaction between 1,3-diynes and EtMgBr to give enynes; in contrast to the stoichiometric reaction, mixtures of stereoisomers are obtained (Scheme 58). Stoichiometric reactions of $ZrEt_2Cp_2$ [a source of $Zr(\eta\text{-}C_2H_4)Cp_2$] with $R^1C{\equiv}CC{\equiv}CR^2$ ($R^1 = R^2 =$ Me, Et, Pr, Ph, $SiMe_3$; $R^1 =$ Ph, $R^2 =$ Bu, Bu^t) gave **261.** With tetradeca-1,13-dien-6,8-diyne, bicyclic **262** is formed.[346] In contrast with monoalkynes, which afford zirconacyclopentadienes, no incorporation of a second molecule of diyne was found, with the exception of $Me_3SiC{\equiv}CC{\equiv}CSiMe_3$.[297] Treatment of zirconacyclopentenes with CO and quenching with iodine gave alkynylcyclopentenones,[347] while reaction of the intermediate with H^+ gave the corresponding enynes.[346] Heating in the presence of PMe_3 afforded alkynylcyclobutenes, while with alkynes, cyclobutenylzirconacyclopentenes **263** are formed.[348] Reactions of bis(alkynyl)silanes with $ZrEt_2Cp_2$ give 2-methylenesilacyclobutenes after hydrolysis, possibly via intermediate **264.**[349]

VII

OTHER REACTIONS OF DIYNES WITH METAL COMPLEXES

There are many examples of reactions of diynes with metal species that give mono or binuclear products containing ligands other than simple σ- or π-bonded diynes. This chemistry is summarized in this section and there are obvious connections with that presented earlier as many of the products are derived from isomerization, rearrangement, and bond-forming reactions of initially formed η^2 diyne complexes.

A. *Mono- and Binuclear Complexes from Metal Carbonyls*

1. *Iron*

Reactions of $PhC{\equiv}CC{\equiv}CPh$ with iron carbonyls [$Fe(CO)_5$, $Fe_2(CO)_9$, or $Fe_3(CO)_{12}$] give isomers of complexes $Fe(CO)_4\{(diyne)_2\}$ (**265**), $Fe_2(CO)_6\{(diyne)_2\}$ (**266**), and $Fe_2(CO)_7\{(diyne)_2\}$ (**267**), to which structures analogous to those found for similar products obtained from C_2Ph_2 were ascribed; all three isomers of the second complex were formed.[160] The reactions of hexa-2,4-diyne and $Fe(CO)_5$ have been described in more detail. UV irradiation of mixtures of the two reagents in thf gave Fe$\{2\eta^1$-C(O)CMe=C(C≡CMe)C(O)$\}$(CO)$_4$ (**268**) as the only product.[350] In cyclohexane, a complex with composition Fe(CO)$_4$(MeC$_2$C$_2$Me)$_2$ was formed, possibly **265a.** The three possible isomers of Fe$_2\{2\eta^1$:η^4-C$_4$Me$_2$(C≡CMe)$_2\}$(CO)$_6$ (**266a–c**) have been obtained from the thermal reaction of the diyne with Fe(CO)$_5$.[351,352] One product of the photochemical reaction between Fe(CO)$_5$ and Me$_3$SiC≡CC≡CSiMe$_3$ in benzene is the fly-over complex Fe$_2\{\mu$-[η^1:η^2-C(C≡CSiMe$_3$)=C(SiMe$_3$)]$_2$CO$\}$(CO)$_6$ (**267a**) rather than the octacarbonyl previously reported.[181,353] The ynamine complex Fe$_2$(μ-PhC$_2$NEt$_2$)(CO)$_7$ reacts with PhC≡CC≡CPh to give three isomers of the ferrole Fe$_2\{\mu$-2η^1:η^4-CPhC(NEt$_2$)CPhC(C≡CPh)$\}$(CO)$_6$ (**266d–e**).[224]

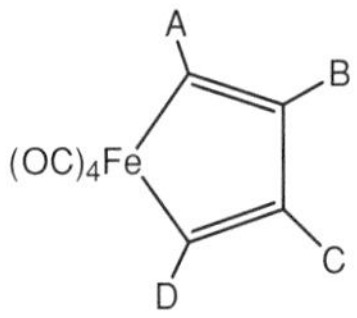

(**265**)

	A	B	C	D
a	C≡CMe	Me	Me	C≡CMe
b	C≡CPh	Ph	Ph	C≡CPh
c	C≡CPh	Ph	C≡CPh	Ph
d	Ph	CPh	C≡CPh	Ph

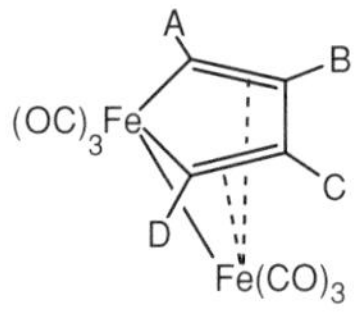

(**266**)

	A	B	C	D
a	Me	C≡CMe	C≡CMe	Me
b	Me	C≡CMe	Me	C≡CMe
c	C≡CMe	Me	Me	C≡CMe
d	Ph	NEt$_2$	Ph	C≡CPh
e	Ph	NEt$_2$	C≡CPh	Ph
f	NEt$_2$	Ph	C≡CPh	Ph

Reduction of $ClCH_2C{\equiv}CC{\equiv}CCH_2Cl$ with zinc in the presence of $Fe_3(CO)_{12}$ gave a mixture of the *syn* and *anti* hexapentaene-$\{Fe_2(CO)_6\}_2$ complexes (**269**). Similar reactions of $ClCMe_2C{\equiv}CC{\equiv}CCMe_2Cl$ afforded the analogous isomeric tetramethyl derivatives, together with $Fe_2(\mu\text{-}\eta^3{:}\eta^3\text{-}Me_2CCCCCCMe_2)(CO)_6$ (**270**).[354]

Reactions of 1,3-diynes with $Fe_2(\mu\text{-}H)(\mu\text{-}PPh_2)(CO)_7$ (prepared in situ from $Na[Fe_2(\mu\text{-}PPh_2)(CO)_8]$ and HBF_4) give $Fe_2(\mu\text{-}PPh_2)\{\mu\text{-}\eta^1{:}\eta^2\text{-}C(C{\equiv}CR){=}CHR\}(CO)_6$ (**271;** R = Me, Ph) by *cis* hydrometallation of only one C≡C triple bond.[355]

Several 1,3-diynes HC≡CC≡CR (R = H, Me, Bu) react with $Fe_2(\mu\text{-}EE')(CO)_6$ (E, E′ = S, Se, Te) in the presence of NaOAc to give complexes in which the diyne has coupled with the chalcogen atoms in $Fe_2\{\mu\text{-}ECH{=}C(C{\equiv}CR)E'\}(CO)_6$ (**272**) or $\{(OC)_6Fe_2(\mu\text{-}E)\}_n\{CHC(C{\equiv}CR)\}$; in most cases, both isomers were obtained. In solution, the SSe derivative slowly disproportionates to the homochalcogen complexes. The uncoordinated C≡C triple bond has been used to prepare heterometallic complexes with $Mo_2(CO)_4Cp_2$, $Co_2(CO)_6$ and $M_3(\mu\text{-}CO)(CO)_9$ (M = Ru, Os) fragments attached.[356–358] Reactions of $Fe_2(\mu\text{-}Se_2)(CO)_6$ with HC≡CC≡CR (R = $SiMe_3$, $SnBu_3$) in the presence of NaOAc/MeOH, or directly with HC≡CC≡CH in lower yield, give $\{Fe_2(CO)_6(\mu\text{-}Se)_2\}_2(\mu\text{-}C_4H_2)$ in which the diene has the *s-trans* conformation.[359]

The STe complex gives a single isomer of $Fe_2\{\mu\text{-}SC(C{\equiv}CMe){=}CHTe\}(CO)_6$, in agreement with EH MO calculations[360] which show the relative energies of isomeric forms to be

$$\begin{aligned} &MeC{\equiv}CC(S){=}CH(Te) < HC(S){=}C(C{\equiv}CMe)(Te) \\ &\quad < MeC(S){=}C(C{\equiv}CH)(Te) < HC{\equiv}CC(S){=}CMe(Te). \end{aligned}$$

2. *Ruthenium*

Several reactions of 1,3-diynes with ruthenium cluster carbonyls have given mono- and bi-nuclear complexes (see Section IV.B.2). Reactions of $Ru(CO)_3(PPh_3)_2$ and PhC≡CC≡CPh, carried out under CO_2, give $Ru(CO)_2(PPh_3)\{\eta^4\text{-}C_4Ph_2(C{\equiv}CPh)_2CO\}$ (**273a,b;** 4/1 ratio) formed by head-to-tail and head-to-head coupling, together with $Ru(CO)(PPh_3)_2\{\eta\text{-}C_4Ph_2(C{\equiv}CPh)_2\}$ (**274a,b; 1/1**). The role of the CO_2 is assumed to aid production of transient *cis*- and *trans*-isomers of $[Ru(CO)_4(PPh_3)_2]^{2+}$ which afford the cyclopentadienone and cyclobutadiene complexes, respectively.[361]

3. *Cobalt*

Cobalt carbonyl complexes react with 1,3-diynes to give a variety of complexes in which two molecules of diyne have coupled to form η-cyclobutadiene ligands; slightly different conditions result in formation of cluster complexes (see Section VII.E.2). In the mixture of complexes obtained from the reaction

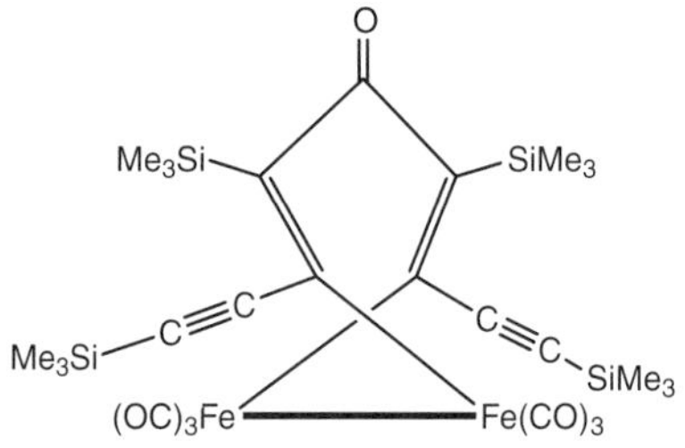

O
Me3Si SiMe3
Me3Si C≡C C≡C SiMe3
(OC)3Fe Fe(CO)3

(**267**)

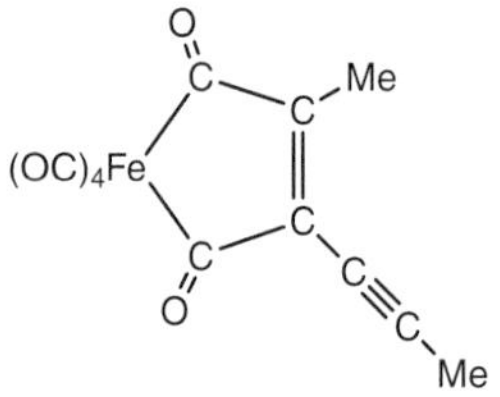

O
C Me
(OC)4Fe C
C
C C
O C
Me

(**268**)

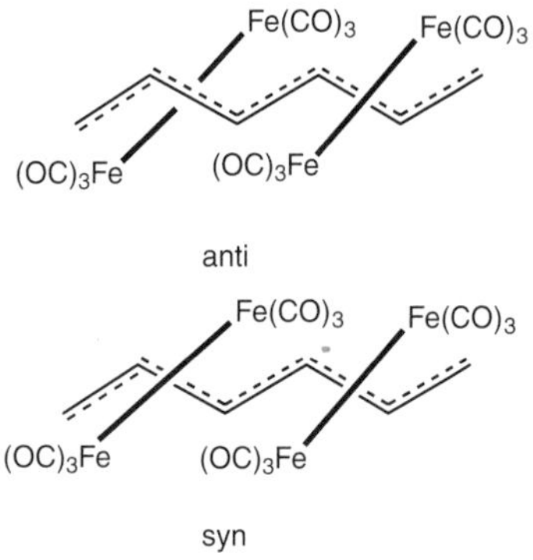

Fe(CO)3 Fe(CO)3
(OC)3Fe (OC)3Fe

anti

Fe(CO)3 Fe(CO)3
(OC)3Fe (OC)3Fe

syn

(**269**)

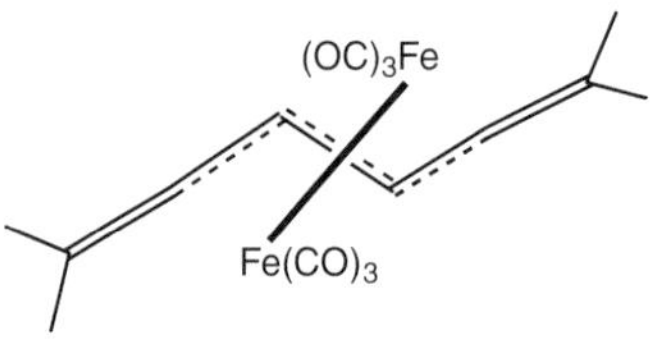

(OC)3Fe
Fe(CO)3

(**270**)

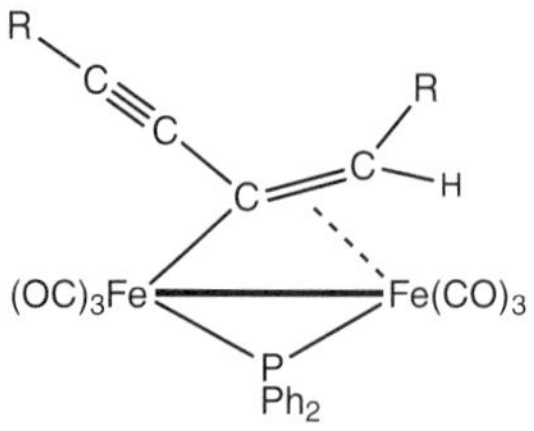

R
C≡C R
C=C H
(OC)3Fe Fe(CO)3
P
Ph2

(**271**)

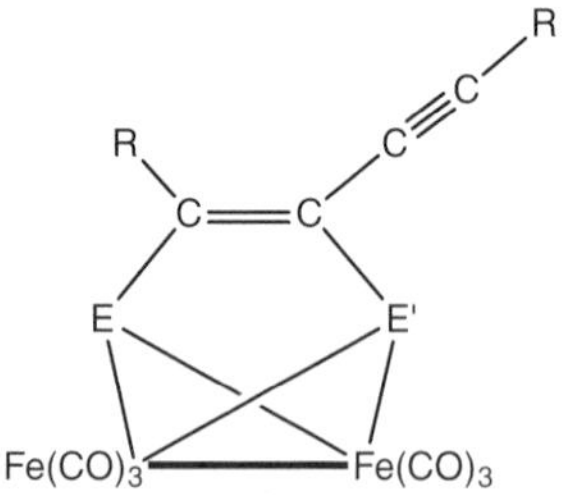

R
R C≡C
C=C
E E'
Fe(CO)3 Fe(CO)3

(**272**)

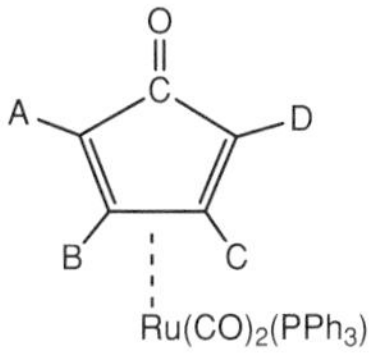

(273)

	A	B	C	D
a	Ph	CCPh	CCPh	Ph
b	Ph	CCPh	Ph	CCPh

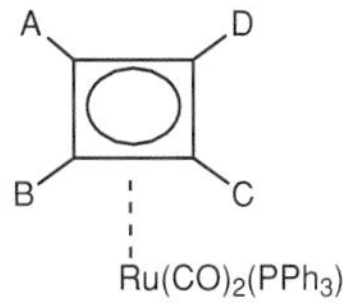

(274)

	A	B	C	D
a	Ph	CCPh	CCPh	Ph
b	Ph	CCPh	Ph	CCPh

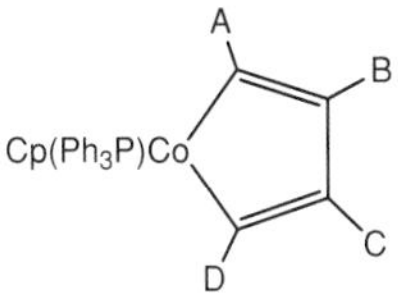

(278)

	A	B	C	D
a	CCPh	Ph	Ph	CCPh
b	Ph	CCPh	Ph	CCPh

of $Me_3SiC{\equiv}CSiMe_3$ with $Co(CO)_2Cp$ (Scheme 59) $Co\{\eta^4\text{-}C_4(SiMe_3)_3(C{\equiv}CSiMe_3)\}Cp$ (**275**) is assumed to arise from $Me_3SiC{\equiv}CC{\equiv}CSiMe_3$, perhaps formed by the metathesis reaction

$$2\,Me_3Si{-}C{\equiv}C{-}SiMe_3 \rightleftharpoons Me_3Si{-}SiMe_3 + Me_3Si{-}C{\equiv}C{-}C{\equiv}C{-}SiMe_3,$$

presumably catalyzed by the cobalt complex. However, no alkyne metathesis was found in the reactions of the diyne itself.[362]

In refluxing decane, a mixture of $Me_3SiC{\equiv}CC{\equiv}CSiMe_3$ and $Co(CO)_2Cp$ gives cyclobutadiene (**275a,b;** Scheme 59) and cyclopentadienone complexes (**276**), which could be protodesilylated with ethanolic KOH.[363,364] Complexes **275** were also obtained by co-dimerization of the diyne with $Me_3SiC{\equiv}CSiMe_3$ or from a mixture of the diyne and $Me_3Si(C{\equiv}C)_3SiMe_3$. The cyclopentadienones were formed by preferential oxidative coupling of two diyne molecules. Two isomeric tris(ethynyl)benzenes (**277a,b**) were also formed, while with a large excess of the diyne, **277c,** was isolated, probably arising from 3% of $Me_3Si(C{\equiv}C)_3SiMe_3$ present in the diyne. The complexes can be protodesilylated and separated to give the 1,2- and 1,3-isomers of $Co\{\eta\text{-}C_4(SiMe_3)_2(C{\equiv}CH)_2\}Cp$. Thermal decomposition of $Co\{\eta\text{-}1,2\text{-}C_4R_2(C{\equiv}CH)_2\}Cp$ occurs via $Co\{\eta\text{-}1,2\text{-}C_4H_2(C{\equiv}CR)_2\}Cp$ to give $RC{\equiv}CC{\equiv}CR$ and C_2H_2.[363]

The 1,2-isomer was flash pyrolyzed (525°C, 10^{-4} Torr) to $Co\{\eta\text{-}1,2\text{-}C_4H_2(C{\equiv}CSiMe_3)_2\}Cp$, which in turn was converted to $Co\{\eta\text{-}1,2\text{-}C_4H_2(C{\equiv}CH)_2\}Cp$. Possible intermediates are indicated in Scheme 60.[363] Flash pyrolysis (0.005 s, 10^{-5} torr, up to 800°C) of **275a** in a quartz tube gives **275b** (the 1,2-diethynylcyclobutadiene rearrangement; Scheme 60), probably to release steric strain of two adjacent $SiMe_3$ groups.[363,364] In general, such pyrolyses of $Co(\eta\text{-}C_4R^1{}_2R^2{}_2)Cp$, derived from 1,3-diynes, have shown that efficient reversion to the alkynes occurs. The rearrangement has $\Delta G^{\#}$ 37 kcal mol^{-1} and occurs by ring opening–rotation–reclosure mechanisms, while the $\Delta G^{\#}$ for activation of decomposition is ~47–50 kcal mol^{-1}. The CoCp bond strength is 64 kcal mol^{-1}.[365] Decomplexation is easier than the competing recyclization reaction. Mutual interconversion of the alkynylcyclobutadiene complexes occurs, but migration of the CoCp residue along the diyne chain does not occur.[365]

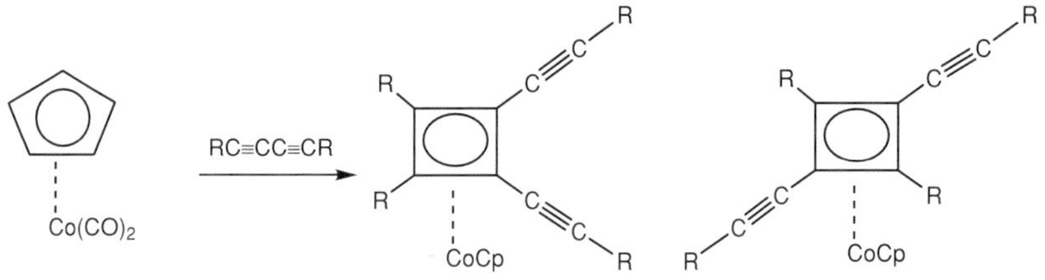

$RC{\equiv}CC{\equiv}CR$

a **b**

(**275**)

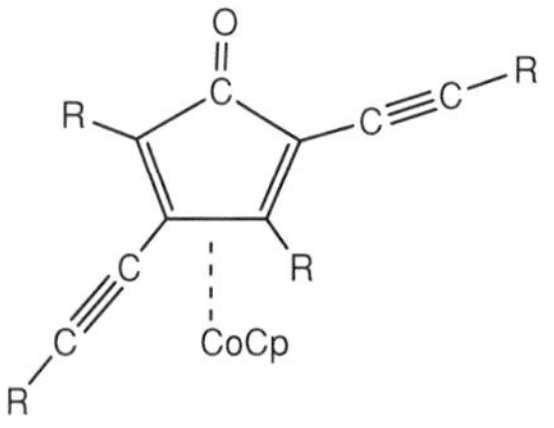

(**276**)

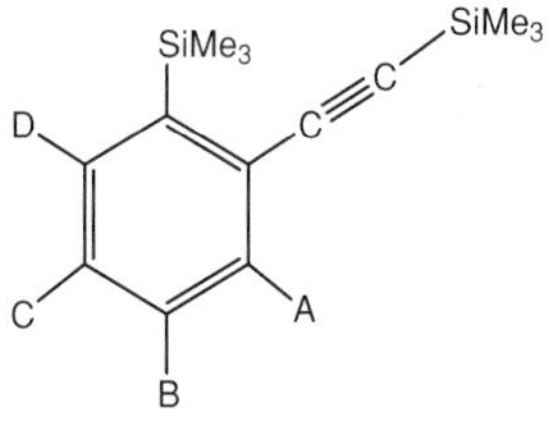

(**277**)

	A	B	C	D
a	$SiMe_3$	$C{\equiv}CSiMe_3$	$SiMe_3$	$C{\equiv}CSiMe_3$
b	$C{\equiv}CSiMe_3$	$SiMe_3$	$C{\equiv}CSiMe_3$	$SiMe_3$
c	$C{\equiv}CSiMe_3$	$SiMe_3$	$C{\equiv}CSiMe_3$	$C{\equiv}CSiMe_3$

SCHEME 59

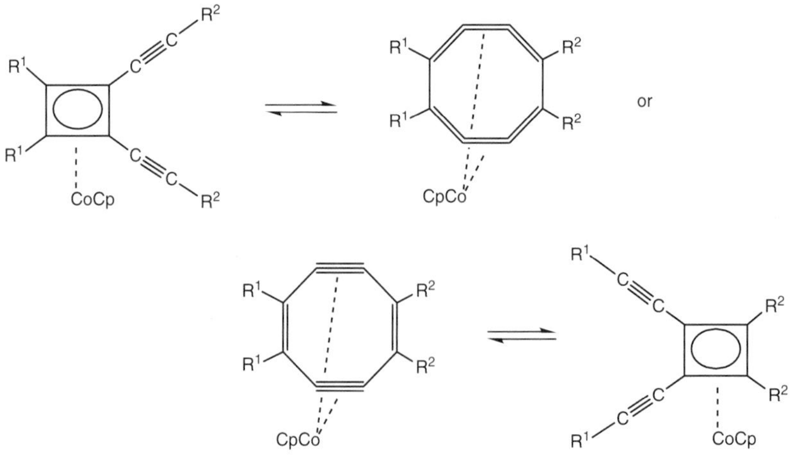

or

SCHEME 60

Three isomers of the cobaltacyclopentadiene, Co$\{(C_4Ph_2)_2\}$$(PPh_3)$Cp (**278**), were obtained from Co$(PPh_3)_2$Cp and PhC≡CC≡CPh, of which the major product was the 2,4-bis(phenylethynyl) compound. The 2,5-isomer was only obtained on heating. Some insoluble polymer was also formed, with average and highest MW 1.2×10^4 and 5.4×10^4, respectively.[366,367] The isomeric composition is dependent upon substituent bulk, as shown by only the 2,4- and 2,5-isomers being obtained with Me_3SiC≡CC≡CSiMe_3 and MeC≡CC≡CMe, respectively. The redox properties of these compounds show that the Me compound is oxidised most easily, and chemical reversibility in MeCN increases Me < SiMe_3 < 2,4-Ph_2.[367] The structures of Co(CR^1CR^2CR^3CR^4)(PPh_3)Cp [**278,** R^1=R^4= C≡CMe, R^2=R^3=Me (**a**); II, R^1=R^3=C≡CPh, R^2=R^4=Ph (**b**)] have been reported.[368]

4. *Rhodium*

The use of several rhodacyclopentadiene complexes in syntheses of polycyclic aromatic compounds has been described by Müller.[369] In general, alkynes displace the rhodium center to give substituted arenes. In this way, complex **279** [from 1,2-$\{$PhC≡CC(O)$\}_2C_6H_4$] reacts with PhC≡CC≡CR (R = Me, Ph) to give **280** (R = Me, Ph) (Scheme 61).[370]

B. *Formation of Unsaturated Carbene Complexes*

1. *Tungsten*

Sequential reactions of Li$(C≡C)_3C(NMe_2)_3$ [formed from Me_3Si$(C≡C)_3$ $C(NMe_2)_3$ and LiBu *in situ*], W$(CO)_5$(thf), and $BF_3(OEt_2)$ afford low yields of a separable mixture of W$\{$=C=C=C=C=C(NMe_2)CH=C$(NMe_2)_2\}(CO)_5$ and W$\{$=C(C≡CSiMe_3)CBu=C=C=C$(NMe_2)_2\}(CO)_5$, the former corresponding to a 1/1 adduct of $NHMe_2$ and W$\{$=C=C=C=C=C=C=C$(NMe_2)_2\}(CO)_5$, nucleophilic attack taking place at C_ε. The carbene complex is formed by attack of the lithium reagent on a W-CO group, followed by abstraction of an NMe_2 group by $BF_3(OEt_2)$.[103]

2. *Iron*

Metal-promoted 1,2-migration of silyl groups in silylalkynes results in the formation of silylvinylidenes which are subsequently readily desilylated.[77,371–373] Alkynyl-substituted silylvinylidenes have been obtained from silylated diynes and the $Fe(N_2)(CO)_2\{P(OMe)_3\}_2$/$\{Fe(CO)_2[P(OMe)_3]_2\}_2(\mu\text{-}N_2)$ reagent[76] and similar species are implicated in the reactions of several Group 8 metal complexes with mono- and bis-trialkylsilyl diynes.[32,89,119]

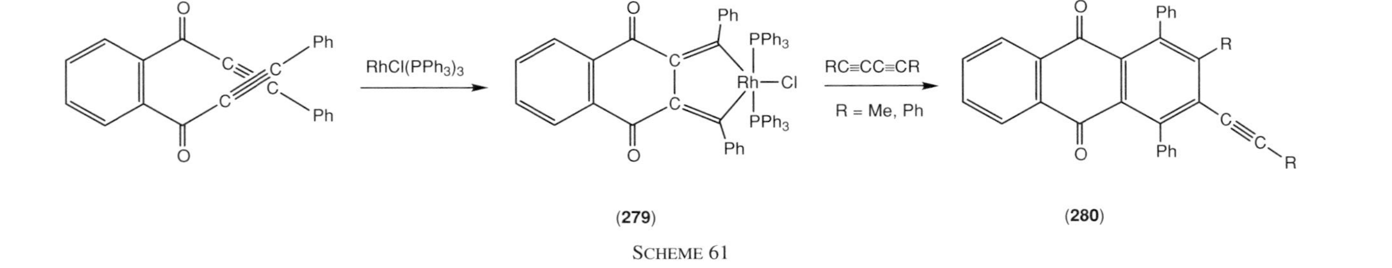

SCHEME 61

Addition of $HC{\equiv}CC{\equiv}CSiMe_3$ to FeCl(dppe)Cp* in MeOH in the presence of $NaBPh_4$ gave allenylidene $[Fe\{=C=C=CMe(OMe)\}(dppe)Cp^*]^+$, possibly via the unobserved intermediate $[Fe(=C=C=C=CH_2)(dppe)Cp^*]^+$ which adds solvent MeOH across the $C_\gamma=C_\delta$ bond.[374]

3. *Ruthenium*

Electrophilic complexes, such as $RuCl_2(PR_3)(\eta\text{-arene})$, react with 1,3-diynes to give metallacumulene intermediates, which are readily attacked by nucleophiles, e.g., MeOH solvent, at C_α or C_γ to give Fischer carbene or allenylidene complexes as the isolated products. If the reactions are carried out in CH_2Cl_2, dark blue solutions are formed, which on addition of R′OH form the violet allenylidenes. Evidence for the presumed cumulated intermediates has been obtained from reaction of $Me_3SiC{\equiv}CC{\equiv}CC(OSiMe_3)(C_6H_4NMe_2\text{-}4)_2$ with $RuCl_2(PMe_3)(\eta\text{-}C_6Me_6)$ which gives the deep blue C_5 cumulene **281.**[121]

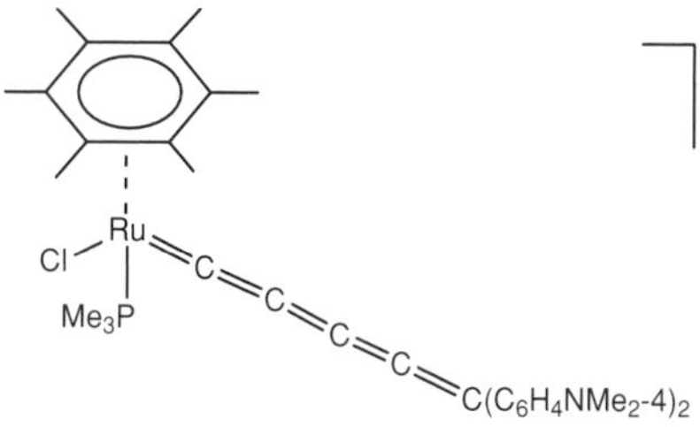

(**281**)

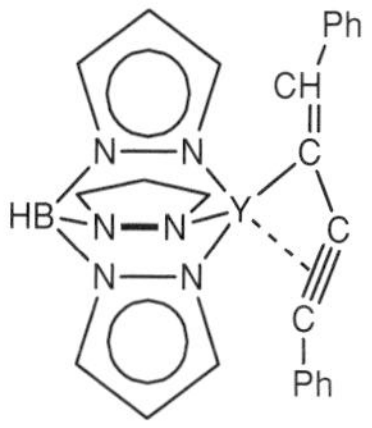

(**290**)

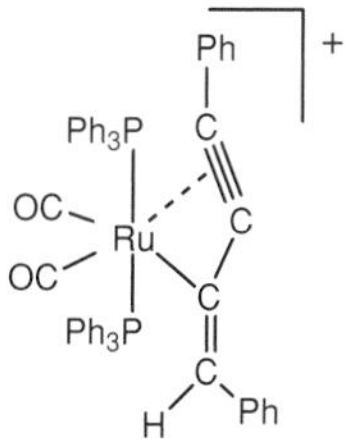

(**291**)

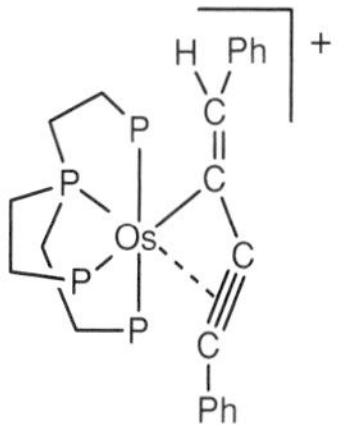

(**292**) P = PPh2

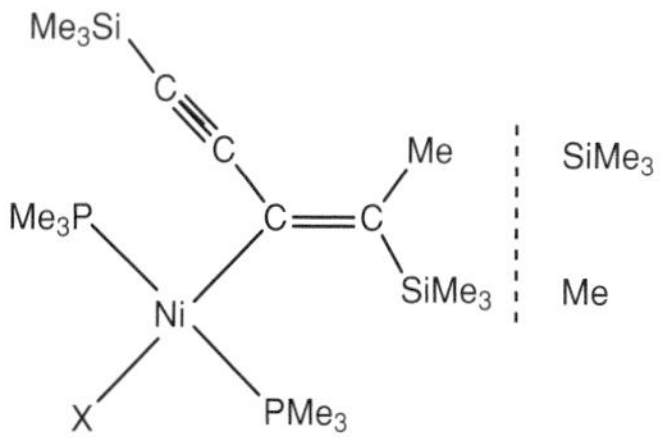

(**293**)

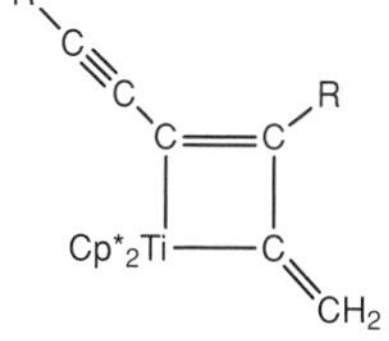

(**294**)

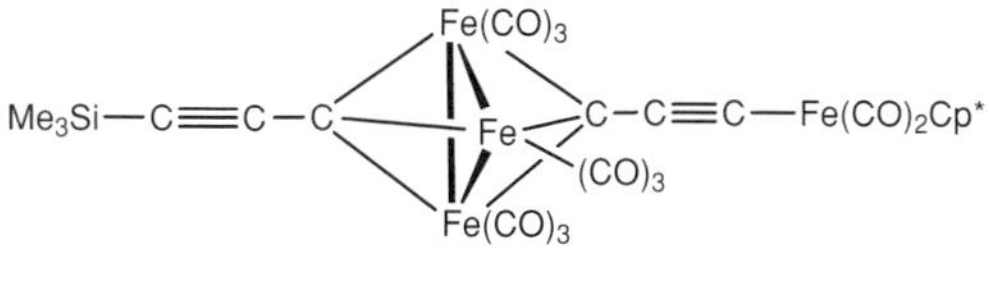

(**300**)

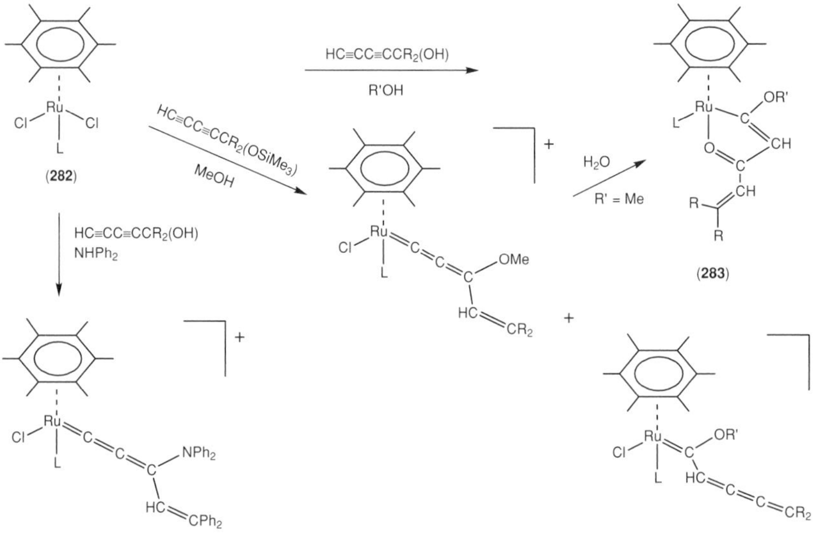

SCHEME 62

Reactions of $HC\equiv CC\equiv CCPh_2(OSiMe_3)$ with $RuCl_2(L)(\eta\text{-}C_6Me_6)$ (**282;** L = PMe_3, PMe_2Ph, $PMePh_2$; Scheme 62) result in elimination of $Si(OH)Me_3$ to give the corresponding pentatrienylidene complexes which rapidly add any nucleophile present. For example, the carbene $[RuCl\{=C(OMe)CH=C=C=CPh_2\}(L)(\eta\text{-}C_6Me_6)]PF_6$ and the alkenylallenylidenes $[RuCl\{=C=C=C(OR)CH=CPh_2\}(L)(\eta\text{-}C_6Me_6)]PF_6$ (R = Et, Pr^i; L = PMe_3, $PMePh_2$) were obtained from the reactions of $HC\equiv CC\equiv CCPh_2(OSiMe_3)$ with $RuCl_2(L)(\eta\text{-}C_6Me_6)$ in the presence of $NaPF_6$ and the appropriate alcohols. Similar derivatives are obtained with secondary amines,[120] $NHPr^i_2$ or $NHPh_2$ (but not more basic amines), or the isolated diynyl complex in the presence of HBF_4, giving $[RuCl\{=C=C=C(NPr^i_2)CH=CR_2\}(PR_3)(\eta\text{-}C_6Me_6)]^+$.[120] Interestingly, the reaction of **282** (L = PMe_3) with $HC\equiv CC\equiv CCPh_2(OH)$ gave the chelate complex $[RuCl\{C(OMe)=CHC(O)CH=CPh_2\}(PMe_3)(\eta\text{-}C_6Me_6)]^+$ (**283**), possibly by dehydration followed by addition of MeOH at C_α and of water at C_γ.[375,376]

Reactions of *cis*-$RuCl_2(dppm)_2$ with $HC\equiv CC\equiv CH$ in the presence of tertiary amines afford *trans*-$[RuCl\{C\equiv CC(NR_3)=CH_2\}(dppm)_2]^+$ [NR_3 = NEt_3, NPr_3, quinuclidine, $EtN(C_2H_4)_2O$, 1,4,7-Me_3-tacn, $NMe_2(CH_2Ph)$, $NMe_2(CH_2C_6H_4OMe\text{-}3)$], rationalized as proceeding by addition of the amine to C_γ of the intermediate butatrienylidene cation **284** (Scheme 63).[94] With 4-$Me_2NC_5H_4N$, the pyridine nitrogen attacks C_γ to give *trans*-$[RuCl\{C\equiv CC(NC_5H_4NMe_2\text{-}4)=CH_2\}(dppm)_2]^+$. Competing reactions of the amines to give *trans*-$[RuCl(NR_3)(dppm)_2]^+$

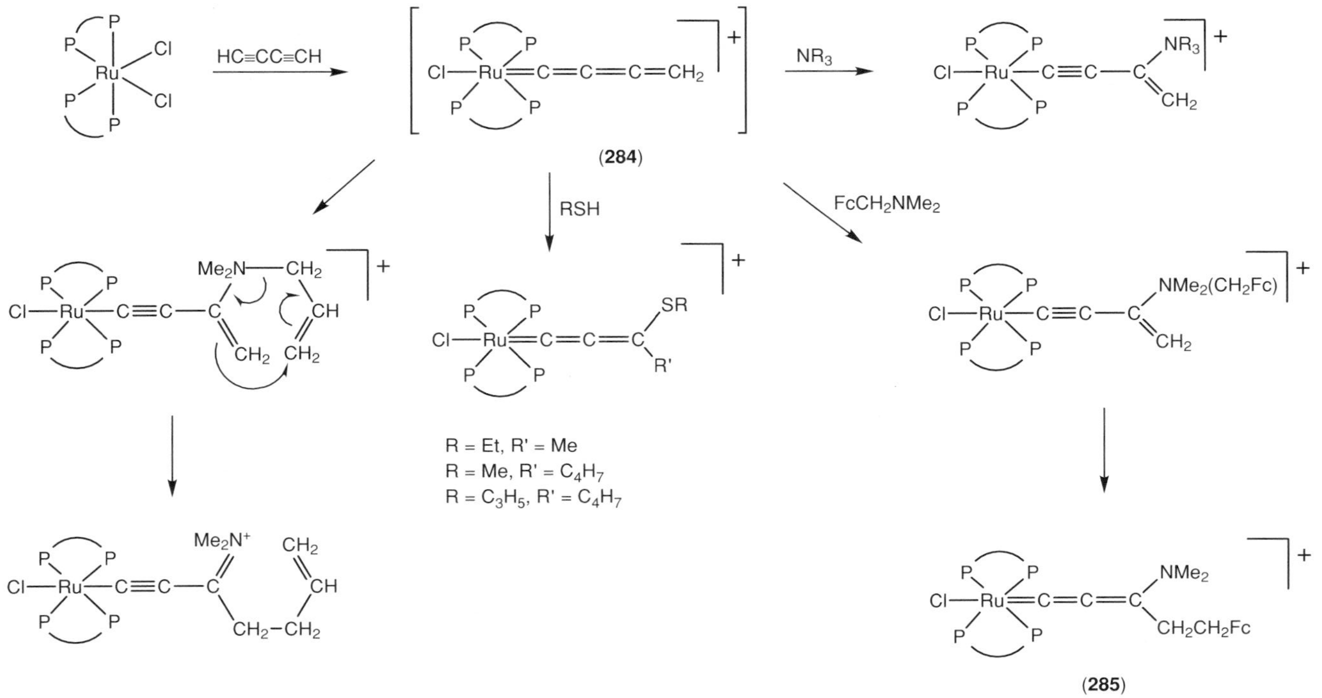

$HC\equiv CC\equiv CH$

(**284**)

NR_3

RSH

$FcCH_2NMe_2$

R = Et, R' = Me
R = Me, R' = C_4H_7
R = C_3H_5, R' = C_4H_7

(**285**)

SCHEME 63

also occur. Reactions with allylamine result in an aza-Cope rearrangement and give functionalized allenylidenes.[377] With $FcCH_2NMe_2$, the first-formed adduct *trans*-$[RuCl\{C{\equiv}CC(NMe_2CH_2Fc){=}CH_2\}(dppm)_2]^+$ apparently rearranges by migration of the $FcCH_2^+$ carbenium ion from nitrogen to carbon to give redox-active *trans*-$[RuCl\{C{\equiv}CC(NMe_2)CH_2CH_2Fc\}(dppm)_2\}]^+$ (**285**).[378] Thioallenylidene complexes have been obtained from *trans*-$[RuCl({=}C{=}C{=}C{=}CH_2)(PP)_2]^+$ (PP = dppm, dppe) and either thiols or allylic thioethers.[379]

Direct reaction of HC≡CC≡CPh with *trans*-$RuCl_2(dppe)_2$ in the presence of MeOH gives *trans*-$[RuCl\{{=}C{=}C{=}C(OMe)CH_2Ph\}(dppe)_2]^+$, which is deprotonated by NEt_3 to *trans*-$RuCl\{C{\equiv}CC(OMe){=}CHPh\}(dppe)_2$ (**286**). Both reactions indicate formation of the butatrienylidene complex as a reactive intermediate.[95] Similarly, protonation of *trans*-RuCl(C≡CC≡CPh)$(dppe)_2$ with CF_3SO_3H results in formation of a bright red intermediate which turns green in seconds, affording *trans*-$[RuCl\{{=}C{=}CHC(O)CH_2Ph\}(dppe)_2]^+$, probably by addition of water to the intermediate *trans*-$[RuCl({=}C{=}C{=}C{=}CHPh)(dppe)_2]^+$. The former is deprotonated to *trans*-$RuCl\{C{\equiv}CC(O)CH_2Ph\}(dppe)_2$.

Reactions of *cis*-$RuCl_2(dppm)_2$ either with $Bu_3SnC{\equiv}CC{\equiv}CCPh_2(OSiMe_3)$, or directly with $HC{\equiv}CC{\equiv}CCPh_2(OSiMe_3)$, both in the presence of $NaPF_6$, with *in situ* deprotonation by NEt_3, give *trans*-$RuCl\{C{\equiv}CC{\equiv}CCPh_2(OSiMe_3)\}(dppm)_2$ (Section II.B.4). Treatment of this complex with HBF_4 in MeOH gives *trans*-$[RuCl\{{=}C{=}C{=}C(OMe)CH{=}CPh_2\}(dppm)_2]^+$, possibly via intermediate formation of *trans*-$[RuCl({=}C{=}C{=}C{=}C{=}CPh_2)(dppm)_2]^+$ which rapidly adds MeOH at C_γ (Scheme 64). The bis-diynyl complex gives *trans*-$[Ru\{{=}C{=}C{=}C(OMe)CH{=}CPh_2\}_2(dppm)_2]^{2+}$ from *trans*-$Ru\{C{\equiv}CC{\equiv}CCPh_2(OSiMe_3)\}_2(dppm)_2$. In CH_2Cl_2, intramolecular cycloaddition of C_γ to an ortho carbon of a phenyl ring in the intermediate cumulene gives *trans*-$[RuCl\{{=}C{=}C{=}C(C_6H_4)CPh{=}CH\}(dppm)_2]^+$ (**287**).[85,380]

The first isolable pentatetraenylidene complex, $[RuCl({=}C{=}C{=}C{=}C{=}CPh_2)(dppe)_2]^+$, was obtained from *cis*-$RuCl_2(dppe)_2$ and $HC{\equiv}CC{\equiv}CCPh_2(OSiMe_3)$ in the presence of $NaPF_6$ and NEt_3 using thf as solvent, via $RuCl\{C{\equiv}CC{\equiv}CCPh_2(OSiMe_3)\}(dppe)_2$ and its subsequent reaction with $[CPh_3]^+$.[89,124]

The cation $[Ru({=}C{=}C{=}C{=}CH_2)(PPh_3)_2Cp]^+$ is obtained directly from buta-1,3-diyne and $[Ru(thf)(PPh_3)_2Cp]^+$[62,92,381–383] and is also formed by protonation of the diynyl $Ru(C{\equiv}CC{\equiv}CH)(PPh_3)_2Cp$.[62] Nucleophiles add readily to C_γ (C_α and C_β are sterically protected by the PPh_3 ligands.) The course of the reaction is apparently decided by the substituent on C(3) of the first-formed vinylacetylide (Scheme 65). Addition of water gives $Ru\{C{\equiv}CC(O)Me\}(PPh_3)_2Cp$ by deprotonation of the intermediate hydroxy-allenylidene. With PPh_3, the vinylphosphonium ethynyl complex $[Ru\{C{\equiv}CC(PPh_3){=}CH_2\}(PPh_3)_2Cp]^+$ is formed, which can be further protonated to the dicationic vinylphosphonium vinylidene $[Ru\{{=}C{=}CHC(PPh_3){=}CH_2\}(PPh_3)_2Cp]^{2+}$. With nucleophiles containing a hydrogen, addition to C_γ is followed by H-shift to C_δ to give a methylallenylidene (also see water above).

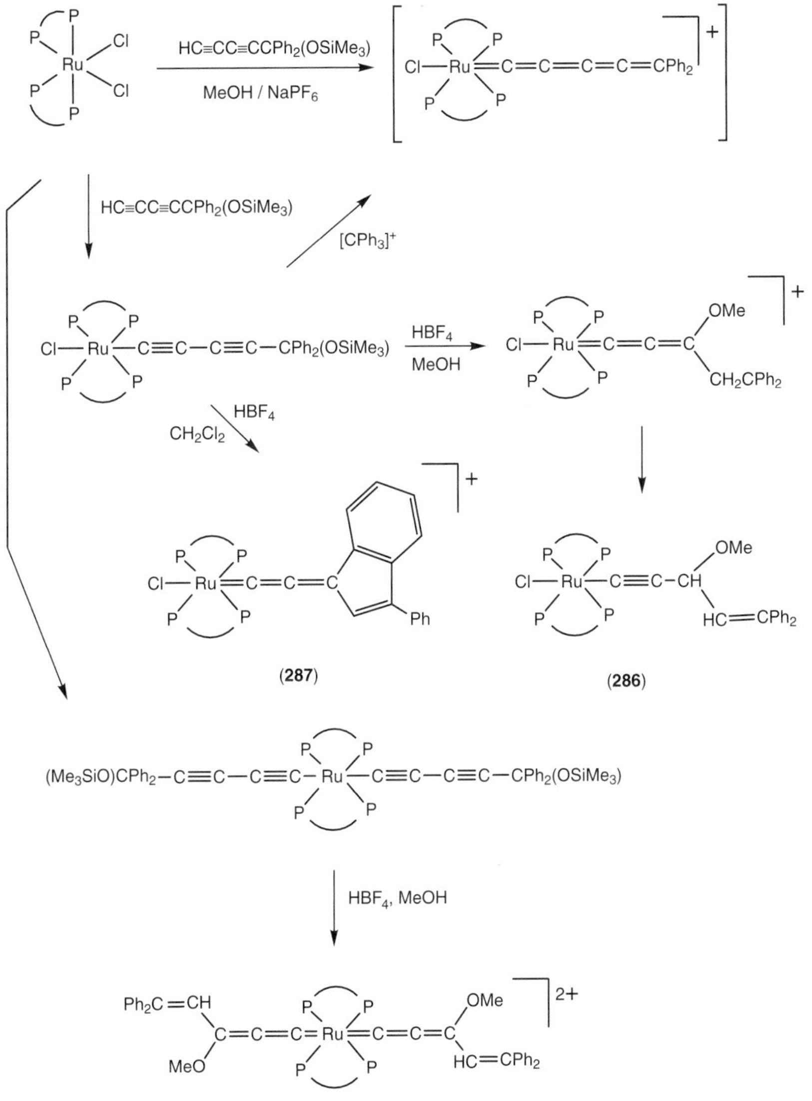

SCHEME 64

Thus with $NHPh_2$, $[Ru\{=C=C=CMe(NPh_2)\}(PPh_3)_2Cp]^+$ is obtained, while with *N*-methylpyrrole, the allenylidene $[Ru\{=C=C=CMe(C_4H_3NMe\text{-}2)\}(PPh_3)_2Cp]^+$ is formed. The latter can be deprotonated (LiBu) to give $Ru\{C{\equiv}CC(C_4H_3NMe\text{-}2){=}CH_2\}(PPh_3)_2Cp$. A slow reaction of the diphenylamino complex with CH_2Cl_2 afforded $Ru(C{\equiv}CCH{=}CHCl)(PPh_3)_2Cp$, possibly by reaction with the

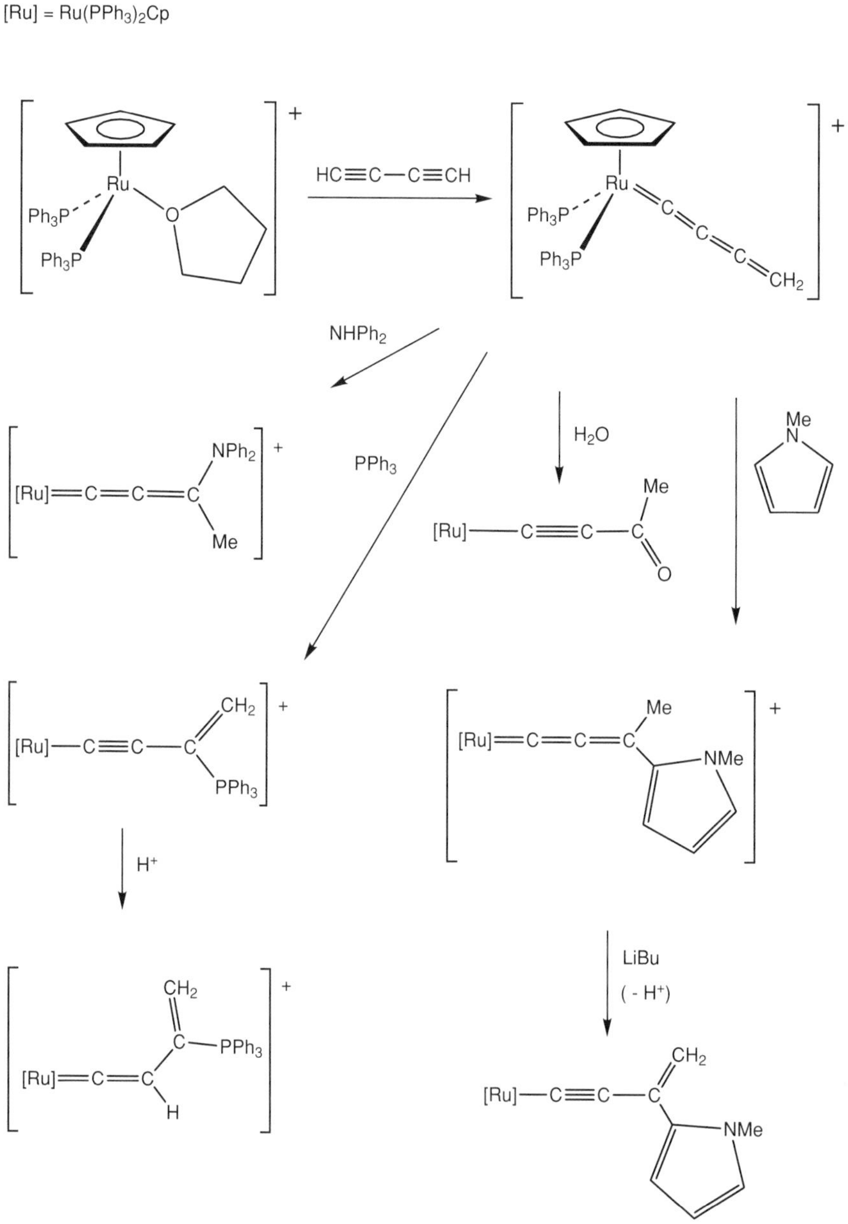

[Ru] = Ru(PPh3)2Cp

[Ru(PPh3)2Cp(O-thf)]+ : Ru, Ph3P, Ph3P, O

HC≡C—C≡CH

[Ru=C=C=C=CH2]+ : Ru, Ph3P, Ph3P, C, C, C, CH2

NHPh2

[[Ru]=C=C=C(NPh2)(Me)]+ : [Ru]═C═C═C, NPh2, Me

PPh3

H2O

[Ru]—C≡C—C(=O)Me : Me, O

Me
N
(N-methylpyrrole)

[[Ru]—C≡C—C(=CH2)PPh3]+ : CH2, PPh3

[[Ru]=C=C=C(Me)(C4H3NMe)]+ : Me, NMe

H+

[[Ru]=C=C(H)—C(=CH2)PPh3]+ : CH2, C—PPh3, H

LiBu
(- H+)

[Ru]—C≡C—C(=CH2)(C4H3NMe) : CH2, NMe

SCHEME 65

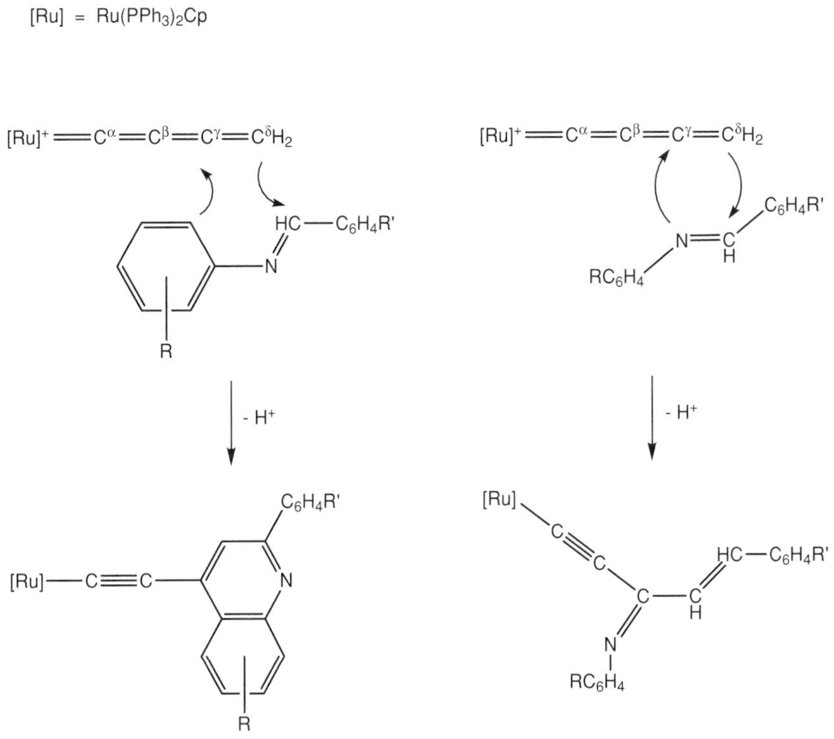

[Ru] = $Ru(PPh_3)_2Cp$

(**288**) (**289**)

SCHEME 66

butatrienylidene.[92,382] Cycloaddition of aromatic imines with $[Ru(C{=}C{=}C{=}CH_2)(PPh_3)_2Cp]^+$ affords ethynylquinoline (**288**) or azabutadienyl complexes (**289**) (Scheme 66). The formation of quinoline products appears to be favored when electron-rich *N*-aryl groups are present.[381]

C. *Formation of Enynyl Complexes*

Many examples of complexes containing enynyl ligands are known from reactions of 1-alkynes with various metal complexes; two coordination sites are necessary for this reaction.[384] Displacement of the ligand often results in a catalytic cycle of head-to-head dimerization of the alkyne. The protonation may occur by solvent, e.g., MeOH, in other cases acid is required, e.g., CF_3CO_2H. Coupling of alkynyl and vinylidene ligands on rhodium has been described.[385] The few reactions of 1,3-diynes to give similar enyl complexes are summarized in the following.

1. *Yttrium*

The η^3-ynenyl complex **290** is formed from $\{Yb(\mu\text{-}H)Tp^{Me, Bu^t}\}_2$ and $Me_3SiC{\equiv}CC{\equiv}CSiMe_3$ by insertion of one C≡C triple bond into the Yb–H bond.[386]

2. *Ruthenium and Osmium*

Addition of acids ($HClO_4$, HBF_4, HPF_6) to $Ru(\eta\text{-}PhC_2C{\equiv}CPh)(CO)_2(PPh_3)_2$ gives $[Ru(\eta^3\text{-}PhC_3C{=}CHPh)(CO)_2(PPh_3)_2]^+$ (**291**) while addition of HCl gives $RuCl\{C(C{\equiv}CPh){=}CHPh\}(CO)(PPh_3)_2$. The latter reacts with CO, followed by removal of chloride with $AgBF_4$ or $AgPF_6$ to give the same cation. With an excess of HCl, *cis*-$RuCl_2(CO)_2(PPh_3)_2$ is formed.[176]

The reaction of PhC≡CC≡CPh with $RuHCl(CO)(PPh_3)_3$ results in mono-insertion and formation of $RuCl\{C(C{\equiv}CPh){=}CHPh\}(CO)(PPh_3)_2$.[387] Reactions of $Hg(C{\equiv}CR)_2$ [R = Ph, tol, Bu, $CMe_2(OH)$] with $RuHCl(CO)(PPh_3)_3$ also provide a useful route to $RuCl\{C({=}CHR)C{\equiv}CR\}$ $(CO)(PPh_3)_2$, although the osmium analogue is unreactive.[387,388] Similarly, insertion of $SiMe_3C{\equiv}CC{\equiv}CSiMe_3$ into the Ru–H bond of $RuHCl(CO)(PPh_3)_3$ gives $RuCl\{C(C{\equiv}CSiMe_3){=}CH(SiMe_3)\}$ $(CO)(PPh_3)_2$, while substitution of one PPh_3 by dppe affords $RuCl\{C(C{\equiv}CSiMe_3){=}CH(SiMe_3)\}(CO)(PPh_3)(dppe)$.[389] However, reaction of $RuHCl(CO)(PPh_3)_3$ with HC≡CC≡CH (from $Me_3SiC{\equiv}CC{\equiv}CSiMe_3$ and $[NBu_4]F/[NH_4]F/H_2O$) gives $\{RuCl(CO)(NH_3)(PPh_3)_2\}_2(\mu\text{-}CH{=}CHCH{=}CH)$; omission of NH_4F gives $\{RuCl(CO)(PPh_3)_2\}_2(\mu\text{-}CH{=}CHCH{=}CH)$. The former reacts with an excess of PEt_3 to give $\{RuCl(CO)(PEt_3)_3\}_2(\mu\text{-}CH{=}CHCH{=}CH)$, whereas CO insertion occurrs in the reaction with Bu^tNC, which affords $[\{Ru(CNBu^t)_3(PPh_3)_2\}_2(\mu\text{-}COCH{=}CHCH{=}CHCO)]^{2+}$.[389]

Reaction of HC≡CPh and $RuH(O_2CCF_3)(CO)(PPh_3)_2$ is presumed to give $Ru(C{\equiv}CPh)(O_2CCF_3)(CO)(PPh_3)_2$ which reacts with excess phenylethyne to give $Ru(O_2CCF_3)\{C(C{\equiv}CPh){=}CHPh\}(CO)(PPh_3)_2$.[390] However, this complex and its Os analogue are better formed from PhC≡CC≡CPh and $MH(O_2CCF_3)(CO)(PPh_3)_2$ by *cis*-1,2 addition of the M-H fragment across one C≡C triple bond; the Os complex is a catalyst for HC≡CPh oligomerization.[391] In the same manner, one of the C≡C triple bonds of (tol)C≡CC≡C(tol) inserts into the Os–H bond of $OsHCl(btd)(CO)(PPh_3)_2$ (btd = 2,1,3-benzothiadiazole) to give the corresponding enynyl complex $OsCl\{C[C{\equiv}C(tol)]{=}CH(tol)\}(btd)(CO)(PPh_3)_2$.[392]

The complex $[OsH(N_2)(pp_3)]BPh_4$ [$pp_3 = P(CH_2CH_2PPh_2)_3$] reacts with RC≡CC≡CR to give (*E*)-$[Os(\eta^3\text{-}RC_3{=}CHR)(pp_3)]BPh_4$ [R = $SiMe_3$, Ph (**292**)] quantitatively. Two isomers of **292** have different bonding modes for the enynyl ligand; both react with CO to give $[Os\{C(C{\equiv}CPh){=}CHPh\}(CO)(pp_3)]^+$.[393] The dinitrogen cation is a catalyst for regio- and stereoselective dimerisation of HC≡CR (R = Ph, $SiMe_3$) to (Z)-RC≡CCH=CHR. It is thought that the intermediate alkynyl(vinylidene)osmium complexes rearrange to η^3-butenynyl derivatives, which undergo σ-bond metathesis with HC≡CR to give the enyne.

3. *Nickel*

Insertion of 1,3-diynes into $NiXMe(PMe_3)_2$ (X = Cl, Br, I) gives the corresponding 1-alkynylalkenyl-nickel complexes **293,** which undergo *E*/*Z* isomerization in solution, the former predominating.[208]

D. *Other Types of Products*

1. *Titanium and Zirconium*

In common with alkynes, the titanium vinylidene $Ti(=C=CH_2)Cp^*_2$, which is formed by elimination of CH_4 or C_2H_4 from $TiMe(CH=CH_2)Cp^*_2$ or Ti (*c*-$CH_2CH_2C=CH_2$)Cp^*_2, respectively, undergoes [2+2] cycloaddition reactions with RC≡CC≡CR (R = Me, Ph, $SiMe_3$, Bu^t) to give the stable titanacyclobutenes **294** in 100% regiospecific yield.[394] The unsymmetrical diyne $Bu^tC≡CC≡CSiMe_3$ gives a 9/1 mixture of the two isomeric products, with the 2-$Bu^tC≡C$ isomer predominating. The products are formed regioselectively with an alkynyl substituent on C_α, this reaction proceeding under electronic control according to the polarity of the C≡C bond reacting with strongly nucleophilic C_α in $Ti^+=C^-=CH_2$, confirmed by ab initio Hartree–Fock calculations.[395]

Presumed intermediate Zr{OCPh=C(C≡CPh)}Cp^*_2 (**295**) formed by trapping of $ZrOCp^*_2$ [generated from $ZrPh(OH)Cp^*_2$ in C_6H_6 or PhMe at 160°C] by PhC≡CC≡CPh, probably undergoes Zr—C bond cleavage to give **296,** which rearranges to the observed enolate Zr(OCPh=CHC=CPhCH$_2$-η-C_5Me_4)Cp* (**297**) (Scheme 67).[396]

2. *Tungsten*

The phosphinidene complex $W(PPh)(CO)_5$, obtained from the 7-phenyl-7-phosphanorbornadiene derivative $W(CO)_5\{PPhC_6H_2Me_2(CO_2Me)_2\}$ in the presence of CuI at 60°C, reacts with 1,3-diynes to give alkynylphosphirene complexes $W(CO)_5$\{PPhCR=C(C≡CR)\} (**298**) (Scheme 68). Insertion of the free C≡CR moiety into the proximal C—P bond affords the 1,2-dihydro-1,2-diphosphetes **299** in 2/1 *cis*/*trans* ratio, the proportion of thermodynamically favored *cis* isomer increasing to 4/1 on heating, presumably by epimerization at a phosphorus center. The formation of a single, sterically unfavorable isomer is accounted for by attack of the phosphinidene on the less hindered C(2) atom.[397,398] Use of other

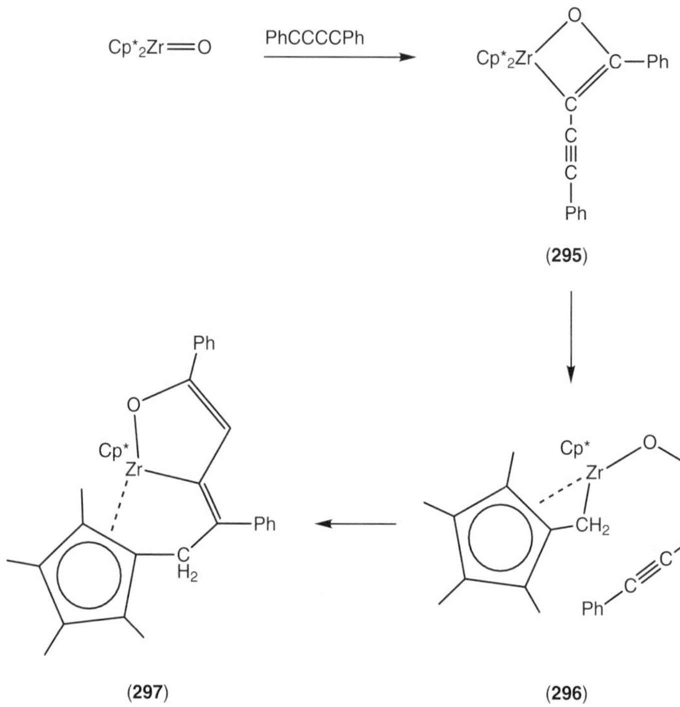

(295)

(297)

(296)

SCHEME 67

phosphinidene precursors allowed mixed diphosphetes to be prepared (**299;** R = Me, CH_2Ph, $CH_2CH{=}CH_2$); however, only **299** (R = Ph, R′ = CH_2OPh) was isolated from the reaction with $(PhO)CH_2C{\equiv}CC{\equiv}CCH_2(OPh)$.

E. *C—C Bond-Breaking Reactions at Metal Centers*

1. *C—C Single Bonds*

As has been previously described, breaking of the central C—C single bond in 1,3-diynes has been reported in several systems, including complexes of groups 3 and 4 (Section VI) and cluster complexes of Group 8 elements (Section IV.D).

2. *C≡C Triple Bonds*

The reaction of $Fe\{(C{\equiv}C)_3SiMe_3\}(CO)_2Cp^*$ with $Fe_2(CO)_9$ gives the bis-μ_3-carbyne complex $Fe_3(\mu_3\text{-}CC{\equiv}CSiMe_3)\{\mu_3\text{-}CC{\equiv}C[Fe(CO)_2Cp^*]\}(CO)_9$ (**300**), in

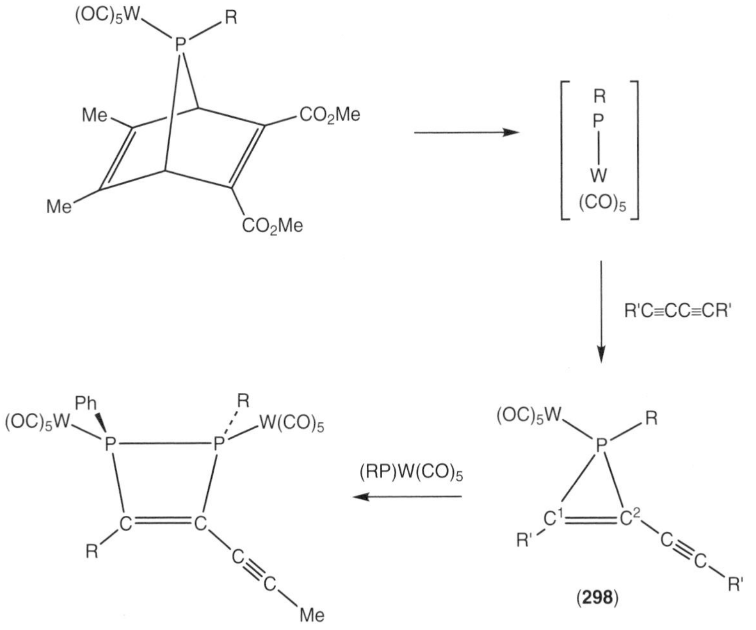

SCHEME 68

reactions which may involve initial formation of a C_2Fe_2 intermediate, which adds the third $Fe(CO)_3$ group with C—C bond breakage. The presence of an electron-rich group, such as $Fe(CO)_2Cp^*$, is necessary, perhaps to stabilize an electron-deficient tetrahedral intermediate; reactions of $Fe_2(CO)_9$ with $Me_3Si(C\equiv C)_3SiMe_3$ give only mixtures of complexes which do not contain μ_3-C ligands.[399]

The reactions of 1,3-diynes with $Co(CO)_2Cp$ to give cyclobutadiene and cyclopentadienone complexes has been previously described (Section VII.A.3). Under different conditions, addition of CoCp units to alkynes results in cleavage of the C≡C triple bond to give several bis-carbyne clusters (Scheme 69).[400] Flash or solution pyrolysis reforms the alkyne without scrambling. Similar complexes were obtained from $Me_3Si(C\equiv C)_3SiMe_3$. The so-called "double-decker" cluster **301-Me,** obtained from the latter reaction, symmetrizes to **302-Me** on FVP or solution pyrolysis at 290°C (Scheme 70). The rearrangement is reversible, as **301-Me** is formed in low yield by FVP of **302-Me.** Also formed in the reaction is **303-Me,** possibly via unobserved intermediate **304-Me.** Analogues of **301–303** were

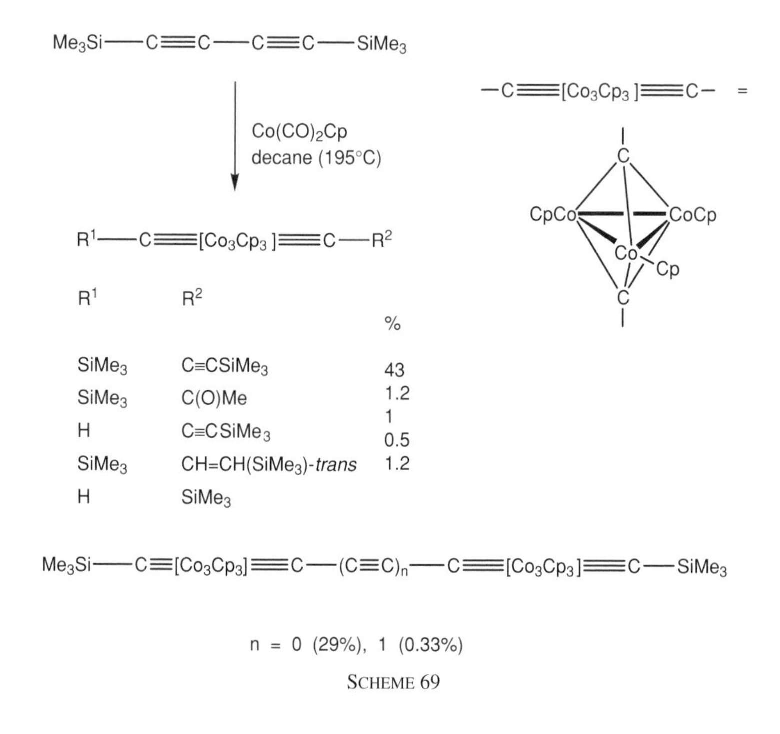

SCHEME 69

obtained from $Co(CO)_2(\eta\text{-}C_5H_4Me)$, while $SiEt_3(C{\equiv}C)_3SiEt_3$ and $Co(CO)_2Cp$ gave **301-Et** and **303-Et** (but no **302-Et**).

No cross-over products were obtained from mixtures of complexes, confirming that isomerization is an intramolecular reaction. Possible mechanisms considered included "breathing" expansion of the Co_3 core, with the carbon chain passing through the center (which would involve the Co ··· Co separation increasing to ca 3.24 Å), or movement of the carbyne fragments about the Co_3 core edges, leading to recombination and cleavage of an alternative C≡C triple bond.[401] Subsequent theoretical explorations of the reaction concluded that the latter mechanism was energetically the most favorable.[402]

F. *Reactions of Metal Complexes with 1,3-Diynes Giving Organic Products*

1. *With Carbene Complexes*

Reactions of Fischer carbene complexes with diynes have been extensively studied as a synthetic approaches to alkynylarenes and biaryls. In general, Cr $\{=C(OR^1)R^2\}(CO)_5$ ($R^1 = Me$, Bu; $R^2 = Ph$, 1-nap, 1-cyclohexenyl) react with

[Co] = CoCp; R = Me, Et

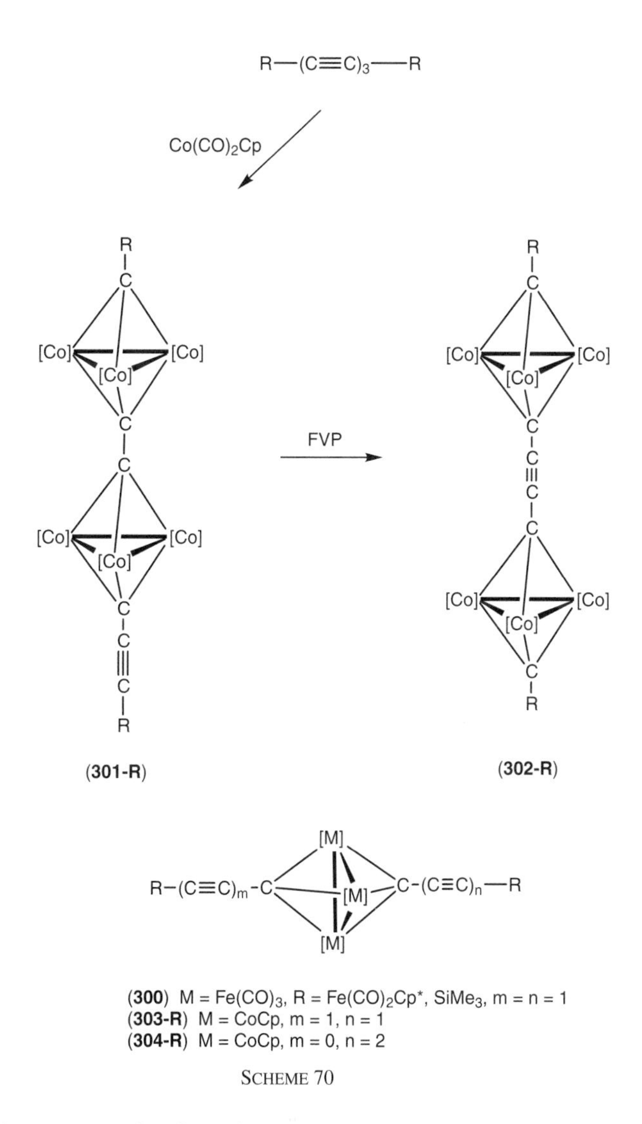

SCHEME 70

1 equiv of $R^3C{\equiv}CC{\equiv}CR^3$ ($R^3 = Pr^i$, Bu^t, Ph) to give alkynylarenes **305** in which R^3 is incorporated regioselectively adjacent to the phenol function, together with the cyclobutenes **306** (Scheme 71). Bis-phenols such as **307a** were obtained from subsequent reactions with a second molecule of carbene complex; other products reported include 2,2′-binaphthol **307b** and indenylnaphthalene **308.**[403]

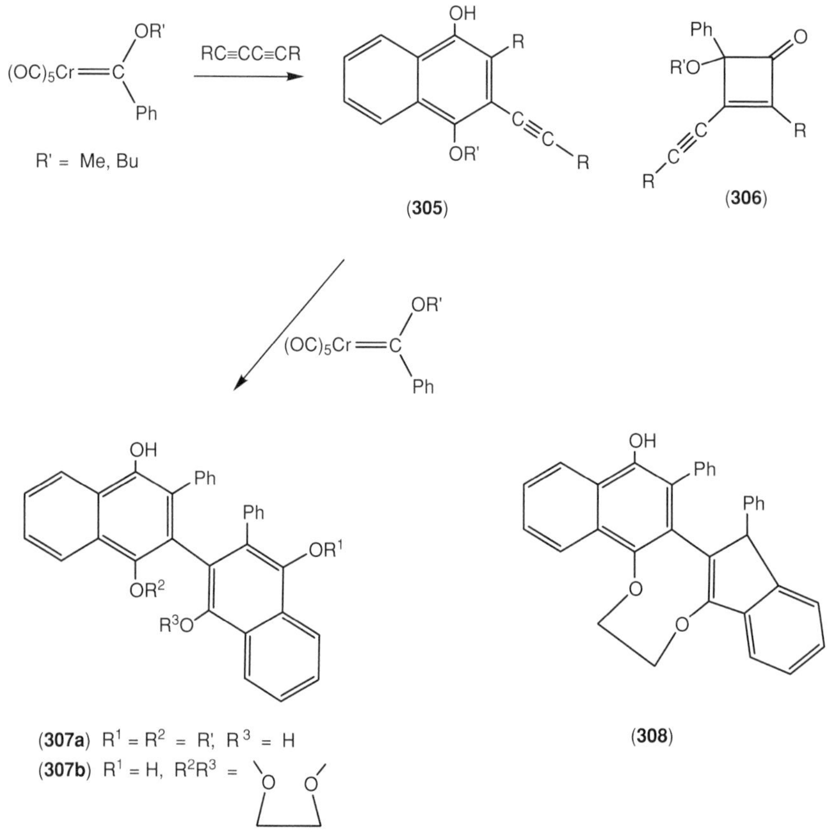

(OC)5Cr=C(OR')Ph
RC≡CC≡CR
R' = Me, Bu
(305)
(306)
(307a) R1 = R2 = R', R3 = H
(307b) R1 = H, R2R3 =
(308)

SCHEME 71

2. *Reactions of Diynes Catalyzed by Metal Complexes*

a. *Oligomerization and Related Reactions.* As found with mono-ynes, several catalyst systems trimerize 1,3-diynes to the corresponding trialkynylbenzenes. An early study described the formation of up to 5% 1,3,5-$Me_3C_6(C{\equiv}CMe)_3$-2,4,6 by trimerization of MeC≡CC≡CMe with Ziegler catalysts ($TiCl_4/AlEt_3$), in addition to polymers. The latter were soluble (catalyst:diyne ratio 0.03, 80°C/1 hr) or insoluble vitreous materials (catalyst:diyne ratio 1, 30°C/1 hr) dependent on reaction conditions.[404] With $Hg\{Co(CO)_4\}_2$ mixtures of 1,3,5- and 1,2,4-isomers were obtained in most cases, although buta-1,3-diyne gave only insoluble polymeric material and HC≡CC≡CPh gave benzene-soluble high-molecular-weight products.[160] 1,2,4-Tris(alkynyl)benzenes are formed in high yield (R = Me, 77%; Ph, 83%) by trimerization of the 1,3-diyne using $Ni(CO)_2(PPh_3)_2$ as catalyst. High yields of the mixture of 1,2,4- and 1,3,5-isomers are formed with $Co(CO)_2Cp$ (R = Ph, 72 and 16%, respectively).[405]

Regioselective syntheses of 1,3,5-unsymmetrically substituted benzenes (**309**) are catalyzed by $Pd(dba)_2/PPh_3$; mixed alkyne/diyne reactants give mixtures containing homocoupled and mixed products (24:21 from HC≡CPh + HC≡CC≡CC_6H_{13}).[406] The probable mechanism involves oxidative addition to the Pd(0) center, insertion of the second diyne into the Pd—H bond, reductive coupling and subsequent π-complexation of this product to Pd(0), followed by Diels–Alder cycloaddition of the third diyne and elimination of product.

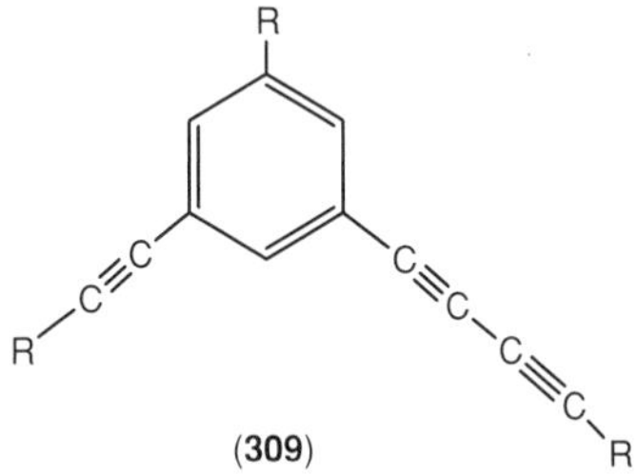

(**309**)

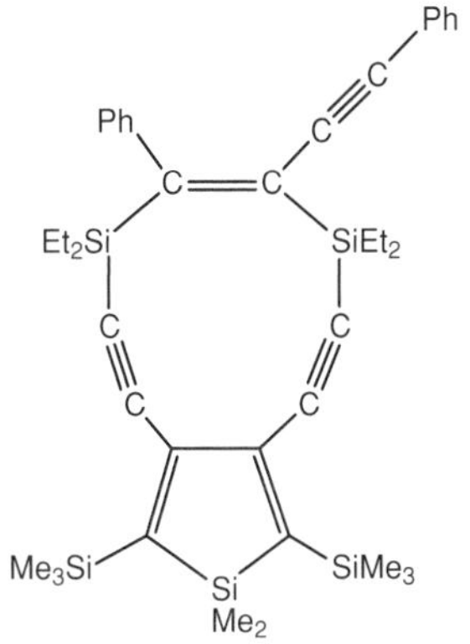

(**311**)

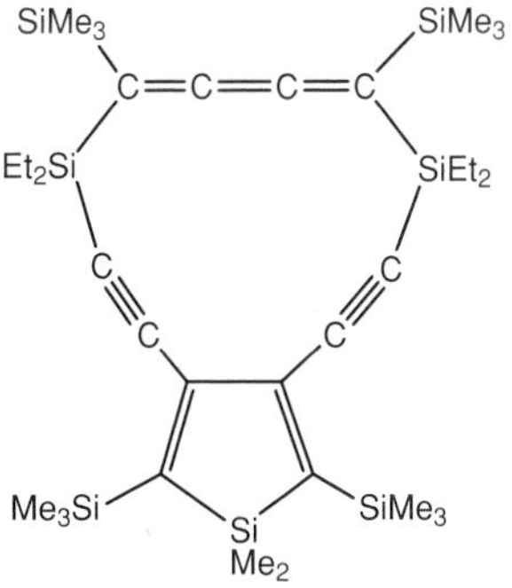

(**312**)

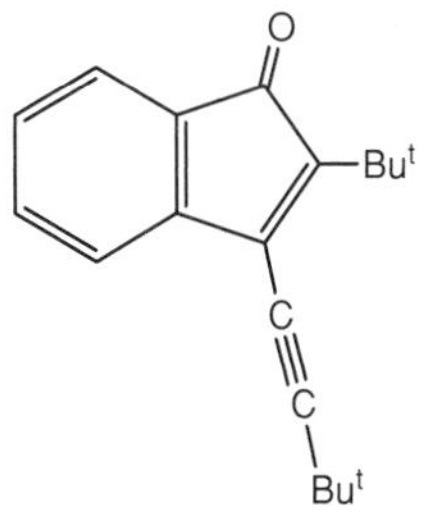

(**315**)

The intramolecular enyne-diyne [4 + 2] cross-benzannulation reaction affords 1,2,4-trisubstituted benzenes, catalysed by $Pd(PPh_3)_4$. Reversible coordination of Pd with enyne and diyne gives a palladacycle, stabilized by coordination to the η^3-propargyl group. Subsequent reductive elimination gives a strained cumulene which the rearranges to the cross-annulation product.[407] Palladium-catalyzed enyne-triyne cross-annulation gives a diynylbenzene.

Trimerization of 1-alkynes to substituted cyclobutadienes occurs in reactions of RhCl(l-alaninate)Cp* with HC≡CR (R = Ph, tol), which afford Rh{η^4-C_4HR_2(C≡CR)}Cp* (**310**) possibly via intermediate dialkynylrhodium(III) complexes. Reductive coupling to an η^2-diyne complex, which coordinates the third molecule of alkyne, is followed by further coupling to the rhodacyclopentadiene and reductive elimination of the cyclobutadiene (Scheme 72).[408]

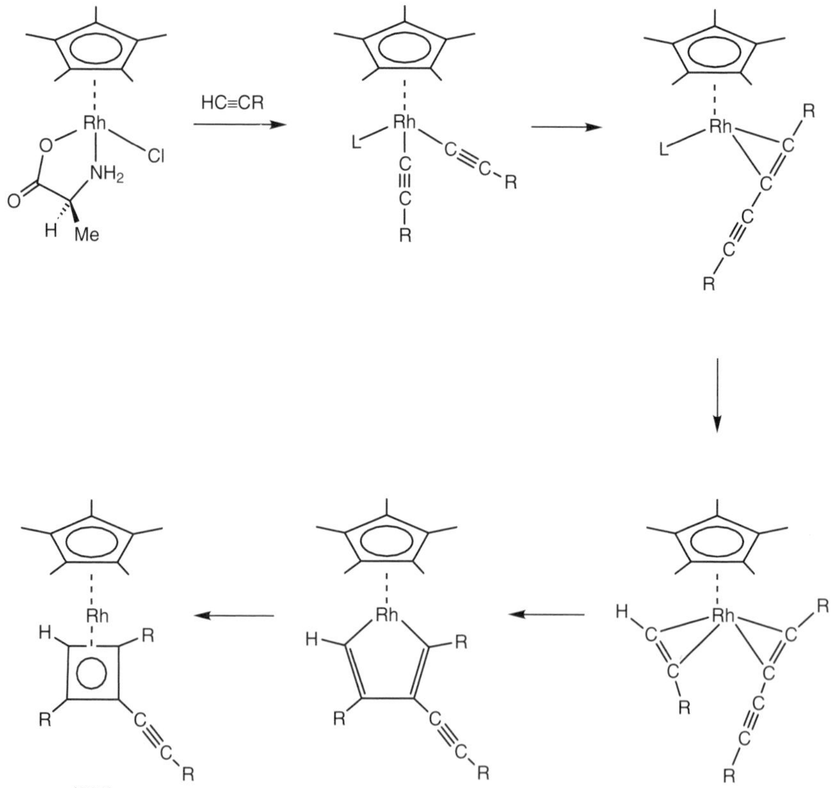

HC≡CR

(**310**)

SCHEME 72

b. *Silylation and stannylation.* *cis*-1,2-Bis-silylation of PhC≡CC≡CPh occurs in the $PdCl_2(PPh_3)_2$-catalyzed reaction with $ClMe_2SiSiMe_2Cl$ to give, after methylation, $Me_3SiCPh{=}C(SiMe_3)C{\equiv}CPh$, whereas with $Me_3SiC{\equiv}CC{\equiv}CSiMe_3$, a mixture of $(Me_3Si)_2C{=}C(SiMe_3)C{\equiv}CSiMe_3$ (major) and $(Me_3Si)_2C{=}C{=}C{=}C(SiMe_3)_2$ is obtained from noncyclic disilanes.[409] Cyclic disilanes containing silacyclopentadiene rings undergo regiospecific reactions with 1,3-diynes, catalyzed by $PdCl_2(PPh_3)_2$. The products depend on the diyne substituents. With PhC≡CC≡CPh, enediyne **311** is formed by inserion of one C≡C triple bond into the Si–Si bond of the disilane. In contrast, $Me_3SiC{\equiv}CC{\equiv}CSiMe_3$ affords 4,9,14-trisilabicyclo[10.3.0]pentadeca-1(15),5,6,7,12-pentaen-2,1-diyne (**312**).[410] Both reactions are considered to proceed via intermediate palladacycles formed by insertion of an alkyne-palladium fragment into the Si–Si bond.

Carbostannylation of RC≡CC≡CR (R = Bu, $SiMe_3$) with $SnBu_3(CH_2CH{=}CH_2)$ is catalyzed by $Ni(cod)_2$ and proceeds stereospecifically to give $SnBu_3\{C(C{\equiv}CR)CRCH_2CH{=}CH_2\}$.[411]

c. *Boration.* Tetraboration of RC≡CC≡CR (R = C_6H_4OMe-4, $SiMe_3$) by $B_2(cat)_2$ (cat = 1,2-$O_2C_6H_4$ or 4-$Bu^tC_6H_3O_2$) is catalyzed by $Pt\{B(cat)\}_2(PPh_3)_2$

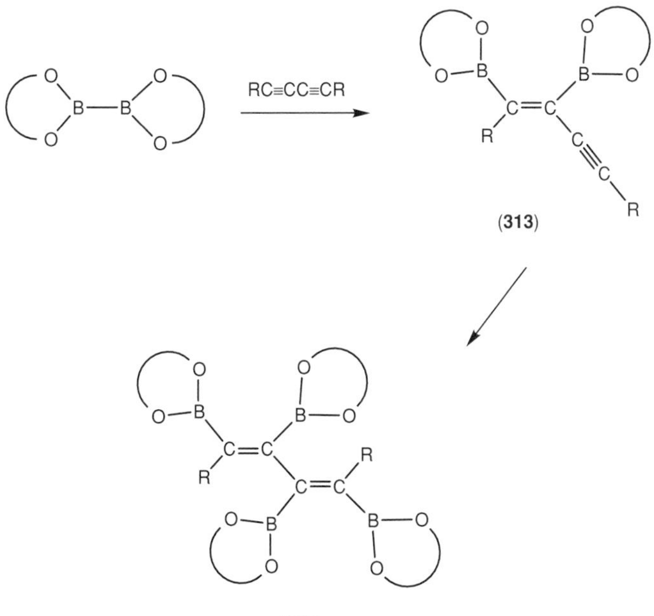

RC≡CC≡CR

(**313**)

(**314**)

SCHEME 73

or $Pt(\eta\text{-}C_2H_4)(PPh_3)_2$ to give (Z,Z)-RC{B(cat)}=C{B(cat)}C{B(cat)}=CR {B(cat)}.[412] The rate of the reaction is reduced by free PPh_3. Mixtures of the bis- and tetrakis-products with unreacted 1,3-diyne are obtained with a deficiency of $B_2(cat)_2$, indicating that rates of the first and second additions are comparable (see Scheme 73). With $Me_3SiC{\equiv}CC{\equiv}CSiMe_3$, addition is very slow, giving other products including tris(boronate) esters by cleavage of the C—Si bond and subsequent diboration of the $C{\equiv}CB(OR_2)$ fragment. With $B_2(pin)_2$ (pin = pinacolato), rapid addition gave *Z*-(4-$MeOC_6H_4$)C{B(pin)}=C{B(pin)} $C{\equiv}C(C_6H_4OMe\text{-}4)$ (**313**), while tetraboration to **314** was slow. Slow reactions with $Me_3SiC{\equiv}CC{\equiv}CSiMe_3$ proceed via *Z*-Me_3SiC{B(pin)}=C{B(pin)}$C{\equiv}CSiMe_3$ to the tetrakis(boronate ester).

d. *Coupling reactions.* The $Pd(OAc)_2$-catalyzed reaction between $Bu^tC{\equiv}CC{\equiv}CBu^t$ and 2-iodobenzaldehyde gave indenone **315** in 58% yield.[413]

e. *Metal derivatives of 1,3-diynes as catalysts.* Hydrosilylation of alkynes (no reaction with $Me_3SiC{\equiv}CSiMe_3$) and 1,3-diynes with $SiHEt_3$, $SiHMe_2Ph$ or SiH_2Ph_2 is catalyzed by $Ni(\eta^2\text{-}HC_2C{\equiv}CH)(PR_3)_2$ (R = Ph, OC_6H_4Me-2) with *cis* addition of the silanes.[414] Disubstituted alkynes also give dienes and trienes together with nonsilylated benzenes. Subsequent 1,4-addition (Bu^t, Ph) to allenes or 3,4-addition (Ph) to the diene occurs with the diynes.

VIII

FORMATION OF FREE DI- AND POLY-YNES BY REACTIONS AT METAL CENTERS

A. *Alkyne-Coupling Reactions*

As has been discussed previously, formation of diyne or diynyl ligands may occur by coupling of alkynyl fragments at metal centers. In what follows, we concern ourselves with the reactions in which coupling of alkynyl fragments gives the free (uncoordinated) diynes or related poly-ynes. In many cases, mechanistic details are not known. The preparation of di- and poly-ynes $R(C{\equiv}C)_nR'$ by metal-catalyzed coupling reactions of terminal, halo- and stannyl-alkynes and the corresponding higher poly-ynes has been achieved by several routes. The various coupling protocols appropriate for sp carbon centers include oxidative coupling of terminal acetylenes (Glaser and Hay reactions) and organometallic alkynyls and coupling between terminal alkynes and bromoalkynes (Cadiot–Chodkiewicz reaction). These and related reactions have been extensively reviewed.[3,67,415–421] Space precludes more than a brief survey of these reactions and here we have chosen to discuss the more recent applications relevant to the present survey.

1. *Titanium and Zirconium*

Chemical ($NiCl_2$ or $AgPF_6$) or electrochemical oxidation of Ti{(C≡C)$_n$Mc}$_2$ $Cp^{Si}{}_2$ (Mc = Fc, Rc) affords Mc(C≡C)$_{2n}$Mc in excellent yield.[97,422] These reactions are thought to occur via initial oxidation of the Mc centers, followed by homolytic intramolecular electron transfer from the Ti–C bonds to the metallocene and subsequent coupling of the Mc(C≡C)$_n$ radicals. Sequential treatment of Si(C≡CPh)$_2$R$_2$ (R = Me, Et, Ph) with $ZrEt_2Cp_2$ [a source of Zr(η-C_2H_4)Cp_2] and I_2 gave high yields of PhC≡CC≡CPh.[349] In contrast to the similar reaction with mono-alkynylsilanes, no incorporation of C_2H_4 is found. Reactions of a mixture of $SiMe_2$(C≡CAr)$_2$ (Ar = Ph and tol) with $ZrEt_2Cp_2$ and I_2 gave only intramolecular coupling products and no PhC≡CC≡C(tol).

2. *Iron, Ruthenium, and Osmium*

Thermolysis of Fe(C≡CSiMe$_3$)(CO)$_2$Cp gives Me$_3$SiC≡CC≡CSiMe$_3$ and {Fe(CO)$_2$Cp}$_2$.[188] Binuclear ruthenium complexes containing two alkynyl groups undergo facile coupling reactions. Thus, reactions of {Ru(μ-SPr^i)(C≡CC$_6$H$_4$R-4)Cp*}$_2$ (R = H, Me) with iodine give {Ru(μ-SPr^i)(I)Cp*}$_2$ and the 1,3-diynes (R = H, 74%; R = Me, 99%).[423] In contrast to its ruthenium congenor, which reacts with Hg(C≡CR)$_2$ to give enynyl complexes, OsHCl(CO)(PPh$_3$)$_3$ acts as a catalyst for the production of 1,3-diynes and elemental mercury. The corresponding enynyl complex is apparently unstable toward β-elimination.[424]

3. *Rhodium*

Catalytic demercuration of Hg(C≡CR)$_2$ by RhCl(CO)(PPh$_3$)$_2$, Rh{H$_2$B(bta)$_2$}(CO)(PPh$_3$)$_2$ (bta = benzotriazolyl), [Rh(PPh$_3$)$_2$([9]aneS$_3$)]PF$_6$, or RhCl(PPh$_3$)$_3$ also affords the 1,3-diynes in excellent yield.[424,425]

4. *Nickel and Palladium*

Homo-coupling of alkynyllithiums to 1,3-diynes occurs in the presence of nickel(II) complexes, as exemplified by reactions of LiC≡CR with NiCl$_2$(PPh$_3$)$_2$ (thf, −78°C) in the presence of ligands (PPh$_3$, dbu, tetramethylguanidine) followed by reductive elimination.[412] The diynes RC≡CC≡CR [R = Bu^t, n-C$_5$H$_{11}$, Ph, C$_2$H$_4$OBn, (CH$_2$)$_n$O(thp) (n = 2–4)] are obtained in 31–73% yield; a degree of oligomerization also accompanies this reaction.[426]

Either heating bis(alkynyl)nickel complexes, or treatment with CO or PMe_3, gives the corresponding 1,3-diynes, apparently via a series of binuclear reactions.[208] Reactions of $LiC{\equiv}CBu^t$ with $PdCl_2(PPh_3)_2$ (2/1 ratio, thf, 22°C) afford $Bu^tC{\equiv}CC{\equiv}CBu^t$ (>95%). However, with a threefold excess of organolithium reagent, competitive formation of the 1,3-diyne and $Li_2[Pd(C{\equiv}CBu^t)_4]$ occurs; the latter complex is formed exclusively from $Li_2[PdCl_4]$ and $LiC{\equiv}CBu^t$ (1/4). The reaction of $PdCl_2(PPh_3)_2$ with $ZnCl(C{\equiv}CBu^t)$ to give the diyne is independent of concentration.[427]

5. *Copper*

The most useful approaches to the synthesis of di- and poly-ynes from terminal alkynes are undoubtedly the copper-catalyzed couplings discovered by Glaser ($CuCl$, NH_4OH, EtOH, O_2),[428] Eglinton [$Cu(OAc)_2$, hot pyridine or quinoline, O_2],[429] and Hay [Cu(I), tmed, O_2].[430] Some of the many applications of these reactions are discussed in the following.

In addition to these, several individual reactions of copper compounds have given diynes. Thus, the reaction of $HC{\equiv}CPh$ with the copper(I) derivative of an N_4-macrocyclic ligand (L) afforded the $PhC{\equiv}CC{\equiv}CPh$ adduct of $[Cu_4(C_2Ph)(L)_2][ClO_4]_3$ and the free diyne.[431] Dimerization of silylated enynes $RCH{=}CHC{\equiv}CSiMe_3$ [$R = SiMe_3$, Ph, 2-thienyl, C_6H_4OMe-4, CH_2CH_2Ph, n-C_6H_{13}, $(CH_2)_2CH{=}CH_2$, $CH{=}CHSiMe_3$] can be achieved by treatment with $\{CuOTf\}_2(\mu\text{-}C_6H_6)$ and $CaCO_3$ in thf-dioxan mixtures at 70°C.[432] Heterogeneous catalytic coupling of $HC{\equiv}CPh$ to $PhC{\equiv}CC{\equiv}CPh$ (>80%) occurs on a Cu–Mg–Al hydroxycarbonate derived from the corresponding hydrotalcite, in the presence of oxygen and NaOH.[433]

IX

LIGANDS CONTAINING DIYNE GROUPS

Application of the coupling reactions described in the previous section has given many organometallic complexes of ligands containing diyne fragments, which may not be directly coordinated to a metal center. The following surveys this area briefly and describes some recent applications of these systems in the construction of novel molecular architectures.

A. *Essays in Molecular Architecture*

The use of ethynyl- or butadiynyl-substituted π-complexes as intermediates for the synthesis of compounds which are typically larger than conventional organic molecules, such as linear rods, star molecules, dendrimers, as well as poly-metallic noncluster molecules, which may be linked into rug-like polymers (cf. the carbon nets of Diederich[151]), is an active contemporary area of research. Approaches to the syntheses of all-carbon networks via per-alkynylated cyclic systems have attracted much interest. Fragments of such nets based on multiply substituted cyclobutadienes can be stabilized by coordination to $Fe(CO)_3$ or CoCp groups, while similar $Mn(CO)_3$ complexes of substituted cyclopentadienyl ligands (cymantrenes) have also been considered to be portions of fullerenyne-metal complexes. An exotic C_{180} fullerenyne constructed around $Mn(CO)_3\{\eta\text{-}C_5(C{\equiv}C\text{-})_5\}$ groups has been suggested as a possible means of stabilizing expanded and *endo*-metallic fullerenes.[51] The solid-state structures of the rigid star-shaped poly-diynyl π-complexes contain large voids leading to low densities; although solvent is expected to be incorporated readily, the resulting crystals also desolvate readily resulting in crystal decomposition, even at low temperatures during attempted X-ray data collection.[49]

The preparation of metal complexes directly from the perethynylated hydrocarbons [e.g., $C_2(C{\equiv}CH)_4$,[434–436] $C_6(C{\equiv}CH)_6$,[437] or $C_6(C{\equiv}CC{\equiv}CH)_6$[438]] has not yet received widespread attention. However, metal complexes with cyclobutadiene, cyclopentadienyl, or benzene ligands are readily functionalized to give poly-ynyl derivatives. Common starting points for these derivatives are pentaalkynyl-cyclopentadienes, penta- or deca-iodoferrocene, $Mn(CO)_3(\eta\text{-}C_5I_5)$ or $Fe(CO)_3(\eta\text{-}C_4I_4)$. Butadiyne linkers separate ferrocenyl or $Co(\eta\text{-}C_4R_3)Cp$ groups in fused organometallic dehydroannulenes obtained by oxidative coupling reactions.[439] Coupling of di- or triynylstannanes with periodo-η-ring complexes of $Cr(CO)_3$, $Mn(CO)_3$, $Fe(CO)_3$, or FeCp has given star-shaped complexes which may be suitable as building blocks for extended carbon networks.[49]

1. *Cyclobutadienes*

a. *$Fe(CO)_3(\eta\text{-}C_4R_4)$.* Stille-type coupling (Stille–Beletskaya reaction) of stannylbuta-1,3-diynes, $Me_3SnC{\equiv}CC{\equiv}CR$ ($R = SiMe_3$, Pr^i, Bu^t, $n\text{-}C_5H_{11}$) with iodocyclobutadiene complexes, using $Pd_2(dba)_3/AsPh_3$ as catalyst, is an efficient route to poly-alkynyl complexes[440,441]; the $SiMe_3$ derivatives can be conventionally protodesilylated to the parent ethynyl compounds. In this way have been made $Fe(CO)_3\{\eta\text{-}C_4(C{\equiv}CH)_4\}$,[45,49,442,443] $Fe(CO)_3(\eta\text{-}C_4H_3C{\equiv}CC{\equiv}CH)$,[46] and $Fe(CO)_3\{\eta\text{-}C_4(C{\equiv}CC{\equiv}CR)_4\}$ ($R = H$, $SiMe_3$).[49] Attempts to desilylate the latter complex ($R = SiMe_3$) gave only insoluble black materials. Binuclear complexes were made similarly, including $\{Fe(CO)_3(\eta\text{-}C_4H_3\text{-})\}_2(\mu\text{-}C_4)$ from $Fe(CO)_3(\eta\text{-}C_4H_3I)$ and $Me_3SnC{\equiv}CC{\equiv}CSnMe_3$,[46] 1,4-$\{Fe(CO)_3[\eta\text{-}C_4(C{\equiv}CSiMe_3)_3\}_2C_4$ (66%) as a hydrocarbon-soluble lemon-yellow solid, stable in the cold (−18°C) from $Fe(CO)_3\{\eta\text{-}C_4I(C{\equiv}CSiMe_3)_3\}$ and $Me_3SnC{\equiv}CC{\equiv}CSnMe_3$,[48] and $Fe(CO)_3\{\eta\text{-}C_4[(C{\equiv}C\text{-}\eta\text{-}C_4H_3)Fe(CO)_3]_4\}$ (18% yield) with some $\{Fe(CO)_3(\eta\text{-}C_4H_3\text{-})\}_2C_4$, from $Fe(CO)_3(\eta\text{-}C_4I_4)$ and $Fe(CO)_3(\eta\text{-}C_4H_3C{\equiv}CSnMe_3)$.[49] Other couplings have given $Fe(CO)_3\{\eta\text{-}C_4[(C{\equiv}CC{\equiv}C\text{-}\eta\text{-}C_5H_4)Mn(CO)_3]_4\}$ from $Fe(CO)_3(\eta\text{-}C_4I_4)$, and $Fe(CO)_3\{\eta\text{-}C_4[(C{\equiv}C)_3Bu^t]_4\}$ from $Me_3Sn(C{\equiv}C)_3Bu^t$.[49]

b. *Co(η-C₄R₄)Cp.* As described earlier (Section VII.A.3), reactions of 1,3-diynes with $Co(CO)_2Cp$ afford η-cyclobutadiene complexes which undergo the 1,2-diethynylcyclobutadiene rearrangement. While oxidative coupling (Hay, 40°C) of $Co\{\eta\text{-}1,2\text{-}C_4H_2(C{\equiv}CH)_2\}Cp$ gave no well-defined product, the more bulky $Co\{\eta\text{-}1,2\text{-}C_4(SiMe_3)_2(C{\equiv}CH)_2\}Cp$ (**316**) was coupled ($CuCl_2$ in refluxing tmed) to cyclic oligomers **317a** (as *syn* and *anti* isomers), **317b** (as four stereoisomers) and **317c** (as three stereoisomers).[439,444] The cyclo-trimer was protodesilylated with $[NBu_4]F$. Similarly, $Co\{\eta\text{-}1,2\text{-}C_4(SiMe_3)_2(C{\equiv}CC{\equiv}CH)_2\}Cp$ (from **316** and $BrC{\equiv}CCH_2OH$) followed by treatment with MnO_2/KOH) affords the orange tetraynyl polymer **318.**[439]

Coupling of **316** with $BrC{\equiv}CSiPr^i_3$ gives $Co\{\eta\text{-}1,2\text{-}C_4(SiMe_3)_2(C{\equiv}CC{\equiv}CSiPr^i_3)_2\}Cp$, which rearranged by FVP at 550°C to $Co\{\eta\text{-}1,2\text{-}C_4(C{\equiv}CSiMe_3)_2(C{\equiv}CSiPr^i_3)_2\}Cp$, together with **319** ($R = SiPr^i_3$).[445] The related complex **319** ($R = H$) was obtained from $\{CpCo[\eta\text{-}1,2\text{-}C_4(SiMe_3)_2(CHO)]\text{-}\}_2C_4$ with dimethyl (1-diazo-2-oxopropyl)phosphonate.[446] While treatment of the tetra-alkynyl complex with $[NMe_4]F$ gave the stable tetraethynyl $Co\{\eta\text{-}C_4(C{\equiv}CH)_4\}Cp$,[445] partial protodesilylation (K_2CO_3/MeOH) to $Co\{\eta\text{-}1,2\text{-}C_4(C{\equiv}CH)_2(C{\equiv}CSiPr^i_3)_2\}Cp$ and oxidative coupling with $Cu(OAc)_2$ in MeCN afforded **320a,b,** also in two and four stereoisomeric forms, respectively.[439] Ring closure is preceded by formation of open oligomers which are conformationally mobile.

Hay coupling of $Co\{\eta\text{-}1,3\text{-}C_4(SiMe_3)_2(C{\equiv}CH)_2\}Cp$ gave a series of oligomers **321** ($n = 0$–7), of which the first two members were protodesilylated; higher oligomers were formed after prolonged reaction times.[442] In the UV/vis spectra, the intensities of bands at ca 348 nm increase with n and are assigned to a diynyl-η-cyclobutadiene interaction. NMR studies showed that there was no significant rotational barrier around the C_4 units, even at −100°C.

Oxidative coupling [$Cu(OAc)_2$/py; Eglinton] of $Co\{\eta\text{-}C_4(C_5H_{10})_2\}(\eta\text{-}C_5H_4C{\equiv}CH)$ gave binuclear **322** in 82% yield; oxidation potentials E_1, E_2 for the two cobalt centers are +0.846, +0.947 V, respectively, suggesting that the two radical centers interact via the butadiynyl chain.[447]

2. *Cyclopentadienyls*

a. *Mn(CO)₃(η-C₅R₅).* Stille-type coupling of polyiodo π-complexes with stannylbuta-1,3-diynes $Me_3SnC{\equiv}CC{\equiv}CR$ ($R = SiMe_3$, Bu^t, Pr^i, n-C_5H_{11}), using $Pd_2(dba)_3/AsPh_3$ as catalyst,[441] have given $Mn(CO)_3\{\eta\text{-}C_5(C{\equiv}CC{\equiv}CR)_5\}$ in low yields only, possibly as a result of steric congestion around the iodines.[49] Attempts to desilylate these complexes gave only insoluble black materials.

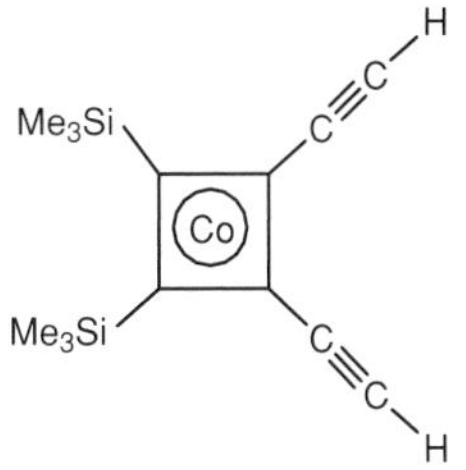

(**316**)

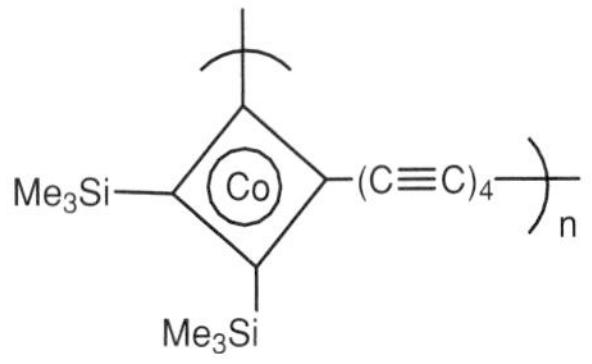

(**318**)

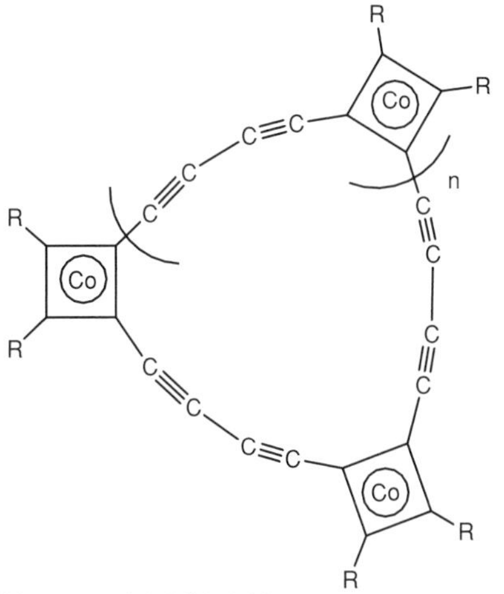

(**317**) R = $SiMe_3$, n = 1 (**a**), 2 (**b**), 3 (**c**)

(**320**) R = $C{\equiv}CSiPr^i_3$, n = 1 (**a**), 2 (**b**)

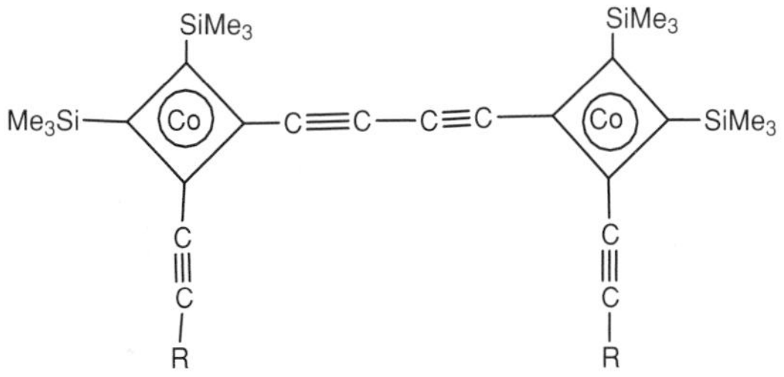

(**319**) R = $C{\equiv}CSiPr^i_3$, H

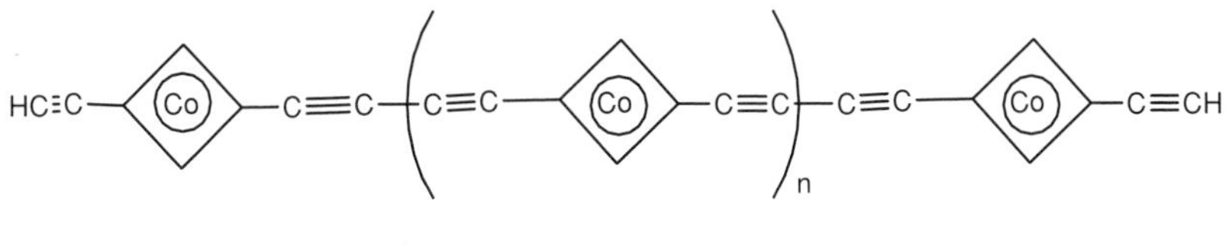

(**321**) n = 0-7

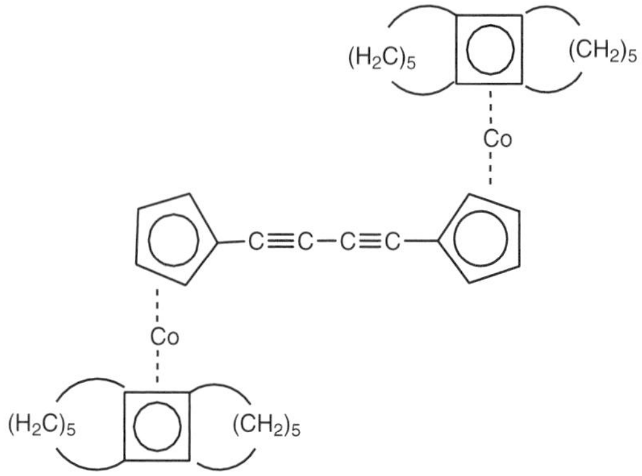

(**322**)

Oxidative coupling (Hay; CuCl/tmeda/acetone) of $Mn(CO)_3\{\eta\text{-}C_5H_3(C{\equiv}CH)_2\text{–}1,2\}$ gave polymeric $\{Mn(CO)_3[\eta\text{-}C_5H_3(C{\equiv}C\text{—})_2]\}_n$ (68%) with n_{av} ca 38; a small amount of presumed cyclic oligomers is also obtained. Co-coupling with $Mn(CO)_3(\eta\text{-}C_5H_4C{\equiv}CH)$ gave a series of oligomers $\{Mn(CO)_3(\eta\text{-}C_5H_4C{\equiv}C\text{—})\}_2\{\mu\text{-}[\text{-}(C{\equiv}C\text{-}\eta\text{-}C_5H_3C{\equiv}C\text{—})Mn(CO)_3]_n\}$ ($n = 1$–5), together with some 1,4-$\{Mn(CO)_3(\eta\text{-}C_5H_4\text{-})\}_2(C{\equiv}CC{\equiv}C)$.[51]

b. *Ferrocenes and ruthenocenes.* $FcC{\equiv}CC{\equiv}CFc$ has been described by several groups[448–451] who made it by the oxidative coupling of $FcC{\equiv}CH$ [$Cu(OAc)_2$/py]; the X-ray structure has been described on at least two occasions.[451,452] A similar route afforded $\{FeCp^{Me}(\eta\text{-}C_5H_3Me\text{-}3\text{-}(C{\equiv}C\text{—})\}_2$.[449] The electrochemistry of $FcC{\equiv}CC{\equiv}CFc$ shows two oxidation waves (ΔE 100 mV).[453] Metalation (LiBu) and iodination of $HC{\equiv}CCFc_2(OMe)$, followed by coupling of the resulting $IC{\equiv}CCFc_2(OMe)$ with $FcC{\equiv}CCu$, afford $FcC{\equiv}CC{\equiv}CCFc_2(OMe)$, which can be readily hydrolyzed to $FcC{\equiv}CC{\equiv}CCFc_2(OH)$. Treatment with HBF_4 gave carbenium cation **323,** for which one canonical form is the diyne shown. Further derivatization with LiFc gives the bis-allenyne $\{Fc_2C{=}C{=}C(C{\equiv}CFc)\text{-}\}_2$.[454] Chemical oxidation ($AgPF_6$) of $FcC{\equiv}CC{\equiv}CMe$ gives $[FcC{\equiv}CC{\equiv}CMe][PF_6]$.[97]

The series $Fc(C{\equiv}C)_nFc$ ($n = 2, 4, 6, 8$) and $Fc(C{\equiv}C)_mPh$ ($m = 2, 4$) were made by coupling of $Fc(C{\equiv}C)_nH$ with $BrC{\equiv}CCO_2H$, decarboxylation, or oxidative coupling.[448] Coupling of $FcC{\equiv}CH$ with *cis*-$CHCl{=}CHCl$, followed by treatment with LDA and quenching with NH_4Cl, gives $FcC{\equiv}CC{\equiv}CH$, which can be oxidatively coupled [$Cu(OAc)_2$/py] to $Fc(C{\equiv}C)_4Fc$.[450]

Dimerization (oxygen with CuCl in pyridine) of *E*- and *Z*-4-$FcCH{=}CHC_6H_4C{\equiv}CH$ gave the *E,E*- and *Z,Z*,-diynes, respectively, in >95% yield.[439,455]

Eglinton coupling of FeCp{*η*-$C_5H_3(C\equiv CH)_2$-1,2} gave both *cis*- and *trans*-{FeCp [*η*-$C_5H_3(C\equiv CC\equiv C\text{-})$]}$_3$ (**324,** $n = 1$), together with some higher oligomers, but no linear polymer. The electrochemistry of the two isomers differ: the *trans* isomer shows three reversible 1-e oxidations (+0.61, +0.74, +0.83 V) while the *cis* isomer shows reversible 1-e (at +0.61 V) and 2-e (+0.78 V) oxidations, i.e., there is weak to moderate electronic communication between the Fc nuclei.[445] Coupling with a mixture of CuCl and $CuCl_2$ in pyridine afforded the tri- and tetramers **324** ($n = 1$ and 2), together with some {CpFe[*η*-1,2-$C_5H_3(C\equiv CH)$-]}$_2C_4$.[439]

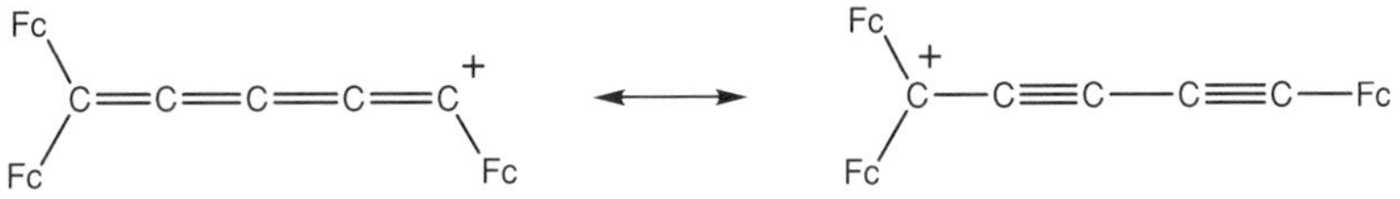

(**323**)

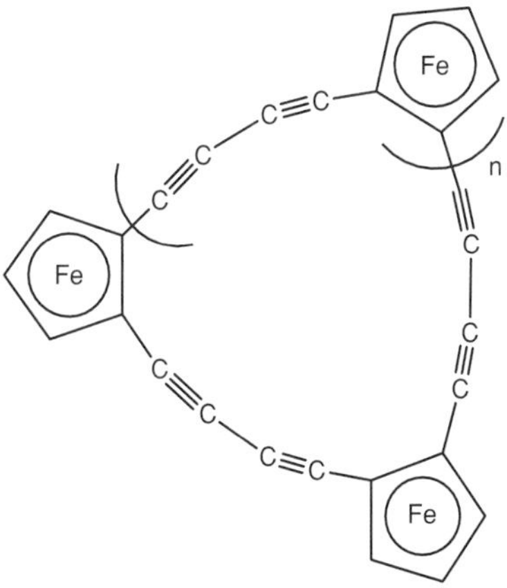

(**324**) n = 1, 2

1,4-Bis(2′,3′,4′,5′-tetramethylruthenocenyl)buta-1,3-diyne is a 19% by-product from the synthesis of RuCp{$C_5Me_4(C\equiv CH)$} from LiBu and RuCp{$C_5Me_4(CH{=}CCl_2)$}.[456]

Pyridyl-poly-ynes containing ferrocenyl end-groups have been prepared, together with their $W(CO)_5$ derivatives. Oxidative coupling of $Fc(C\equiv C)_nH$ ($n =$ 1, 2) with $HC\equiv CC_5H_4N$-4, using $Cu(OAc)_2$–CuI in MeOH–pyridine gave $Fc(C\equiv C)_{n+1}C_5H_4N$-4 ($n = 1$, 15%; 2, 35%), together with $Fc(C\equiv C)_nFc$ ($n = 2$, 15%; 4, 20%, respectively).[457]

3. η^6-*Arene Complexes*

Oxidative coupling [$Cu(OAc)_2$/py] of $Cr(CO)_3(\eta^6\text{-PhC}{\equiv}\text{CH})$ gave $\{Cr(CO)_3\}_2(\mu\text{-}\eta^6{:}\eta^6\text{-PhC}{\equiv}\text{CC}{\equiv}\text{CPh})^{96\%}$; because of the oxygen sensitivity of the phenylethyne complex, the Glaser coupling conditions were unsuccessful.[458] Cadiot–Chodkiewicz coupling of $Cr(CO)_3(\eta\text{-PhC}{\equiv}\text{CH})$ with BrC≡CR (R = $C_6H_4NO_2$-4, $C_6H_4NMe_2$-4, Fc) gives the corresponding diynes $Cr(CO)_3(\eta\text{-PhC}{\equiv}\text{CC}{\equiv}\text{CR})$. The NMe_2 complex shows large solvatochromism in the π-π^* transition [$\Delta\lambda$($CHCl_3$/dmso) −158 (NO_2), −235 cm^{-1} (NMe_2)] and intra-ligand CT band [$\Delta\lambda$($CHCl_3$/dmso) −194 (NO_2), +421 cm^{-1} (NMe_2)]; it has a large NLO hyperpolarizability (β°_{333} 34).[459]

B. *Diynyl-Containing Hydrocarbyl Ligands*

1. *Alkyls and Aryls*

Complexes $M(CH_2C{\equiv}CC{\equiv}CMe)(CO)_nCp$ [**325;** M = Mo, $n = 3$; M = Fe, $n = 2$ (Scheme 74)] were obtained from the carbonyl anions and 1-chlorohexa-2,4-diyne.[460] Subsequent chemistry involves protonation (HBF_4) to cationic allene or diene complexes, or addition of MeOH to give allylic derivatives, which are formed with concomitant insertion of CO. The latter can also be obtained from the cationic species and NaOMe. The allene-iron cation reacts with $NHEt_2$ to form an ynenyl complex. The luminescent complex $\{Re(CO)_3(5,5'\text{-Bu}^t_2\text{-bpy})\}_2(\mu\text{-C}{\equiv}\text{CC}_6H_4C{\equiv}CC{\equiv}CC_6H_4C{\equiv}C)$ has been reported.[461]

Oxidative coupling (CuCl/tmeda/O_2, CH_2Cl_2) of *trans*-$Pt(C{\equiv}CSiMe_2C{\equiv}CH)_2(PBu_3)_2$ gives polymeric $\{Pt(PBu_3)_2(C{\equiv}CSiMe_2C{\equiv}CC{\equiv}CSiMe_2C{\equiv}C)\}_n$, which reacts with $Co_2(CO)_8$ to give either dark red (1/1) or green (1/3) derivatives according to reagent stoichiometry, both of which are soluble in benzene. In the latter, both C≡C triple bonds of the diyne are coordinated to cobalt, the Pt-C≡C system being sterically hindered by the PBu_3 ligands. While it is an insulator, doping with I_2 gave a black material which increased conductivity up to about 20% iodine incorporation.[462] Protodesilylation of the tetraethynylethene complex *trans*-$Pt\{C{\equiv}CC(C{\equiv}CSiPr^i_3){=}C(C{\equiv}CC_6H_3Bu^t_2\text{-}3,5)_2\}_2(PEt_3)_2$ followed by oxidative coupling (Hay conditions) gave orange-red metallacycle **326,** in which two opposite edges of the "square" are diynyl units. Significant electronic communication around the perimeter is evidenced by the UV-vis spectrum; the end absorption extends beyond 500 nm. Irradiation at 450 nm produces fluorescence at 543 nm.[463]

Other molecular squares (**327**) have been obtained from *cis*-$Pt(C{\equiv}CC{\equiv}CH)_2(PR_3)_2$ and *cis*-$PtX_2(PR'_3)_2$ (X = Cl, OTf) in reactions carried out under high dilution conditions.[464,465] These reactions have been extended to condensations with

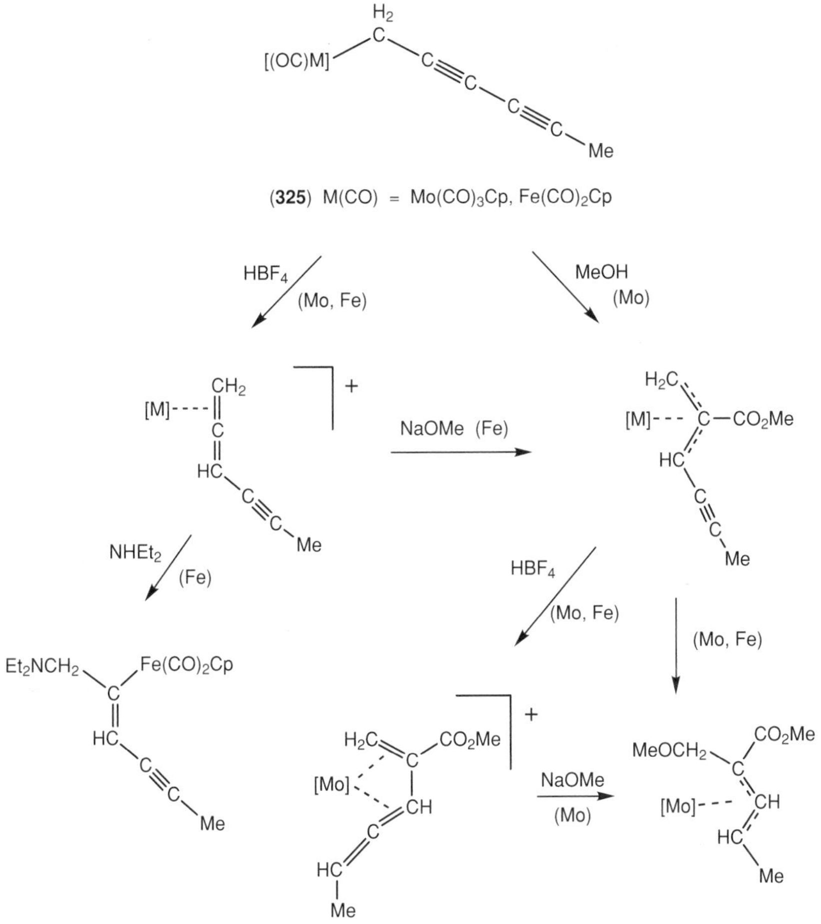

(**325**) M(CO) = $Mo(CO)_3Cp$, $Fe(CO)_2Cp$

SCHEME 74

the analogous *trans*-Pt complexes or [ppn][$Au(C{\equiv}CC{\equiv}CH)_2$] to give even larger squares (**328** and **329**), or with other metal systems, such as $\{AuCl\}_2(\mu$-dppm) to give rectangle **330.**

The platinadehydrobenzo[19]annulene **331** is formed by reaction of the $SnMe_3$ derivative with *trans*-$PtCl_2(PEt_3)_2$ and CuI (Scheme 72). The molecule is not planar, maximum deviations from the mean plane of the $C_{30}H_{12}$ macrocycle being between 0.484 Å above to 0.283 Å below. The UV/vis absorption spectrum suggests that electron delocalization occurs throughout the macrocycle.[466]

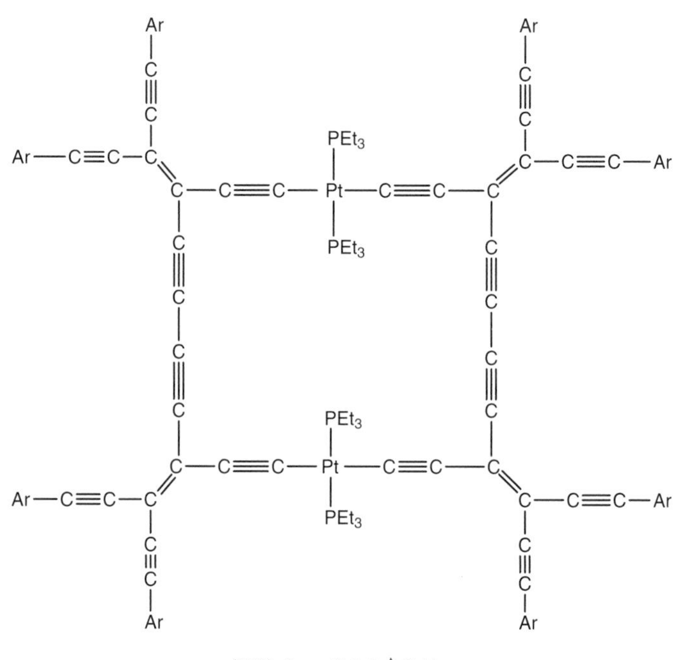

($\mathbf{326}$) Ar = 3,5-$Bu^t_2C_6H_3$

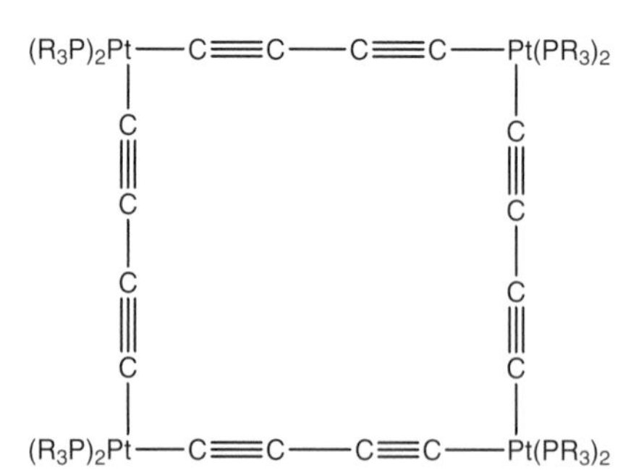

($\mathbf{327}$) PR_3 = PEt_3, PBu_3, dcype, dppe, dppp

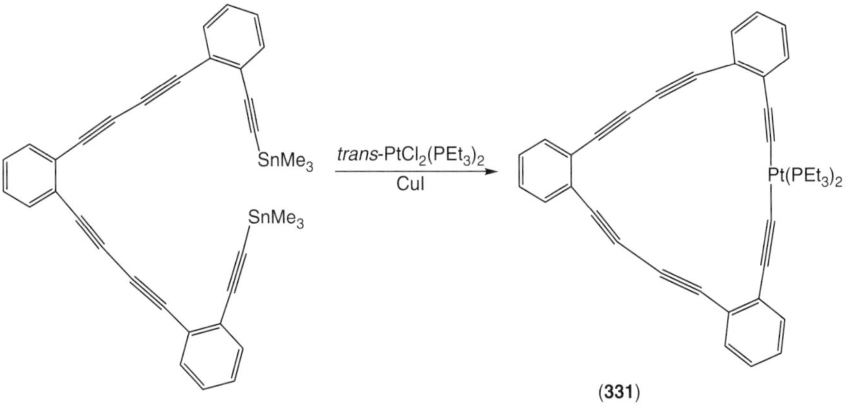

$SnMe_3$

$SnMe_3$

trans-$PtCl_2(PEt_3)_2$

CuI

$Pt(PEt_3)_2$

(**331**)

SCHEME 75

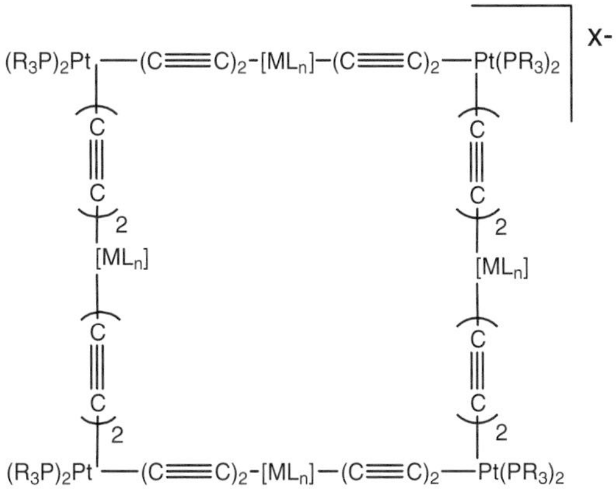

$(R_3P)_2Pt$ —$(C{\equiv}C)_2$-$[ML_n]$-$(C{\equiv}C)_2$—$Pt(PR_3)_2$ X-

$(C{\equiv}C)_2$ — $[ML_n]$ — $(C{\equiv}C)_2$

$(C{\equiv}C)_2$ — $[ML_n]$ — $(C{\equiv}C)_2$

$(R_3P)_2Pt$ —$(C{\equiv}C)_2$-$[ML_n]$-$(C{\equiv}C)_2$—$Pt(PR_3)_2$

(**328**) $ML_n = Pt(PBu_3)_2$, $(PR_3)_2$ = dcype, x = 0

(**329**) ML_n = Au, $(PR_3)_2$ = dppe, x = 4

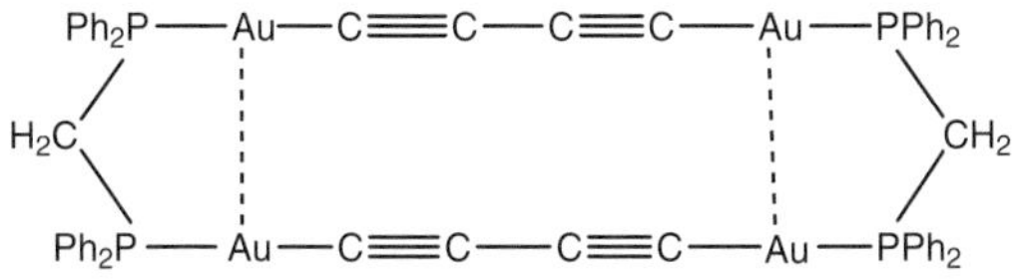

Ph_2P—Au—C≡C—C≡C—Au—PPh_2

H_2C CH_2

Ph_2P—Au—C≡C—C≡C—Au—PPh_2

(**330**)

2. *Diynyl Carbenes*

Successive treatment of HC≡CC≡CBu^t with LiBu, $Cr(CO)_6$, and $[Et_3O]BF_4$ gave the first diynyl carbene complex Cr{=C(OEt)(C≡CC≡CBu^t)}$(CO)_5$, which reacts with $NH(CH_2Ph)_2$ to give Cr{=C(OEt)[CH=C(C≡CBu^t)[N$(CH_2Ph)_2$]]}$(CO)_5$.[467]

Lithiation of W{=C(NMe_2)(C≡CH)}$(CO)_5$ and subsequent reaction with CuI and BrC≡$CSiMe_3$ give a low yield of W{=C(NMe_2)(C≡CC≡$CSiMe_3$)}$(CO)_5$. Protodesilylation (KF/MeOH/thf) affords the butadiynylcarbene W{=C(NMe_2)(C≡CC≡CH)}$(CO)_5$, which with LiBu and $SnClBu_3$ gives W{=C(NMe_2)(C≡CC≡$CSnBu_3$)}$(CO)_5$.[66] Some subsequent reactions with metal halides are described in Section II.

Conversion of M{=C(NMe_2)C≡CH}$(CO)_5$ (M = Cr, W) to the iodoalkynyl carbenes (with LiBu, then I_2) or the $SnBu_3$ derivatives (with LiBu, then $SnClBu_3$), followed by $PdCl_2(NCMe)_2$-catalyzed coupling of these two complexes gave binuclear {$M(CO)_5$}$_2${μ-[=C(NMe_2)C≡CC≡CC(NMe_2)=]}.[468] Spectroscopic data indicate that allenyl–alkynyl- and pentatetraenyl–carbene resonance forms contribute little at best to the structure. Similarly, the carbenes W{=C(NMe_2)(C≡C)$_n$H}$(CO)_5$ ($n = 2, 3$) could be transformed into {$W(CO)_5$}$_2${μ-[=C(NMe_2)(C≡C)$_{2n}$C(NMe_2)=]}. While the iodoalkynyl complexes slowly decompose at r.t., the alkynediyl-bridged dicarbene complexes are stable.

The complex W{=C(NMe_2)(C≡C)$_3SiMe_3$}$(CO)_5$ was obtained from W{=C(NMe_2)C≡CX}$(CO)_5$ by coupling with IC≡CC≡$CSiMe_3$ either after lithiation and addition of CuI (X = H), or by Stille coupling with $PdCl_2(NCMe)_2$ as catalyst (X = $SnBu_3$)[116]; protodesilylation occurred with KF in MeOH/thf. The structures are hybrids of resonance forms **N–S,** dipolar **P–R** being more important. The ^{13}C NMR spectra show alternating δ values along the chain.

(N) R_2N(C=[M])–C≡C–C≡C–C≡C–R' ⟷ **(Q)** R_2N(C–[M]$^-$)=C=C=C=C$^+$–C≡C–R'

(O) R_2N^+=C([M]$^-$)–C≡C–C≡C–C≡C–R' ⟷ **(R)** R_2N(C–[M]$^-$)=C=C=C=C=C=C$^+$–R'

(P) R_2N(C–[M]$^-$)=C=C$^+$–C≡C–C≡C–R' ⟷ **(S)** R_2N(C–[M]$^-$)=C=C=C=C=C=C=R'$^+$

Deprotonation of the parent complex occurred with LiBu; the lithio derivative was converted to the $SnBu_3$ and $Ru(CO)_2Cp$ complexes, or the Hg derivative containing two C_7 chains. The IR ν(CO) spectra of the Ru and Hg compounds are at significantly lower wavenumbers than the others, suggesting that form **S** is more important.

Alternation of ^{13}C chemical shifts occurs along the C_n chain, with $\Delta\delta(C_\beta\text{-}C_\alpha) > \Delta\delta(C_\delta\text{-}C_\gamma) > \Delta\delta(C_\varepsilon\text{-}C_\delta)$, while in the UV/vis spectra, the lowest energy absorption moves to longer wavelengths with increasing chain length. The latter observation suggests that while the HOMO energy remains relatively constant, the energy of the LUMO decreases (by ca 15–20 kJ mol^{-1} for addition of a C_4 unit). Solvatochromic effects decrease with increasing chain length.[468]

C. *Tertiary Phosphines Containing Diynyl Groups*

Interaction of $PPh_2(C{\equiv}CC{\equiv}CR)$ (R = Bu^t, Ph; from Cadiot–Chodkiewicz coupling of $RC{\equiv}CBr$ and $HC{\equiv}CCH_2OH$, followed by lithiation and reaction with $PClPh_2$) with $Ru_3(CO)_{12}$ gave $Ru_3(CO)_{11}(PPh_2C{\equiv}CC{\equiv}CR)$. The transformations occurring upon thermolysis are discussed in Section III. An extensive study of the chemistry of $PPh_2C{\equiv}CC{\equiv}CPPh_2$ (bdpp) has been reported. Simple *P*-donor complexes $\{M(CO)_5\}_2(\mu\text{-bdpp})$ were obtained from $M(CO)_5(L)$ (M = Mo, L = MeCN; M = W, L = thf), $\{Fe(CO)_4\}_2(\mu\text{-bdpp})$ from $Fe_2(CO)_9$, $\{RuCl(PPh_3)Cp\}_2(\mu\text{-bdpp})$ from $RuCl(PPh_3)_2Cp$, and $\{AuCl\}_2(\mu\text{-bdpp})$ from AuCl (thiodiglycol).[211]

The previously described Mo, W, and Fe complexes reacted with $Co_2(CO)_8$ to give black compounds in which one $Co_2(CO)_6$ moiety coordinates to one of the two C≡C triple bonds. Complexes containing a single $Pt(PPh_3)_2$ group coordinated to the diyne ligand were obtained from $Pt(\eta\text{-}C_2H_4)(PPh_3)_2$.[211] Addition of $SePh(SiMe_3)$ to a mixture of CuCl and bdpp in thf gave the diynylphosphine-bridged complex $\{Cu(SePh)\}_2(\mu\text{-bdpp})_3$.[493]

Reactions of bdpp with metal cluster carbonyls have given "barbell" shaped molecules. With $M_n(CO)_{11}(NCMe)$ [$M_n = Ru_3$, Os_3, $Re_3(\mu\text{-H})_3$, $Ru_4(\mu\text{-H})_4$], the bis-cluster complexes $\{M_n(CO)_{11}\}_2(\mu\text{-bdpp})$ were obtained in 38–68% yield; with an excess of bdpp, monocluster derivatives $M_n(CO)_{11}(bdpp)$ [$M_n = Re_3(\mu\text{-H})_3$, Os_3] are formed. From these, mixed-metal complexes $\{Os_3(CO)_{11}\}(\mu\text{-bdpp})\{M_n(CO)_{11}\}$ [$M_n = Ru_3$, $Re_3(\mu\text{-H})_3$] and $\{Re_3(\mu\text{-H})_3(CO)_{11}](\mu\text{-bdpp})\{Ru_3(CO)_{11}\}$ were obtained by combination with the appropriate cluster carbonyl precursors.[260] Thermolysis of $\{M_3(CO)_{11}\}_2(\mu\text{-bdpp})$ (M = Ru, Os) in refluxing toluene gave $\{Ru_3(\mu\text{-PPh}_2)(CO)_9\}_2(\mu_3{:}\mu_3\text{-}C_4)$ together with $Ru_4(\mu\text{-H})\{\mu_4\text{-PPh}(C_6H_4)C_2C{\equiv}CPPh_2[Ru_3(CO)_{12}]\}(CO)_{12}$ (**332**) in the former case. Recrystallization of **332** from MeOH-containing solvents gave $Ru_4\{\mu_4\text{-CCHC}_2[Ru_2(\mu\text{-PPh}_2)(CO)_6]\}(\mu_3\text{-OMe})(\mu\text{-PPh}_2)(CO)_{10}$ (**333**).

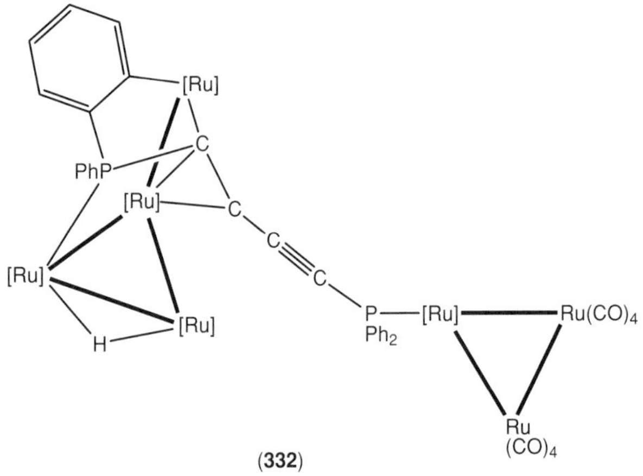

(**332**)

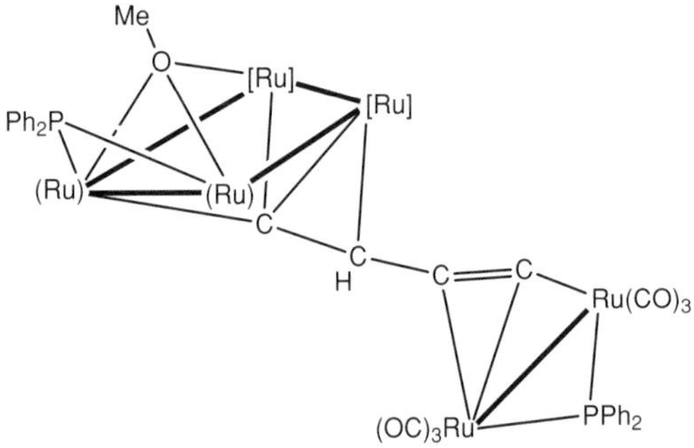

(**333**)

D. *Applications of Coupling Reactions to Inorganic Complexes*

1. *Diynyl Pyridines*

Elegant and extensive work on photoactive molecular scale wires and their use for information transfer, using complexes containing alkynyl- and diynyl bridged polypyridyl and porphyrin ligands, has been reviewed recently.[37,469] While most work has involved single C≡C units separating the complex moieties, several examples containing diynyl units are known. Use of a bis-alkynyl-platinum(II) unit as spacer results in formation of a relay or insulator as a result of the low-lying orbitals on the Pt center. Electronic tuning can also be achieved using polycyclic aromatic hydrocarbons.

1,4-Bis(4-pyridyl)buta-1,3-diyne (dbp) was prepared by oxidative coupling of 4-ethynylpyridine.[470] Complexes containing the *N*-bonded ligand are usually formed: interest has centered on electron transfer, and NLO properties were demonstrated in addition with η^1 complexes. Most investigations have been directed toward the synthesis of either binuclear complexes, as with $\{cis\text{-}W(CO)_4(L)\}_2(\mu\text{-dpb})$ [L = CO, PEt_3, PPh_3, $P(OEt)_3$, $P(OPh)_3$] obtained from dpb and $W(CO)_4(L)(thf)$,[471] $\{fac\text{-}ReCl(CO)_3(L)\}_2(\mu\text{-dpb})$ [L = PPh_3, $P(OMe)_3$] and $[\{fac\text{-}Re(CO)_3(bpy)\}_2(\mu\text{-dpb})][PF_6]_2$,[472] or polymeric Co, Ni, and Cu compounds.[473] Reaction of $Mo_2\{\mu\text{-}\eta^2\text{-}(4\text{-}NC_5H_4)C_2C{\equiv}C(C_5H_4N\text{-}4)\}(CO)_4Cp_2$ with $W(CO)_5(thf)$ gave deep red $Mo_2\{\mu\text{-}C_4(C_5H_4N[W(CO)_5]\text{-}4)_2\}(CO)_4Cp_2$.[175] Self-assembly of $ReCl(CO)_5$ and dpb gives the molecular square *cyclo*-$\{Re(\mu\text{-dpb})(Cl)(CO)_3\}_4$,[474] while triangles and squares containing dpb as edges bridging Pd(en) groups have also been obtained; significant amounts of oligomeric products are also formed.[475]

The complex $[Cu_2(dpb)_3(NCMe)_2][PF_6]_2$, obtained from the diyne and $[Cu(NCMe)_4]PF_6$, consists of a network of interpenetrating ladder polymers containing tetrahedral Cu(I), with the diyne moiety bridging Cu(I) centers and forming the steps of the ladder. Adjacent symmetry-related ladders have π-π interactions [3.484 Å].[476] A ribbon-like one-dimensional polymer is formed by the reaction of $AgNO_3$ with $2\text{-}NC_5H_4C{\equiv}CC{\equiv}CC_5H_4N\text{-}2$, in which the silver atom is coordinated by two pyridines and a chelating NO_3 anion.[477] The $Cd(CN)_2$ complex of dpb also forms an extended framework structure, darkening in light probably resulting from polymerization within the crystals.[478]

2. *Bipyridines and Terpyridines*

Dimerization of $4\text{-}(4\text{-}HC{\equiv}CC_6H_4)$-terpy ($CuCl/tmed/O_2$) gave the disubstituted buta-1,3-diyne.[479] Rigid-rod complexes of the type $[Ru(terpy)\{4\text{-terpy-}(C{\equiv}C)_2\text{-terpy-}4\}]^{2+}$, $[\{Ru(terpy)[terpy\text{-}(C_6H_4\text{-}4)_m\text{-}4']\}_2(\mu\text{-}C_4)]^{4+}$ ($m = 0$, 1) and $[\{(terpy)Ru[terpy\text{-}(C{\equiv}C)_2\text{-terpy}]\}_2Fe\}]^{6+}$ were obtained from the corresponding bridged terpy ligands and $RuCl_2(dmso)(terpy)$, followed by $FeSO_4$ for the mixed Ru_2Fe system.[480] Addition of alkyne groups to $[Ru(terpy)_2]^{2+}$ lowers the energy of the MLCT triplet state, with the result that excitation into the MLCT transition may result in luminescence. Light-induced electron transfer takes place over a distance of 15–18 Å, indicating that there is efficient electronic communication between the terminal unit and the C_n bridge, with the energy of the bridge decreasing with increasing length.[481] Removal of the second Ru system or insertion of the C_6H_4 groups in the chain results in shorter lifetimes.[16]

The CVs of alkynylbipyridyl-ruthenium complexes, such as $[Ru(bpy)_2(5\text{-}HC{\equiv}C\text{-bpy})]^{2+}$ and the related terpy complex, $[Ru(terpy)(4\text{-}HC{\equiv}C\text{-terpy})]^{2+}$, contain reversible metal-centered oxidation waves and several ligand-dependent reduction waves. Polymerization involves activation of the alkyne bridge, with electrochemical and MO studies showing the LUMO delocalized over both the alkyne and the π-radical anions of the metal complexes. Coupling at the alkyne centre leads to greater distortions of the bpy ligands than for the terpy ligands and this suggests the dianion adopts a [3]-cumulene structure before coupling.[482] Coupling of 4-HC≡CC≡C-terpy with *trans*-$PtCl_2(PBu_3)_2$ (CuI/$NHPr^i{}_2$) gives the *trans*-bis(diynyl) complex.[134]

3. *Porphyrins*

μ-Diynyl-porphyrins have been reported and complexes with a range of metal ions have been studied electrochemically. Spectroelectrochemical results indicate that the C_4 chain is involved in strong cooperative effects in the reduced species.[483] Fast electron transfer occurs between the porphyrin nucleus and quinone substituents.[484] Diyne-linked oligomeric and polymeric porphyrins have interesting electro-optical and NLO properties as a result of extended interactions between the porphyrin nuclei, which may extend over long distances.[485] Unsymmetrical porphyrin oligomers have been obtained using a bis(diynyl)binaphthyl spacer.[486]

Platinum derivatives of zinc *meso*-ethynylporphyrins have been made by conventional reactions; one product was the 1,3-diynydiyl-linked bis-porphyrin {*trans*-$(Ph_3P)_2ClPt$}C≡C{Zn(porph)}C≡CC≡C{Zn(porph)}C≡C{$PtCl(PPh_3)_2$-*trans*}, which was identified by MALDI mass spectrometry through the ion [HC≡C{Zn(porph)}C_4{Zn(porph)}C≡CH + Na]$^+$ at *m/z* 1197.[487]

The synthesis of linear and cyclic porphyrin oligomers by Glaser coupling and their use to enclose molecules and to carry out reactions in the cavities have been reviewed.[488] Conjugated porphyrin ladders have been assembled from the diynyl-linked zinc porphyrins by bridging zinc atoms in adjacent polymeric strands with bidentate ligands such as 1,4-pyrazine.[36] A similar trimer catalyses acyl transfer from *N*-acetylimidazole to 4-hydroxymethylpyridine, the reaction being inhibited by *N*-donor ligands on the zinc atoms.[489] Substantial acceleration of Diels–Alder reactions also occurs in the cavity, presumably as a result of favorable alignment of reactants by coordination to the zinc atoms.[490]

4. *β-Diketonates*

Coupling of $Ru(acac)_2${$(OCMe)_2C$(C≡CH)} with CuI/$PdCl_2(PPh_3)_2$/NEt_3 gave the diyne {$Ru(acac)_2[(OCMe)_2C$}$_2$(μ-C≡CC≡C), while in the presence of HC≡CFc coupling with $Cu(OAc)_2$/NEt_3 gave $Ru(acac)_2${$(OCMe)_2C$(C≡CC≡CFc)}.[491] Coupling under Hay conditions gives instead a green chain polymer containing $Ru(acac)_3$ units linked by diyndiyl bridges. Electrochemical results indicate strong Ru–Ru interactions between neighboring units in the polymer.[492]

X

SPECTROSCOPIC PROPERTIES, ELECTRONIC STRUCTURE, AND REDOX BEHAVIOR

A. *Infrared Spectra*

Numerous measurements of $\nu(C{\equiv}C)$ frequencies and vibrational data associated with other supporting ligands such as $\nu(CO)$ have been used to characterize most of the complexes included in this review and, in fewer cases, to probe the electronic effects of σ-bonded poly-ynyl $[(C{\equiv}C)_nR]$ ligands. However, as theoretical studies of the vibrational spectra of poly-ynes are surprisingly scarce,[494] conclusions drawn from these observations are based more on empirical rationalisation. Several detailed comparisons of data pertaining to complexes containing $C{\equiv}CC{\equiv}CH$ ligands,[115] and extensive series of compounds $Re\{(C{\equiv}C)_nR\}(NO)(PPh_3)Cp^*$ ($n = 2$–10; R = H, $SiMe_3$, $SiEt_3$, tol),[29,144] $Fe(C{\equiv}CC{\equiv}CR)(L_2)(\eta^5\text{-}C_5R'_5)$ [R = H, $SiMe_3$; $L_2 = (CO)_2$, dppe; R′ = Me, Ph],[27] and $Pt\{(C{\equiv}C)_nR\}(tol)(PR'_3)_2$ ($n = 2, 3$; R = H, $SiEt_3$; R′ = Ph, tol)[147] have been made in an effort to identify trends that might provide insight into the electronic structure of these species. A detailed complementary IR and Raman study of polymers with $\{\text{-}Pt(PR_3)_2(C{\equiv}C)\text{-}\}_n$ units has also been reported.[495]

For the majority of buta-1,3-diynyl complexes $M(C{\equiv}CC{\equiv}CH)L_n$ only one $\nu(C{\equiv}C)$ band is found in their IR spectra (see Table I). Notable exceptions are $Fe(C{\equiv}CC{\equiv}CH)(dppe)Cp^*$,[27] $Mo(C{\equiv}CC{\equiv}CH)(CO)(dppe)Cp$,[107] and $Re(C{\equiv}CC{\equiv}CH)(NO)(PPh_3)Cp^*$[87] In the rhenium series, the number of $\nu(C{\equiv}C)$ bands generally reflects the number of $C{\equiv}C$ moieties in the ynyl ligand, with the extinction coefficients of the most intense bands increasing with chain length.[29,144] Complexes containing $SiMe_3$-capped diynyl ligands show one more $\nu(C{\equiv}C)$ band than those with the corresponding terminal diynyl ligands. Thus, three vibrational modes are observed for the carbonyl ligands and $\nu(C{\equiv}C)$ bands of $Fe(C{\equiv}CC{\equiv}CSiMe_3)(CO)_2Cp^*$, this being attributed to the coupling of one of the expected normal modes with another unspecified oscillator.[27]

The electronic effect of the $(C{\equiv}C)_nR$ ligand has been estimated from the $\nu(CO)$ and $\nu(NO)$ stretching frequencies of co-ligands in some cases. For $Fe\{(C{\equiv}C)_nH\}(CO)_2Cp^*$, average values of the symmetric and anti-symmetric $\nu(CO)$ bands are 1995 cm^{-1} ($n = 1$) and 2002 cm^{-1} ($n = 2$).[115] Similar trends are found for *trans*-$Ru\{(C{\equiv}C)_nR\}_2(CO)_2(PEt_3)_2$ [R = $SiMe_3$: $n = 1$, $\nu(CO)$ 1986 cm^{-1}; $n = 2$, $\nu(CO)$ 2002 cm^{-1}; R = H: $n = 1$, $\nu(CO)$ 1987 cm^{-1}; $n = 2$, $\nu(CO)$ 2002 cm^{-1}]. For $Re\{(C{\equiv}C)_nR\}(NO)(PPh_3)Cp^*$ (R = $SiEt_3$, H), values of $\nu(NO)$ increase from 1627 to 1637 cm^{-1} ($n = 1$) to an apparent limit of 1660–1662 cm^{-1} ($n = 5, 6$). However, the poly-ynyl ligands are poor π-accepting ligands because of their high lying π^* levels.[2,157] As n increases, the π MO energies increase while the π^* MO energies decrease, i.e., the HOMO–LUMO gap decreases. The trends described here are consistent with a progressively greater σ withdrawing effect of the longer chain poly-ynyl ligands, rather than indicating a progressive decrease in metal $\rightarrow$ π-ligand back-bonding interactions. This is consistent with the increase in inductive σ withdrawing effects based on acidity measurements of the free poly-ynes.[496]

B. *NMR Spectra*

The ^{13}C NMR data for the diynyl ligands in $\{L_nM\}C\equiv CC\equiv CH$ can usually be assigned unequivocally on the basis of *J*(CH) coupling constants [and *J*(CP) in the case of metal centres bearing phosphine co-ligands].[87] Coupling constants to ^{117}Sn, ^{119}Sn,[66] and ^{183}W[104] nuclei have also been employed. In general, values for *J*(CH) decrease in the order ^1J(CH) > ^2J(CH) > ^3J(CH) > ^4J(CH), while the ^4J(CP) values are larger than ^3J(CP). The chemical shift sequence $\delta(C_\alpha) > \delta(C_\beta) > \delta(C_\gamma) > \delta(C_\delta)$ is frequently found and is often assumed when assigning the ^{13}C resonances in cases where diagnostic coupling constants are absent. However, this ordering is by no means universal and where possible, assignments based on the magnitudes of coupling constants are preferable. In the iron series, the Fe–C_α resonances move downfield as the electron density at the metal centre increases [assignments based on *J*(CH) and *J*(CP) values].[27] Comparison of the butadiynyl and $SiMe_3$-substituted complexes shows that the C_4 chain is a good electronic communicator, and there is a ^6J(HP) coupling between the terminal H and ^{31}P nuclei in *trans*-Ru$(C\equiv CC\equiv CH)_2(CO)_2(PEt_3)_2$.[118]

The chemical shifts of the carbons of the butadiynyl chains are sensitive to the nature of the ancillary ligands about the metal center as well as the end-capping substituent, which has been taken as further evidence of there being effective interactions through the diynyl π-system.[27] Thus, within the series of structurally comparable complexes Fe$(C\equiv CC\equiv CR)(L_2)(\eta^5\text{-}C_5R'_5)$ [R = H, SiMe$_3$; $L_2 = (CO)_2$, dppe; R′ = Me, Ph] the chemical shift of C_α is found to move progressively to lower field as the electron density at the metal center increases (see Table I). The chemical shifts are also sensitive to the nature of the R group, a small shift in $\delta(C_\alpha)$ to higher field being found when $SiMe_3$ replaces H. The chemical shift of C_β is less affected by the terminal groups, while $\delta(C_\gamma)$ and $\delta(C_\delta)$ are more affected by the nature of the R group than the ML_n fragment. It has been suggested the deshielding of C_α in Fe$(C\equiv CC\equiv CH)(CO)_2Cp^*$ may be indicative of a degree of cumulenic structure in these species, resulting from back-donation from the metal to the ligand. The higher field shift of C_δ may then be attributed to the increased electron density at this centre.[115] However, this interpretation seems to be at odds with the results of IR measurements, which have given little evidence for such a structure.

For the rhenium series Re{$(C{\equiv}C)_nR$}(NO)(PPh_3)Cp* ($n = 2$–6; R = H, $SiMe_3$, $SiEt_3$; not all combinations) the Re—C≡C signals show a steady downfield shift with increasing n. The influence of the R group diminishes with n [δ_C: Re$(C{\equiv}C)_2$$SiMe_3$/H 102.1–105.8; Re$(C{\equiv}C)_3$$SiMe_3$/$SiEt_3$/H 112.3–113.6; Re$(C{\equiv}C)_4$$SiMe_3$/H 117.2–117.6; Re$(C{\equiv}C)_5$$SiMe_3$/$SiEt_3$ 122.7–122.9].[29] A limiting value of δ_C 133–138 was estimated graphically. The C≡CSi resonances show a similar trend toward δ_C 88–90, while, as found for the Fe series, the Re–C≡C signals are less sensitive to the nature of R. The ^2J(CP) values are similar across the series, indicating little change in the Re–C bond order.[29,31] The interior C≡C carbons tend to a limit near δ_C 65 in a manner similar to that found for other long-chain poly-ynes.[497–499]

C. *Electronic Spectra*

Progression from alkynyl to diynyl and higher poly-ynyl ligands results in the maximum absorption progressing to higher wavelengths as a result of the diminished HOMO–LUMO gap, and spectra of increasing complexity are observed. The longest wavelength bands in these species have relatively low intensities, which suggests that these bands arise from symmetry-forbidden transitions, possibly between the HOMO with high metal d character and unoccupied orbitals with appreciably more poly-yne π^* character. In each case λ_{max} is highly dependent on the metal–ligand fragment, as expected from the nature of the HOMO which features appreciable metal and C≡C character (see following). In contrast, the longest wavelength bands in poly-ynes with carbon,[500] silicon,[501] or hydrogen endcaps[502] are the most intense and are assigned to symmetry-allowed $\pi \rightarrow \pi^*$ transitions.[29] Within a given series of compounds based upon a single metal–ligand end-group {ML_x}$(C{\equiv}C)_nR$, there are only minor variations with the nature of R.

Given the relatively small amount of electronic spectral data available at the present time, and the large variation in the spectral profiles found with different metal end-groups, we have chosen to present the available data in a periodic fashion. We also note that extensive optical studies have been described for polymeric materials with poly-yndiyl repeat units[503,504] although further discussion is not appropriate here.

1. *Tungsten*

Comparison of the UV/Vis spectra of several complexes W{=C(NMe_2)$(C{\equiv}C)_n$[M$(CO)_2$Cp]}$(CO)_5$ ($n = 1, 2$) reveals the expected bathochromic shift of MLCT bands with increasing n, e.g., for M = Fe, from 390 ($n = 1$) to 416 nm ($n = 2$); for Ru, from 387 to 414 nm (both in PhMe). There is also a moderate solvatochromic shift toward shorter wavelengths in more polar solvents, e.g., $\Delta\nu$(PhMe/dmf) = 970 cm^{-1} for M = Ru.[66]

2. *Rhenium*

Intense absorption bands at 404 and 416 nm in Re(C≡CC≡CR)$(CO)_3$(Bu^t_2-bpy) (R = H, Ph, respectively) have been assigned to spin-allowed $d\pi$(Re) → π^*(Bu^t_2-bpy) MLCT transitions, the C≡CC≡CPh ligand having better σ- and π-donating abilities than the H analogue.[80] Excitation at $\lambda > 400$ nm gave strong orange luminescence from the ^{3}MLCT excited state, the emissions being at higher energies than found for the corresponding alkynyl complexes (R = H: 620, 670 nm; R = Ph: 625, 688 nm).

The spectra of Re$\{(C{\equiv}C)_nR\}$(NO)(PPh_3)Cp* (R = H, $SiMe_3$, $SiEt_3$, tol) each contain a series of strong absorption bands between 300 and 400 nm that show some dependence on the nature of the end cap.[29,143,144] The lowest energy absorption maxima shift to longer wavelengths and become more intense with increasing chain length. This is in agreement with the notion that these transitions are from occupied orbitals with increasing C≡C character and that the transitions themselves become more $\pi \rightarrow \pi^*$ in nature.

3. *Nickel and Platinum*

The electronic spectra of M$(C{\equiv}CC{\equiv}CR)_2(PEt_3)_2$ (M = Ni, Pt) have been studied in conjunction with the related alkynyls. For the lowest energy band, which is assigned to charge transfer within the M–C≡C moiety, the diynyl groups fall within the series R = H < Me , CH_2F < CH=CH_2 < C≡CH < Ph < C≡CMe.[65] In general the diynyl complexes give rise to more complex spectra, with a small blue shift of the absorption maxima found in MeOH when compared with Et_2O, possibly due to H-bonding effects.

Trends in the UV/vis spectra of *trans*-Pt$\{(C{\equiv}C)_nR\}$(tol)$\{P(tol)_3\}_2$ ($n = 2$, R = H; $n = 3$, R = $SiEt_3$, H; $n = 4$, R = $SiEt_3$)[147] follow the generalizations stated previously. Interestingly, λ_{max} of the longer members of the series (325–340 nm, ε = 5000–122,000 M^{-1} cm^{-1}, in CH_2Cl_2) are remarkably similar to those of the purely organic poly-yne $Bu^t(C{\equiv}C)_{10}Bu^t$ (320 nm, ε = 345,000 $M^{-1}cm^{-1}$), suggesting a more limited contribution of the Pt d-type orbitals to the frontier orbital structure of these molecules. This probably arises from the poor energy match between, and hence limited mixing of, the low-lying filled $Pt(PR_3)_2$ d orbitals and the C≡C π-type orbitals of similar symmetry.

4. *Mercury*

The bis(carbene) complex Hg$\{C{\equiv}CC{\equiv}CC(NMe_2){=}W(CO)_5\}_2$ shows a maximum wavelength absorption at 444 nm in toluene.[66]

D. *Electronic Structure*

The electronic structure of Fe(C≡CC≡CH)$(CO)_2$Cp has been probed using He(I) and He(II) photoelectron spectroscopy (PES), which allows for ligand π-bonding effects to be distinguished from σ-bonding and charge potential effects. Relatively sharp and intense absorptions were assigned with the help of the interaction diagram for $[Fe(CO)_2Cp]^+$ and $[C{\equiv}CC{\equiv}CH]^-$ fragments which indicates that the important interactions are the Fe—C σ bond (formed from donation from diynyl σ_{sp} orbital into empty metal d_{z^2} orbitals) and between occupied diynyl π e set of levels and occupied metal dπ orbitals. For this electron-poor metal center the diynyl ligand behaves as relatively good π-donor, as a result of the filled–filled (or four-electron, two-orbital) π-interactions of the ligand with the metal, the HOMO containing considerable diynyl π-character. This is more pronounced than similar orbitals in the analogous ethynyl complex.[505] Conversely, the π-acceptor character of C≡CC≡CH is negligible, the empty π^*-orbitals of the $[C{\equiv}CC{\equiv}CH]^-$ fragment lying 13.9 eV above the occupied orbitals of similar symmetry.[157]

However, for complexes with more electron-rich metal centers a cross-over in the electronic behavior of the C≡CC≡CH ligand from π-donor to predominantly π-acceptor may occur as the occupied metal orbital energies begin to approach those of the π^*-levels of the C≡CC≡CH fragment.[505] While this remains to be tested by experiment in diynyl systems, similar interactions have been observed for alkynyl complexes.[506] Ligand substituents can also have a strong influence on the metal–ligand π-interactions due to perturbations in the ligand-based orbitals: C≡CC≡CH and C≡CBu^t were both found to be superior π-donors to C≡CH.[505] The FeCCCC—H system was considered to be a five-centered poly-yne and it was noted that introduction of a second metal atom in place of H would give a six-centered poly-yne with the potential for electronic communication between the metal centers via the π-system.[157] Experimental justification for this suggestion is now well established.[25]

An alternative view based on ^{13}C NMR, IR, and structural data is of the C≡CC≡CH ligand as a π-acceptor.[115] The deshielding of the C_α signal and the relatively higher field shift of C_δ is interpreted in terms of a contribution from a cumulenylidene-type structure arising from back-donation from the metal center assisted by the electron-rich Cp* ligand. However, the effects of deshielding in the butadiynyl complexes are significantly less than are observed for ligands such as vinylidene. Average ν(CO) values for Fe(C≡CC≡CH)$(CO)_2$Cp* (2002 cm^{-1}) and Fe(C≡CH)$(CO)_2$Cp* (1995 cm^{-1}) also suggest that C≡CC≡CH is a better π-acceptor than C≡CH. In addition, comparison of the structures of Fe($C_\alpha{\equiv}C_\beta C_\gamma{\equiv}C_\delta$H)$(CO)_2$Cp and Fe($C_\alpha{\equiv}C_\beta$H)$(CO)_2$Cp indicate shorter Fe—$C_\alpha$ [1.907(4) vs. 1.921(3) Å] and longer C_α—C_β distances [1.207(5) vs. 1.173(4) Å] in the case of the diynyl complex, while the C_β—C_γ separation [1.378(6) Å] is comparable to that in buta-1,3-diyne [1.3284(2) Å]. These structural features are in accord with the electron-accepting ability of the C≡CC≡CH ligand and a contribution from the cumulene structure deduced from the δ_C and $\nu(CO)$ data.

These inconsistencies may be due to different evaluation methods, as the $\nu(CO)$ and structural parameters reflect the net electronic effects of the ligand, while the PES results are based on metal band splitting of the π-type orbitals. In this connection, it is worth noting that the relative σ-bonding capabilities of a ligand may also influence the observed stretching frequencies.[507] The interpretation of M—C and C≡C bond lengths is not conclusive, the length of a C≡C bond being an unreliable measure of bond order.[2,505]

As more studies on the nature of the bonding of M—C≡CR and M—C≡CC≡CR systems are performed, the most recent and favoured interpretation appears to be that based upon filled–filled interactions between M(π)-C≡C(π) orbitals, electronic interactions being transmitted between the metal center and the capping group at the other end of the $(C{\equiv}C)_n$ chain by perturbations in the ligand π-system. These interactions have been found in many poly-yndiyl complexes[25,27–29,33] and in heteronuclear systems such as $Co_2(\mu\text{-dppm})\{\mu\text{-}\eta^2\text{-}Me_3SiC_2(C{\equiv}C)_n[Ru(PPh_3)_2Cp]\}(CO)_4$ ($n = 1, 2$).[194] In the latter, both the Ru and Co_2C_2 fragments possess filled π-type valence orbitals that are of the appropriate energy and symmetry to interact with occupied frontier orbitals of the poly-ynyl moiety. In this manner the electronic effects of each organometallic fragment are efficiently transmitted between the termini of the poly-ynyl chain. The extensive mixing of metal and carbon character in these systems results in efficient electronic interactions between metal-based remote sites when linked via a poly-yndiyl spacer.

Simple EH MO calculations on the model complexes $\{Co_2(CO)_6\}_2(\mu\text{-}\eta^2{:}\mu\text{-}\eta^2\text{-}HC_2C_2H)$ and $\{Ni_2Cp_2\}_2(\mu\text{-}\eta^2{:}\mu\text{-}\eta^2\text{-}HC_2C_2H)$ showed that the LUMOs are M—M antibonding with 10% contribution from diyne π-orbitals. For the Ni_4 complex, there is also a significant contribution from Cp ligands (ca 20%), suggesting possible further delocalization of the extra electron introduced upon reduction to the mono-anion.[508]

E. *Redox Properties*

On the basis of the few diynyl systems for which sufficient electrochemical data have been collected to enable meaningful comparisons with other functional groups to be made, it appears that the diynyl ligand behaves as a moderate electron-donating group, albeit not as strong a donor as the alkynyl ligand. A useful indication of this behavior is found in the oxidation half-wave potentials of the series FcH (E° 0.49 V), FcC≡CH (0.130 V) and FcC≡CC≡CH (0.190 V).[123] Values in the alkyl/alkynyl/diynyl series $Re(Me)(PPh_3)(NO)Cp^*$ (E_{pa} 0.32 V), $Re(C{\equiv}CMe)(PPh_3)(NO)Cp^*$ (E_{pa} 0.40 V) and $Re(C{\equiv}CC{\equiv}CMe)(PPh_3)(NO)Cp^*$

(E_{pa} 0.52 V) also suggest that the diynyl ligand is less effective as an electron donor than the corresponding alkynyl ligand. However, as the frontier orbitals of $L_nM{-}(C{\equiv}C)_nR$ systems are often derived from extensive mixing of both metal and carbon fragment orbitals a straightforward comparison of ynyl and non-ynyl electrode potentials may not be entirely appropriate in all cases.

The formal electrode potentials of diynyl complexes are greatly influenced by the nature of the other supporting ligands (Table V). For example, the cyclic voltamograms (CVs) of the iron complexes $Fe(C{\equiv}CC{\equiv}CSiMe_3)(CO)_2(\eta^5\text{-}C_5R_5)$

TABLE V

SOME ELECTROCHEMICAL DATA FOR POLY-YNYL COMPLEXES, $\{L_nM\}$ $(C{\equiv}CC{\equiv}CR)_n$

ML_n	R	n	E° (V)	Notes	Reference
$TiCp_2$	Fc	2	Ti(III/IV) −1.48 Fe(II/III) +0.21 (2e, irreversible)	*a*	97
$TiCp^{Si}{}_2$	Fc	2	Ti(III/IV) −1.53 Fe(II/III) +0.16 (2e, irreversible)	*a*	97
$TiCp^{Si}{}_2$	$SiMe_3$	2	Ti(III/IV) = −1.43	*a*	97
$Fe(CO)_2Cp^*$	$SiMe_3$	1	+1.15 (irreversible)	*b*	27
$Fe(CO)_2(C_5Ph_5)$	$SiMe_3$	1	−0.97 (irreversible)	*b*	27
$Fe(dppe)Cp^*$	$SiMe_3$	1	+0.00, $i_c/i_a = 1$	*b*	27
$Fe(CO)_2Cp^*$	H	1	−1.30 (irreversible)	*b*	27
$Fe(CO)_2(C_5Ph_5)$	H	1	−0.87 (irreversible)	*b*	27
$Fe(dppe)Cp^*$	H	1	+0.00, $i_c/i_a = 0.48$	*b*	27
$Re(NO)(PPh_3)Cp^*$	Me	1	+0.52 (irreversible)	*c*	87
trans-$Ru(C{\equiv}CC_6H_4NO_2)$ $(dppe)_2$	Fc	1	Ru(II/III) +0.318 Fe(II/III) −0.094	*d*	123
trans-$RuCl(dppe)_2$	Fc	1	Ru(II/III) +0.295 Fe(II/III) −0.165	*d*	123
trans-$Ru(dppe)_2$	Fc	2	Ru(II/III) +0.404 Fe(II/III) +0.15 Fe(II/III) −0.124	*d*	123
trans-$Ru(C{\equiv}CFc)(dppe)_2$	Fc	1	Ru(II/III) +0.460 Fe(II/III) −0.15 Fe(II/III) −0.300	*d*	123

[a] CH_2Cl_2, 0.1M $[NBu_4]ClO_4$ (temperature, electrodes and Fc/Fc^+ redox couple unspecified).

[b] CH_2Cl_2, 0.1M $[NBu_4]PF_6$, 20°C, Pt working electrode, SCE reference electrode, 100 mV s^{-1}, Fc/Fc^+ +0.46 V.

[c] CH_2Cl_2, 0.1M $[NBu_4]BF_4$, r.t., Pt working, Ag-wire pseudo reference electrode, 100 mV s^{-1}, Fc/Fc^+ +0.56 V.

[d] CH_2Cl_2, 0.1M $[NBu_4]PF_6$, 20°C, Pt working electrode, SCE reference electrode, 200 mV s^{-1}, Fc/Fc^+ +0.49 V.

(R = Ph, Me) each exhibit a single wave.[27] The complex containing the very electron-withdrawing C_5Ph_5 ligand shows an irreversible Fe^{II}/Fe^{I} reduction at −0.97 V while in the case of the C_5Me_5 analogue an irreversible oxidation was found at +1.15 V. The nature of the end-cap also plays a role in the stability of the odd-electron species. This is thought to be due primarily to the degree of steric protection afforded by the end-cap, and by way of illustration it is worth noting that in contrast to the silyl-capped complexes mentioned earlier, Fe(C≡CC≡CH)(dppe)Cp* gives an oxidation wave ($E^\circ = 0.00$ V) that is only partially reversible (all measured vs. SCE, Pt electrodes, as CH_2Cl_2 solutions containing 0.1 M $[NBu_4]PF_6$, at a Pt working electrode).

Many studies of the redox properties of transition metal carbonyl clusters show that these species are structurally flexible electron reservoirs with tunable redox properties dependent on the coordination sphere of the metal framework. Consequently, multimetallic systems featuring redox-active metal cluster cores bridged by ligands derived from poly-ynes that might mediate electronic interactions have come under scrutiny as part of a global search for molecular materials that display useful electronic communication between remote sites. Electrochemical techniques, together with UV/vis and NIR spectral information, have been used to probe the nature of these interactions in model systems. The electrochemical responses of the various cluster systems derived from poly-yne ligands are described in more detail in the following.

Comparisons of electrochemical results obtained from different laboratories are complicated by differing preferences for solvents, electrolytes, working electrode surfaces and, most significantly, reference electrodes. Rather than relating the data to a common electrochemical reference, we cite the data as given in the original literature together with an indication of the conditions employed. Furthermore, we have included the formal electrode potential of any internal standard (typically ferrocene or decamethylferrocene) used in the study where available.

1. *Titanium*

The CV of $Ti(C{\equiv}CC{\equiv}CSiMe_3)_2Cp^{Si}{}_2$ shows a reversible reduction wave at −1.43 V (CH_2Cl_2 vs FcH/FcH^+). No oxidation wave was observed within the potential range of the solvent.[97] By comparison with bis(alkynyl)titanocene complexes, the reduction is assigned to the Ti^{IV}/Ti^{III} couple,[422] although on the basis of EHMO calculations of similar bis(alkynyl)titanocene models it is likely that these processes actually involve orbitals with considerable Ti/Cp/diynyl character.[509,510] The analogous metallocenyl derivatives $Ti(C{\equiv}CC{\equiv}CFc)_2(\eta^5\text{-}C_5H_4R)_2$ (R = H, $SiMe_3$) give similar reversible 1-e reduction waves under CV conditions (CH_2Cl_2, 0.1 M $[NBu_4]ClO_4$, 100 mV/s, vs FcH/FcH^+, R = H, $E_{1/2} = -1.48$ V; R = $SiMe_3$, $E_{1/2} = -1.53$ V).[97]

However, in contrast to the $SiMe_3$-capped bis(diynyl) complexes, the CVs of the bis(metallocenediynyl) complexes $Ti(C{\equiv}CC{\equiv}CMc)_2(\eta^5\text{-}C_5H_4R)_2$ show several oxidation waves. Simultaneous oxidation of both Fc moieties (R = H, E_{pa} = +0.21 V; R = $SiMe_3$, $E_{1/2}$ = +0.16 V) is followed by two unresolved 1-e waves arising from the oxidation of each Fc moiety of the reductive elimination product $Fc(C{\equiv}C)_4Fc$. Intramolecular electron transfer from the two Ti–C bonds to each $Fc^{\bullet}$ center follows the initial oxidation to give $[Ti(\eta^5\text{-}C_5H_4R)_2]^{2+}$ and two $FcC{\equiv}CC{\equiv}C^{\bullet}$ radicals; the latter couple to generate the tetrayne.[97] The mixed metallocenyl complex $Ti(C{\equiv}CC{\equiv}CFc)(C{\equiv}CC{\equiv}CRc)Cp^{Si}{}_2$ behaves similarly. The Fc/Fc^+ couple is observed at +0.190 V, followed by $Rc^{II/III}$ near +0.310 V which overlaps the Fc/Fc^+ wave from $Fc(C{\equiv}C)_4Rc$ generated *in situ* and finally the 2-e Rc/Rc^+ oxidation in the tetrayne.[97] Preparative scale oxidations of $Ti\{(C{\equiv}C)_mFc\}\{(C{\equiv}C)_nMc\}Cp^{Si}{}_2$ (*m*, *n* = 1 or 2; Mc = Fc or Rc) with 2 equiv of $AgPF_6$ afforded $Fc(C{\equiv}C)_m(C{\equiv}C)_nMc$ in good yield. In the case of $Ti(C{\equiv}CC{\equiv}CFc)_2Cp_2$, oxidation with 2 equiv of $AgPF_6$ lead to the reductive coupling of the diynyl ligands to give 1,8-bis(ferrocenyl)octatetrayne, $Fc(C{\equiv}C)_4Fc$, with the isolated trinuclear $[\{Ti(C{\equiv}CC{\equiv}CFc)_2Cp_2\}_2Ag][PF_6]$ being implicated as a likely intermediate.[277]

2. *Niobium*

The CV of $Nb(\eta^2\text{-}Me_3SiC_2C{\equiv}CSiMe_3)Cp^{Si}{}_2$ shows a reversible 1-e oxidation at −0.15 V (vs SCE) to give the 16-e cation, possibly solvated by thf, and an irreversible 1-e reduction process at −1.68 V.[165]

3. *Rhenium*

The compound $Re(C{\equiv}CC{\equiv}CMe)(NO)(PPh_3)Cp^*$ is oxidized at a more positive potential (+0.520 V) than is the analogous propynyl $Re(C{\equiv}CMe)(NO)(PPh_3)Cp^*$ (+0.400 V).[87]

4. *Iron*

The carbonyl complexes $Fe(C{\equiv}CC{\equiv}CR)(CO)_2(\eta^5\text{-}C_5R'_5)$ (R = H, $SiMe_3$; R′ = Me, Ph) each feature a single irreversible redox process (Table V), the nature (reduction or oxidation) of which is governed by the electronic effects of the $\eta^5\text{-}C_5Ph_5$ (electron-withdrawing) or $\eta^5\text{-}C_5Me_5$ (electron-donating) ligands. In the latter series, replacement of CO groups by the bulky electron-donating dppe ligand stabilizes the 17-e oxidized form. The stability of the oxidized species is further enhanced by the presence of a bulky group on the diynyl ligand, and the CV of $Fe(C{\equiv}CC{\equiv}CSiMe_3)(dppe)Cp^*$ displays a fully reversible 1-e oxidation wave.[27] Similar trends to those found for rhenium complexes are found for the complexes $Fe(C{\equiv}CR)(dppe)Cp^*$ (R = Ph, E° = −0.130 V; R = Bu^t,

$E^\circ = -0.280$ V vs FcH/FcH$^+$ +0.470 V)[511] and Fe(C≡CC≡CR)(dppe)Cp* (R = SiMe$_3$, $E^\circ = +0.00$ V; R = H, $E^\circ = +0.00$ V vs FcH/FcH$^+$ +0.460 V).[27]

5. *Ruthenium*

In comparison with *trans*-RuCl$_2$(dppe)$_2$ ($E_{1/2}$ +0.017 V, 0.1 M [NBu$_4$]PF$_6$ in CH$_2$Cl$_2$, Pt electrodes, 293 K, 200 mV/s, vs FcH/FcH$^+$) and Fc(C≡C)$_n$H ($n = 1$, $E_{1/2} = +0.130$ V; $n = 2$, $E_{1/2} = +0.190$ V), the ferrocenyl moieties in *trans*-RuX(C≡CC≡CFc)(dppe)$_2$ are oxidised at remarkably low potentials (X = Cl, $E_{1/2} = -0.344$ V; C≡CC$_6$H$_4$NO$_2$-4, $E_{1/2} = -0.094$ V; C≡CFc, $E_{1/2} = -0.300$ V, -0.015 V; C≡CC≡CFc, $E_{1/2} = -0.124$ V, +0.015 V), which was taken as an indication of the strong electron-donating ability of the Ru(dppe)$_2$ fragment.[123] Oxidation waves at relatively high positive potentials were assigned to the Ru$^{II/III}$ couple (X = Cl, $E_{1/2} = +0.377$ V; C≡CC$_6$H$_4$NO$_2$-4, $E_{1/2} = +0.318$ V; C≡CFc, $E_{1/2} = +0.460$ V; C≡CC≡CFc, $E_{1/2} = +0.404$ V). An alternative explanation based upon significant orbital rearrangement in these complexes when compared with the models was not explored. The two ferrocenyl moieties in *trans*-Ru(C≡CC≡CFc)$_2$(dppm)$_2$ are oxidized at different potentials ($\Delta E = 139$ mV) indicating moderate electronic interactions occurring between them via the (C≡C)$_2$Ru(C≡C)$_2$ bridge. Similar effects of greater magnitude have been observed with alkynyl bridges.[123,512,513]

The CV of Ru(acac)$_2${(OCMe)$_2$C(C≡CC≡CFc)} showed two reversible 1-e oxidations and a 1-e reduction which were assigned to Fc/Fc$^+$ (+0.195 V), Ru^{III}/Ru^{IV} (+0.660 V), and Ru^{II}/Ru^{III} couples (-1.157 V), respectively.[492] The diyne {Ru(acac)$_2$[(OCMe)$_2$C]}$_2$(μ-C≡CC≡C) shows two 1-e oxidation and two 1-e reduction steps, both as overlapping pairs, suggesting stepwise processes occurring via mixed Ru^{II}/Ru^{III} and Ru^{III}/Ru^{IV} valence states.

6. *Cobalt*

The diyne complexes Co$_2$(μ-η^2-RC$_2$C≡CR)(CO)$_6$ (R = Ph, Fc) give irreversible reduction waves even at 213 K which indicates that fast chemical reactions follow the electrochemical production of the corresponding radical anions [Co$_2$(μ-η^2-RC$_2$C≡CR)(CO)$_6$]$^-$. The ESR spectra of the anion radical generated *in situ* were not consistent with the presence of two different Co centers. In the case of the ferrocenyl-substituted complex, two distinct oxidation waves separated by 70 mV are observed, which indicates a modest degree of interaction between the Fc cores through the cluster.[190]

At ambient temperatures, the primary CV processes observed for {Co$_2$(CO)$_6$}$_2$(μ-η^2:μ-η^2-RC$_2$C$_2$R) (R = Ph, Fc), which contain two chemically equivalent Co$_2$(μ-alkyne)(CO)$_6$ redox centers, are an apparent irreversible 2-e reduction, with an

$$Co_2(\mu\text{-}\eta^2\text{-}RC_2R)(CO)_6 \underset{-e}{\overset{e,\ E_1}{\rightleftharpoons}} [Co_2(\mu\text{-}\eta^2\text{-}RC_2R)(CO)_6]^{\bullet-}$$

$$\downarrow +CO$$

$$[Co(RC_2R)(CO)_3]^- \underset{e,\ E_2}{\rightleftharpoons} [Co(RC_2R)(CO)_3]^{\bullet} + [Co(CO)_4]^-$$

SCHEME 76

additional apparent 2-e oxidation in the case of R = Fc. The chemical reactions which follow reduction are related to loss of $[Co(CO)_4]^-$[190] with activation energies of ca 10 kcal mol^{-1} (Scheme 76).[514] However, at -80°C these chemical reactions are slowed sufficiently to allow observation of two reversible reduction processes ($\Delta E^\circ = 220$ mV) for $\{Co_2(CO)_6\}_2(\mu\text{-}\eta^2\text{:}\mu\text{-}\eta^2\text{-}PhC_2C_2Ph)$ which clearly indicates a moderate intercluster electronic interaction. Improved reversibility is also found in experiments conducted in thf, a solvent that is better able to solvate and therefore stabilise the intermediate radical species.[515]

Comparable reduction processes in the ferrocenyl analogue $\{Co_2(CO)_6\}_2(\mu\text{-}\eta^2\text{:}\mu\text{-}\eta^2\text{-}FcC_2C_2Fc)$ are more difficult to resolve and two irreversible reductions with $\Delta E^\circ \sim 140$ mV were observed at -40°C. The Fc/Fc^+ oxidations were readily resolved into two reversible oxidation couples ($\Delta E^\circ = 40$ mV) at -20°C. The decrease in separations between the twin Fc/Fc^+ couples in the series FcC≡CC≡CFc (100 mV)/$Co_2(\mu\text{-}\eta^2\text{-}FcC_2C{\equiv}CFc)(CO)_6$ (70 mV)/$\{Co_2(CO)_6\}_2(\mu\text{-}\eta^2\text{:}\mu\text{-}\eta^2\text{-}FcC_2C_2Fc)$ (40 mV) is consistent with decreasing electronic interactions between the ferrocenyl centres. Insertion of aromatic spacer groups further decreases the interactions between the cluster cores, as evidenced by the diminished separations between the Co_2C_2-centered reduction potentials. The likely influence of solvation effects and the relatively small values of ΔE° hamper efforts to distinguish the roles of the aromatic spacer and structural effects in the electronic communication observed in this series.[190]

The phosphite-substituted complexes $Co_2(\mu\text{-}\eta^2\text{-}RC_2C{\equiv}CR)(CO)_{6\text{-}n}\{P(OMe)_3\}_n$ (R = Ph, Fc; $n = 1$, 2, 3), $\{Co_2(CO)_5[P(OMe)_3]\}\{Co_2(CO)_{6\text{-}n}[P(OMe)_3]_n\}(\mu\text{-}\eta^2\text{:}\mu\text{-}\eta^2\text{-}RC_2C_2R)$ ($n = 0, 1, 2$) and $\{Co_2(CO)_4[P(OMe)_3]_2\}\{Co_2(CO)_6\}(\mu\text{-}\eta^2\text{:}\mu\text{-}\eta^2\text{-}RC_2C_2R)$, which feature 0/1, 1/1 and 1/2 substitution patterns at each Co center, give rise to complicated CV responses during the cathodic sweep as a result of undefined ECE processes.[192] The oxidation processes were more amenable to study, and resulted in the formation of the radical cations $[C_2Co_2(CO)_{6\text{-}n}\{P(OMe)_3\}_n]^{\bullet+}$ as the primary products. The oxidation half-wave potentials decrease and the chemical reversibility of the processes increases as the number of phosphite ligands per redox center increases. For $Co_2(\mu\text{-}\eta^2\text{-}FcC_2C{\equiv}CFc)(CO)_{6\text{-}n}\{P(OMe)_3\}_n$ the ferrocenyl centers are oxidized before the Co_2 moiety and at rather more positive potentials than the phenyl-substituted analogues.

The related compounds $Co_2(\mu\text{-dppm})(\mu\text{-}\eta^2\text{-}RC_2C{\equiv}CR')(CO)_4$ (R = R′ = H, Ph, Fc, $SiMe_3$, $C{\equiv}CSiMe_3$; R = $SiMe_3$, R′ = H, $C{\equiv}CSiMe_3$) give well-behaved electrochemical responses at higher temperatures.[192,194,278] A systematic shift of ν(CO) to higher energy with each oxidation (2030, 2059, 2084 cm^{-1} for 0/1+/2+) indicates that the odd electron is delocalized in the mono-oxidized cation radical. Two new bands at 450 and 835 nm were observed in the spectrum of the monocation, the latter being lost on conversion to the dication.[192] Coordination of a second dppm ligand as in $Co_2(\mu\text{-dppm})_2(\mu\text{-}\eta^2\text{-}PhC_2C{\equiv}CPh)(CO)_2$ results in a significant increase of electron density at the cluster and a reversible oxidation wave at very low potential is observed.[192]

Two oxidation processes separated by 448 mV are found in the CV of the complex $\{Co_2(\mu\text{-dppm})(CO)_4\}_2(\mu\text{-}\eta^2{:}\mu\text{-}\eta^2\text{-}PhC_2C_2Ph)$, and square-wave voltametry confirmed the full reversibility of the electrochemical events. Similar results have been observed in electrochemical experiments conducted in thf at ambient and lower temperatures.[515] The magnitude of ΔE° shows a small solvent dependence, consistent with the operation of a predominantly through-bond mechanism for the electronic coupling phenomenon in these systems, together with a small residual through-space component.[515] The different values of ΔE° found in the cathodic and anodic sweeps are also consistent with a predominantly through-bond interaction and EH MO calculations on the model complexes $Co_2(\mu\text{-}\eta^2\text{-}HC_2C{\equiv}CH)(CO)_6$ and $Co_2(\mu\text{-dppm})(\mu\text{-}\eta^2\text{-}HC_2C{\equiv}CH)(CO)_4$ indicate that the HOMO and LUMO have significantly different orbital compositions.[508]

The CV of $\{Co_2(CO)_6\}\{Co_2(CO)_4(\mu\text{-bma})\}(\mu\text{-}\eta^2{:}\mu\text{-}\eta^2\text{-}PhC_2C_2Ph)$ (**52**) contains two reversible reduction waves (E −0.51, −0.63 V) for the bma π-system and an irreversible reduction (−1.32 V) associated with the $Co_2(CO)_6$ moiety which is apparently unaffected by the presence of the $Co_2(CO)_4$(bma) moiety.[195] In light of the other results described earlier, and given the apparent multielectron nature of the irreversible reduction, we suggest that the nature of any electron interaction between the cluster moieties in this system remains unclear.

The electrochemical responses of poly-ynyl cluster-based systems $Co_3(\mu_3\text{-}CC{\equiv}CR)(CO)_9$ [R = $C{\equiv}CSiMe_3$,[516] $Co_3(\mu_3\text{-}C)(CO)_9$, $Co_3(\mu_3\text{-}CC{\equiv}C)(CO)_9$[516,517]] have also been investigated. The monocluster displays a chemically reversible reduction at −0.49 V (vs Ag/AgCl in CH_2Cl_2, FcH/FcH^+ +0.68V) followed by irreversible formation of a dianion near −1.3 V. The bis-cluster compounds also undergo two reduction processes, which were chemically reversible at low temperature. Spectroscopic studies suggest that the radical anions may isomerize to a form which contains bridging carbonyl ligands.[516] Coordination of a $Co_2(CO)_6$ unit to one of the C≡C moieties in $\{Co_3(CO)_9\}_2$

(μ_3,η^1:μ_3,η^1-CC≡CC≡CC) afforded $\{Co_3(CO)_9\}_2\{\mu_3,\eta^1:\mu_3,\eta^1:\mu,\eta^2$-$CC_2[Co_2(CO)_6]C$≡CC} which showed electrochemical responses characteristic of independent redox centers.[516]

7. *Nickel*

In thf the complexes Ni_2 (μ-η^2-$PhC_2R)Cp_2$ (including R = Ph, C≡CPh) undergo irreversible oxidation processes near +0.7 V (vs SCE, FcH*/FcH*+ +0.11 V, FcH/FcH$^+$ +0.56 V) which results in the formation of deposits on the electrode surface. The anodic sweep indicates the presence of a reversible reduction near −1.30 V attributed to a Ni_2-centered reduction and the formation of $[Ni_2(\mu$-η^2-$PhC_2R)Cp_2]^{\bullet -}$ Further reduction results in decomposition of the complexes, and the liberation of the alkyne or diyne ligand, as evidenced by two characteristic alkyne/diyne reductions at very negative potentials.[508]

For $\{Ni_2Cp_2\}_2(\mu$-η^2:μ-η^2-$PhC_2C_2Ph)$, two well-resolved, reversible metal-centered reductions are found at −1.26 and −1.93 V (vs. FcH*).[508,514] The separation of these waves by 670 mV (comproportionation constant $K_c = 2.1 \times 10^{11}$) indicates the thermodynamic stability of the odd-electron species. The first reduction occurs at a potential very similar to that of the mono-complexed species $\{Ni_2Cp_2\}(\mu$-η^2-PhC_2C≡CPh). On the basis of electrochemical evidence, electronic interactions between the Ni_2Cp_2 moieties in $\{Ni_2Cp_2\}_2(\mu$-η^2:μ-η^2-$PhC_2C_2Ph)$ are greater than those between the $Co_2(CO)_6$ moieties in $\{Co_2(CO)_6\}_2(\mu$-η^2:μ-η^2-$PhC_2C_2Ph)$ [ΔE° 220 mV in CH_2Cl_2,[190] 350 mV (solvent not given),[514] 400 mV (thf, GCE, −30°C)[515]]. As a crystal structure of the nickel complex is not available, variations in structural parameters which may change the degree of π-overlap between the metal centers in the two complexes could not be evaluated directly.[508]

Contrary to the usual observations that redox processes generally become more favorable as the number of redox sites and the degree of interaction between them increase,[21] addition of the second electron-donating Ni_2Cp_2 moiety appears to counter this effect. Attempts to resolve the problem of electron delocalization in the radical anion $[\{Ni_2Cp_2\}_2(\mu$-η^2-$PhC_2C_2Ph)]^{\bullet -}$ using ESR spectroscopy were inconclusive.[508] Electrolytic reduction of $\{Ni_2Cp_2\}(\mu$-η^2-$PhC_2Ph)$, $\{Ni_2Cp_2\}(\mu$-η^2-PhC_2C≡CPh) or $\{Ni_2Cp_2\}_2(\mu$-η^2:μ-η^2-$PhC_2C_2Ph)$ *in situ* each gave singlet ESR resonances devoid of other features. At 140 K, frozen solutions gave well-resolved anisotropic spectra indicative of axial symmetry for $\{Ni_2Cp_2\}(\mu$-η^2-$PhC_2Ph)$ and $\{Ni_2Cp_2\}(\mu$-η^2-PhC_2C≡CPh), with a small measure of splitting of the perpendicular component in $\{Ni_2Cp_2\}_2(\mu$-η^2:μ-η^2-$PhC_2C_2Ph)$. It was concluded that in solution, the odd electron in the latter is delocalized over two nonequivalent thermally accessible Ni sites, but at 140 K, the odd electron is trapped at a single site in all three complexes. There is linear relationship between the reduction potentials of $Ni_2(\mu$-η^2-$RC_2R')Cp_2$ and δ_H(Cp).

8. *Heterometallic Complexes*

Electrochemical studies of complexes $Co_2(\mu\text{-dppm})\{\mu\text{-}RC_2\text{—}Y\text{—}C{\equiv}C[ML_n]\}(CO)_4$ [$ML_n = Ru(PPh_3)_2Cp$, Y = bond, C≡C] reveal large electronic interactions between the mononuclear fragment and the Co_2(alkyne) cluster core. The parent cobalt complexes $Co_2(\mu\text{-dppm})\{\mu\text{-}\eta^2\text{-}Me_3SiC_2(C{\equiv}C)_nSiMe_3\}(CO)_4$ ($n = 1, 2$) exhibit 1-e reduction and 1-e oxidation processes, which become more reversible in thf at −30°C. Upon attachment of the ML_n fragment, the reduction shifts to more negative half-wave potentials, while two oxidation processes, both with half-wave potentials less positive than the parent cobalt complex or model $\{ML_n\}C{\equiv}CPh$ complexes, were observed.[194] This behavior was interpreted with the aid of DFT and Electron Localization Function (ELF) studies. The LUMO of the heterometallic complex, which is predominantly Co—Co antibonding in character, lies at a higher energy than in the cobalt model complex. The HOMO and SOMO both contain appreciable Ru, C≡C, and Co_2C_2 character and are delocalized over the entire molecule. Thus, the electrochemical oxidation processes in these systems cannot be interpreted in terms of independent oxidations of the ML_n and cluster core fragments.

The mixed complex $\{Co_2(CO)_6\}\{Ni_2Cp_2\}(\mu\text{-}\eta^2{:}\mu\text{-}\eta^2\text{-}PhC_2C_2Ph)$ gives two reduction processes which are correlated with the formation of a Co_2-centered monoanion followed by reduction of the Ni_2 moiety to give the dianion. Chemical reactions were suppressed at low temperatures (−20°C) and fast scan rates (10 V/s). The shift in reduction potentials relative to the Co_2-centered reduction of $Co_2(\mu\text{-}\eta^2\text{-}PhC_2C{\equiv}CPh)(CO)_6$ and the second reduction of $\{Ni_2Cp_2\}_2(\mu\text{-}\eta^2{:}\mu\text{-}\eta^2\text{-}PhC_2C_2Ph)$ indicate that the $Co_2(CO)_6$ fragment acts as an electron-withdrawing group while the Ni_2Cp_2 group is more electron-donating and that these systems interact through the diyne ligand.[508]

XI

METALLADIYNES AND RELATED COMPLEXES

Formally, substitution of one or more diyne carbon atoms by a metal center leads to metalladiynes which may possess rod-like linear or branched structures according to the geometry about the metal atom. In addition to the intrinsic interest of these unusual highly unsaturated systems, the combination of M dπ and C pπ orbital fragments in the molecular scaffold suggests that metalladiynes may also be viable building blocks for the construction of molecular scale wires and other metal-containing oligomeric species. This section considers complexes containing conjugated M≡CC≡CR, RC≡MC≡CR′, RC≡M=M≡CR and M $\underline{4}$ MC≡CR moieties.

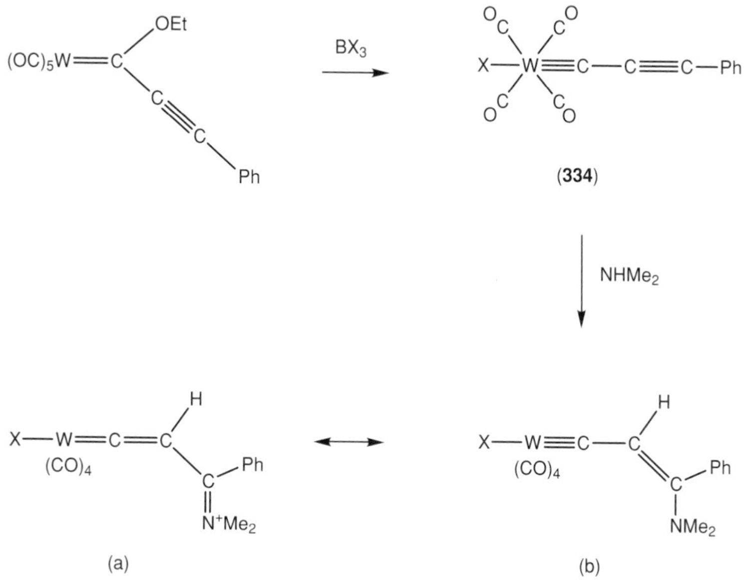

SCHEME 77

A. *Complexes M≡CC≡CR*

Following the synthesis of metal carbyne complexes, the first metalladiyne derivative was prepared by treatment of W{=C(OEt)C≡CPh}$(CO)_5$ with BX_3 (X = Cl, Br, I) (pentane, −45°C) to give *trans*-W(≡CC≡CPh)(X)$(CO)_4$ (**334;** Scheme 77) in good yields (30–60%). Subsequent reactions with $NHMe_2$ give W{≡CCH=CPh(NMe_2)}(X)$(CO)_4$ by addition to the C≡C triple bond, the structure of which indicates a contribution from the vinylidene resonance form.[518]

Treatment of $M(CO)_6$ (M = Mo, W) with LiC≡CBu^t gives the acylate $[M\{C(O)C{\equiv}CBu^t\}(CO)_5]^-$; subsequent reactions with $(CF_3CO)_2O$, followed by tmeda, give M(≡CC≡CBu^t)$(CO)_2(O_2CCF_3)$(tmeda).[519] Related complexes with bpy (Mo) or py (W) have also been described. The metal-bonded carbon resonates at δ 245–252 (for the tmeda complexes). The bpy or py complexes (but not tmeda) react with NaCp to give M(≡CC≡CBu^t)$(CO)_2$Cp, and the tmed complex with K[Tp] or K[Tp′] gives Mo(≡CC≡CBu^t)$(CO)_2$Tp(Tp′), although the tungsten complex could not be prepared in this way; the py precursor was used instead. Compared with the carbyne complexes containing saturated substituents, the M≡C resonance is considerably deshielded, appearing between δ 253 and 275.

Reactions of M(≡CC≡CBu^t)$(CO)_2$Cp with $Co_2(CO)_8$ afforded the cluster complexes $Co_2M(\mu_3$-CC≡$CBu^t)(CO)_8$Cp (**335,** M = Mo, W; Scheme 78), which exist

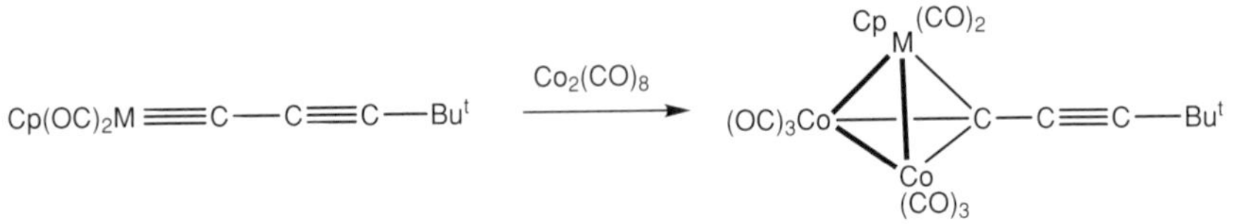

$Cp(OC)_2M{\equiv}C{-}C{\equiv}C{-}Bu^t \xrightarrow{Co_2(CO)_8}$ $(OC)_3Co$, $Cp\,M(CO)_2$, $Co(CO)_3$ cluster $C{-}C{\equiv}C{-}Bu^t$

(**335**) M = Mo, W

SCHEME 78

in solution as mixtures of rotamers differentiated by the *distal* or *proximal* orientations of the $M(CO)_2Cp$ fragment relative to the CCo_2 fragment. The Co_2Mo complex reacts with dppm to give $Co_2Mo(\mu_3\text{-}CC{\equiv}CBu^t)(\mu\text{-dppm})(CO)_6Cp$, in which the dppm bridges the two Co atoms. In contrast, the presence of the bulky Tp′ ligand precludes cluster formation, the $Co_2(CO)_6$ now being attached to the C≡C triple bond to give $Co_2\{\mu\text{-}Bu^tC_2C{\equiv}[Mo(CO)_2Tp']\}(CO)_6$. The Tp complex also reacts with dppm to give $Co_2\{\mu\text{-}Bu^tC_2C{\equiv}[Mo(CO)_2Tp]\}(\mu\text{-dppm})(CO)_4$.[519]

A moderate yield of $Mo_2W(\mu_3\text{-}CC{\equiv}CBu^t)(CO)_6Cp_3$ is obtained from the reaction between $W({\equiv}CC{\equiv}CBu^t)(CO)_2Cp$ and $\{Mo(CO)_3Cp\}_2$. In solution a mixture of unsymmetrical and symmetrical isomers is present, the latter having two equivalent Cp groups. Attachment of the Mo_2 fragment to the C≡C triple bond occurs with $W({\equiv}CC{\equiv}CBu^t)(CO)_2Tp$ to give $Mo_2\{\mu\text{-}Bu^tC_2C{\equiv}[W(CO)_2Tp]\}(CO)_4Cp_2$, which also exhibits dynamic behavior in solution.

B. *Complexes RC≡MC≡CR′*

Internally metallated diynes −C≡M−C≡C− have been prepared from reactions between alkynyllithiums and $W({\equiv}CH)(OTf)(dmpe)_2$ which give *trans*-$HC{\equiv}W(C{\equiv}CR)(dmpe)_2$ (**336,** R = H, $SiMe_3$, Ph, $C_6H_4C{\equiv}CPr$-4).[520] In the parent compound, the $^5J(HH)$ coupling constant (0.8 Hz) suggests a degree of electron delocalization over the HCWCCH chain [cf. 2.2 Hz in HC≡C≡CH[521]]. The $SiMe_3$ compound has an essentially linear C≡W−C≡C−Si chain [W−C 2.246(6), W≡C 1.801(7), C≡C 1.228(9) Å], these values suggesting π-conjugation between the W≡C and C≡C triple bonds, which is further supported by the electronic spectra. The band between 23,470 and 24,810 cm^{-1} found in $W({\equiv}CH)(X)(dmpe)_2$ (X = Bu, I, Cl), assigned to the $d_{xy} \rightarrow \pi^*(W{\equiv}C)$ transition, is found at lower energies for **336** (20,240–22,270 cm^{-1}, the red shift increasing with increasing conjugation). Mixing of $\pi^*(W{\equiv}C)$ and $\pi^*(C{\equiv}C)$ orbitals probably stabilizes the former.

Metathesis of $W_2(OBu^t)_6$ with one C≡C triple bond of substituted 1,4-diethynylbenzenes has given carbyne complexes which can be converted into *trans*-$WCl({\equiv}CC_6H_4C{\equiv}CH)(dmpe)_2$.[522] Functionalization via the W−Cl and ≡CH groups affords metalladiynes such as *trans*-$W\{C{\equiv}C(tol)\}({\equiv}CC_6H_4C{\equiv}CSiPr^i_3)$

$(dmpe)_2$ (**337**), for which spectroscopic and structural data indicate extended π-conjugation. The $n \rightarrow \pi^*$ transition is shifted to 15,870 cm^{-1} from 16,780 cm^{-1} for the chloro complex.

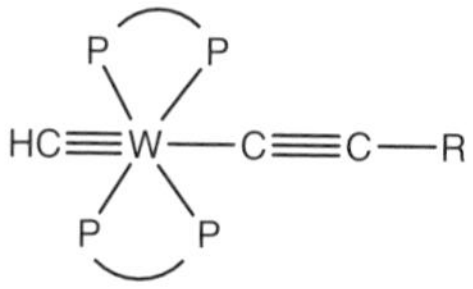

(**336**) P-P = dmpe

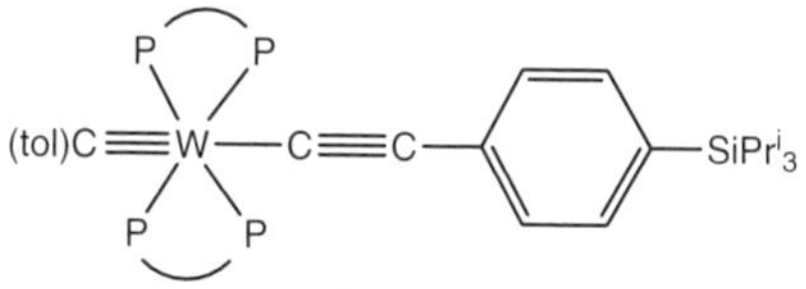

(**337**) P-P = dmpe

C. *Complexes RC≡M=M≡CR*

Four examples of complexes containing dimetalladiynes have been described. Reactions of $CH_2{=}CH(OEt)$ with *syn*-$Re(\equiv CBu^t)(=CHBu^t)\{OCMe(CF_3)_2\}_2$ in thf afford $Re(\equiv CBu^t)\{=CH(OEt)\}\{OCMe(CF_3)_2\}_2(thf)_2$ (**338,** Scheme 79), but

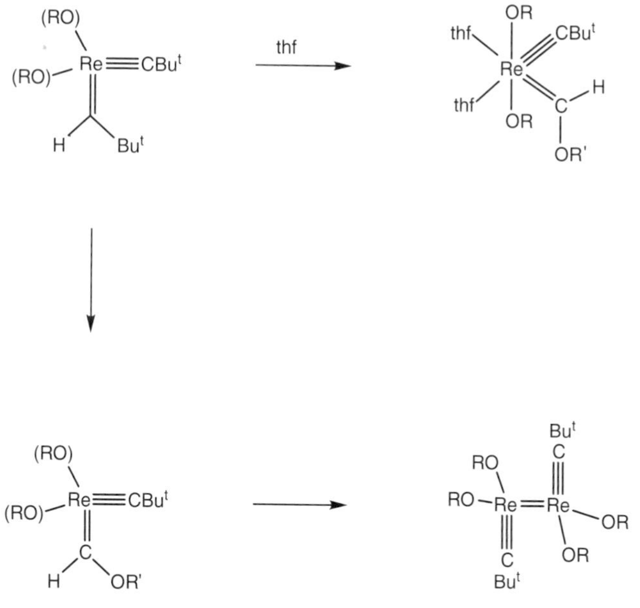

(**338**) R' = Et, $SiMe_3$

(**339**) R = $CMe(CF_3)_2$, Bu^t

SCHEME 79

in benzene or CH_2Cl_2 give Re(≡CBu^t){CH(OEt}{OCMe(CF_3)$_2$}$_2$ which rapidly (minutes) converts to {Re(≡CBu^t)[OCMe(CF_3)$_2$]$_2$}$_2$ (**339**).[523] The related complexes {Re(≡CBu^t)(OR)$_2$}$_2$ (R = Bu^t, CMe$_2$Ph) are also mentioned. These molecules have a staggered ethane-like geometry with bent —C≡Re=Re≡C— systems (angles at Re, 90°). The reaction of Re(≡CBu^t){=CH(OEt)}{OCMe(CF_3)$_2$}$_2$ (thf)$_2$ with *syn*-Re(≡CBu^t)(=CHBu^t){OCMe(CF_3)$_2$}$_2$ gives **339** directly, possibly via a dimetallacycle and/or dimetallatetrahedrane.

Several complexes containing the Ru_2 unit bridged by bidentate ligands and containing axial alkynyl groups are known. In these, the Ru—Ru bond orders range between 1 and 2.5, and so do not fall strictly within the scope of this survey.[524–526] For example, a large excess of LiC≡CPh reacts with Ru_2(μ-form)$_4$Cl to give intermediate anions [Ru_2(μ-form)$_4$(C≡CPh)$_2$]$^-$ (form = diarylformamidinate, ArNCH-NAr; Ar = Ph, 3- and 4-ClC$_6$H$_4$, 3,4- and 3,5-Cl$_2$C$_6$H$_3$, 3-CF$_3$C$_6$H$_4$) which dissociates one C≡CPh group on purification. The Ru—Ru and C≡C bonds are colinear (linear Ru—Ru-C≡C) in contrast to bis-adducts. The Ru—Ru separation is very sensitive to crystal packing effects, e.g., values of 2.369, 2.431(1) Å for two independent molecules of Ph complex. The complexes show two 1-e redox processes consistent with Ru_2^{4+}/Ru_2^{5+}/Ru_2^{6+} oxidation states, and become more difficult to oxidize Ru_2^{5+} with increasing electron-withdrawing power of aryl substituents. Substitution of Cl by C≡CPh shifts $E_{1/2}$ cathodically for Ru_2^{5+}/Ru_2^{4+} by 200 and 700 mV for first and second C≡CPh groups, respectively. The compounds are paramagnetic (three unpaired e, ground-state $\sigma^2\pi^4\delta^2\delta^{*1}\pi^{*2}$), with UV-vis spectra containing a well-resolved peak at ca 530 nm with two shoulders between 380 and 590 nm. The IR ν(C≡C) bands are between 2031 and 2045 cm^{-1}, considerably lower than the bis-adducts (ca 2100 cm^{-1}), and decrease with increasing electron-accepting power of the substituents (linear correlation with Hammett constants). Changes may occur by a σ-donor effect, $d_\pi \rightarrow \pi^*$(C≡C) back-donation, or π(C≡C) $\rightarrow d_\pi$ donation. In contrast with CO and cyano complexes, where $d_\pi \rightarrow \pi^*$(CX) back-donation occurs, here the changes in ν(C≡C) result from electron donation from the alkynyl group to the Ru_2 center.

D. *Complexes M $\underline{4}$ M—C≡CR*

Blue complexes M_2(C≡CR)$_4$(PMe$_3$)$_4$ (**340,** M = Mo, W; R = Me, Pr^i, Bu^t, Ph, SiMe$_3$) were obtained from reactions between M_2Cl_4(PMe$_3$)$_4$ and LiC≡CR in dme. Their thermal stability increases Ph < alkyl < SiMe$_3$ and Mo < W. Curiously, the W—SiMe$_3$ compound has not been obtained, an unidentified maroon complex being formed in its place.[527,528] Only the Mo/SiMe$_3$ complex shows ν(C≡C) at 1991 cm^{-1}. Most complexes are highly disordered in the crystal among three axial directions, although X-ray data indicate the compounds have D_{2d} symmetry.

The M $\underline{4}$ M quadruple bond is retained as shown by ν(MM) ca 362 cm^{-1}. The UV-vis spectra contain intense absorptions for the $^1(\delta \rightarrow \delta^*)$ transition at lower energy than for analogues with simpler ligands. Vibronic fine structure arising from ν(MoC) rather than ν(MoMo) has also been resolved; resonance Raman spectra also indicate an enhancement of the former vibration. These data are consistent with frontier orbital mixing between M_2 [δ,δ^*] and CCR [π,π^*] which both have π-symmetry.

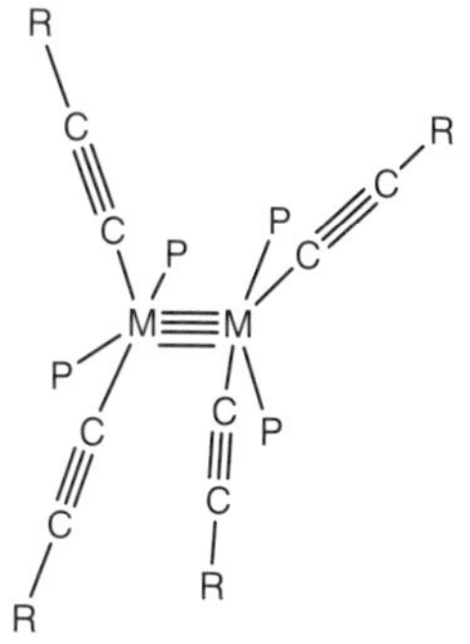

(**340-M/R**)

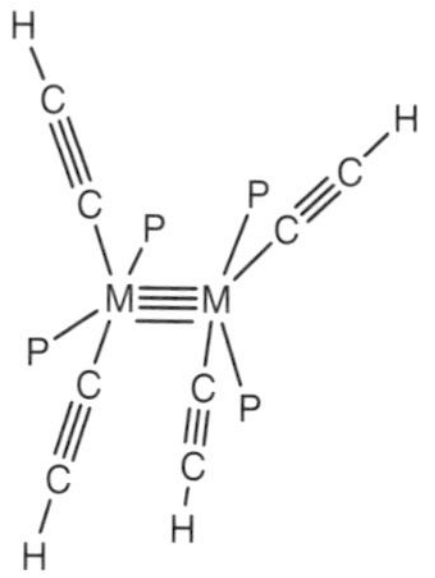

(**341**)

Mo-Mo 2.134(1), Mo-C 2.153, 2.161(4),
C≡C 1.174, 1.183(6) Å

Although reactions between LiC≡CH(en) or MgCl(C≡CH) and $Mo_2Cl_4(PMe_3)_4$ gave no tractable products, protodesilylation of $Mo_2(C{\equiv}CSiMe_3)_4(PMe_3)_4$ with $[NBu_4][HF_2]$ gave $Mo_2(C{\equiv}CH)_4(PMe_3)_4$ (**341**).[527] The X-ray structure shows D_{2d} symmetry, with expected bond lengths with no shortening of the Mo—C bond consistent with π(Mo—C) bonding. However, the ^{1}H NMR spectrum shows long-range ^4J(HP) coupling to the ethynyl proton, while the $^1(\delta \rightarrow \delta^*)$ absorption (ν_{max} 15,150 cm^{-1}, ε 4550) is both red-shifted and of ca 2.5 times the intensity of the similar band in $Mo_2Me_4(PMe_3)_4$. At low temperatures, vibronic structure of this band has a 400-cm^{-1} progression, corresponding to ν(MoC); substituted derivatives show a 360-cm^{-1} progression, assigned to the ν(MoC) + ν(MoMo) combination. The parent compound is thus electronically different from the substituted compounds, and is similar to $C_2(C{\equiv}CH)_4$. This is also demonstrated by its sensitivity to irradiation at 15,150 cm^{-1} (substituted compounds are stable under these conditions).

The presence of π(CC)-δ(MM)-π(CC) conjugation in dimetallapoly-ynes of the type $M_2(CCR)_4(PMe_3)_4$ has been inferred from Raman and electronic spectral data. In turn this may lead to donor (D)-acceptor (A) interactions in complexes of the type $L_2(DCC)_2M \underline{4} M(CCA)_2L_2$. The synthesis of asymmetric quadruply-bonded M_2 complexes is rare, but potential precursor $W_2Cl_2(C{\equiv}CMe)_2(PMe_3)_4$ has been obtained as a single isomer from LiC≡CMe and $W_2Cl_4(PMe_3)_4$.[528] On the basis of relative *trans* effects of Cl and CCR and because the latter is capable of π-back-bonding, reaction of initially formed $W_2Cl_3(C{\equiv}CMe)(PMe_3)_4$ would be expected to proceed by substitution of the chloride *trans* to the C≡CMe group. In the crystal, the W—C(sp) distances are 0.19 Å shorter than the W—C(sp^3) distances in $[W_2Me_8]^{4-}$, compared with the difference in covalent radii of 0.08 Å. The C≡C bonds are 1.21 Å, not significantly longer than conjugated triple bonds in organic molecules. In the ^{1}H NMR spectrum, the ≡CMe group (δ 2.95, septet) displays ^5J(HP) and ^6J(HW) couplings, again consistent with the presence of π(CC)-δ(MM)-π(CC) conjugation. The $^1(\delta \rightarrow \delta^*)$ band is at 13,765 cm^{-1}, intermediate between those found for the tetrachloro and tetrapropynyl complexes.

In $Mo_2(C{\equiv}CR)_4(PMe_3)_4$ (**340-Mo**), δ(MM) and δ^*(MM) orbitals have π-symmetry with respect to π^*(C≡C) orbitals; suitable design of ligands and photochemical studies allow determination of the role of π-back-bonding in the M—CCR bond. The complex $Mo_2(C{\equiv}CSiMe_3)_4(PMe_3)_4$ (**340-Mo/Si**) shows a reversible 1-e reduction wave at −2.13 V (vs FcH/[FcH]$^+$).[506] Chemical reduction ($K[C_{10}H_8]$) afforded [K(crypt-222)][$Mo_2(C{\equiv}CSiMe_3)_4(PMe_3)_4$] ([K][**340-Mo/Si**]) which is instantly oxidized in air. The ν(C≡C) bands for **340-Mo/Si** and [K][**340-Mo/Si**] are at 1991 and 1954 cm^{-1}, respectively. Electronic spectra of these compounds at 10 K contain vibronically structured $^1(\delta \rightarrow \delta^*)$ and $^2(\delta \rightarrow \delta^*)$ absorptions; the latter is red-shifted by ca 7000 cm^{-1} from the former as a result of larger spin-pairing energy contributions. Extensive spectroscopic data (UV/vis, Raman, NMR) provide direct evidence for M → CCR back-bonding, the former containing vibronic progressions corresponding to ν(MoMo)/ν(MoC) modes, with the high-energy edge of each band containing a feature with 0—0 spacing of 1970 (**340-Mo/Si**) or 1890 cm^{-1} ([K][**340-Mo/Si**]), assigned to ν(CC).

Similar studies of the vibrational modes of $M_2(C{\equiv}CR)_4(PMe_3)_4$ (M = Mo, W; R = H, Me, Bu^t, $SiMe_3$) have been made in conjunction with X-ray structural data, which are independent of R.[529] The three observed vibronic progressions originate from ν(MoMo), ν(MoC), and λ(MoCC) modes, which are strongly mixed. However, there is negligible mixing of the ν(MoMo) and ν(CC) modes, the latter being highly localized. These findings again substantiate the presence of π(CC)-δ(MM)-π(CC) conjugation.

E. *Dimetalladiynes, M≡CC≡M*

These complexes, exemplified by $(Bu^tO)_3W{\equiv}CC{\equiv}W(OBu^t)_3$, will be described in a later article.[530]

F. *Heteroatom Versions of Diyne Ligands*

1. *C≡CC≡N and RC≡CC≡N (R = H, CN)*

The cyanoethynyl ligand has been found in *cis*-Pt(C≡CCN)(CN)$(PPh_3)_2$, formed by photochemical rearrangement of Pt{η^2-$C_2(CN)_2$}$(PPh_3)_2$,[531] and in Fe(C≡CCN)(CO)(L)Cp (L = CO, PPh_3)[532] and Co(C≡CCN){CH=CH(CN)}(L)Cp [L = PPh_3, η^2-$C_2(SiMe_3)_2$].[533] Reactions of Me_3SnC≡CCN with $[NEt_4]_2[MCl_4]$ (M = Ni, Pd, Pt) give square planar $[NEt_4]_2$[M(C≡CC≡N)$_4$]; in the case of M = Pt, the Cl/alkynyl exchange is catalyzed by $PdCl_2(PPh_3)_2$.[534] The IR spectra showed a decrease of ca 69 cm^{-1} in ν(CN) compared with $[M(CN)_4]^{2-}$, while ν(C≡C) values of 2039–2047 cm^{-1} compare with 2062 cm^{-1} in HC≡CC≡N. These data, together with bond lengths in the Ni anion of Ni–C (1.856 Å), C≡C (1.203 Å), C–C (1.373 Å), and C≡N (1.148 Å), indicate that there is only a small contribution from the M=C=C=C=N resonance form. The group electronegativity of C≡CCN is estimated at $\chi = 3.17$, which indicates that it is one of the best π-acceptor ligands in the alkynyl series.[535]

Cyanoethyne, HC≡CC≡N, and dicyanoethyne, N≡CC≡CC≡N, are two highly activated alkynes which readily form η^2 complexes with tungsten (as a 4-e donor),[536] cobalt,[537] rhodium and iridium,[538] or platinum (as a 2-e donor)[539]; the Ni_2Cp_2 adduct of $C_2(CN)_2$[533] and the $Co_2(CO)_6$ adduct of Fe(C≡CCN)$(CO)_2$Cp have been reported.[532] Insertion reactions of these alkynes into M–H [M = Ta(η^2-C_2R_2)Cp_2,[540] M′Cp_2 (M′ = Mo, W),[541] ReCp_2,[540] Fe(CO)(L)Cp (L = CO, PPh_3, 1/2dppe)[542] and M–S bonds [M = W$(CO)_3$Cp,[543] Fe$(CO)_2$Cp[532]] have been described. Cycloaddition of $C_2(CN)_4$ to give pentacyanobutadienyl complexes is known.[532]

2. *Isocyano-Alkylidynes*

Isocyano-alkylidynes have been used as ligands for extended π-systems. In some complexes $L_2(OC)_2$ClW{≡CC_6H_4(C≡CC_6H_4)$_n$N≡C}ML_m (n = 0, 1 L_2 = tmeda, dppe; M = Re, Pd, Pt), photo-induced electron transfer along the chain has been demonstrated (but not for aromatic-free systems).[544]

3. *N≡CC≡N*

The high electronegativity of the cyano group ($\chi = 3.32$) will result in this ligand being an even stronger π-acceptor than cyanoethyne with significantly different chemistry. Extensive comparisons of alkynyl and cyano complexes have been made.[2]

4. *P≡CC≡P*

Theoretical studies of 1,4-diphosphabutadiyne conclude that while the molecule is thermodynamically stable, with structure P≡C–C≡P ↔ P–C≡C–P, it has a low kinetic stability and is likely to polymerize readily.[545] Possible stabilization by coordination to $Cr(CO)_5$ suggests side-on coordination to be preferred over end-on, while double side-on coordination to two $Cr(CO)_5$ or with one C≡P bond bridging a single $Co_2(CO)_6$ moiety, offers even more stabilization.

XII

POLYMER AND MATERIALS CHEMISTRY

As surmized in a recent review article there are literally hundreds of polymeric systems featuring organometallic complexes within a conjugated organic backbone.[38] Given recent reviews of these systems, here we shall restrict discussion to the various polymeric species and ceramic materials derived from diyne complexes and from coordination of metal fragments to polymers featuring C≡CC≡C repeat units.

A. *Materials from Diyne Complexes*

Thermolysis of $Mo_2(\mu\text{-}Me_3SiC_2C{\equiv}CSiMe_3)(CO)_3Cp$ or $\{W_2(CO)_4Cp_2\}_2(\mu\text{-}\eta^2{:}\mu\text{-}\eta^2\text{-}Me_3SiC_2C_2SiMe_3)$ gave black metallocarbide ceramic materials with some free metal and carbon.[546] The complex $\{Co_2(CO)_6\}_2(\mu\text{-}\eta^2{:}\mu\text{-}\eta^2\text{-}Me_3SiC_2C_2SiMe_3)$ has been shown to be unstable in methanol solutions, affording a black insoluble electrically conducting polymer which precipitated over 24 h.[162] Microanalytical data suggest that this material is a polyacetylene with most triple bonds being attached to $Co_2(CO)_6$ fragments. Pyrolysis of **90** (Section III) (800°C, 6 hr) gives powders containing well-formed carbon onions and multiwalled nanotubes. Most of the cobalt is deposited amorphously in discrete patches or in crystalline form inside the tubes and at the tips.[221] At lower temperatures **90** loses CO and is converted to graphitic material.

B. *Coordination of Metal Groups to Poly-yne-Containing Polymers*

The pyrolysis of several transition metal-containing organosilicon-diyne oligomers has been investigated as a method of preparing multiphase SiC-X or GeC-X ceramics.[547] Conventional methods, such as polycondensation of Li_2C_4 with dichlorosilanes $SiCl_2R_2$ ($R_2 = Me_2$, MePh, Ph_2), were used to prepare the precursor poly[(silylene)diynes], poly[(germylene)diynes] and their $Co_2(CO)_6$ derivatives. Room-temperature reactions with $Co_2(CO)_8$ gave $-\{SiR_2C_2[Co_2(CO)_6]C{\equiv}C\}_n-$, in which up to three Si environments were observed, corresponding to the three combinations of C≡C triple bond coordination in the $-(C{\equiv}CC{\equiv}C)SiR_2(C{\equiv}CC{\equiv}C)-$ sequence. In the case of the $SiPh_2$ polymer the most hindered $-\{C{\equiv}CC_2[Co_2(CO)_6]SiR_2C_2[Co_2(CO)_6]C{\equiv}C\}-$ arrangement was not observed.

Similar polymers containing a single coordinated C≡C moiety in the repeat units of poly[(methylphenylgermylene)diyne] and poly[(2,5-diphenyl-1-silacyclopenta-diene-1,1-diyl)diyne] were also prepared.

Pyrolysis of the $Co_2(CO)_6$ derivatives of poly(diorganosilylene)diynes at temperatures up to ca 1400°C gave multiphase Si–M–C ceramics retaining most of the Si and Co. X-ray powder diffraction of the resulting ceramic material indicated the presence of Co_2Si and graphitic carbon, rather than crystalline β-SiC.[548] TGA results suggest that at lower pyrolysis temperatures carbonyl groups are incorporated into the carbon matrix. Compared with the pure poly(diorganosilylene)-diynes, the cobalt derivatives form ceramic phases at lower temperatures, indicating that the cobalt may act as a catalyst for this process. Thermolysis of poly {1,1′-bis(diorganosilylethynyl)ferrocenes}, $-\{C{\equiv}CSiRR'\text{-}Fc'\text{-}SiRR'C{\equiv}C\}_n-$ (R, R = Me, Ph; R = Me, R′ = Ph) between 350 and 390°C results in a slow cross-linking of the C≡C triple bonds. Pyrolysis at 400–800°C gave black ceramic powders containing all of the Si and Fe present in the precursors with both β-SiC and $Fe_xSi_yC_z$ phases being identified in the X-ray powder patterns.[548]

Both $Mo(CO)_4(cod)$ and $Fe_2(CO)_9$ react with the coupling product from Li_2C_4 and 1-chloro-2,5-dimethylsilacyclopentadiene to give oligomers containing $Mo(CO)_4$ and $Fe(CO)_3$ groups, respectively, although not all cyclopentadiene groups were complexed.[547]

Reactions of poly(phenylenediyne) with $Pt(\eta\text{-}C_2H_4)(PPh_3)_2$ gave toluene-soluble oligomers which on heating to 600°C gave Pt-doped glassy carbon in which 0.1–1 atom-% metal is incorporated into an sp^2-carbon framework as particles with average diameter ca 16 Å. The materials are catalysts for electroreduction of H^+ in $HClO_4$ with activities similar to that of electroformed platinum microparticles (of ca 600 Å diameter).[549] Mixing PtO_2 with poly(1,3-phenylenebuta-1,3-diynyl) followed by thermal treatment (600°C) resulted in incorporation of platinum in oxidation states 0, II, and IV.

Cobalt octacarbonyl reacts with polydiynes obtained by ^{60}Co γ-irradiation to give metallated products in which 50% of the available alkyne moieties are coordinated. Reactions with $\{M(CO)_2Cp\}_2$ (M = Mo, W) also gave partly characterized polymeric metallated products, together with significant amounts of $\{M(CO)_3Cp\}_2$.[550]

XIII

PROGNOSIS

The synthetic chemistry associated with the preparation of complexes containing diyne or diynyl ligands and their longer chain analogues is now well established. In many cases, metal cluster reagents react with systems containing multiple C≡C moieties to give products which are similar to the products obtained from reactions of simple mono-alkynes. Derivatives are known for virtually all metals and it is possible to design rational syntheses for many complexes by employing one or more of the reactions types described earlier. However, much of the chemistry associated with diynes bearing electron-withdrawing groups such as CO_2Me, CF_3, and CN is conspicuously absent from the work reported to date. Metal-based reagents such as $M(C{\equiv}CC{\equiv}CR)(L)$ (M = Cu, Ag; L = phosphine) are yet to be fully exploited, although the alkyne chemistry of these species is rich in structural and chemical diversity.

A major challenge for the future lies in the systematic syntheses of metal complexes designed with a view to performing specific functions which result from their molecular shapes and/or electronic structures. Poly-yne ligands provide a fairly rigid rod-like structure, which when coupled with the geometric control possible about metal centers provides great promise for the assembly of supramolecular species with preconceived and controlled shapes. In addition, the varying degrees of orbital mixing which occur with different metal centers suggest potential applications to electronics. Their NLO and magnetic exchange properties remain to be fully explored, while the concept of molecular wires has been demonstrated. More work in these areas is required to determine how the combinations of frontier orbital overlap and electrostatic effects can be combined to transmit electronic information. Careful work aimed at constructing systems featuring symmetry and energetically well-matched, or deliberately mismatched, poly-yne and metal orbitals is required. In addition, studies directed toward gaining an understanding of the diynyl ligand/semiconductor surface interface need to be addressed.

The use of metal complexes of diynes as reagents in organic chemistry also promises many new developments, and the variety of alkyne-coupling reactions unearthed in recent times points toward novel synthetic methodologies involving early transition metals for the preparation of diynes and related unsaturated systems. However, a great deal of work remains to be done in this area if these new reactions are to compete with the well-established copper- and palladium-based reactions.[415] Investigations of the reactions of yne and ynyl ligands with nucleophilic and electrophilic reagents will no doubt continue to generate surprising products.

Ligands comprised of heteroatomic C_nY_m chains are a natural extension of the work on ynyl and yne ligands and are intriguing synthetic targets. Computational studies of complexes with short-chain carbon—boron and carbon—phosphorus ligands have been performed, and suggest that synthesis of molecules containing these ligands is an achievable aim. Chemically, one possible synthon is $R_3P^+C{\equiv}CC{\equiv}CB^-R'_3$, one example of which, $Ph_2MePC{\equiv}CC{\equiv}CB(CH_2Ph)_3$, is already known.[551] To our knowledge, reagents of this type have not yet been applied to transition metal chemistry.

A further aspect of this chemistry is in relation to the preparation of extended carbon networks and related systems. Imaginative consideration of various as yet unknown forms of carbon, both molecular and polymeric, has included the concept of metal complexes of π-systems which carry various di- and poly-ynyl fragments (Section XI), which may in turn be converted to two- and three-dimensional networks and cages.[53,151] Indeed, as we write, carbon networks containing triangular motifs (so-called polytriangle-n-ynes) are suggested to have negative Poisson ratios, becoming wider when stretched and thinner when compressed.[552]

Finally, the study of gaseous species containing ligand-free metal-diynyl fragments, such as TiCCCCH[553] and FeCCCCH[554] is still in its infancy, and it will be interesting to see if these molecules inhabit circum- or interstellar space. The literature is expanding rapidly and we have no doubt that, while many advances have already been made toward all these goals, this area will reward with many surprises in the future.

XIV

APPENDIX: ABBREVIATIONS

[9]aneS_3	1,4,7-Trithiacyclononane
bdpp	1,4-Bis(diphenylphosphino)buta-1,3-diyne
bma	3,4-Bis(diphenylphosphino)maleic anhydride
bpy	2,2′-Bipyridyl
bta	Benzotriazole
btd	2,1,3-Benzothiadiazole
c.v.e.	Cluster valence electrons
cat	Catecholate ($C_6H_4O_2$-1,2 or 4,5-$Bu^t_2C_6H_2O_2$-1,2)
cod	1,5-Cyclooctadiene
Cp*	η-C_5Me_5
Cp^R	η-C_5H_4R
CV	Cyclic voltamogram/voltametry
dbp	1,4-Bis(4-pyridyl)buta-1,3-diyne
dbu	1,8-Diazabicyclo[5.4.0]undec-7-ene
dcype	1,2-Bis(dicyclohexylphosphino)ethane
DFT	Density functional theory
dippe	1,2-Bis(di-isopropylphosphino)ethane
dippp	1,3-Bis(di-isopropylphosphino)propane
dmpe	1,2-Bis(dimethylphosphino)ethane
dppe	1,2-Bis(diphenylphosphino)ethane
dppm	Bis(diphenylphosphino)methane
EH MO	Extended Hückel molecular orbital
ELF	Electron localization function
Fc	Ferrocenyl

FcH*	Decamethylferrocene, $Fe(\eta\text{-}C_5Me_5)_2$
Fv	Fulvalenyl, $\eta^5{:}\eta^5\text{-}C_5H_4C_5H_4$
GCE	Glassy carbon electrode
LDA	$LiNPr^i_2$
Mc	Metallocenyl
nap	Naphthyl
NLO	Nonlinear optical
OTf	Triflate, trifluoromethanesulfonate
PES	Photoelectron spectroscopy
pin	Pinacolinate
porph	Porphyrin dianion
pp_3	$P(CH_2CH_2PPh_2)_3$
ppn	bis(triphenylphosphine)iminium cation, $[N(PPh_3)_2]$
(Ru)	$Ru(CO)_2$
[Ru]	$Ru(CO)_3$
Rc	Ruthenocenyl
SCE	Standard calomel electrode
SOMO	Singly occupied molecular orbital
tacn	1,4,7-Triazacyclononane
tcne	Tetracyanoethene
terpy	2:2′,6′:6″-Terpyridyl
tmeda	*N,N,N′,N′*-Tetramethyldiaminoethane
tol	p-Tolyl, $C_6H_4Me\text{-}4$
Tp	Hydrotris(pyrazolyl)borate
Tp′	Hydrotris(3,5-dimethylpyrazolyl)borate
$Tp^{R,R'}$	Hydrotris(3-R-5-R′-pyrazolyl)borate

XV

ADDENDUM

The following up-dates this review to mid-2001: the area is very active and it has not been possible to include more than a general indication of the content of these later papers, which are generally arranged in the order of the above sections.

I. Recent review topics include carbon-rich acetylenic materials as molecular scaffolding,[558] studies of luminescent di- and poly-ynyl-rhenium complexes containing $Re(CO)_3(bpy)$ groups[559] and Group 4 diynyl "tweezer" complexes in the context of other tweezer and related molecules.[560]

II.A. Treatment of Z-CH(OMe)=CHCHC≡CSi(OR)$_3$ (R = Me, Pr^i, Bu^t) with LiNPr$^i_2$ affords LiC≡CC≡CSi(OR)$_3$, several reactions of which are reported.[561]

II.B. Syntheses of Cu(C≡CC≡CEt) {(Bu^{t}C≡C)$_2$TiCp$^{Si}{}_2$},[562] Mo(C≡CC≡CH)(CO)(dppe)(η-C$_5$H$_4$CO$_2$Me), and its Co$_2$(CO)$_6$ complex.[563] Complexes Re{C≡CC≡CAr$_2$(OMe)}(NO)(PPh$_3$)Cp* [CAr$_2$(OMe) = 9-methoxyfluorenyl and 2,7-dibromo- and -dichloro- derivatives] have been made by lithiating Re(C≡CC≡CH)(NO)(PPh$_3$)Cp* and reaction with the 9-fluorenones, en route to the corresponding pentatetraenylidenes [Re(=C=C=C=C=CAr$_2$)(NO)(PPh$_3$)Cp*]$^+$.[564] Electrochemistry and spectroscopic properties of binuclear {Me$_3$SiC≡CC≡C[Rh$_2$(ap)$_4$]}$_n$ (ap = 2-anilinopyridinate) complexes.[565]

II.C. The luminescent properties of Re{(C≡C)$_3$R}(CO)$_3$(bpy-Bu$^t_2$) (R = Ph, SiMe$_3$) have been compared with those of analogous mono- and di-ynyl complexes.[566]

III.B. While equimolar amounts of Pt(C≡CPh)$_2$(dppf) and [Au(PPh$_3$)]OTf react to give the enynyl complex Pt{η^3-PhCCC≡CPh[Au(PPh$_3$)]}(dppf), excess alkynylplatinum complex is converted to Pt(η^2-PhC$_2$C≡CPh)(dppf). The latter, which is also formed from the diyne and Pt(η-C$_2$H$_4$)(dppf), reacts directly with [Au(PPh$_3$)]$^+$ to give the enynyl complex.[567]

III.C. Addition of VCp$_2$ to Bu^t(C≡C)$_4$Bu^t gives successively mono- and divanadium complexes, attached to the C^3—C^4 or to the C^1—C^2 and C^7—C^8 fragments, respectively, indicating the movement of the VCp$_2$ group along the carbon chain. X-ray structures of both complexes and the {VCp}$_2$ complex of Ph(C≡C)$_4$Ph are given.[568] Several examples of Cr(CO)$_3$(η-PhC≡CC≡CR) (R = C$_6$H$_4$NO$_2$-4, C$_6$H$_4$NMe$_2$-4, Fc) have been prepared.[569] Reactions of RC≡CC≡CR (R = Me, Ph) with Ru$_2$(μ-dppm)$_2$(μ-CO)(CO)$_4$ afford Ru$_2$(μ-dppm)$_2$(μ-RC$_2$C≡CR)(CO)$_4$.[570] Oxidative coupling of the alkynyl groups in Co(C≡CR)$_2${(PPh$_2$CH$_2$)$_3$CMe} (R = Ph, Bu^t, SiMe$_3$) occurs with [FcH]$^+$ to give Co(η^2-RC$_2$C≡CR){(PPh$_2$CH$_2$)$_3$CMe}.[571]

Treatment of Co$_2$(μ-RC$_2$C≡CH)(μ-dppm)(CO)$_4$ (R = H, SiMe$_3$) with RuCl(PPh$_3$)$_2$Cp in the presence of NH$_4$PF$_6$ gives the vinylidene, which can be deprotonated *in situ* (NaOMe) to give deep green Co$_2${μ-RC$_2$C≡C[Ru(PPh$_3$)$_2$Cp]}(μ-dppm)(CO)$_4$. Electrochemical, spectroscopic, and theoretical studies suggest that the HOMO contains significant contributions from RuC≡C and C$_2$Co$_2$ orbitals and is delocalized over the molecule, with accumulation of negative charge on the C$_2$Co$_2$ center.[572] Syntheses and electrochemical studies of Co$_2$(Me$_3$SiC$_2$C≡CSiMe$_3$)(CO)$_{6-n}$(L)$_n$ [L = PMe$_3$ (n = 1,2), PMePh$_2$ (n = 2), L$_2$ = dppm, (PPh$_2$)$_2$NH (n = 2)]; the normally readily reducible C$_2$Co$_2$ center becomes readily oxidizable when phosphine ligands replace CO.[573] Cationic Co$_2$(CO)$_6$ derivatives of various

1,3-diynes bearing N- or S-centers (X-ray structures of SMe_2, 3-picoline, and neutral OH complexes) are described.[574] Reactions of the SMe_2 dication with N-, P-, or S-nucleophiles proceed in high yield, the diyne salts being considered to be rigid masked electrophiles.

IV.B. The reaction of $Ru_3(\mu\text{-dppm})(CO)_{10}$ with FcC≡CC≡CFc results in coupling of the diyne to give $Ru_3\{\mu_3\text{-FcCC(C≡CFc)CFcCC}_2\text{Fc}\}(\mu\text{-dppm})(\mu\text{-CO})(CO)_5$.[575] Enynyl complexes have been obtained from RC≡CC≡CR and $Ru_3(\mu\text{-H})(\mu\text{-dmpz})(CO)_{10}$ (R = Me).[576] Metalation of a phenyl ring of $Ru_3(\mu\text{-H})(\mu\text{-N=CPh}_2)(CO)_{10}$ occurs in reactions with RC≡CC≡CR (R = Me, CH_2OPh, Ph) to give butatriene, enyne, allenyl, or allyl ligands, the latter two incorporating an N=CPh(C_6H_4) fragment.[577] Several Ru_4 clusters containing di- or tri-hydrogenated diyne ligands were obtained from $Ru_4(\mu\text{-H})_4(CO)_{12}$ and RC≡CC≡CR (R = Me, $SiMe_3$, Ph).[578]

Open Os_3 clusters containing $\mu_3\text{-}\eta^4$-diyne ligands have been obtained from $Os_3(CO)_{10}(NCMe)_2$ and RC≡CC≡CR [R = 2-C_4H_3S,[579] Fc[580]]. The latter reaction also gives $Os_3(\mu_3\text{-FcC}_2\text{C≡CFc})(\mu\text{-CO})(CO)_9$. Electronic communication between the Fc groups is decreased in the closed Os_3 cluster, but increased in the open cluster.

$Os_4(\mu\text{-H})_4(CO)_{12}$ gives the enynyl cluster $Os_4(\mu\text{-H})_3\{\mu\text{-}\eta^2\text{-(Z)-FcCCHC≡CFc}\}(CO)_{11}$.[581] Reactions of $Os_3(\mu_3\text{-FcC}_2\text{C≡CFc})(\mu\text{-CO})(CO)_9$ with water afforded enynyl complexes $Os_3(\mu_3\text{-}\eta^3\text{-FcC}_3\text{CHFc})(\mu\text{-OH})(CO)_9$ (E and Z isomers; open Os_3 clusters) and $Os_3(\mu\text{-H})(\mu_3\text{-E-FcC}_3\text{CHFc})(CO)_9$; ΔE values for the oxidations of the two Fc groups are larger than in the diyne complex.[582] Reaction of Me_3SiC≡CC≡$CSiMe_3$ with $Os_3(\mu\text{-H})_2(CO)_{10}$ gives $Os_3(\mu\text{-H})\{\mu\text{-CHC(SiMe}_3)\text{C≡CSiMe}_3\}(CO)_{10}$ via a hydride shift and 1,2-migration of an $SiMe_3$ group. An excess of the diyne affords $Os_3(\mu_3\text{-}\eta^2\text{-Me}_3\text{SiC}_2\text{C≡CSiMe}_3)(\mu\text{-CO})(CO)_9$. The mono-proto-desilylated complex reacts with $Co_2(CO)_8$ to give known $Os_3\{\mu_3\text{-}\eta^2\text{-Me}_3\text{SiC}_2\text{C}_2\text{H[Co}_2\text{(CO)}_6]\}(\mu\text{-CO})(CO)_9$ together with $Os_3(\mu\text{-H})\{\mu_3\text{-C}_2\text{C}_2\text{SiMe}_3[\text{Co}_2\text{(CO)}_6]\}(CO)_9$.[583] Reactions of $Os_3(\mu_3\text{-FcC}_2\text{C≡CFc})(CO)_{10}$ with $Pt(cod)_2$ give $Os_3Pt(\mu_4\text{-FcC}_2\text{C}_2\text{Fc})(CO)_9(cod)$ (butterfly, Pt in wing-tip) and $Os_3Pt_2(\mu_5\text{-FcC}_2\text{C}_2\text{Fc})(CO)_{10}(cod)$ (bow-tie, Pt knot).[584]

IV.C. Tail-to-tail coupling of alkynyl groups occurs with $W(C≡CPh)(CO)_3Cp^*$ and $Fe_3(\mu_3\text{-E})_2(CO)_9$ (E = S, Se, Te) to give $Fe_3W_2(\mu_4\text{CCPhCPhC})(\mu_3\text{-E})_2(CO)_6Cp^*_2$.[585] Butenynyl and butatrienyl complexes are formed by coupling HC≡CAr on $Ru_3(\mu\text{-H})\{\mu_3\text{-NS(O)MePh}\}(CO)_9$,[586] while $PtRu_3(\mu_4\text{-PhCCCCHBu}^t)(\mu_4\text{-Te})(\mu\text{-TePr}^i)(CO)_6(dppe)$ is obtained from $PtRu_3(\mu\text{-H})(\mu_4\text{-C}_2\text{Bu}^t)(CO)_9(dppe)$ and $PhC≡CTePr^i$.[587]

V.B. Reactions of $W(C{\equiv}CC{\equiv}CH)(CO)_3Cp$ with $Ru_3(CO)_{10}(NCMe)_2$ or $Ru_3(\mu\text{-dppm})(CO)_{10}$ have given the expected η^2-diyne and hydrido-diynyl complexes. $Ru_3(\mu\text{-H})\{\mu_3\text{-}C_2C{\equiv}C[W(CO)_3Cp]\}(CO)_9$ reacts with $Fe_2(CO)_9$ or $Ru_3(CO)_{12}$ to give complexes containing C_4 ligands bridging Ru_2M and M_2W (M = disordered Fe/Ru or Ru, respectively) clusters. With $Co_2(CO)_8$, migration of hydride from the cluster to the C_4 unit gives the μ_3:μ_3-ethynylvinylidene derivative containing CoRuW and RuM_2 (M = disordered Co/Ru) clusters.[588]

VI.B. The X-ray structure of $Ti(\eta^2\text{-}PhC_2C{\equiv}CPh)(ttp)$ [ttp = tetra(4-tolyl)porphyrin] is reported.[589] Variable temperature NMR studies of $\{TiCp_2\}_2(\mu\text{-}\eta^2{:}\eta^2\text{-}Bu^tC_2C_2SiMe_3)$ revealed dynamic behavior that probably involves central C—C bond cleavage to form two μ-C≡CR ligands which allow exchange of metal centers on the diyne ($\Delta G^{\#}$ 63 kJ mol^{-1}). This process is related to the C—C single bond metathesis reactions.[590] PhC≡CC≡CPh reacts with $Zr(C_6H_4)Cp_2$ to give zirconacyclocumulene $Zr(PhC{=}C{=}C{=}CPhC_6H_4)Cp_2$.[591]

The reaction of $TiCl_2Cp^*_2$ with Mg and $Bu^t(C{\equiv}C)_3Bu^t$ gives the symmetrical complex $Ti(\eta^2\text{-}Bu^tC{\equiv}CC_2C{\equiv}CBu^t)Cp^*_2$ (X-ray) while, in contrast, $ZrCl_2Cp^*_2$ gives $Zr(\eta^4\text{-}Bu^tC_4C{\equiv}CBu^t)\,Cp^*_2$ (X-ray).[592] The variable temperature NMR spectra of the zirconium complex indicate that the $ZrCp^*_2$ group slides along the poly-yne chain, probably via an η^2 bonded intermediate. Thermolysis of $Ti\,(\eta^2\text{-}Me_3SiC_2C{\equiv}CSiMe_3)(\eta\text{-}C_5Me_4R)_2$ (R = CH_2Ph, Ph, C_6H_4F-4) gives products formed by double activation and reaction of C—H bonds; in the presence of $Bu^tC{\equiv}CC{\equiv}CBu^t$, further coupling occurs to give "doubly tucked-in" compounds. Reactions with HCl afford $TiCl_2Cp^R_2$ (Cp^R = the modified η^5 ligands).[593] Coupling of σ-C≡CR ligands on $ZrCp_2$ centers to give methylenecyclopropenes is induced by $B(C_6F_5)_3$.[594]

VII.C. Enynyl $RuCl\{C(C{\equiv}CSiMe_3){=}CHSiMe_3\}(CO)(PPh_3)_2$, obtained from $Me_3SiC{=}CC{=}CSiMe_3$ and $RuHCl(CO)(PPh_3)_2$, is converted to $RuCl\{C(C{\equiv}CSiMe_3){=}CHSiMe_3\}(CO)(PPh_3)(dppe)$ with dppe; with HC≡CC≡CH, binuclear $\{RuCl(CO)(PPh_3)_2(L)\}_2(\mu\text{-}CH{=}CHCH{=}CH)$ (L = -, NH_3, PEt_3, PPh_3) were obtained.[595] The reaction of PhC≡CC≡CPh with $RuH(S_2CNEt_2)(CO)(PPh_3)_2$ gives $Ru\{C(C{\equiv}CPh){=}CHPh\}\,(S_2CNEt_2)(CO)(PPh_3)_2$.[596]

VII.F. $NiX_2(PPh_3)_2$ (X = Cl, Br) catalyzes the synthesis of $C_6Et_5(C{\equiv}CEt)$ from EtC≡CC≡CEt and $Zr(C_4Et_4)Cp_2$[597] and the cyclo-addition of C_2Et_2, EtCN, and PhC≡CC≡CPh to give 2,3,4-Et_3-5-PhC≡C-6-Ph-pyridine.[598] RuCl(cod)Cp* catalyzes the reaction of $Me_3SiC{\equiv}CC{\equiv}CSiMe_3$ with $SiMe_3CHN_2$ to give alkynyldiene $CH(SiMe_3){=}C(SiMe_3)C(C{\equiv}CSiMe_3){=}CHSiMe_3$.[599]

VIII. Symmetrization of terminal alkynes to the corresponding diynes (43–67% yield) occurs in reactions with $TiCl_4/NEt_3$.[600] Reactions of $Ti(C{\equiv}CR)_2Cp^{Si}_2$ (R = Fc, $SiMe_3$, Ph, or mixed Ph/$SiMe_3$) with MCl_2 (M = Pd, Pt, Cu) or $AuCl_3$

gives $TiCl_2Cp^{Si}{}_2$ and RC≡CC≡CR.[601] Oxidation of Cp_2Zr-alkenyl/alkynyl complexes with $VOCl(OPr^i)_2$ gives the corresponding diynes.[602] 1,3-Diynes are side products in reactions of PhC≡CX (X = Cl, I) with $Pd(PPh_3)_4$.[603]

IX.A. Conventional coupling and desilylation reactions of Co{η-C_4(C≡CH)$_2$-(C≡CSiPr$^i{}_3$)$_2$}Cp have given large concave organometallic hydrocarbons containing the Co{η-C_4(C≡C-)$_4$}Cp core.[604] Similar precursors have been converted to dehydrobenzannulenes containing Co{η-C_4(SiR_3)$_2$}Cp and related groups.[605] Syntheses of various poly-ferrocenyl cumulenes involve the corresponding diynes, such as $Fc_2C(OMe)(CH_2)_n$C≡CC≡CCH_2CFc_2(OMe) (n = 0, 1), Fc_2C=CHC≡CC≡CCH≡CFc_2, Fc_2C^+C≡CC≡CCH=CFc_2.[606]

IX.C. The spectra and electrochemistry of molecular rectangles containing Ph_2C≡CC≡$CPPh_2$ and RuCl(tpy) groups have been described.[607]

IX.D. Several one-, two- and three-dimensional polymeric complexes derived from (py)C≡CC≡C(py) (py = 2- or 4-C_5H_4N) containing copper(I) or silver(I) have been prepared, but involve only N-coordination.[608] Syntheses and electrochemical properties of butadiynyl-linked metallo-phthalocyanines, including heterodimetallic complexes containing push-pull substituents, and molecular diyads containing homo- or hetero-metallic phthalocyanines (Zn or Zn/Co); also a square nickel porphyrin tetramer linked by diyndiyl edges.[609, 610] A cyclic dimer of a zinc porphyrin with substituted phenoxymethyldiynyl edges forms an inclusion complex with C_{60}.[611] The preparation of *trans*-{$(Et_3P)_2$PhPt}$_2${μ-C≡CSC≡CC≡CSC≡C) is reported.[612]

References

(1) Nast, R. *Coord. Chem. Rev.* **1982,** *47,* 89.
(2) Manna, J.; John, K. D.; Hopkins, M. D. *Adv. Organomet. Chem.* **1995,** *38,* 79.
(3) Stang, P. J.; Diederich, F. (Eds.) *Modern Acetylene Chemistry;* VCH: Weinheim, 1995.
(4) Hagihara, N.; Sonogashira, K.; Takahashi, S. *Adv. Polym. Sci.* **1981,** *41,* 149.
(5) Takahashi, S.; Morimoto, H.; Murata, E.; Kataoka, S.; Sonogashira, K.; Hagihara, N. *J. Polym. Sci., Polym. Chem. Ed.* **1982,** *20,* 565.
(6) Rouke, J. P.; Bruce, D. W.; Marder, T. B. *J. Chem. Soc. Dalton Trans.* **1995,** 317.
(7) (a) Whittall, I. R.; McDonagh, A. M.; Humphrey, M. G. *Adv. Organomet. Chem.* **1998,** *42,* 291. (b) Whittall, I. R.; McDonagh, A. M.; Humphrey, M. G.; Samoc, M. *Adv. Organomet. Chem.* **1999,** *43,* 349.
(8) (a) Stang, P. J.; Olenyuk, B. *Acc. Chem. Res.* **1997,** *30,* 502. (b) Stang, P. J. *Chem. Eur. J.* **1998,** *4,* 19.
(9) Crabtree, R. H. *The Organometallic Chemistry of the Transition Metals,* 2nd ed.; Wiley: New York, 1994, p. 388.
(10) Cafryn, A. J. M.; Nicholas, K. M. *Comprehensive Organometallic Chemistry II;* Abel, E. W.; Stone, F. G. A.; Wilkinson, G., Eds.; Pergamon: Oxford, 1995, Vol. 12, ch. 7.1, p. 685.
(11) Schore, N. E. *Comprehensive Organometallic Chemistry II;* Abel, E. W.; Stone, F. G. A.; Wilkinson, G., Eds.; Pergamon: Oxford, 1995, Vol. 12, ch. 7.2, p. 703.

(12) Beck, W.; Niemer, B.; Wieser, M. *Angew. Chem.* **1993,** *105,* 969; *Angew. Chem. Int. Ed. Engl.* **1993,** *32,* 923.
(13) Lang, H. *Angew. Chem.* **1994,** *106,* 569; *Angew. Chem. Int. Ed. Engl.* **1994,** *33,* 547.
(14) Bunz, U. H. F. *Angew. Chem.* **1996,** *108,* 1047; *Angew. Chem. Int. Ed. Engl.* **1996,** *35,* 969.
(15) Astruc, D. *Electron Transfer and Radical Processes in Transition Metal Chemistry;* VCH: New York, 1995.
(16) Benniston, A. C.; Grosshenny, V.; Harriman, A.; Ziessel, R. *Angew. Chem.* **1994,** *106,* 1956; *Angew. Chem. Int. Ed. Engl.* **1994,** *33,* 1884.
(17) Grosshenny, V.; Harriman, A.; Ziessel, R. *Angew. Chem.* **1995,** *107,* 1211; *Angew. Chem. Int. Ed. Engl.* **1995,** *34,* 1100.
(18) Schumm, J. S.; Pearson, D. L.; Tour, J. M. *Angew. Chem.* **1994,** *106,* 1445; *Angew. Chem. Int. Ed. Engl.* **1994,** *33,* 1360.
(19) Aviram, A. (Ed.) *Molecular Electronics: Science and Technology;* Am. Inst. Phys., Conf. 262, 1992.
(20) Bunz, U. H. F. *Angew. Chem.* **1994,** *106,* 1127; *Angew. Chem. Int. Ed. Engl.* **1994,** *33,* 1073.
(21) Ward, M. D. *Chem. Ind.* **1996,** *568;* Ward, M. D. *Chem. Soc. Rev.* **1995,** *24,* 121.
(22) Aviram, A.; Ratner, M. *Chem. Phys. Lett.* **1974,** *29,* 277.
(23) Metzger, R. M. *Acc. Chem. Res.* **1999,** *32,* 950.
(24) Wong, A.; Kang, P. C. W.; Tagge, C. D.; Leon, D. R. *Organometallics* **1990,** *9,* 1992.
(25) Le Narvor, N.; Toupet, L.; Lapinte, C. *J. Am. Chem. Soc.* **1995,** *117,* 7129.
(26) Coat, F.; Lapinte, C. *Organometallics* **1996,** *15,* 477.
(27) Coat, F.; Guillevic, M.-A.; Toupet, L.; Paul, F.; Lapinte, C. *Organometallics* **1997,** *16,* 5988.
(28) Guillemot, M.; Toupet, L.; Lapinte, C. *Organometallics* **1998,** *17,* 1928.
(29) Dembinski, R.; Bartik, T.; Bartik, B.; Jaeger, M.; Gladysz, J. A. *J. Am. Chem. Soc.* **2000,** *122,* 810.
(30) Brady, M.; Weng, W.; Zhou, Y.; Seyler, J. W.; Amoroso, A. J.; Arif, A. M.; Böhme, M.; Frenking, G.; Gladysz, J. A. *J. Am. Chem. Soc.* **1997,** *119,* 775.
(31) Bartik, T.; Weng, W.; Ramsden, J. A.; Szafert, S.; Falloon, S. B.; Arif, A. M.; Gladsyz, J. A. *J. Am. Chem. Soc.* **1998,** *120,* 11071.
(32) Bruce, M. I.; Low, P. J.; Costuas, K.; Halet, J.-F.; Best, S. P.; Heath, G. A. *J. Am. Chem. Soc.* **2000,** *122,* 1949.
(33) Kheradmandan, S.; Heinze, K.; Schmalle, H. W.; Berke, H. *Angew. Chem. Int. Ed.* **1999,** *38,* 2270.
(34) Paul, F.; Lapinte, C. *Coord. Chem. Rev.* **1998,** *178–180,* 431.
(35) Lin, V. S.-Y.; DiMagno, S. G.; Therien, M. J. *Science* **1994,** *264,* 1105.
(36) Anderson, H. L. *Inorg. Chem.* **1994,** *33,* 972.
(37) Ziessel, R.; Hissler, M.; El-Ghayoury, A.; Harriman, A. *Coord. Chem. Rev.* **1998,** *178–180,* 1251.
(38) Kingsborough, R. P.; Swager, T. M. *Prog. Inorg. Chem.* **1999,** *48,* 123.
(39) Manners, I. *Angew. Chem.* **1996,** *108,* 1712; *Angew. Chem. Int. Ed. Engl.* **1996,** *35,* 1603, but esp. p. 1616.
(40) Diederich, F.; Faust, R.; Gramlich, V.; Seiler, P. *J. Chem. Soc. Chem. Commun.* **1994,** 2045 and refs therein.
(41) Diederich, F. *Pure Appl. Chem.* **1999,** *71,* 265.
(42) Anthony, J.; Boldi, A. M.; Rubin, Y.; Hobi, M.; Gramlich, V.; Knobler, C. B.; Seiler, P.; Diederich, F. *Helv. Chim. Acta* **1995,** *78,* 13.
(43) Martin, R. E.; Diederich, F. *Angew. Chem. Int. Ed.* **1999,** *38,* 1350.
(44) Wiegelmann, J. E. C.; Bunz, U. H. F.; Schiel, P. *Organometallics* **1994,** *13,* 4649.
(45) Altmann, M.; Bunz, U. H. F. *Makromol. Rapid Commun.* **1994,** *15,* 785.
(46) Wiegelmann, J. E. C.; Bunz, U. H. F. *Organometallics* **1993,** *12,* 3792.

(47) Bunz, U. H. F.; Wiegelmann-Kreiter, J. C. *Chem. Ber.* **1996,** *129,* 785.
(48) Wiegelmann-Kreiter, J. E. C.; Bunz, U. H. F. *Organometallics* **1995,** *14,* 4449.
(49) Bunz, U. H. F.; Enkelmann, V. *Organometallics* **1994,** *13,* 3823.
(50) Bunz, U. H. F.; Enkelmann, V. *Angew. Chem.* **1993,** *105,* 1712; *Angew. Chem. Int. Ed. Engl.* **1993,** *32,* 1653.
(51) Bunz, U. H. F.; Enkelmann, V.; Beer, F. *Organometallics* **1995,** *14,* 2490.
(52) Bunz, U. H. F.; Enkelmann, V.; Rader, J. *Organometallics* **1993,** *12,* 4745.
(53) Bunz, U. H. F.; Rubin, Y.; Tobe, Y. *Chem. Soc. Rev.* **1999,** *28,* 107.
(54) Blenkiron, P.; Enright, G. D.; Carty, A. J. *Chem. Commun.* **1997,** 483.
(55) Low, P. J.; Enright, G. D.; Carty, A. J. *J. Organomet. Chem.* **1998,** *565,* 279.
(56) Low, P. J.; Udachin, K. A.; Enright, G. D.; Carty, A. J. *J. Organomet. Chem.* **1999,** *578,* 103.
(57) Adams, R. D.; Bunz, U. H. F.; Fu, W.; Riodl, G. *J. Organomet. Chem.* **1999,** *578,* 55.
(58) Shostakovskii, M. F.; Bogdanova, A. V. *The Chemistry of Diacetylenes;* Wiley: New York, 1974.
(59) Brefort, J. L.; Corriu, R. J. P.; Gerbier, P.; Guerin, C.; Henner, B.; Jean, A.; Kuhlmann, T.; Garnier, F.; Yassar, A. *Organometallics* **1992,** *11,* 2500.
(60) Holmes, A. B.; Jennings-White, C. L. D.; Schulthess, A. H.; Kinde, B.; Walton, D. R. M. *J. Chem. Soc. Chem. Commun.* **1979,** 840.
(61) Kim, J. P.; Masai, H.; Sonogashira, N.; Hagihara, N. *Inorg. Nucl. Chem. Lett.* **1970,** *6,* 181.
(62) Bruce, M. I.; Hinterding, P.; Tiekink, E. R. T.; Skelton, B. W.; White, A. H. *J. Organomet. Chem.* **1993,** *450,* 209.
(63) Zweifel, G.; Rajagopalan, S. *J. Am. Chem. Soc.* **1985,** *107,* 700.
(64) Stracker, E. C.; Zweifel, G. *Tetrahedron Lett.* **1990,** *31,* 6815.
(65) Masai, H.; Sonogashira, K.; Hagihara, N. *Bull. Chem. Soc. Jpn.* **1971,** *44,* 2226.
(66) Hartbaum, C.; Fischer, H. *Chem. Ber./Recueil* **1997,** *130,* 1063.
(67) Sonogashira, K. *Comprehensive Organic Synthesis;* Trost, B. M.; Fleming, I., Eds.; Pergamon: Oxford, 1991, Vol. 3, ch. 2.5, p. 551.
(68) (a) Sonogashira, K.; Yatake, T.; Tohda, Y.; Takahashi, S.; Hagihara, N. *J. Chem. Soc. Chem. Commun.* **1977,** *291.* (b) Sonogashira, K.; Fujikura, Y.; Yatake, T.; Takahashi, S.; Hagihara, N. *J. Organomet. Chem.* **1978,** *145,* 101.
(69) Bruce, M. I.; Hall, B. C.; Low, P. J.; Smith, M. E.; Skelton, B. W.; White, A. H. *Inorg. Chim. Acta* **2000,** *300–302,* 633.
(70) Bruce, M. I.; Humphrey, M. G.; Matisons, J. G.; Roy, S. K.; Swincer, A. G. *Aust. J. Chem.* **1984,** *37,* 1955.
(71) Werner, H.; Gervert, O.; Steinert, P.; Wolf, J. *Organometallics* **1995,** *14,* 1786.
(72) Lass, R. W.; Steinert, P.; Wolf, J.; Werner, H. *Chem. Eur. J.* **1996,** *2,* 19.
(73) Eastmond, R.; Walton, D. R. M. *Tetrahedron* **1972,** 4591.
(74) Mitchell, T. N. *Metal Catalysed Cross-Coupling Reactions;* Diederich, F.; Stang, P. J., Eds.; Wiley-VCH: Weinheim, 1998, ch. 4, p. 167.
(75) Farina, V.; Krishnamurthy, V.; Scott, W. J. *The Stille Reaction,* Wiley: New York, 1998.
(76) Gauss, C.; Veghini, D.; Berke, H. *Chem. Ber./Recueil* **1997,** *130,* 183.
(77) Schneider, D.; Werner, H. *Angew. Chem.* **1991,** *103,* 710; *Angew. Chem. Int. Ed. Engl.* **1991,** *30,* 700.
(78) Rappert, T.; Nürnberg, O.; Werner, H. *Organometallics* **1993,** *12,* 1359.
(79) Werner, H.; Lass, R. W.; Gervert, O.; Wolf, J. *Organometallics* **1997,** *16,* 4077.
(80) Yam, V. W.-W.; Chong, S. H.-F.; Cheung, K.-K. *Chem. Commun.* **1998,** 2121.
(81) Bruce, M. I.; Hall, B. C.; Kelly, B. D.; Low, P. J.; Skelton, B. W.; White, A. H. *J. Chem. Soc. Dalton Trans.* **1999,** 3719.
(82) Le Bozec, H.; Dixneuf, P. H. *Izvest. Akad. Nauk. Ser. Khim.* **1995,** 827; *Russ. Chem. Bull.* **1995,** *44,* 801.

(83) Crescenzi, R.; Lo Sterzo, C. *Organometallics* **1992,** *11,* 4301.
(84) Viola, E.; Lo Sterzo, C.; Crescenzi, R.; Frachey, G. *J. Organomet. Chem.* **1995,** *493,* 55.
(85) Touchard, D.; Pirio, N.; Fettouhi, M.; Ouahab, L.; Dixneuf, P. H. *Organometallics* **1995,** *14,* 5263.
(86) Pirio, N.; Touchard, D.; Dixneuf, P. H.; Fettouhi, M.; Ouahab, L. *Angew. Chem.* **1992,** *104,* 664; *Angew. Chem. Int. Ed. Engl.* **1992,** *31,* 651.
(87) Weng, W.; Bartik, T.; Brady, M.; Bartik, B.; Ramsden, J. A.; Arif, A. M.; Gladysz, J. A. *J. Am. Chem. Soc.* **1995,** *117,* 11922.
(88) (a) Bruce, M. I.; Swincer, A. G. *Adv. Organomet. Chem.* **1983,** *22,* 59. (b) Bruce, M. I. *Chem. Rev.* **1991,** *91,* 197.
(89) Péron, D.; Romero, A.; Dixneuf, P. H. *Gazz. Chim. Ital.* **1994,** *124,* 497.
(90) Romero, A.; Peron, D.; Dixneuf, P. H. *J. Chem. Soc. Chem. Commun.* **1990,** 1410.
(91) Bruce, M. I. *Chem. Rev.* **1998,** *98,* 2797.
(92) Bruce, M. I.; Hinterding, P.; Low, P. J.; Skelton, B. W.; White, A. H. *Chem. Commun.* **1996,** 1009.
(93) Winter, R. F.; Hornung, F. M. *Organometallics* **1997,** *16,* 4248.
(94) Winter, R. F.; Hornung, F. M. *Organometallics* **1999,** *18,* 4005.
(95) Haquette, P.; Touchard, D.; Toupet, L.; Dixneuf, P. H. *J. Organomet. Chem.* **1998,** *565,* 63.
(96) Lang, H.; Blau, S.; Pritzkow, H.; Zsolnai, L. *Organometallics* **1995,** *14,* 1850.
(97) (a) Hayashi, Y.; Osawa, M.; Kobayashi, K.; Wakatsuki, Y. *Chem. Commun.* **1996,** 1617. (b) Hayashi, Y.; Osawa, M.; Wakatsuki, Y. *J. Organomet. Chem.* **1997,** *542,* 241. (c) Hayashi, Y.; Osawa, M.; Wakatsuki, Y. in *Hyper-Structures of Molecules,* Sasabe, H. (ed.), Gordon & Breach: Amsterdam, 1999, p. 35.
(98) Lang, H.; Weber, C. *Organometallics* **1995,** *14,* 4415.
(99) Lang, H.; Wu, I.-Y.; Weinmann, S.; Weber, C.; Nuber, B. *J. Organomet. Chem.* **1997,** *541,* 157.
(100) Oberthur, M.; Hillebrand, G.; Arndt, P.; Kempe, R. *Chem. Ber./Receuil.* **1997,** *130,* 789.
(101) Roth, G.; Fischer, H. *Organometallics* **1996,** *15,* 1139.
(102) Roth, G.; Fischer, H.; Meyer-Friedrichsen, T.; Heck, J.; Houbrechts, S.; Persoons, A. *Organometallics* **1998,** *17,* 1511.
(103) Roth, G.; Fischer, H. *Organometallics* **1996,** *15,* 5766.
(104) Bruce, M. I.; Ke, M.; Low, P. J.; Skelton, B. W.; White, A. H. *Organometallics* **1998,** *17,* 3539.
(105) Yamamoto, J. H.; Low, P. J.; Carty, A. J. unpublished results.
(106) Huang, T.-K.; Chi, Y.; Peng, S.-M.; Lee, G.-H.; Wang, S.-L.; Liao, F.-L. *Organometallics* **1995,** *14,* 2164.
(107) Moreno, C.; Gómez, J. L.; Medina, R.-M.; Macazaga, M.-J.; Arnanz, A.; Lough, A.; Farrar, D. H.; Delgado, S. *J. Organomet. Chem.* **1999,** *579,* 63.
(108) Cambridge, J.; Choudhury, A.; Friend, J.; Garg, R.; Hill, G.; Hussain, Z. I.; Lovett, S. M.; Whiteley, M. W. *J. Organomet. Chem.* **1999,** *577,* 249.
(109) Bruce, M. I.; Ke, M.; Low, P. J. *Chem. Commun.* **1996,** 2405.
(110) Bruce, M. I.; Smith, M. E.; Hall, B. C. unpublished work.
(111) Bruce, M. I.; Smith, M. E. unpublished work.
(112) Chi, Y. personal communication; see also Shiu, C.-W.; Su, C.-J.; Pin, C.-W.; Chi, Y.; Peng, P. S.-M.; Lee, G.-H. *J. Organomet. Chem.* **1997,** *545–546,* 151.
(113) Yamamoto, J. H.; Low, P. J.; Chi, Y.; Carty, A. J. unpublished results.
(114) Weng, W.; Bartik, T.; Gladysz, J. A. *Angew. Chem.* **1994,** *106,* 2272; *Angew. Chem. Int. Ed. Engl.* **1994,** *33,* 2199.
(115) Akita, M.; Chung, M.-C.; Sakurai, A.; Sugimoto, S.; Terada, M.; Tanaka, M.; Moro-oka, Y. *Organometallics* **1997,** *16,* 4882.
(116) Hartbaum, C.; Fischer, H. *J. Organomet. Chem.* **1999,** *578,* 186.
(117) Sun, Y.; Taylor, N. J.; Carty, A. J. *J. Organomet. Chem.* **1992,** *423,* C43.

(118) Sun, Y.; Taylor, N. J.; Carty, A. J. *Organometallics* **1992,** *11,* 4293.
(119) Bruce, M. I.; Hall, B. C.; Kelly, B. D.; Low, P. J.; Skelton, B. W.; White, A. H. *J. Chem. Soc. Dalton Trans.* **1999,** 3719.
(120) Peron, D.; Romero, A.; Dixneuf, P. H. *Organometallics* **1995,** *14,* 3319.
(121) Touchard, D.; Dixneuf, P. H. *Coord. Chem. Rev.* **1998,** *178–180,* 409.
(122) Dahlenburg, L.; Weiss, A.; Bock, M.; Zahl, A. *J. Organomet. Chem.* **1997,** *541,* 465.
(123) Lebreton, C.; Touchard, D.; Le Pichon, L.; Daridor, A.; Toupet, L.; Dixneuf, P. H. *Inorg. Chim. Acta* **1998,** *272,* 188.
(124) Touchard, D.; Haquette, P.; Daridor, A.; Toupet, L.; Dixneuf, P. H. *J. Am. Chem. Soc.* **1994,** *116,* 11157.
(125) Touchard, D.; Haquette, P.; Pirio, N.; Toupet, L.; Dixneuf, P. H. *Organometallics* **1993,** *12,* 3132.
(126) Gevert, O.; Wolf, J.; Werner, H. *Organometallics* **1996,** *15,* 2806.
(127) Kovacik, I.; Laubender, M.; Werner, H. *Organometallics* **1997,** *16,* 5607.
(128) Kim, P. J.; Masai, H.; Sonogashira, K.; Hagihara, N. *Inorg. Nucl. Chem. Lett.* **1970,** *6,* 181.
(129) Gallagher, J. F.; Butler, P.; Manning, A. R. *Acta Crystallogr.* **1998,** *C54,* 342.
(130) Fujikura, Y.; Sonogashira, K.; Hagihara, N. *Chem. Lett.* **1975,** 1067.
(131) Lewis, J.; Lin, B.; Raithby, P. R. *Trans. Met. Chem.* **1995,** *20,* 569.
(132) Sonogashira, K.; Ohga, K.; Takahashi, S.; Hagihara, N. *J. Organomet. Chem.* **1980,** *188,* 237.
(133) Bruce, M. I.; Hall, B. C.; Low, P. J.; Skelton, B. W.; White, A. H. unpublished work.
(134) Harriman, A.; Hissler, M.; Ziessel, R.; De Cian, A.; Fisher, J. *J. Chem. Soc., Dalton Trans* **1995,** 4067.
(135) Curtis, R. F.; Taylor, J. A. *J. Chem. Soc. C* **1971,** 186.
(136) Rubin, Y.; Knobler, C. B.; Diederich, F. *J. Am. Chem. Soc.* **1990,** *112,* 1607.
(137) (a) Vicente, J.; Chicote, M. T.; Cayuelas, J. A.; Fernandez-Baeza, J.; Jones, P. G.; Sheldrick, G. M.; Espinet, P. *J. Chem. Soc. Dalton Trans* **1985,** 1163. (b) Vicente, J.; Chicote, M.-T. *Inorg. Synth.* **1998,** *32,* 172.
(138) Mori, S.; Iwakura, H.; Takechi, S. *Tetrahedron Lett.* **1988,** *29,* 5391.
(139) Krieger, M.; Gould, R. O.; Neumüller, B.; Harms, K.; Dehnicke, K. *Z. Anorg. Allg. Chem.* **1998,** *624,* 1434.
(140) (a) Jones, G. E.; Kendrick, D. A.; Holmes, A. B. *Org. Synth.* **1987,** *65,* 52. (b) Rubin, Y.; Lin, S. S.; Knobler, C. B.; Antony, J.; Boldi, A. M.; Diederich, F. *J. Am. Chem. Soc.* **1991,** *113,* 6943.
(141) Aizoh, S.; Akita, M.; Moro-oka, Y. *Organometallics* **1999,** *18,* 3241.
(142) LePichon, L. Ph.D. Thesis, Université de Rennes I, 1999.
(143) Bartik, B.; Dembinski, R.; Bartik, T.; Arif, A. M.; Gladysz, J. A. *New J. Chem.* **1997,** *21,* 739.
(144) Dembinski, R.; Lis, T.; Szafert, S.; Mayne, C. L.; Bartik, T.; Gladysz, J. A. *J. Organomet. Chem.* **1999,** *578,* 229.
(145) Chodkiewicz, W.; Cadiot, P. *C.R. Acad. Sci. Fr. Ser. C* **1955,** *241,* 1055.
(146) Bartik, T.; Bartik, B.; Brady, M.; Dembinski, R.; Gladysz, J. A. *Angew. Chem.* **1996,** *108,* 467; *Angew. Chem. Int. Ed. Engl.* **1996,** *35,* 414.
(147) Peters, T. B.; Bohling, J. C.; Arif, A. M.; Gladysz, J. A. *Organometallics* **1999,** *18,* 3261.
(148) Leroux, F.; Stumpf, R.; Fischer, H. *Eur. J. Inorg. Chem.* **1998,** *37,* 1225.
(149) March, J. *Advanced Organic Chemistry;* Wiley: New York, 4th ed., 1992; pp. 21–22.
(150) Tanimoto, M.; Kuchitsu, K.; Morino, Y. *Bull. Chem. Soc. Jpn* **1971,** *44,* 386.
(151) Diederich, F.; Rubin, Y. *Angew. Chem.* **1992,** *104,* 1123; *Angew. Chem. Int. Ed. Engl.* **1992,** *31,* 1101.
(152) Goroff, N. S. *Acc. Chem. Res.* **1996,** *29,* 77.
(153) Marder, T. B. personal communication.
(154) Marder, T. B.; Lesley, G.; Yaun, Z.; Fyfe, H. B.; Chow, P.; Stringer, G.; Jobe, I. R.; Taylor, N. J.; Williams, I. D.; Kurtz, S. K. *Materials for Nonlinear Optics. Chemical Perspectives;* ACS Symp. Ser. 455, 1991, Ch. 40.

(155) Greenfield, H.; Sternberg, H. W.; Friedel, R. A.; Wotiz, J. H.; Markby, R.; Wender, I. *J. Am. Chem. Soc.* **1956,** *78,* 120.
(156) Tilney-Bassett, J. F. *J. Chem. Soc.* **1961,** 577.
(157) Lichtenberger, D. L.; Renshaw, S. K.; Wong, A.; Tagge, C. D. *Organometallics* **1993,** *12,* 3522.
(158) Schager, F.; Bonrath, W.; Pörschke, K.-R.; Kessler, M.; Krüger, C.; Seevogel, K. *Organometallics* **1997,** *16,* 4276.
(159) Ajayi-Obe, T.; Armstrong, E. M.; Baker, P. K.; Prakash, S. *J. Organomet. Chem.* **1994,** *468,* 165.
(160) Hübel, W.; Merenyi, R. *Chem. Ber.* **1963,** *96,* 930.
(161) Ustynyuk, N. A.; Vinogradova, V. N.; Korneva, V. N.; Kravtsov, D. N.; Andrianov, V. G.; Struchkov, Yu. T. *J. Organomet. Chem.* **1984,** *277,* 285.
(162) Magnus, P.; Becker, D. P. *J. Chem. Soc. Chem. Commun.* **1985,** 640.
(163) Choukroun, R.; Donnadieu, B.; Malfant, I.; Haubrich, S.; Frantz, R.; Guerin, C.; Henner, B. *Chem. Commun.* **1997,** 2315.
(164) Rodewald, D.; Schulzke, C.; Rehder, D. *J. Organomet. Chem.* **1995,** *498,* 29.
(165) Garcia-Yebra, C.; Carrero, F.; Lopez-Mardomingo, C.; Fajardo, M.; Rodriguez, A.; Antinolo, A.; Otero, A.; Lucas, D.; Mugnier, Y. *Organometallics* **1999,** *18,* 1287.
(166) Castellano, B.; Solari, E.; Floriani, C.; Re, N.; Chiesi-Villa, A.; Rizzoli, C. *Chem. Eur. J.* **1999,** *5,* 722. See also Floriani, C.; Floriani-Moro, R., *Adv. Organomet. Chem.,* **2001,** *47,* 167.
(167) Kersting, M.; Dehnicke, K.; Fenske, D. *J. Organomet. Chem.* **1986,** *309,* 125.
(168) Stahl, K.; Weller, F.; Dehnicke, K. *Z. Anorg. Allg. Chem.* **1984,** *518,* 175.
(169) Stahl, K.; Dehnicke, K. *J. Organomet. Chem.* **1986,** *316,* 85.
(170) Werth, A.; Dehnicke, K.; Fenske, D.; Baum, G. *Z. Anorg. Allg. Chem.* **1990,** *591,* 125.
(171) Templeton, J. L.; Ward, B. C. *J. Am. Chem. Soc.* **1980,** *102,* 3288.
(172) Bruce, M. I.; Low, P. J.; Werth, A.; Skelton, B. W.; White, A. H. *J. Chem. Soc. Dalton Trans* **1996,** 1551.
(173) Lin, J. T.; Yang, M.-F.; Tsai, C.; Wen, Y.S. *J. Organomet. Chem.* **1998,** *564,* 257.
(174) Ruffolo, R.; Kainz, S.; Gupta, H. K.; Brook, M. A.; McGlinchey, M. J. *J. Organomet. Chem.* **1997,** *547,* 217.
(175) Lang, H.; Blau, S.; Rheinwald, G.; Zsolnai, L. *J. Organomet. Chem.* **1995,** *494,* 65.
(176) Alcock, N. W.; Hill, A. F.; Melling, R. P.; Thompsett, A. R. *Organometallics* **1993,** *12,* 641.
(177) Klein, H. F.; Schwind, M.; Flörke, U.; Haupt, H.-J. *Inorg. Chim. Acta* **1993,** *207,* 79.
(178) (a) Dickson, R. S.; Fraser, P. J. *Adv. Organomet. Chem.* **1974**, 12, 323. (b) Kemmitt, R.D.W.; Russell, D.R. *Comprehensive Organometallic Chemistry,* Wilkinson, G.; Stone, F. G. A.; Abel, E. W., Eds.; Pergamon: Oxford, 1982, Vol. 5, ch. 34, p. 192.
(179) Nicholas, K. M. *Acc. Chem. Res.* **1987,** *20,* 207.
(180) Karpov, M. G.; Tunik, S. P.; Denisov, V. R.; Starova, G. L.; Nikol'skii, A. B.; Dolgushin, F. M.; Yanovsky, A. I.; Rybinskaya, M. I. *J. Organomet. Chem.* **1995,** *485,* 219.
(181) Pannell, K. H.; Crawford, G. M. *J. Coord Chem.* **1973,** *2,* 251.
(182) Dickson, R. S.; Tailby, G. R. *Aust. J. Chem.* **1969,** *22,* 1143.
(183) Lewis, J.; Lin, B.; Khan, M. S.; Al-Mandhury, R. A.; Raithby, P. R. *J. Organomet. Chem.* **1994,** *484,* 161.
(184) Draper, S. M.; Delamesiere, M.; Champeil, E.; Turamley, B.; Byrne, J. J.; Long, C. *J. Organomet. Chem.* **1999,** *589,* 157.
(185) Peyronel, G.; Ragni, A.; Trogu, E. F. *Gazz. Chim. Ital.* **1967,** *97,* 1327.
(186) Diederich, F.; Rubin, Y.; Chapman, O. L.; Goroff, N. S. *Helv. Chim. Acta* **1994,** *77,* 1441.
(187) Johnson, B. F. G.; Lewis, J.; Raithby, P. R.; Wilkinson, D. A. *J. Organomet. Chem.* **1991,** *408,* C9.
(188) Lang, H.; Blau, A.; Rheinwald, G. *J. Organomet. Chem.* **1995,** *492,* 81.
(189) Bruce, M. I.; Skelton, B. W.; Smith, M. E.; White, A. H. *Aust. J. Chem.* **1999,** *52,* 431.

(190) Duffy, N.; McAdam, J.; Nervi, C.; Osella, D.; Ravera, M.; Robinson, B.; Simpson, J. *Inorg. Chem. Acta* **1996,** *247,* 99.
(191) Brook, M. A.; Ramacher, B.; Dallaire, C.; Gupta, H. K.; Ulbrich, D.; Ruffolo, R. *Inorg. Chim. Acta* **1996,** *250,* 49.
(192) McAdam, C. J.; Duffy, N. W.; Robinson, B. H.; Simpson, J. *Organometallics* **1996,** *15,* 3935.
(193) Bruce, M. I.; Low, P. J. unpublished results.
(194) Low, P. J.; Rousseau, R.; Lam, P.; Udachin, K. A.; Enright, G. D.; Tse, J. S.; Wayner, D. D. M.; Carty, A. J. *Organometallics* **1999,** *18,* 3885.
(195) Yang, K.; Martin, J. A.; Bott, S. G.; Richmond, M. G. *Organometallics* **1996,** *15,* 2227.
(196) Constable, E. C.; Housecroft, C. E.; Johnston, L. A. *Inorg. Chem. Commun.* **1998,** *1,* 68.
(197) Bruce, M. I.; Halet, J.-F.; Kahal, S.; Low, P. J.; Skelton, B. W.; White, A. H. *J. Organomet. Chem.* **1999,** *578,* 155.
(198) Isobe, M.; Hosokawa, S.; Kira, K. *Chem. Lett.* **1996,** 473.
(199) Chen, X.-N.; Zhang, J.; Wu, S.-L.; Yin, Y.-Q.; Wang, W.-L.; Sun, J. *J. Chem. Soc. Dalton Trans* **1999,** 1987.
(200) (a) Davies, J. E.; Hope-Weeks, L. J.; Mays, M. J.; Raithby, P. R. *Chem. Commun.* **2000,** 1411. (b) Mays, M. J. private communication.
(201) Werner, H.; Gevert, O.; Haquette, P. *Organometallics* **1997,** *16,* 803.
(202) (a) Schwab, P.; Werner, H. *J. Chem. Soc. Dalton Trans* **1994,** 3415, (b) Werner, H.; Schwab, P.; Heinemann, A.; Steinert, P. *J. Organomet. Chem.* **1995,** *496,* 207.
(203) Heyns, J. B. B.; Stone, F. G. A. *J. Organomet. Chem.* **1978,** *160,* 337.
(204) Bonrath, W.; Pörschke, K.-R.; Wilke, G.; Angermund, K.; Krüger, C. *Angew. Chem.* **1988,** *100,* 853; *Angew. Chem. Int. Ed. Engl.* **1988,** *27,* 833.
(205) Maekawa, M.; Munakata, M.; Kuroda-Sawa, T.; Hachiya, K. *Polyhedron* **1995,** *14,* 2879.
(206) Rosenthal, U.; Pulst, S.; Arndt, P.; Baumann, W.; Tillack, A.; Kempe, R. *Z. Naturforsch.* **1995,** 50b, 368; 377.
(207) Rosenthal, U.; Pulst, S.; Kempe, R.; Pörschke, K.-R.; Goddard, R.; Proft, B. *Tetrahedron* **1998,** *54,* 1277.
(208) (a) Klein, H.-F.; Beck-Hemetsberger, H.; Reitzel, L.; Rodenhauser, B.; Cordier, G. *Chem. Ber.* **1989,** *122,* 43. (b) Klein, H.-F.; Petermann, A. *Inorg. Chim. Acta* **1997,** *261,* 187. (c) Klein, H.-F.; Heiden, M.; He, M.; Jung, T.; Röhr, C. *Organometallics* **1997,** *16,* 2003.
(209) Tripathy, P. B.; Renoe, B. W.; Adzamli, K.; Roundhill, D. M. *J. Am. Chem. Soc.* **1971,** *93,* 4406.
(210) Babaeva, A. B.; Beresneva, T. I.; Kharitonov, Yu.Ya. *Dokl. Akad. Nauk SSSR* **1967,** *175,* 591.
(211) Adams, C. J.; Bruce, M. I.; Horn, E.; Tiekink, E. R. T. *J. Chem. Soc. Dalton Trans* **1992,** 1157.
(212) Yamazaki, S.; Deeming, A. J.; Speel, D. M. *Organometallics* **1998,** *17,* 775.
(213) Maekawa, M.; Munakata, M.; Kuroda-Sawa, T.; Hachiya, K. *Inorg. Chim. Acta* **1995,** *231,* 213; 236, 181.
(214) Maekawa, M.; Munataka, M.; Kuroda-Sowa, T.; Hachiya, K. *Inorg. Chim. Acta* **1995,** *233,* 1.
(215) Casey, C. P.; Chung, S.; Ha, Y.; Powell, D. R. *Inorg. Chim. Acta* **1997,** *265,* 127.
(216) Tilney-Bassett, J. F. *J. Chem. Soc.* **1963,** 4784.
(217) Choukroun, R.; Lorber, C.; Donnadieu, B.; Henner, B.; Frantz, R.; Guerin, C. *Chem. Commun.* **1999,** 1099.
(218) Rubin, Y.; Knobler, C. B.; Diederich, F. *J. Am. Chem. Soc.* **1990,** *112,* 4966.
(219) Adams, R. D.; Bunz, U. H. F.; Fu, W.; Nguyen, L. *J. Organomet. Chem.* **1999,** *578,* 91.
(220) Guo, L.; Hrabusa III, J. M.; Tessier, C. A.; Youngs, W. J.; Lattimer, R. *J. Organomet. Chem.* **1999,** *578,* 43.
(221) Dosa, P. I.; Erben, C.; Iyer, V. S.; Vollhardt, K. P. C.; Wasser, I. M. *J. Am. Chem. Soc.* **1999,** *121,* 10430.
(222) Butler, G.; Eaborn, C.; Pidcock, A. *J. Organomet. Chem.* **1981,** *210,* 403.
(223) Hoffmann, R. *Tetrahedron* **1966,** *22,* 521.

(224) Daran, J. C.; Gilbert, E.; Gouygou, M.; Halut, S.; Heim, B.; Jeannin, Y. *J. Cluster Sci.* **1994,** *5,* 373.
(225) Bruce, M. I. *Comprehensive Organometallic Chemistry;* Wilkinson, G.; Stone, F. G. A.; Abel, E. W., Eds.; Pergamon, Exeter, 1982; Vol. 4, ch. 32.5, p. 843.
(226) Deeming, A. J. *Comprehensive Organometallic Chemistry II;* Abel, E. W.; Stone, F. G. A.; Wilkinson, G., Eds.; Elsevier, 1995; Vol. 7, ch. 12, p. 683.
(227) Smith, A. K. *Comprehensive Organometallic Chemistry II;* Abel, E. W.; Stone, F. G. A.; Wilkinson, G., Eds.; Elsevier, 1995; Vol. 7, ch. 13, p. 747.
(228) Tunik, S. P.; Grachova, E. V.; Denisov, V. R.; Starova, G. L.; Nikol'skii, A. B.; Dolgushin, F. M.; Yanovsky, A. I.; Struchkov, Yu. T. *J. Organomet. Chem.* **1997,** *536–537,* 339.
(229) Koridze, A. A.; Zdanovich, V. I.; Andrievskaya, N. V.; Siromakhova, Yu.; Petrovski, P. V.; Ezernitskaya, M. G.; Dolgushin, F. M.; Yanovsky, A. I.; Struchkov, Yu. T. *Izv. Akad. Nauk, Ser. Khim.* **1996,** 1261; *Russ. Chem. Bull.* **1996,** *45,* 1200.
(230) Bruce, M. I.; Zaitseva, N. N.; Skelton, B. W.; White, A. H. *Inorg. Chim. Acta* **1996,** *250,* 129.
(231) Bruce, M. I.; Zaitseva, N. N.; Skelton, B. W.; White, A. H. *J. Organomet. Chem.* **1997,** *536–537,* 93.
(232) Bruce, M. I.; Skelton, B. W.; White, A. H.; Zaitseva, N. N. *Inorg. Chem. Commun.* **1998,** *1,* 134.
(233) Bruce, M. I.; Skelton, B. W.; White, A. H.; Zaitseva, N. N. *Aust. J. Chem.* **1996,** *49,* 155.
(234) Lau, C. S.-W.; Wong, W.-T. *J. Chem. Soc. Dalton Trans.* **1999,** 2511.
(235) Bruce, M. I.; Zaitseva, N. N.; Skelton, B. W.; White, A. H. *Izv. Akad. Nauk, Ser. Khim.* **1998,** 1012; *Russ. Chem. Bull.* **1998,** *47,* 983.
(236) Bruce, M. I.; Skelton, B. W.; White, A. H.; Zaitseva, N. N. *J. Organomet. Chem.* **1998,** *558,* 197.
(237) Lavigne, G.; Nombel, P.; Lugan, N. unpublished work cited in Lavigne, G. *Eur. J. Inorg. Chem.* **1999,** 917.
(238) Corrigan, J. F.; Doherty, S.; Taylor, N. J.; Carty, A. J. *Organometallics* **1993,** *12,* 1365.
(239) Eichele, K.; Wasylishen, R. E.; Corrigan, J. F.; Taylor, N. J.; Carty, A. J. *J. Am. Chem. Soc.* **1995,** *117,* 6961.
(240) Corrigan, J. F.; Doherty, S.; Taylor, N. J.; Carty, A. J. *Organometallics* **1992,** *11,* 3160.
(241) Corrigan, J. F.; Taylor, N. J.; Carty, A. J. *Organometallics* **1994,** *13,* 3778.
(242) Ang, H. G.; Ang, S. G.; Du, S. *J. Chem. Soc. Dalton Trans.* **1999,** 2963.
(243) Blenkiron, P.; Corrigan, J. F.; Pilette, D.; Taylor, N. J.; Carty, A. J. *Can. J. Chem.* **1996,** *74,* 2349.
(244) Blenkiron, P.; Enright, G. D.; Low, P. J.; Corrigan, J. F.; Taylor, N. J.; Chi, Y.; Saillard, J.-Y.; Carty, A. J. *Organometallics* **1998,** *17,* 2447.
(245) Bobbie, B. J.; Taylor, N. J.; Carty, A. J. *Chem. Commun.* **1991,** 1511.
(246) Blenkiron, P.; Taylor, N. J.; Carty, A. J. *J. Chem. Soc. Chem Commun.* **1995,** 327.
(247) Bruce, M. I.; Zaitseva, N. N.; Skelton, B. W.; White, A. H. *Aust. J. Chem.* **1998,** *51,* 165.
(248) Bruce, M. I.; Low, P. J.; Zaitseva, N. N.; Kahlal, S.; Halet, J.-F.; Skelton, B. W.; White, A. H. *J. Chem. Soc. Dalton Trans.* **2000,** 2939.
(249) Bruce, M. I.; Skelton, B. W.; White, A. H.; Zaitseva, N. N. *J. Chem. Soc. Dalton Trans.* **1996,** 3151.
(250) MacLaughlin, S. A.; Taylor, N. J.; Carty, A. J. *Organometallics* **1984,** *3,* 392.
(251) Adams, C. J.; Bruce, M. I.; Skelton, B. W.; White, A. H. *J. Organomet. Chem.* **1999,** *589,* 213.
(252) Aime, S.; Deeming, A. J. *J. Chem. Soc. Dalton Trans.* **1983,** 1807.
(253) Aime, S.; Tiripicchio, A.; Tiripicchio-Camellini, M.; Deeming, A. J. *Inorg. Chem.* **1981,** *20,* 2027.
(254) Deeming, A. J.; Felix, M. S. B.; Bates, P. A.; Hursthouse, M. B. *J. Chem. Soc. Chem. Commun.* **1987,** 461.
(255) Deeming, A. J.; Felix, M. S. B.; Nuel, D. *Inorg. Chim. Acta* **1993,** *213,* 3.
(256) Worth, G. H.; Robinson, B. H.; Simpson, J. *Organometallics* **1992,** *11,* 501.

(257) Huang, T.-K.; Chi, Y.; Peng, S.-M.; Lee, G.-H. *Organometallics* **1999,** *18,* 1675.
(258) Waterman, S. M.; Humphrey, M. G.; Tolhurst, V.-A.; Bruce, M. I.; Low, P. J.; Hockless, D. C. R. *Organometallics* **1998,** *17,* 5789.enlargethispage-9pt
(259) Chung, M.-C.; Sakurai, A.; Akita, M.; Moro-oka, Y. *Organometallics* **1999,** *18,* 4684.
(260) Adams, C. J.; Bruce, M. I.; Horn, E.; Skelton, B. W.; Tiekink, E. R. T.; White, A. H. *J. Chem. Soc. Dalton Trans.* **1993,** *3299,* 3313.
(261) Bruce, M. I.; Zaitseva, N. N.; Skelton, B. W.; White, A. H. *Polyhedron* **1995,** *14,* 2647.
(262) Carty, A. J.; Hogarth, G.; Enright, G.; Frapper, G. *Chem. Commun.* **1997,** 1883.
(263) Davies, J. E.; Mays, M. J.; Raithby, P. R.; Sarveswaran, K. *Angew. Chem.* **1997,** *109,* 2784; *Angew. Chem. Int. Ed. Engl.* **1997,** *36,* 2668.
(264) Chi, Y.; Carty, A. J.; Blenkiron, P.; Delgado, E.; Enright, G. D.; Wang, W.; Peng, S.-M.; Lee, G.-H. *Organometallics* **1996,** *15,* 5269.
(265) Delgado, E.; Chi, Y.; Wang, W.; Hogarth, G.; Low, P. J.; Enright, G. D.; Peng, S.-M.; Lee, G.-H.; Carty, A. J. *Organometallics* **1998,** *17,* 2936.
(266) Low, P. J.; Udachin, K. A.; Enright, G. D.; Carty, A. J. unpublished work.
(267) Jeannin, S.; Jeannin, Y.; Robert, F.; Rosenberger, C. *Inorg. Chem.* **1994,** *33,* 243.
(268) Bruce, M. I.; Koutsantonis, G. A.; Tiekink, E. R. T. *J. Organomet. Chem.* **1991,** *407,* 391.
(269) Yasufuku, K.; Yamazaki, H. *Bull. Chem. Soc. Jpn.* **1972,** *45,* 2664.
(270) Deeming, A. J.; Donovan-Mtunzi, S.; Hardcastle, K. I. *J. Chem. Soc. Dalton Trans.* **1986,** 543.
(271) Carty, A. J.; Taylor, N. J.; Smith, W. F. *J. Chem. Soc. Chem. Commun.* **1979,** 750.
(272) Bruce, M. I.; Zaitseva, N. N.; Skelton, B. W.; White, A. H. *J. Cluster Sci.* **1996,** *7,* 109.
(273) Low, P. J.; Udachin, K. A.; Enright, G. D.; Carty, A. J. unpublished results.
(274) Etches, S. J.; Hart, I. J.; Stone, F. G. A. *J. Chem. Soc. Dalton Trans.* **1989,** 2281.
(275) Lotz, S.; van Rooyen, P. H.; Meyer, R. *Adv. Organomet. Chem.* **1995,** *37,* 219.
(276) Varga, V.; Mach, K.; Hiller, J.; Thewalt, U. *J. Organomet. Chem.* **1996,** *506,* 109.
(277) Hayashi, Y.; Osawa, M.; Kobayashi, K.; Sato, T.; Sato, M.; Wakatsuki, Y. *J. Organomet. Chem.* **1998,** *569,* 169.
(278) Snaith, T. J.; Low, P. J. unpublished work.
(279) Blenkiron, P.; Pilette, D.; Corrigan, J. F.; Taylor, N. J.; Carty, A. J. *J. Chem. Soc. Chem. Commun.* **1995,** 2165.
(280) Cherkas, A. A.; Randall, L. H.; MacLaughlin, S. A.; Mott, G. N.; Taylor, N. J.; Carty, A. J. *Organometallics* **1988,** *7,* 969.
(281) Blenkiron, P.; Enright, G. D.; Taylor, N. J.; Carty, A. J. *Organometallics* **1996,** *15,* 2855.
(282) Forsyth, C. M.; Nolan, S. P.; Stern, C. L.; Marks, T. J.; Rheingold, A. L. *Organometallics* **1993,** *12,* 3618.
(283) Evans, W. J.; Keyer, R. A.; Ziller, J. W. *Organometallics* **1993,** *12,* 2618.
(284) Evans, W. J.; Keyer, R. A.; Ziller, J. W. *Organometallics* **1990,** *9,* 2628.
(285) Nolan, S. P.; Stern, D.; Marks, T. J. *J. Am. Chem. Soc.* **1989,** *111,* 7844.
(286) Evans, W. J.; Keyer, R. A.; Zhang, H.; Atwood, J. L. *J. Chem. Soc. Chem. Commun.* **1987,** 837.
(287) Heeres, H. J.; Nijhoff, J.; Teuben, J. H.; Rogers, R. D. *Organometallics* **1993,** *12,* 2609.
(288) Lee, L.; Berg, D. J.; Bushnell, G. W. *Organometallics* **1995,** *14,* 5021.
(289) Ohff, A.; Pulst, S.; Lefeber, C.; Peulecke, N.; Arndt, P.; Burlakov, V. V.; Rosenthal, U. *Synlett* **1996,** 111.
(290) Pellny, P.-M.; Burlakov, V. V.; Baumann, W.; Spannenberg, A.; Rosenthal, U. *Z. Anorg. Allg. Chem.* **1999,** *625,* 910.
(291) Rosenthal, U.; Pellny, P.-M.; Kirchbauer, F. G.; Burlakov, V. V. *Acc. Chem. Res.* **2000,** *33,* 119.
(292) Takagi, K.; Rousset, C. J.; Negishi, E. *J. Am. Chem. Soc.* **1991,** *113,* 1440.
(293) (a) Choukroun, R.; Cassoux, P. *Acc. Chem. Res.* **1999,** *32,* 494. (b) Choukroun, R.; Zhao, J.; Lorber, C.; Cassoux, P.; Donnadieu, B. *Chem. Commun.* **2000,** 1511.

(294) Rosenthal, U.; Ohff, A.; Michalik, M.; Görls, H.; Burlakov, V. V.; Shur, V. B. *Angew. Chem.* **1993,** *105,* 1228; *Angew. Chem. Int. Ed. Engl.* **1993,** *32,* 1193.
(295) Rosenthal, U.; Ohff, A.; Baumann, W.; Kempe, R.; Tillack, A.; Burlakov, V. V. *Organometallics* **1994,** *13,* 2903.
(296) Metzler, N.; Nöth, H. *J. Organomet. Chem.* **1993,** *454,* C5.enlargethispage-9pt
(297) Rosenthal, U.; Ohff, A.; Baumann, W.; Kempe, R.; Tillack, A.; Burlakov, V. V. *Angew. Chem.* **1994,** *106,* 1678; *Angew. Chem. Int. Ed. Engl.* **1994,** *33,* 1605.
(298) Kirmaier, L.; Weidenbruch, M.; Marsmann, H.; Peters, K.; von Schnering, H. G. *Organometallics* **1998,** *17,* 1237.
(299) (a) Ostendorf, D.; Kirmaier, L.; Saak, W.; Marsmann, H.; Weidenbruch, M. *Eur. J. Inorg. Chem.* **1999,** 2301. (b) Ostendorf, D.; Saak, W.; Weidenbruch, M.; Marsmann, H. *Organometallics,* **2000,** *14,* 4938.
(300) Pellny, P.-M.; Peubecke, N.; Burlakov, V. V.; Baumann, W.; Spannenberg, A.; Rosenthal, U. *Organometallics* **2000,** *19,* 1198.
(301) Pellny, P.-M.; Burlakov, V. V.; Peulecke, N.; Baumann, W.; Spannenberg, A.; Kempe, R.; Francke, V.; Rosenthal, U. *J. Organomet. Chem.* **1999,** *578,* 125.
(302) Pavan Kumar, P. N. V.; Jemmis, E. D. *J. Am. Chem. Soc.* **1988,** *110,* 125.
(303) Jemmis, E. D.; Giju, K. T. *Angew. Chem.* **1997,** *109,* 633; *Angew. Chem. Int. Ed. Engl.* **1997,** *36,* 606.
(304) Jemmis, E. D.; Giju, K. T. *J. Am. Chem. Soc.* **1998,** *120,* 6952.
(305) Burlakov, V. V.; Ohff, A.; Lefeber, C.; Tillack, A.; Baumann, W.; Kempe, R.; Rosenthal, U. *Chem. Ber.* **1995,** *128,* 967.
(306) Rosenthal, U.; Ohff, A.; Tillack, A.; Baumann, W.; Görls, H. *J. Organomet. Chem.* **1994,** *468,* C4.
(307) Pulst, S.; Arndt, P.; Heller, B.; Baumann, W.; Kempe, R.; Rosenthal, U. *Angew. Chem.* **1996,** *108,* 1175; *Angew. Chem. Int. Ed. Engl.* **1996,** *35,* 1112.
(308) Teuben, J. H.; de Liefde Meier, H. J. *J. Organomet. Chem.* **1969,** *17,* 87.
(309) Sekutowski, D. G.; Stucky, G. D. *J. Am. Chem. Soc.* **1976,** *98,* 1376.
(310) Sebald, A.; Fritz, P.; Wrackmeyer, B. *Spectrochim. Acta* **1985,** *41A,* 1405.
(311) Burlakov, V. V.; Peulecke, N.; Baumann, W.; Spannenberg, A.; Kempe, R.; Rosenthal, U. *J. Organomet. Chem.* **1997,** *536–537,* 293.
(312) (a) Wood, G. L.; Knobler, C. B.; Hawthorne, M. F. *Inorg. Chem.* **1989,** *28,* 382. (b) Rosenthal, U.; Görls, H. *J. Organomet. Chem.* **1992,** *439,* C6.
(313) Lang, H.; Seyferth, D. *Z. Naturforsch.* **1990,** *45b,* 212.
(314) Rosenthal, U.; Görls, H. *J. Organomet. Chem.* **1992,** *439,* C36.
(315) Erker, G.; Frömberg, W.; Benn, R.; Mynott, R.; Angermund, K.; Krüger, C. *Organometallics* **1989,** *8,* 911.
(316) Hsu, D. P.; Davis, W. M.; Buchwald, S. L. *J. Am. Chem. Soc.* **1993,** *115,* 10394.
(317) Rosenthal, U.; Pulst, S.; Arndt, P.; Ohff, A.; Tillack, A.; Baumann, W.; Kempe, R.; Burlakov, V. V. *Organometallics* **1995,** *14,* 2961.
(318) Kempe, R.; Ohff, A.; Rosenthal, U. *Z. Kristallogr.* **1995,** *210,* 707.
(319) Erker, G.; Frömberg, W.; Mynott, R.; Gabor, B.; Krüger, C. *Angew. Chem.* **1986,** *98,* 456; *Angew. Chem. Int. Ed. Engl.* **1986,** *25,* 463.
(320) Lang, H.; Keller, H.; Imhof, W.; Martin, S. *Chem. Ber.* **1990,** *123,* 417.
(321) Heshmatpour, F.; Wocadlo, S.; Massa, W.; Dehnicke, K. *Acta Crystallogr.* **1995,** *C51,* 2225.
(322) Pulst, S.; Arndt, P.; Heller, B.; Baumann, W.; Kempe, R.; Rosenthal, U. *Angew. Chem.* **1996,** *108,* 1175; *Angew. Chem. Int. Ed. Engl.* **1996,** *35,* 1112.
(323) Pulst, S.; Kirchbauer, F. G.; Heller, B.; Baumann, W.; Rosenthal, U. *Angew. Chem.* **1998,** *120,* 2029; *Angew. Chem. Int. Ed.* **1998,** *37,* 1925.

(324) Pellny, P.-M.; Burlakov, V. V.; Baumann, W.; Spannenberg, A.; Kempe, R.; Rosenthal, U. *Organometallics* **1999,** *18,* 2906.
(325) Pellny, P.-M.; Peulecke, N.; Burlakov, V. V.; Tillack, A.; Baumann, W.; Spannenberg, A.; Kempe, R.; Rosenthal, U. *Angew. Chem.* **1997,** *109,* 2728; *Angew. Chem. Int. Ed. Engl.* **1997, 36,** 2615.
(326) Cuenca, T.; Gomez, R.; Gomez-Sal, P.; Rodriguez, G. M.; Royo, P. *Organometallics* **1992,** *11,* 1229.
(327) Lang, H.; Blau, S.; Nuber, B.; Zsolnai, L. *Organometallics* **1995,** *14,* 3216.
(328) Cano, A.; Cuenca, T.; Galakhov, M.; Rodriguez, G. M.; Royo, P.; Cardin, C. J.; Convery, M. A. *J. Organomet. Chem.* **1995,** *493,* 17.
(329) Ashe III, A. J.; Al-Ahmad, S.; Kampf, J. W. *Organometallics* **1999,** *18,* 4234.
(330) Pellny, P.-M.; Kirchbauer, F. G.; Burlakov, V. V.; Baumann, W.; Spannenberg, A.; Rosenthal, U. *Chem. Eur. J.* **2000,** *6,* 81.
(331) Troyanov, S. I.; Varga, V.; Mach, K. *Organometallics* **1993,** *12,* 2820.
(332) Pellny, P.-M.; Kirchbauer, F. G.; Burlakov, V. V.; Spannenberg, A.; Mach, K.; Rosenthal, U. *Chem. Commun.* **1999,** 2505.
(333) Pellny, P.-M.; Kirchbauer, F. G.; Burlakov, V. V.; Baumann, W.; Spannenberg, A.; Rosenthal, U. *J. Am. Chem. Soc.* **1999,** *121,* 8313.
(334) Horacek, M.; Stepnicka, P.; Gyepes, R.; Cisarova, I.; Polasek, M.; Mach, K.; Pellny, P.-M.; Burlakov, V. V.; Baumann, W.; Spannenberg, A.; Rosenthal, U. *J. Am. Chem. Soc.* **1999,** *121,* 10638.
(335) Pellny, P.-M.; Burlakov, V. V.; Baumann, W.; Spannenberg, A.; Horácek, M.; Stepnicka, P.; Mach, K.; Rosenthal, U. *Organometallics* **2000,** *19,* 2816.
(336) Schottek, J.; Erker, G.; Fröhlich, R. *Eur. J. Inorg. Chem.* **1998,** 551.
(337) Danjoy, C.; Zhao, J.; Donnadieu, B.; Legros, J.-P.; Valade, L.; Choukroun, R.; Zwick, A.; Cassoux, P. *Chem. Eur. J.* **1998,** *4,* 1100.
(338) Pulst, S.; Arndt, P.; Baumann, W.; Tillack, A.; Kempe, R.; Rosenthal, U. *J. Chem. Soc. Chem. Commun.* **1995,** 1753.
(339) Röttger, D.; Erker, G. *Angew. Chem.* **1997,** *109,* 840; *Angew. Chem. Int. Ed. Engl.* **1997,** *36,* 813.
(340) Ahlers, W.; Temme, B.; Erker, G.; Fröhlich, R.; Zippel, F. *Organometallics* **1997,** *16,* 1440.
(341) Ahlers, W.; Erker, G.; Fröhlich, R.; Peuchert, U. *Chem. Ber./Recueil.* **1997,** *130,* 1069.
(342) Choukroun, R.; Donnadieu, B.; Zhao, J.-S.; Cassoux, P.; Lepetit, C.; Silvi, B. *Organometallics* **2000,** *19,* 1901.
(343) Temme, B.; Erker, G.; Fröhlich, R.; Grehl, M. *Angew. Chem.* **1994,** *106,* 1570; *Angew. Chem. Int. Ed. Engl.* **1994,** *33,* 1480.
(344) Erker, G.; Röttger, D. *Angew. Chem.* **1993,** *105,* 1691; *Angew. Chem. Int. Ed. Engl.* **1993,** *32,* 1623.
(345) Kempe, R.; Pulst, S.; Spannenberg, A.; Rosenthal, U. *Z. Kristallogr. New Cryst. Struct.* **1998,** *213,* 215.
(346) Takahashi, T.; Aoyagi, K.; Denisov, V.; Suzuki, N.; Choueiry, D.; Negishi, E. *Tetrahedron Lett.* **1993,** *34,* 8301.
(347) Takahashi, T.; Xi, Z.; Nishihara, Y.; Huo, S.; Kasai, K.; Aoyagi, K.; Denisov, V.; Negishi, E. *Tetrahedron* **1997,** *53,* 9123.
(348) Liu, Y.; Sun, W.-H.; Nakajima, K.; Takahashi, T. *Chem. Commun.* **1998,** 1133.
(349) Takahashi, T.; Xi, Z.; Obora, Y.; Suzuki, N. *J. Am. Chem. Soc.* **1995,** *117,* 2665.
(350) Pettersen, R. C.; Levenson, R. A. *Acta Crystallogr.* **1976,** *B32,* 723.
(351) Young III, F. R.; O'Brien, D. H.; Pettersen, R. C.; Levenson, R. A.; von Minden, D. L. *J. Organomet. Chem.* **1976,** *114,* 157.
(352) Pettersen, R. C.; Cihonsky, J. L.; Young III, F. R.; Levenson, R. A. *Chem. Commun.* **1975,** 370.
(353) Pettersen, R. C.; Cash, G. G. *Inorg. Chim. Acta* **1979,** *34,* 261.

(354) Iyoda, M.; Kuwatani, Y.; Oda, M.; Tatsumi, K.; Nakamura, A. *Angew. Chem.* **1991,** *103,* 1697; *Angew. Chem. Int. Ed. Engl.* **30,** 1670.
(355) MacLaughlin, S. A.; Doherty, S.; Taylor, N. J.; Carty, A. J. *Organometallics* **1992,** *11,* 4315.
(356) Mathur, P.; Hossain, Md.M.; Rheingold, A. L. *J. Organomet. Chem.* **1996,** *507,* 187.
(357) Mathur, P.; Dash, A. K.; Hossain, Md. M.; Satyanarayana, C. V. V.; Rheingold, A. L.; Liable-Sands, L. M.; Yap, G. P. A. *J. Organomet. Chem.* **1997,** *532,* 189.
(358) Mathur, P.; Dash, A. K. *J. Cluster Sci.* **1998,** *9,* 131.
(359) Mathur, P.; Dash, A. K.; Hossain, Md.M.; Satyanarayana, C. V. V.; Vergjese, B. *J. Organomet. Chem.* **1996,** *506,* 307.
(360) Mathur, P.; Hossain, Md.M.; Datta, S. N.; Kondru, R.-K.; Bhadbhade, M. M. *Organometallics* **1994,** *13,* 2532.
(361) Yamazaki, S.; Taira, Z. *J. Organomet. Chem.* **1999,** *578,* 61.
(362) Fritch, J. R.; Vollhardt, K. P. C.; Thompson, M. R.; Day, V. W. *J. Am. Chem. Soc.* **1979,** *101,* 2768.
(363) Fritch, J. R.; Vollhardt, K. P. C. *J. Am. Chem. Soc.* **1978,** *100,* 3643.
(364) Fritch, J. R.; Vollhardt, K. P. C. *Organometallics* **1982,** *1,* 590.
(365) Fritch, J. R.; Vollhardt, K. P. C. *Angew. Chem.* **1979,** *91,* 439; *Angew. Chem. Int. Ed. Engl.* **1979,** *18,* 409.
(366) Ohkubo, A.; Aramaki, K.; Nishihara, H. *Chem. Lett.* **1993,** 271.
(367) Shimura, T.; Ohkubo, A.; Aramaki, K.; Uekusa, H.; Fujita, T.; Ohba, S.; Nishihara, H. *Inorg. Chim. Acta* **1995,** *230,* 215.
(368) Fujita, T.; Uekusa, H.; Ohkubo, A.; Shimura, T.; Aramaki, K.; Nishihara, H.; Ohba, S. *Acta Crystallogr.* **1995,** *C51,*.
(369) Müller, E. *Synthesis* **1974,** 761.
(370) Müller, E.; Beissner, C.; Jäkle, H.; Langer, E.; Muhm, H.; Odenigbo, G.; Sauerbier, M.; Segnitz, A.; Streichfuss, D.; Thomas, R. *Liebigs Ann. Chem.* **1971,** *754,* 64.
(371) Sakurai, H.; Nakadaira, Y.; Hosomi, A.; Eriyama, Y.; Hirama, K.; Kabuto, C. *J. Am. Chem. Soc.* **1984,** *106,* 8315; Sakurai, H.; Fujii, T.; Sakamoto, K. *Chem. Lett.* **1992,** 339.
(372) Werner, H.; Baum, M.; Schneider, D.; Windmüller, B. *Organometallics* **1993,** *13,* 1089.
(373) (a) Connelly, N. G.; Orpen, A. G.; Rieger, A. L.; Rieger, P. H.; Scott, C. J.; Rosair, G. M. *J. Chem. Soc. Chem. Commun.* **1992,** 1293. (b) Bartlett, I. M.; Connelly, N. G.; Martín, A. J.; Orpen, A. G.; Paget, T. J.; Rieger, A. L.; Rieger, P. H. *J. Chem. Soc. Dalton Trans.* **1999,** 691.
(374) Guillaume, V.; Thominot, P.; Coat, F.; Mari, A.; Lapinte, C. *J. Organomet. Chem.* **1998,** *565,* 75.
(375) Romero, A.; Vegas, A.; Dixneuf, P. H. *Angew. Chem.* **1990,** *102,* 210; *Angew. Chem. Int. Ed. Engl.* **1990,** *29,* 215.
(376) Romero, A.; Vegas, A.; Dixneuf, P. H. *Anal. Quim. Int. Ed.* **1996,** *92,* 299.
(377) Winter, R.; Hoznung, F. M. *Organometallics* **1997,** *16,* 4248.
(378) Winter, R. F. *Chem. Commun.* **1998,** 2209.
(379) Winter, R. F. *Eur. J. Inorg. Chem.* **1999,** 2121.
(380) Pirio, N.; Touchard, D.; Toupet, L.; Dixneuf, P. H. *Chem. Comm.* **1991,** 980.
(381) Bruce, M. I.; Hinterding, P.; Ke, M.; Low, P. J.; Skelton, B. W.; White, A. H. *J. Chem. Soc. Chem. Commun.* **1997,** 715.
(382) Bruce, M. I.; Hinterding, P.; Low, P. J.; Skelton, B. W.; White, A. H. *J. Chem. Soc. Dalton Trans.* **1998,** 467.
(383) Bruce, M. I.; Ke, M.; Kelly, B. D.; Low, P. J.; Smith, M. E.; Skelton, B. W.; White, A. H. *J. Organomet. Chem.* **1999,** *590,* 184.
(384) (a) Gotzig, J.; Otto, H.; Werner, H. *J. Organomet. Chem.* **1985,** *287,* 247. (b) Field, L. D.; Messerle, B. A.; Smernik, R. J.; Hambley, T. W.; Turner, P. *J. Chem. Soc. Dalton Trans.* **1999,** 2557 and references cited therein.

(385) Schäfer, M.; Mahr, N.; Wolf, J.; Werner, H. *Angew. Chem.* **1993,** *105,* 1377; *Angew. Chem. Int. Ed. Engl.* **1993,** *32,* 1315.
(386) Ferrence, G. M.; McDonald, R.; Takats, J. *Angew. Chem.* **1999,** *111,* 2372; *Angew. Chem. Int. Ed.* **1999,** *38,* 2233.
(387) Hill, A. F.; Melling, R. P. *J. Organomet. Chem.* **1990,** *396,* C22.
(388) Hill, A. F.; Harris, M. C. J.; Melling, R. P. *Polyhedron* **1992,** *11,* 781.
(389) Xia, H. P.; Yeung, R. C. Y.; Jia, G. *Organometallics* **1998,** *17,* 4763.
(390) Dobson, A.; Moore, D. S.; Robinson, S. D.; Hursthouse, M. B.; New, L. *Polyhedron* **1985,** *4,* 1119.
(391) Dobson, A.; Moore, D. S.; Robinson, S. D. *J. Organomet. Chem.* **1979,** *177,* C8.
(392) Hill, A. F.; Wilton-Ely, J. D. E. T. *J. Chem. Soc. Dalton Trans.* **1998,** 3501.
(393) Barbaro, P.; Bianchini, C.; Peruzzini, M.; Polo, A.; Zanobini, F.; Frediani, P. *Inorg. Chim. Acta* **1994,** *220,* 5.
(394) Beckhaus, R.; Sang, J.; Wagner, T.; Ganter, B. *Organometallics* **1996,** *15,* 1176.
(395) Beckhaus, R.; Sang, J.; Englert, U.; Bohme, U. *Organometallics* **1996,** *15,* 4731.
(396) Carney, M. J.; Walsh, P. J.; Hollander, F. J.; Bergman, R. G. *J. Am. Chem. Soc.* **1989,** *111,* 8751.
(397) Huy, N. H. T.; Ricard, L.; Mathey, F. *Organometallics* **1997,** *16,* 4501.
(398) Wang, B.; Nguyen, K. A.; Srinivas, G. N.; Watkins, C. L.; Menzer, S.; Spek, A. L.; Lammertsma, K. *Organometallics* **1999,** *18,* 796.
(399) Akita, M.; Sakurai, A.; Moro-oka, Y. *Chem. Commun.* **1999,** 101.
(400) Fitch, J. R.; Vollhardt, K. P. C. *Angew. Chem.* **1980,** *92,* 570; *Angew. Chem. Int. Ed. Engl. 19,* 559.
(401) Allison, N. T.; Fritch, J. R.; Vollhardt, K. P. C.; Walborsky, E. C. *J. Am. Chem. Soc.* **1983,** *105,* 1384.
(402) Clauss, A. D.; Shapley, J. R.; Winkler, C. N.; Hoffmann, R. *Organometallics* **1984,** *3,* 619.
(403) Bao, J.; Wulff, W. D.; Fumo, M. J.; Grant, E. B.; Heller, D. P.; Whitcomb, M. C.; Yeung, S.-M. *J. Am. Chem. Soc.* **1996,** *118,* 2166.
(404) Hubert, A. J.; Dale, J. *J. Chem. Soc.* **1965,** 3160.
(405) Chalk, A. J.; Jerussi, R. A. *Tetrahedron Lett.* **1972,** 61.
(406) Takeda, A.; Ohno, A.; Kadota, I.; Gevorgyan, V.; Yamamoto, Y. *J. Am. Chem. Soc.* **1997,** *119,* 4547.
(407) Gevorgyan, V.; Takeda, A.; Yamamoto, Y. *J. Am. Chem. Soc.* **1997,** *119,* 11313.
(408) Lamata, M. P.; San Jose, E.; Carmona, D.; Lahoz, F. J.; Atencio, R.; Oro, L. A. *Organometallics* **1996,** *15,* 4852.
(409) Kusumoto, T.; Hiyama, T. *Tetrahedron Lett.* **1987,** *28,* 1807; *Bull. Chem. Soc. Jpn.* **1990,** *63,* 3103.
(410) Tsutsui, S.; Toyoda, E.; Hamaguchi, T.; Ohshita, J.; Kanetani, F.; Kunai, A.; Naka, A.; Ishikawa, M. *Organometallics* **1999,** *18,* 3792.
(411) Shirakawa, E.; Yamasaki, K.; Yoshida, H.; Hiyama, T. *J. Am. Chem. Soc.* **1999,** *121,* 10221.
(412) Lesley, G.; Nguyen, P.; Taylor, N. J.; Marder, T. B.; Scott, A. J.; Clegg, W.; Norman, N. C. *Organometallics* **1996,** *15,* 5137.
(413) Larock, R. C.; Doty, M. J.; Cacchi, S. *J. Org. Chem.* **1993,** *58,* 4579.
(414) Tillack, A.; Pulst, S.; Baumann, W.; Baudisch, H.; Kortus, K.; Rosenthal, U. *J. Organomet. Chem.* **1997,** *532,* 117.
(415) Siemsen, P.; Livingston, R. C.; Diederich, F. *Angew. Chem. Int. Ed.* **2000,** *39,* 2632.
(416) Brandsma, L. *Preparative Acetylenic Chemistry;* Elsevier: Amsterdam, 1971.
(417) Brandsma, L.; Verkruijsse, H. D. *Synthesis of Acetylenes, Allenes and Cumulenes;* Elsevier: Amsterdam, 1981.
(418) Noyes, A. A.; Turker, C. W. *J. Am. Chem. Soc.* **1897,** *19,* 123.
(419) Taylor, R. J. K.; Ed. *Organocopper Reagents—A Practical Approach;* Oxford UP: Oxford, 1994.

(420) March, J. *Advanced Organic Chemistry;* Wiley: New York, 4th ed., 1992, 715.
(421) Viehe, H. G. *Chemistry of Acetylenes;* Dekker: New York, 1969, p. 597.
(422) Back, S.; Pritzkow, H.; Lang, H. *Organometallics* **1998,** *17,* 41.
(423) Matsuzaka, H.; Hirayama, Y.; Nishio, M.; Mizobe, Y.; Hidai, M. *Organometallics* **1993,** *12,* 36.
(424) Hill, A. F.; Wilton-Ely, J. D. E. T. *Organometallics* **1997,** *16,* 4517.
(425) Bedford, R. B.; Hill, A. F.; Thompsett, A. R.; White, A. J. P.; Williams, D. J. *Chem. Commun.* **1996,** 1059.
(426) Smith, E. H.; Whittall, J. *Organometallics* **1994,** *13,* 5169.
(427) Negishi, E.; Akiyoshi, K.; Takahashi, T. *J. Chem. Soc. Chem. Commun.* **1987,** 477.
(428) Glaser, C. *Ber. Dtsch. Chem. Ges.* **1869,** *2,* 422; *Am. Chem. Pharm.* **1870,** *154,* 137.
(429) Eglinton, G.; Galbraith, A. R. *Chem. Ind. (London)* **1956,** 737.
(430) Hay, A. S. *J. Org. Chem.* **1962,** *27,* 3320.
(431) Drew, M. G. B.; Esho, F. S.; Nelson, S. M. *J. Chem. Soc. Chem. Commun.* **1982,** 1347.
(432) Babudri, F.; Fiandanese, V.; Marchese, G.; Punzi, A. *J. Organomet. Chem.* **1998,** *566,* 251.
(433) Auer, S. M.; Schneider, M.; Baiker, A. *J. Chem. Soc. Chem. Commun.* **1995,** 2057.
(434) Rubin, Y.; Knobler, C. B.; Diederich, F. *Angew. Chem.* **1991,** *103,* 708; *Angew. Chem. Int. Ed. Engl.* **1991,** *30,* 698.
(435) Hopf, H.; Kreitzer, M.; Jones, P. G. *Chem. Ber.* **1991,** *124,* 1471.
(436) Koshunov, S. I.; Haumann, T.; Boese, R.; de Meijere, A. *Angew. Chem.* **1993,** *105,* 426; *Angew. Chem. Int. Ed. Engl.* **1993,** *32,* 401.
(437) Dierks, R.; Armstrong, J. C.; Boese, R.; Vollhardt, K. P. C. *Angew. Chem.* **1986,** *98,* 270; *Angew. Chem. Int. Ed. Engl.* **1986,** *25,* 268.
(438) Boese, R.; Green, J. R.; Mittendorf, J.; Mohler, D. L.; Vollhardt, K. P. C. *Angew. Chem.* **1992,** *104,* 1643; *Angew. Chem. Int. Ed. Engl.* **1992,** *31,* 1643.
(439) Bunz, U. H. F.; Roidl, G.; Altmann, M.; Enkelmann, V.; Shimizu, K. D. *J. Am. Chem. Soc.* **1999,** *121,* 10719.
(440) Beletskaya, I. P. *J. Organomet. Chem.* **1983,** *250,* 551.
(441) Farina, V.; Krishnan, B. *J. Am. Chem. Soc.* **1991,** *113,* 9585.
(442) Altmann, M.; Enkelmann, V.; Beer, F.; Bunz, U. H. F. *Organometallics* **1996,** *15,* 394.
(443) Altmann, M.; Bunz, U. H. F. *Angew. Chem.* **1995,** *107,* 603; *Angew. Chem. Int. Ed. Engl.* **1995,** *34,* 569.
(444) Altmann, M.; Friedrich, J.; Beer, F.; Reuter, R.; Enkelmann, V.; Bunz, U. H. F. *J. Am. Chem. Soc.* **1997,** *119,* 1472.
(445) Altmann, M.; Roidl, G.; Enkelmann, V.; Bunz, U. H. F. *Angew. Chem.* **1997,** *109,* 1133; *Angew. Chem. Int. Ed. Engl.* **1997,** *36,* 1107.
(446) Roidl, G.; Enkelmann, V.; Adams, R. D.; Bunz, U. H. F. *J. Organomet. Chem.* **1999,** *578,* 144.
(447) Schimanke, H.; Gleiter, R. *Organometallics* **1998,** *17,* 275.
(448) Schlögl, K.; Steyrer, W. *Monatsh. Chem.* **1963,** *94,* 376.
(449) Abram, T. S.; Watts, W. E. *Synth. React. Inorg. Met.-Org. Chem.* **1976,** *6,* 31.
(450) Yuan, Z.; Stringer, G.; Jobe, I. R.; Kreller, D.; Scott, K.; Koch, L.; Taylor, N. J.; Marder, T. B. *J. Organomet. Chem.* **1993,** *452,* 115.
(451) Rodriguez, J.-G.; Onate, A.; Martin-Villamil, R. M.; Fonseca, I. *J. Organomet. Chem.* **1996,** *513,* 71.
(452) Shklover, V. E.; Ovchinnikov, Yu. E.; Zamaev, I. A.; Struchkov, Yu. T.; Dement'ev, V. V.; Frunze, T. M.; Antipova, B. A. *J. Organomet. Chem.* **1989,** *378,* 235.
(453) LeVanda, C.; Bechgaard, K.; Cowan, D. O. *J. Org. Chem.* **1976,** *41,* 2700.
(454) Bildstein, B.; Schweiger, M.; Kopacka, H.; Ongania, K.-H.; Wurst, K. *Organometallics* **1998,** *17,* 2414.
(455) Rodriguez, J.-G.; Gayo, M.; Fonseca, I. *J. Organomet. Chem.* **1997,** *534,* 35.
(456) Sato, M.; Iwai, A.; Watanabe, M. *Organometallics* **1999,** *18,* 3208.

(457) Lin, J. T.; Wu, J. J.; Li, C.-S.; Wen, Y. S.; Lin, K.-J. *Organometallics* **1996,** *15,* 5028.

(458) Müller, T. J. J.; Lindner, H. J. *Chem. Ber.* **1996,** *129,* 607.

(459) Müller, T. J. J.; Netz, A.; Ansorge, M.; Schmalzlin, E.; Brauchle, C.; Meerholz, K. *Organometallics* **1999,** *18,* 5066.

(460) Giulieri, F.; Benaim, J. *J. Organomet. Chem.* **1984,** *276,* 367.

(461) Yam, V. W.-W.; Lo, K. K.-W.; Wong, K. M.-C. *J. Organomet. Chem.* **1999,** *578,* 3.

(462) Matsumoto, T.; Kotani, S.; Shiina, K.; Sonogashira, K. *Appl. Organomet. Chem.* **1993,** *7,* 613.

(463) Faust, R.; Diederich, F.; Gramlich, V.; Seiler, P. *Chem. Eur. J.* **1995,** *1,* 111.

(464) AlQuaisi, S. M.; Galat, K. J.; Chai, M.; Ray III, D. G.; Rinaldi, P. L.; Tessier, C. A.; Youngs, W. J. *J. Am. Chem. Soc.* **1998,** *120,* 12149.

(465) Bruce, M. I.; Hall, B. C.; Skelton, B. W.; White, A. H. unpublished work.

(466) Pak, J. J.; Weakley, T. J. R.; Haley, M. M. *Organometallics* **1997,** *16,* 4505.

(467) Funke, F.; de Meijere, A.; cited in Pohl, E.; Kneisel, B. O.; Herbst-Irmer, R.; de Meijere, A.; Funke, F.; Stein, F. *Acta Crystallogr.* **1995,** *C51,* 2503.

(468) Hartbaum, C.; Mauz, E.; Roth, G.; Weissenbach, K.; Fischer, H. *Organometallics* **1999,** *18,* 2619.

(469) Harriman, A.; Ziessel, R. *Coord. Chem. Rev.* **1998,** *171,* 331.

(470) Ciana, L. D.; Haim, A. *J. Heterocyclic Chem.* **1984,** *21,* 607.

(471) Lin, J. T.; Sun, S.-S.; Wu, J. J.; Lee, L.; Lin, K.-J.; Huang, Y. F. *Inorg. Chem.* **1995,** *34,* 2323.

(472) Lin, J. T.; Sun, S.-S.; Wu, J. J.; Liaw, Y.-C.; Lin, K.-J. *J. Organomet. Chem.* **1996,** *517,* 217.

(473) Allan, J. R.; Barrow, M. J.; Beaumont, P. C.; Macindoe, L. A.; Milburn, G. H. W.; Werninck, A. R. *Inorg. Chim. Acta* **1988,** *148,* 85.

(474) Sun, S.-S.; Lees, A. J. *Inorg. Chem.* **1999,** *38,* 4181.

(475) Fujita, M.; Sasaki, O.; Mitsuhashi, T.; Fujita, T.; Yazaki, J.; Yamaguchi, K.; Ogura, K. *Chem. Commun.* **1996,** 1535.

(476) Blake, A. J.; Champness, N. R.; Khlobystov, A.; Lemenovskii, D. A.; Li, W.-S.; Schröder, M. *Chem. Commun.* **1997,** 2027.

(477) Richardson, C.; Steel, P. *Inorg. Chem. Commun.* **1998,** *1,* 260.

(478) Abrahams, B. F.; Hardie, M. J.; Hoskins, B. F.; Robson, R.; Sutherland, E. E. *J. Chem. Soc. Chem. Commun.* **1994,** 1049.

(479) Grosshenny, V.; Ziessel, R. *J. Chem. Soc. Dalton Trans.* **1993,** 817.

(480) Grosshenny, V.; Ziessel, R. *J. Organomet. Chem.* **1993,** *453,* C19.

(481) Grosshenny, V.; Harriman, A.; Ziessel, R. *Angew. Chem.* **1995,** *107,* 2921; *Angew. Chem. Int. Ed. Engl.* **1995,** *34,* 2705.

(482) Grosshenny, V.; Harriman, A.; Gisselbrecht, J.-P.; Ziessel, R. *J. Am. Chem. Soc.* **1996,** *118,* 10315.

(483) (a) Arnold, D. P.; Heath, G. A. *J. Am. Chem. Soc.* **1993,** *115,* 12197. (b) Stranger, R.; McGrady, J. E.; Arnold, D. P.; Lane, I.; Heath, G. A. *Inorg. Chem.* **1996,** *35,* 7791.

(484) Imahori, H.; Tanaka, Y.; Odaka, T.; Sakata, Y. *Chem. Lett.* **1993,** 1215.

(485) Anderson, H. L. *Chem. Commun.* **1999,** 2323.

(486) Vidal-Ferran, A.; Muller, C. M.; Sanders, J. K. *J. Chem. Soc. Chem. Commun.* **1994,** 2657.

(487) Ferri, A.; Polzonetti, G.; Licoccia, S.; Paolesse, R.; Favretto, D.; Traldi, P.; Russo, M. V. *J. Chem. Soc. Dalton Trans.* **1998,** 4063.

(488) Anderson, S.; Anderson, H. L.; Sanders, J. K. M. *Acc. Chem. Res.* **1993,** *26,* 469.

(489) (a) Mackay, L. G.; Wylie, S.; Sanders, J. K. M. *J. Am. Chem. Soc.* **1994,** *116,* 3141. (b) Anderson, H. L.; Sanders, J. K. M. *Angew. Chem.* **1990,** *102,* 1478; *Angew. Chem. Int. Ed. Engl.* **1990,** *29,* 1400.

(490) Walter, C. J.; Anderson, H. L.; Sanders, J. K. M. *J. Chem. Soc. Chem. Commun.* **1993,** 458.

(491) Hoshino, Y.; Hagihara, Y. *Inorg. Chim. Acta* **1999,** *292,* 64.

(492) Kasahara, Y.; Hoshino, Y.; Kajitani, M.; Shimizu, K.; Sato, G. P. *Organometallics* **1992,** *11,* 1968.

(493) Semmelmann, M.; Fenske, D.; Corrigan, J. F. *J. Chem. Soc. Dalton Trans.* **1998,** 2541.
(494) Dudev, T.; Galabov, B. *Spectrochim. Acta* **1997,** *53A,* 2053.
(495) Markwell, R. D.; Butler, I. S.; Kakkar, A. K.; Khan, M. S.; Al-Zakwani, Z. H.; Lewis, J. *Organometallics* **1996,** *15,* 2331.
(496) Eastmond, R.; Johnson, T. R.; Walton, D. R. M. *J. Organomet. Chem.* **1973,** *50,* 87.
(497) Morris, D. G. *The Chemistry of Functional Groups. Supplement C. Acetylene Compounds;* Patai, S.; Rappoport, Z., Eds.; John Wiley: Chichester, 1983; Chapter 25.
(498) Schermann, G.; Grösser, T.; Hampel, F.; Hirsch, A. *Chem. Eur. J.* **1997,** *3,* 1105.
(499) Lagow, R. J.; Kampa, J. J.; Wei, H.-C.; Battle, S. L.; Genge, J. W.; Laude, D. A.; Harper, C. J.; Bau, R.; Stevens, R. C.; Haw, J. F.; Munson, E. *Science* **1995,** *267,* 362.
(500) Jones, E. R. H.; Lee, H. H.; Whiting, M. C. *J. Chem. Soc.* **1960,** 3483.
(501) Eastmond, R.; Johnson, T. R.; Walton, D. R. M. *Tetrahedron* **1972,** *28,* 4601.
(502) Grutter, M.; Wyss, M.; Fulara, J.; Maier, J. P. *J. Phys. Chem. A* **1998,** *102,* 9785.
(503) Johnson, B. F. G.; Kakkar, A. K.; Khan, M. S.; Lewis, J.; Dray, A. E.; Friend, R. H.; Wittmann, F. *J. Mater. Chem.* **1991,** *1,* 485.
(504) Lewis, J.; Khan, M. S.; Kakkar, A. K.; Johnson, B. F. G.; Marder, T. B.; Fyfe, H. B.; Wittmann, F.; Friend, R. H.; Dray, A. E. *J. Organomet. Chem.* **1992,** *425,* 165.
(505) Lichtenberger, D. L.; Gruhn, N. E.; Renshaw, S. K. *J. Molec. Struct.* **1997,** *405,* 79.
(506) John, K. D.; Stoner, T. C.; Hopkins, M. D. *Organometallics* **1997,** *16,* 4948.
(507) Hall, M. B.; Fenske, R. F. *Inorg. Chem.* **1972,** *11,* 768.
(508) Osella, D.; Rossetti, R.; Nervi, C.; Ravera, M.; Moretta, M.; Fiedler, J.; Pospisil, L.; Samuel, E. *Organometallics* **1997,** *16,* 695.
(509) Köhler, K.; Silverio, S. J.; Hyla-Kryspin, I.; Gleiter, R.; Zsolnai, L.; Driess, A.; Huttner, G.; Lang, H. *Organometallics* **1997,** *16,* 4970.
(510) Knight, E. T.; Myers, L. K.; Thompson, M. E. *Organometallics* **1992,** *11,* 3691.
(511) Connelly, N. G.; Pilar Gamasa, M.; Gimeno, J.; Lapinte, C.; Lastra, E.; Maher, J. P.; Le Narvor, N.; Rieger, A. L.; Rieger, P. H. *J. Chem. Soc. Dalton Trans.* **1993,** 2575.
(512) Colbert, M. C. B.; Lewis, J.; Long, N. J.; Raithby, P. R.; White, A. J. P.; Williams, D. J. *J. Chem. Soc. Dalton Trans.* **1997,** 99.
(513) Jones, N. D.; Wolf, M. O.; Giaquinta, D. M. *Organometallics* **1997,** *16,* 1352.
(514) Osella, D.; Milone, L.; Nervi, C.; Ravera, M. *J. Organomet. Chem.* **1995,** *488,* 1.
(515) Osella, D.; Milone, L.; Nervi, C.; Ravera, M. *Eur. J. Inorg. Chem.* **1998,** 1473.
(516) Worth, G. H.; Robinson, B. H.; Simpson, J. *Organometallics* **1992,** *11,* 3863.
(517) Osella, D.; Gambino, O.; Nervi, C.; Ravea, M.; Bertolino, D. *Inorg. Chem. Acta* **1993,** *206,* 155.
(518) Fischer, E. O.; Kalder, H. J.; Köhler, F. H. *J. Organomet. Chem.* **1974,** *81,* C23.
(519) Hart, I. J.; Hill, A. F.; Stone, F. G. A. *J. Chem. Soc. Dalton Trans.* **1989,** 2261.
(520) Manna, J.; Geib, S. J.; Hopkins, M. D. *J. Am. Chem. Soc.* **1992,** *114,* 9199.
(521) Snyder, E. I.; Roberts, J. D. *J. Am. Chem. Soc.* **1962,** *84,* 1582.
(522) John, K. D.; Hopkins, M. D. *Chem. Commun.* **1999,** 589.
(523) Torecki, R.; Schrock, R. R.; Vale, M. G. *J. Am. Chem. Soc.* **1991,** *113,* 3610.
(524) Chakravarty, A. R.; Cotton, F. A. *Inorg. Chim. Acta* **1986,** *113,* 19.
(525) (a) Lin, C.; Ren, T.; Valente, E. J.; Zubkowski, J. D. *J. Chem. Soc., Dalton Trans.* **1998,** 571. (b) Lin, C.; Ren, T.; Valente, E. J.; Zubkowski, J. D. *J. Organomet. Chem.* **1999,** *579,* 114.
(526) (a) Bear, J. L.; Han, B.; Huang, S.; Kadish, K. M. *Inorg. Chem.* **1996,** *35,* 3012. (b) Bear, J. L.; Li, Y.; Han, B.; Van Caemebecke, E.; Kadish, K. M. *Inorg. Chem.* **1997,** *36,* 5449 and references therein.
(527) (a) Stoner, T. C.; Dallinger, R. F.; Hopkins, M. D. *J. Am. Chem. Soc.* **1990,** *112,* 5651. (b) Stoner, T. C.; Schaefer, W. P.; Marsh, R. E.; Hopkins, M. D. *J. Cluster Sci.* **1994,** *5,* 107. (c) Stoner, T. C.; Geib, S. J.; Hopkins, M. D. *J., Am. Chem. Soc.* **1992,** *114,* 4201.
(528) Stoner, T. C.; Geib, S. J.; Hopkins, M. D. *Angew. Chem.* **1993,** *105,* 457; *Angew. Chem. Int. Ed. Engl.* **1993,** *32,* 409.

(529) John, K. D.; Miskowski, V. M.; Vance, M. A.; Dallinger, R. F.; Wang, L. C.; Geib, S. J.; Hopkins, M. D. *Inorg. Chem.* **1998,** *37,* 6858.
(530) Low, P. J.; Bruce, M. I. in preparation.
(531) Baddley, W. H.; Panattoni, C.; Bandoli, G.; Clemente, D. A.; Belluco, U. *J. Am. Chem. Soc.* **1971,** *93,* 5590.
(532) Kergoat, R.; Kubicki, M. M.; Gomes de Lima, L. C.; Scordia, H.; Guerchais, J. E.; L'Haridon, P. *J. Organomet. Chem.* **1989,** *367,* 143.
(533) Kergoat, R.; Gomes de Lima, L. C.; Jegat, C.; Le Berre, N.; Kubicki, M. M.; Guerchais, J. E.; L'Haridon, P. *J. Organomet. Chem.* **1990,** *389,* 71.
(534) Zhou, Y.; Arif, A. M.; Miller, J. S. *Chem. Commun.* **1996,** 1881.
(535) Bratsch, S. *J. Chem. Ed.* **1988,** *65,* 223.
(536) Kiplinger, J. L.; Arif, A. M.; Richmond, T. G. *Inorg. Chem.* **1995,** *34,* 399.
(537) Dickson, R. S.; Yawney, D. B. W. *Aust. J. Chem.* **1968,** *21,* 1077.
(538) (a) McClure, G. L.; Baddley, W. H. *J. Organomet. Chem.* **1971,** *27,* 155. (b) Kirchner, R. M.; Ibers, J. A. *J. Am. Chem. Soc.* **1973,** *95,* 1095.
(539) McClure, G. L.; Baddley, W. H. *J. Organomet. Chem.* **1970,** *25,* 261.
(540) (a) Herberich, G. E.; Barlage, W. *J. Organomet. Chem.* **1987,** *331,* 63. (b) Herberich, G. E.; Englert, U.; Hoeveler, M.; Savvopoulos, I. *J. Organomet. Chem.* **1990,** *399,* 35.
(541) (a) Scordia, H.; Kergoat, R.; Kuibicki, M. M.; Guerchais, J. E. *J. Organomet. Chem.* **1983,** *249,* 371. (b) Kubicki, M. M.; Kergoat, R.; Guerchais, J. E.; L'Haridon, P. *J. Chem. Soc. Dalton Trans.* **1984,** 1791.
(542) Cariou, M.; Etienne, M.; Guerchais, J. E.; Kergoat, R.; Kubicki, M. M. *J. Organomet. Chem.* **1987,** *327,* 393.
(543) Kubicki, M. M.; Kergoat, R.; Scordia, H.; Gomes de Lima, L. C.; Guerchais, J. E.; L'Haridon, P. *J. Organomet. Chem.* **1988,** *340,* 41.
(544) Mayr, A.; Yu, M. P. Y.; Yam, V. W.-W. *J. Am. Chem. Soc.* **1999,** *121,* 1760.
(545) Bickelhaupt, F. M.; Bickelhaupt, F. *Chem. Eur. J.* **1999,** *5,* 162.
(546) Lang, H.; Blau, S.; Rheinwald, G.; Wildermuth, G. *J. Organomet. Chem.* **1995,** *489,* C17.
(547) Corriu, R. J. P.; Devylder, N.; Guerin, C.; Henner, B.; Jean, A. *Organometallics* **1994,** *13,* 3194.
(548) Corriu, R. J. P.; Devylder, N.; Guerin, C.; Henner, B.; Jean, A. *J. Organomet. Chem.* **1996,** *509,* 249.
(549) (a) Pocard, N. L.; Alsmeyer, D. C.; McCreery, R. L.; Neenan, T. X.; Callstrom, M. R. *J. Am. Chem. Soc.* **1992,** *114,* 769. (b) Pocard, N. L.; Alsmeyer, D. C.; McCreery, R. L.; Neenan, T. X.; Callstrom, M. R. *J. Mater. Chem.* **1992,** *2,* 771.
(550) Agh-Atabay, N. M.; Lindsell, W. E.; Preston, P. N.; Tomb, P. J.; Lloyd, A. D.; Rangel-Rojo, R.; Spruce, G.; Wherrett, B. S. *J. Mater. Chem.* **1992,** *2,* 1241.
(551) Bestmann, H. J.; Hadawi, D.; Behl, H.; Brenner, M.; Hampel, F. *Angew. Chem.* **1993,** *105,* 1198; *Angew. Chem. Int. Ed. Engl.* **1993,** *32,* 1205.
(552) Grima, J. N.; Evans, K. E. *Chem. Commun.* **2000,** 1531.
(553) Wang, X.-B.; Ding, C.-F.; Wang, L.-S. *J. Phys. Chem. A* **1997,** *101,* 7699.
(554) (a) Petrie, S.; Becker, H.; Baranov, V.; Bohme, D. K. *Astrophys. J.* **1997,** *476,* 191. (b) Fan, J.; Lou, L.; Wang, L.-S. *J. Chem. Phys.* **1995,** *102,* 2701.
(555) Fischer, H.; Scheck, P. A. *Chem. Commun.* **1999,** 1031.
(556) Davey, A. P.; Cardin, D. J.; Byrne, H. J.; Blau, W. *NATO ASI Ser. Ser E* **1991,** *194,* 391.
(557) Werner, H.; Flugel, R.; Windmüller, B. *Chem. Ber./Recueil.* **1997,** *130,* 493.
(558) Diederich, F. *Chem. Commun.* **2001,** 219.
(559) Yam, V. W.-W. *Chem. Commun.* **2001,** 789.
(560) Lang, H.; George, D. S. A.; Rheinwald, G. *Coord. Chem. Rev.* **2000,** *206–207,* 101.
(561) (a) Brunel, L.; Chaplais, G.; Dutremez, S. G.; Guerin, C.; Henner, B. J. L.; Tomberli, V. *Organometallics* **2000,** *19,* 2516. (b) See also Corriu, R. J. P.; Guerin, C.; Henner, B. J. L.; Jolivet, A. *J. Organomet. Chem.* **1997,** *530,* 39.

(562) Frosch, W.; Back, S.; Müller, H.; Köhler, K.; Driess, A.; Schiemenz, B.; Huttner, G.; Lang, H. *J. Organomet. Chem.* **2001,** *619,* 99.
(563) Moreno, C.; Arnanz, A.; Delgado, S. *Inorg. Chim. Acta* **2001,** *312,* 139.
(564) Szafert, S.; Haquette, P.; Falloon, S. B.; Gladysz, J. A. *J. Organomet. Chem.* **2000,** *604,* 52.
(565) Bear, J. L.; Han, B.; Wu, Z.; Van Caemelbecke, E.; Kadish, K. M. *Inorg. Chem.* **2001,** *40,* 2275.
(566) Yam, V. W.-W.; Cong, S. H.-F.; Ko, C.-C.; Cheung, K.-K. *Organometallics* **2000,** *19,* 5092.
(567) Ursini, C. V.; Dias, G. H. M.; Horner, M.; Bortoluzzi, A. J.; Morigaki, M. K. *Polyhedron* **2000,** *19,* 2261.
(568) Choukroun, R.; Donnadieu, B.; Lorber, C.; Pellny, P.-M.; Baumann, W.; Rosenthal, U. *Chem. Eur. J.* **2000,** *6,* 4505.
(569) Müller, T. J. J.; Netz, A.; Ansorge, M.; Schmälzlin, E.; Bräuchle, C.; Meerholz, K. *Organometallics* **1999,** *18,* 5066.
(570) Kuncheria, J.; Mirza, H. A.; Vittal, J. J.; Puddephatt, R. J. *J. Organomet. Chem.* **2000,** *593–594,* **77.**
(571) Rupp, E.; Huttner, G.; Lang, H.; Heinze, K.; Büchner, M.; Hoveystreydt, E. R. *Eur. J. Inorg. Chem.* **2000,** 1953.
(572) (a) Snaith, T. J.; Low, P. J.; Rousseau, R.; Puschmann, H.; Howard, J. A. K. *J. Chem. Soc. Dalton Trans.* **2001,** 292. (b) See also Low, P. J.; Rousseau, R.; Lam, P.; Udachin, K. A.; Enright, G. D.; Tse, J. S.; Wayner, D. D. M.; Carty, A. J. *Organometallics* **1999,** *18,* 3885.
(573) Marcos, M. L.; Macazaga, M. J.; Medina, R. M.; Moreno, C.; Castro, J. A.; Gomez, J. L.; Delgado, S.; Gonzalez-Velasco, J. *Inorg. Chim. Acta* **2001,** *312,* 249.
(574) Amouri, H.; Da Silva, C.; Malezieux, B.; Andres, R.; Vaissermann, J.; Gruselle, M. *Inorg. Chem.* **2000,** *39,* 5053.
(575) Koridze, A. A.; Zdanovich, V. I.; Yu, V.; Lagunova, I. I.; Ptukhova, F. M.; Dolgushin, *Izv. Akad. Nauk, Ser. Khim.* **2000,** 1324; *Russ. Chem. Bull.* **2000,** *49,* 1321.
(576) Cabeza, J. A.; Moreno, M.; Riera, V.; de J. Rosales-Hoz, M. *Inorg. Chem. Commun.* **2001,** *4,* 57.
(577) Cabeza, J. A.; Grepioni, F.; Moreno, M.; Riera, V. *Organometallics* **2000,** *19,* 5424.
(578) Clarke, L. P.; Davies, J. E.; Raithby, P. R.; Shields, G. P. *J. Chem. Soc. Dalton Trans.* **2000,** 4527.
(579) Adams, C. J.; Clarke, L. P.; Martin-Castro, A. M.; Raithby, P. R.; Shields, G. P. *J. Chem. Soc. Dalton Trans.* **2000,** 4015.
(580) Adams, R. D.; Qu, B. *Organometallics* **2000,** *19,* 2411.
(581) Adams, R. D.; Qu, B. *J. Organomet. Chem.* **2001,** *619,* 271.
(582) Adams, R. D.; Qu, B. *Organometallics* **2000,** *19,* 4090.
(583) Clarke, L. P.; Davies, J. E.; Raithby, P. R.; Rennie, M.-A.; Shields, G. P.; Sparr, E. *J. Organomet. Chem.* **2000,** *609,* 169.
(584) Adams, R. D.; Qu, B. *J. Organomet. Chem.* **2001,** *620,* 303.
(585) Mathur, P.; Ahmed, M. O.; Dash, A. K.; Walawalkar, M. G. *J. Chem. Soc. Dalton Trans.* **1999,** 1795.
(586) Ferrand, V.; Gambs, C.; Derrien, N.; Bolm, C.; Stoeckli-Evans, H.; Süss-Fink, G. *J. Organomet. Chem.* **1997,** *549,* 275.
(587) Farrugia, L. J.; MacDonald, N.; Peacock, R. D. *Polyhedron* **1998,** *17,* 2877.
(588) Bruce, M. I.; Low, P. J.; Zaitseva, V. N.; Kahlal, S.; Halet, J.-F.; Skelton, B. W.; White, A. H. *J. Chem. Soc. Dalton Trans.* **2000,** 2939.
(589) Chen, J.; Guzei, I. A.; Woo, L. K. *Inorg. Chem.* **2000,** *39,* 3715.
(590) Baumann, W.; Pellny, P.-M.; Rosenthal, U. *Mag. Resn. Chem.* **2000,** *38,* 515.
(591) Bredeau, S.; Delmas, G.; Pirio, N.; Richard, P.; Donnadieu, B.; Meunier, P. *Organometallics* **2000,** *19,* 4463.
(592) Pellny, P.-M.; Burlakov, V. V.; Arndt, P.; Baumann, W.; Spannenberg, A.; Rosenthal, U. *J. Am. Chem. Soc.* **2000,** *122,* 6317.

(593) Kupfer, V.; Thewalt, U.; Tislerova, I.; Stepnicka, P.; Gyepes, R.; Kubista, J.; Horacek, M.; Mach, K. *J. Organomet. Chem.* **2001,** *620,* 39.
(594) (a) Ahlers, W.; Erker, G.; Fröhlich, R.; Peuchert, U. *J. Organomet. Chem.* **2000,** *578,* 115. (b) Venne-Dunker, S.; Ahlers, W.; Erker, G.; Fröhlich, R. *Eur. J. Inorg. Chem.* **2000,** 1671.
(595) Xia, H. P.; Yeung, R. C. Y.; Jia, G. *Organometallics* **1998,** *17,* 4762.
(596) Bedford, R. B.; Cazin, C. S. J. *J. Organomet. Chem.* **2000,** *598,* 20.
(597) Takahashi, T.; Tsai, F.-Y.; Li, Y.; Nakajima, K.; Kotora, M. *J. Am. Chem. Soc.* **1999,** *121,* 11093.
(598) Takahashi, T.; Tsai, F.-Y.; Kotora, M. *J. Am. Chem. Soc.* **2000,** *122,* 4994.
(599) Le Paih, J.; Dérien, S.; Özdemir, I.; Dixneuf, P. H. *J. Am. Chem. Soc.* **2000,** *122,* 7400.
(600) Bharathi, P.; Periasamy, M. *Organometallics* **2000,** *19,* 5511.
(601) Frosch, W.; Back, S.; Köhler, K.; Lang, H. *J. Organomet. Chem.* **2000,** *601,* 226.
(602) Ishikawa, T.; Ogawa, A.; Hirao, T. *J. Organomet. Chem.* **1999,** *575,* 76.
(603) Weigelt, M.; Becher, D.; Poetsch, E.; Bruhn, C.; Steinborn, D. *Z. Anorg. Allg. Chem.* **1999,** *625,* 1542.
(604) Laskoski, M.; Roidl, G.; Smith, M. D.; Bunz, U. H. F. *Angew. Chem. Int. Ed.* **2001,** *40,* 1460.
(605) Bunz, U. H. F.; Roidl, G.; Adams, R. D. *J. Organomet. Chem.* **2000,** *600,* 56.
(606) (a) Bildstein, B.; Skibar, W.; Schweiger, M.; Kopacka, H.; Wurst, K. *J. Organomet. Chem.* **2001,** *622,* 135. (b) Bildstein, B.; Schweiger, M.; Angleitner, H.; Kopacka, H.; Wurst, K.; Ongania, K.-H.; Fontani, M.; Zanello, P. *Organometallics* **1999,** *18,* 4286. (c) Bildstein, B. *Coord. Chem. Rev.* **2000,** *206–207,* 369.
(607) Xu, D.; Hong, B. *Angew. Chem. Int. Ed.* **2000,** *39,* 1826.
(608) Maekawa, M.; Konaka, H.; Suenaga, Y.; Kuroda-Sowa, T.; Munakata, M. *J. Chem. Soc. Dalton Trans.* **2000,** 4160.
(609) (a) Maya, E. M.; Vazquez, P.; Torres, T. *Chem. Eur. J.* **1999,** *5,* 2004. (b) Maya, E. M.; Vazquez, P.; Torres, T.; Gobbi, L.; Diederich, F.; Pyo, S.; Echegoyen, L. *J. Org. Chem.* **2000**, *65,* 823.
(610) Sugiura, K.; Fujimoto, Y.; Sakata, Y. *Chem. Commun.* **2000,** 1105.
(611) Tashiro, K.; Aida, T.; Zheng, J.-Y.; Kinbara, K.; Saigo, K.; Sakamoto, S.; Yamaguchi, K. *J. Am. Chem. Soc.* **1999,** *121,* 9477.
(612) Zhang, H.; Lee, A. W. M.; Wong, W.-Y.; Yuen, M. S. M. *J. Chem. Soc. Dalton Trans.* **2000,** 3675.

ADVANCES IN ORGANOMETALLIC CHEMISTRY, VOL. 48

Organoelement Chemistry of Main-Group Porphyrin Complexes

PENELOPE J. BROTHERS*

Department of Chemistry
The University of Auckland
Private Bag 92019, Auckland, New Zealand

I

INTRODUCTION

Porphyrin complexes containing coordinated main group elements have never received the attention enjoyed by their transition metal cousins, perhaps because they lacked their obvious relevance in biology.[1,2] Transition metal porphyrin complexes are now finding applications in areas as diverse as new materials, pharmaceutical agents, and catalysts. The publication in the late 1980s of three reviews dealing specifically with the organometallic chemistry of porphyrin complexes was important as it heralded the establishment of this area as a significant subdiscipline within porphyrin chemistry.[3–5] These reviews focused mainly on transition

*E-mail: p.brothers@auckland.ac.nz

0065-3055/01 $35.00

metals, although some limited advances in main group organometallic porphyrin chemistry were already evident, primarily involving the Groups 13 and 14 metals aluminum and indium, and to a lesser extent gallium, thallium, silicon, germanium, and tin. From these discussions it was apparent that the main group chemistry was unevenly developed, particularly for the light-sensitive Group 14 complexes, and that few structural data were available. Over a decade has elapsed since then, and organometallic porphyrin complexes have again been reviewed, both in a multi-volume series[6] and in a companion review to this one, which focused specifically on the organometallic chemistry of transition metal porphyrin complexes.[7] It is now apparent that significant advances in the organometallic chemistry of main group porphyrins have occurred in the interim, and with the Group 15 elements phosphorus, arsenic, and antimony now also represented in this category, this area has now matured sufficiently to warrant a review in its own right.

This review encompasses compounds containing elements from Groups 13, 14, and 15 coordinated to a porphyrin ligand where an element—carbon bond is also present. One of the key features of the chemistry of the elements from Groups 13–15 is the variation in both size and electronegativity of the elements, encompassing nonmetals, semi-metals, and metallic elements. It is remarkable that a porphyrin ligand with its four nitrogen donors circumscribing a hole of radius close to 2.0 Å can serve to coordinate this diverse set of elements. Currently, these range from boron, which is sufficiently small that two atoms can be accommodated by one porphyrin ligand[8,9] to the heavy elements thallium, lead, and bismuth which reside well out of the N_4 plane in their porphyrin complexes. Group 16 or 17 porphyrins are currently unknown, save for a preliminary report of a tellurium porphyrin.[10] Most of the chemistry discussed in the review will concern complexes specifically containing the porphyrin ligand; and the two most commonly employed ligands, the dianions of octaethylporphyrin (OEP) and tetraphenylporphyrin (TPP) are shown in Fig. 1, along with the porphine skeleton and its numbering scheme. In recent times, related macrocycles (Fig. 1) such as the porphyrin isomer porphycene, corrole (containing one less carbon atom in the skeleton), phthalocyanines, and subphthalocyanines have been investigated,[11] and relevant examples of complexes containing these ligands will be included. The very small element boron is a special case, and boron subphthalocyanine complexes (which contain only three isoindoline rings, Fig. 1) are also featured. The commonly used porphyrin ligands and their abbreviations are given in Table I.

An important recent development in main group porphyrin chemistry has been the improved syntheses and structural characterization of the Group 1 porphyrin complexes $M_2(Por)L_n$ (where M = Li, Na, K; L = coordinating solvent).[12,13] These have been significant in advancing the chemistry of early transition metal porphyrins through their role as precursors to the Groups 3, 4, and 5 metal porphyrin halide complexes via simple salt elimination routes.[14] Surprisingly, despite the similarities often drawn between the halides and oxides of the Groups 3, 4, and

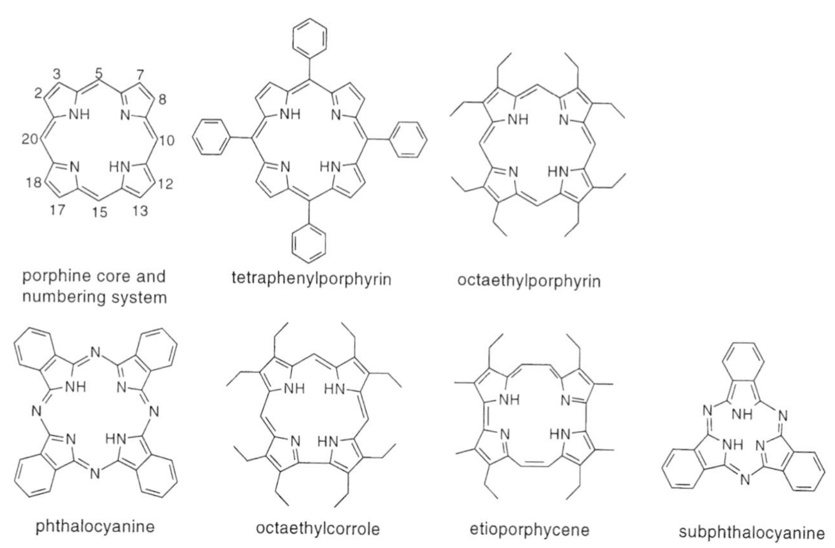

FIG. 1. Porphyrins, tetrapyrrole macrocycles, phthalocyanine, and subphthalocyanine.

5 transition metals and those of the Groups 13, 14, and 15 elements, the Group 1 porphyrins, especially $Li_2(Por)L_n$, have not figured largely in the synthesis of main group porphyrin complexes.

A selection of "inorganometallic" complexes will also be discussed, in which the axial ligand in a Group 13 or 14 porphyrin complex is a transition metal ML_n

TABLE I

ABBREVIATIONS FOR PORPHYRIN DIANIONS AND TETRAPYRROLE ANIONS

Abbreviations for porphyrin dianions	
Por	porphyrin dianion, unspecified
OEP	2,3,7,8,12,13,17,18-octaethylporphyrin
TPP	5,10,15,20-tetraphenylporphyrin
TTP	5,10,15,20-tetra-*p*-tolylporphyrin
OETPP	2,3,7,8,12,13,17,18-octaethyl-5,10,15,20-tetraphenylporphyrin
TAP	5,10,15,20-tetra-*p*-methoxyphenylporphyrin
T*p*ClPP	5,10,15,20-tetra-*p*-chlorophenylporphyrin
TMP	5,10,15,20-tetramesitylporphyrin
Abbreviations for anions of tetrapyrrole macrocycles	
Pc	phthalocyanine (dianion)
SubPc	subphthalocyanine (dianion)
EtioPc	2,7,12,17-tetraethyl-3,6,13,16-tetramethylporphycene (dianion)
OEC	2,3,7,8,12,13,17,18-octaethylcorrole (trianion)
OETAP	octaethyltetraazaporphyrin (dianion)

fragment.[15] In this case the metal–carbon bonds are found in the axial ligands rather than directly associated with the main group element coordinated to the porphyrin. However, they comprise a significant class of main group porphyrin complexes, and the Group 14 examples in particular are interesting because they formally contain Ge^{II}(Por) or Sn^{II}(Por) groups which can be viewed as carbenoid ligands coordinated to the ML_n moiety.

Overall, this review will focus on synthesis, structure, and chemical reactivity. An overview of the porphyrin as a supporting ligand in organometallic complexes was given in the companion review article.[7] The porphyrin ligand is a planar, dianionic macrocycle with four nitrogen donors in a square planar arrangement, with a hole size of radius close to 2.0 Å. This fairly rigid coordination environment generally (but not always) results in *trans* coordination of the two axial ligands in six-coordinate complexes. Structural variations within the 24-atom porphyrin core are possible, with fully planar, saddle-shaped, or ruffled distortions among the common arrangements.[16] Porphyrin complexes have characteristics which are particularly suited to some spectroscopic methods. The macrocycle itself absorbs strongly in the visible and near UV regions. A "normal" UV-visible spectrum exhibits a very intense Soret band in the 400- to 500-nm region, corresponding to the porphyrin π-π^* transition, and less intense Q-bands in the 500- to 700-nm region. Some main group porphyrin complexes show "hyper"-type UV-visible spectra with an additional band in the Soret region, corresponding to a metal a_{2u} (p_z) to porphyrin e_g (π^*) transition. The ratio of intensities of the two bands in the Soret region gives useful information about the electronic properties of the axial ligands. The aromatic ring current associated with the porphyrin macrocycle has an effect on the NMR chemical shifts of axial ligands, which usually exhibit marked upfield shifts in diamagnetic complexes. Electrochemical features of main group porphyrin complexes have been discussed in a recent review.[17] Exhaustive compilations of physical data for organometallic porphyrin complexes have not been included in this review, with the exception of X-ray data. Few crystal structures of organometallic main group porphyrin complexes were available the last time this area was reviewed, and tables collecting selected data for structurally characterized complexes are included. Sketches which use an oval shape to represent the porphyrin ligand are given in the schemes and equations, and a representative group of complete structures are shown, using examples taken from the Cambridge Structural Database. For simplicity, the convention of using "M" to denote the coordinated element in metalloporphyrin complexes has been followed, although several of the elements under discussion are in fact nonmetals (B, Si, P) or semimetals (Ge, As, Sb).

The complexes discussed in the review generally comprise five- or six-coordinate complexes containing σ-alkyl or aryl ligands. This contrast with transition metal organometallic porphyrin complexes which provide, in addition to σ-bonded complexes, examples of carbene and π-complexes (alkene, alkyne, cyclopentadienyl and allyl) and metal—metal multiply bonded dimers. However, this situation does parallel that observed in the broader field of main group chemistry, where the development of both heteronuclear and homonuclear multiply bonded complexes of the heavier elements is only now emerging as an established area,[18] and is not yet reflected in main group porphyrin chemistry. Another significant difference between main group and transition metal porphyrin chemistry is the very limited redox chemistry in the former. In the latter, the energies of the metal *d* orbitals and the frontier orbitals of the porphyrin ligand are often quite close, and the redox chemistry of the porphyrin and the metal become entwined. This is not the case for the main group complexes. For the lighter elements in Groups 13 and 14 (B, Al, Si) only one oxidation state is expected to be readily accessible. Well-characterized, low-valent main group porphyrin complexes have been established only for the heavier elements Tl, Sn, Pb, Sb, and Bi. In general, electrochemical processes are most likely to be restricted to oxidation or reduction at the porphyrin ring, although this is in some cases accompanied by subsequent chemical reactions at the coordinated element. Recent developments in main group porphyrin chemistry have begun to exploit the limited redox chemistry, by harnessing photophysical properties of the porphyrin ligand that are not compromised by redox changes at the central element, a role that has long been played by zinc in transition metal porphyrin chemistry.

The review has been organized using a group by group approach, considering porphyrin complexes of the Groups 13, 14, and 15 elements in turn. Several themes have emerged which cut across these group boundaries. Redox chemistry at the coordinated element is of central importance in transition metal organometallic porphyrin chemistry, and the corresponding chemistry of the main group elements is distinguished by its absence. Electrochemical redox processes are largely confined to the porphyrin ring, and redox chemistry of the central element features only for the heavier elements tin and antimony. All of the organoelement main group porphyrin complexes show electronic absorptions which are shifted to lower energy relative to simple halide or anion-substituted counterparts, reflecting the electron donating character of the alkyl or aryl substituents. The Groups 13 and 14 complexes M(Por)R and M(Por)R_2, respectively, exhibit photolabile M—C bonds which are subject to homolytic cleavage upon irradiation. In contrast the Group 15 porphyrins are inert to photochemical bond cleavage. Small molecule activation by insertion into M—C bonds during photolysis is also observed for the Groups 13 and 14 complexes, notably CO_2 activation by Al(Por)R; CO_2, SO_2, and O_2 activation by Ga(Por)R and O_2 activation for Ge(Por)R_2. Coordination of the small, nonmetallic elements silicon and phosphorus (and to a lesser extent germanium and arsenic) results in ruffling deformations of the porphyrin ligand, and structural and electronic effects relating to this phenomenon are discussed.

II

BORON

Boron, with a covalent radius of 0.85 Å, is too small to coordinate to a porphyrin ligand through all four nitrogen atoms. There are two possible solutions to this problem, either to use a contracted porphyrin-type ligand or to coordinate more than one boron atom to a single porphyrin, and both of these have been realized.

Only two structurally characterized boron porphyrin complexes are known. Both contain two boron atoms per porphyrin ligand, with each boron atom coordinated to two porphyrin nitrogens, representing structural types that are unique in porphyrin chemistry (Fig. 2). B_2OF_2(TpClPP) contains an F—B—O—B—F fragment threaded through the hole in the porphyrin in an asymmetric fashion, with one boron lying approximately in the N_4 plane and the other displaced above it. The related complex $B_2O(OH)_2$(TTP) contains OH groups in place of the F groups.[8] $B_2O_2(BCl_3)_2$(TpClPP) contains a four-membered B_2O_2 ring coordinated in the porphyrin cavity, with the plane of the B_2O_2 ring orthogonal to that of the porphyrin, and the two ring oxygen atoms each coordinated to a BCl_3 acceptor group.[9] Both complexes show remarkable in-plane tetragonal elongation of the porphyrin rings, with one N ··· N distance between adjacent nitrogen atoms over 1.0 Å longer than the other. B_2OF_2(TpClPP) and $B_2O_2(BCl_3)_2$(TpClPP) were prepared from the reactions of $BF_3 \cdot OEt_2$ or $BCl_3 \cdot MeCN$, respectively, with the free base porphyrin in the presence of trace water. The first organoboron porphyrin derivative, B_2O(Ph)(OH)(TTP), was prepared from $PhBCl_2$ with H_2TTP under similar conditions.[19] The complex contains a Ph—B—O—B—OH group with the same arrangement as in B_2OF_2(Por) with the phenyl group attached to the out-of-plane boron atom (Fig. 2).[19] The tetragonal elongation observed in both boron porphyrin structures suggested that a B—B single bonded fragment might be accommodated within the ligand. This occurs in a very new organoboron porphyrin derivative, $B_2{}^nBu_2$(TTP), which represents a further novel structural type. This complex was prepared from the reactive diboron halide B_2Cl_4 with

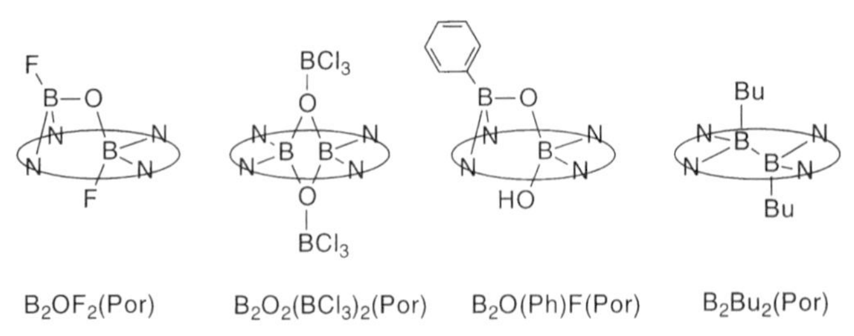

FIG. 2. Structural types for diboron porphyrin complexes.

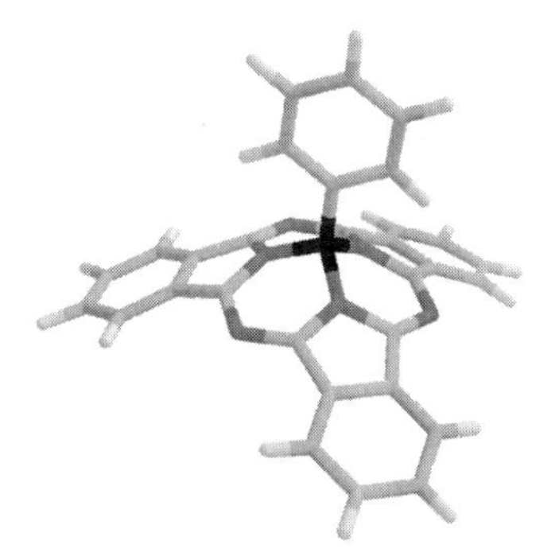

FIG. 3. Molecular structure of phenylboron subphthalocyanine $B(SubPc)(C_6H_5)$.[20]

H_2TPP, followed by treatment with butyl lithium. The B—B distance is 1.71 Å (Fig. 2).[19] These early results indicate a rich chemistry for organoboron porphyrin complexes which should yield further examples without precedent in porphyrin chemistry.

Subphthalocyanines are relatives of phthalocyanines comprised of only three isoindoline groups with boron as the central atom. They have C_{3v} symmetry and a delocalized 14 π electron aromatic system. The simplest SubPc complexes are prepared by the reaction of boron trihalides with phthalonitrile, giving B(SubPc)X.[20,21] One problem with these complexes is their relative insolubility, and this can be alleviated by the use of alkyl substitutents on the periphery.[22] An alternative is to prepare organoboron derivatives by using BPh_3 with phthalonitrile to give B(SubPc)Ph.[21,23] An X-ray crystal structure of this complex shows the characteristic bowl shape and tetrahedral geometry of the boron (Fig. 3).[20] A butyl derivative, $B(SubPc)^nBu$, has been synthesized using Bu_2BBr with phthalonitrile, and a related subnaphthalocyanine complex, B(SubNc)Ph, has also been reported.[21] In addition to experiment, the properties of these complexes have been investigated by molecular orbital calculations.[24]

III
ALUMINUM

Relatively few organometallic aluminum porphyrin complexes have been reported, unlike the heavier Group 13 elements for which more extensive series of compounds have been reported. However, the reaction chemistry of the aluminum porphyrins has been much more extensively studied and exhibits features not replicated by the heavier elements. For this reason the aluminum porphyrin complexes are discussed separately.

A. *Synthesis, Spectroscopy, Structure, and Electrochemistry*

The first organometallic aluminum porphyrin complex to be reported was Al(TPP)Et, formed from the elimination of ethane in the reaction of $AlEt_3$ with H_2TPP in dichloromethane.[25] At the time, this represented a relatively unusual approach to porphyrin metallation, for which the reaction of a metal salt with free base porphyrin in the presence of base was the more usual approach.[1,2] The development of the chemistry of aluminum porphyrins, particularly Al(TPP)Et and Al(TPP)Me, was advanced during the 1980s primarily in the research group of Inoue, who utilized these complexes as initiators in a range of polymerization reactions. It was not until 1990 that a systematic study of the synthesis and properties of organoaluminum porphyrins appeared, including an X-ray crystal structure of Al(OEP)Me.[26] Organoaluminum porphyrin complexes and references to their syntheses are given in Table II.

Eight Al(Por)R complexes (Por = OEP, TPP; R = Me, ^{n}Bu, C_6H_5, C_6F_4H) were prepared from the reaction of Al(Por)Cl with an alkyl or aryllithium reagent.[26] Mass spectra of the compounds showed low intensities for the molecular ion peaks, consistent with facile cleavage of the Al—C bond. The chloride complexes Al(Por)Cl show normal porphyrin UV-visible spectra, whereas the organometallic complexes Al(Por)R show a split Soret band typical of hyperporphyrin spectra.

TABLE II

ALUMINUM, GALLIUM, INDIUM, AND THALLIUM PORPHYRIN COMPLEXES, M(Por)R

Porphyrin	R	Reference
Aluminum		
TPP, OEP	Me, Et, ^{n}Bu, C_6H_5, C_6F_4H	25, 26
Gallium		
TPP, OEP	Me, Et, ^{n}Bu, ^{t}Bu, C_6H_5, *p*-$C_6H_4CH_3$, CH=CHPh, C=CPh	63–65
TPP	$CH_2(CH_2)_3CH{=}CH_2$, CH(*cyclo*-C_5H_9), $CH_2CH{=}CH_2$, $CH_2C(Me){=}CH_2$, C≡CPr	65–67
TAP[a]	$CH{=}CH_2$	67
Indium		
TPP, OEP	Me, Et, ^{i}Pr, ^{n}Bu, ^{t}Bu, C_6H_5, CH=CHPh, C≡CPh, C_6F_5, C_6F_4H	68–70
OETPP	C_6H_5	71
EtioPc	C_6H_5	72, 73
$^{t}Bu_4Pc$, R_8Pc[b]	C_6H_5, C_6F_5, *p*-C_6H_4F, *p*-$C_6H_4CF_3$, *m*-$C_6H_4CF_3$	74
Thallium		
TPP, OEP	Me, norbornenyl, C_6H_5, C_6F_4H, C_6F_5, *p*-C_6H_4OMe	62, 75
TpyP[c]	Me	76

[a] TAP = dianion of tetra-*p*-anisylporphyrin.
[b] R_8Pc, $R = C_5H_{11}$.
[c] TpyP = dianion of tetra-4-pyridylporphyrin.

The ratio of the intensity of the two Soret bands is related to the electron-donating ability of the σ-bonded ligand, and the series showed, not surprisingly, that the n-butyl and C_6F_4H ligands are the strongest and weakest donors, respectively. Overall, the aluminum porphyrins show little coupling between the axial ligand and the properties of the porphyrin ring. For example, for either the OEP or the TPP series the chemical shifts of the porphyrin protons are very similar for Al(Por)Cl and Al(Por)R. The ^{1}NMR spectra of the axial ligands show the marked upfield shifts induced by the porphyrin ring current. An electrochemical study on Al(TPP)R and Al(OEP)R for R = Me, ^{n}Bu and C_6H_5 revealed two reversible reductions at the porphyrin ring, and an irreversible oxidation followed by two reversible oxidations (ECE mechanism). The first oxidation was assigned to the formation of $[Al(Por)R]^+$, which then underwent Al—C cleavage to form $[Al(Por)]^+$ and R·, and the two subsequent reversible oxidations occurred at the porphyrin ring in $[Al(Por)]^+$.[26]

A methylaluminum complex of the porphyrin isomer etioporphycene, Al (EtioPc)Me, has recently been prepared by the reaction of $AlMe_3$ with H_2EtioPc.[27] The overall properties of the complex, including electrochemical behavior, are very similar to those of Al(OEP)Me or Al(TPP)Me. The methyl group chemical shifts in Al(Por)Me appear at −6.48, −6.03, and −5.85 ppm for OEP, TPP, and EtioPc, respectively. The least pronounced upfield chemical shift observed in the EtioPc complex was attributed to a reduced ring current in the porphycene relative to a porphyrin. Both Al(OEP)Me and Al(EtioPc)Me show five-coordinate geometry in their crystal structures (Table III), with the Al atom displaced approximately 0.5 Å above the mean N_4 plane. Despite the rectangular N_4 coordination environment in the porphycene complex the Al—N(av) and Al—C distances are quite similar to those in the porphyrin complex.[26,27]

TABLE III

SELECTED DATA FOR STRUCTURALLY CHARACTERIZED GROUP 13 PORPHYRIN COMPLEXES

	M—C bond length/Å	M—N_{av} bond length/Å	M—N_4 plane/Å	Other		Reference
Al(OEP)(CH_3)	1.942(3)	2.033(3)	0.47	N—Al—C	103.2(5)°	26
Al(EtioPc)(CH_3)	1.960(2)	1.998	0.54			27
Ga(TPP){CH_2(*cyclo*-C_5H_9)}	1.992(6)	2.100	0.58	Ga—C—C	119.2(4)°	66
Ga(TPP)(CH=CH_2)	1.971(5)	2.073	0.53	Ga—C=C	125.0(5)°	67
				C=C	1.301(8) Å	
Ga(TPP)(C≡CPr)	1.949(6)	2.047	0.52	Ga—C≡C	178.8(6)°	65
				C≡C—C	174.8(0)°	
In(TPP)(CH_3)	2.132(15)	2.21	0.78			84
In(EtioPc)(C_6H_5)	2.148(3)	2.174	0.85			73
Tl(TPP)(CH_3)	2.147(12)	2.29	0.98			62
Tl(TPP)(norbornenyl)[a]	2.09	2.29	0.9			62

[a] Data averaged over two independent molecules in cell. Accuracy low.

The methylaluminum complex of a sterically crowded 5,10,15,20-tetrakis (2′-phenylphenyl)porphyrin derivative (which bears a phenyl group on the *ortho* positions of each peripheral phenyl group in TPP) can exist as a number of atropisomers formed by rotation about the porphyrin—phenyl bond. For example, in the $\alpha\alpha\alpha\alpha$ atropisomer all of the pendant phenyl groups are on the same face of the porphyrin, and the aluminum methyl group can then occupy either this face or the opposite, less sterically crowded face. A conformational study on these atropisomers showed, surprisingly, preferential isomerization to the more crowded isomer in which the Al—Me group is on the same face as the pendant phenyl groups. It was proposed that CH-π interactions between the axial methyl group and the pendant phenyl groups were occurring which favored the more hindered isomer. In support of this, the less hindered isomers were favored when the axial methyl group was replaced by Cl, or the pendant phenyl group was replaced by OMe groups.[28]

B. *Chemical and Photochemical Reactivity*

The organoaluminum porphyrins (primarily the methyl and ethyl derivatives) will react with protic reagents HX to eliminate methane or ethane and give Al(Por)X (X = OH, OR, O_2CR, SR). This occurs, for example, for H_2O,[29] HCl,[27] alcohols,[30,31] carboxylic acids,[32,33] thiols,[34] and in an unusual case, with the OH proton in $[M(OEP)(Me)(OH)]^+$ (M = P, As, Sb).[35] The reaction with alcohols has been studied in more detail, and is accelerated by irradiation with visible light, either at the Soret or at the Q-band wavelengths. For example, the reaction of Al(TPP)Et with the hindered alcohol 2,6-di-t-butyl-4-methoxyphenol did not occur at all even after a week in the dark, but proceeded readily upon irradiation. Isosbestic behavior was observed in the UV-visible spectrum, indicating that only two species were involved, Al(TPP)Et and Al(TPP)OAr.[36,37] The reactions with thiols were similarly accelerated.[37] A conclusion from this study was that the reactions are nonhomolytic, and that they occur by nucleophilic attack of the Al—C bond on the protic reagent. This conclusion was based on the fact that in the dark the rates of the reactions increase with more acidic substrates (ROH $<$ ArOH $<$ RCO_2H), and the rates also increase by addition of electron-donating bases such as 1-methylimidazole.[37] These assertions, however, were not directly tested by experiment.

The photochemical acceleration of a variety of other reactions involving Al(Por)R is observed, and as a result, the mechanism of this photochemical activation is important and was the subject of two studies. In the first, a spin trap (tributylnitrosobenzene) was added to Al(TPP)Et in the dark, and slow production of Et· (trapped as ArN(Et)O·) was observed by EPR spectroscopy. Upon irradiation the amount of this product increased and a signal corresponding to the Al—O bonded compound ArNOAl(TPP) was formed, indicating that Al—C bond homolysis was occurring.[38] This was supported by theoretical calculations consistent with initial formation of a photoexcited π-π^* triplet state in the porphyrin ligand, followed by energy transfer from the excited porphyrin π^* orbital to the σ^* orbital of the Al—C bond, resulting in homolytic dissociation.[39]

Aluminum porphyrin enolates are formed from the conjugate addition of Al(TPP)Et to α,β-unsaturated ketones, shown, for example, for t-butyl vinyl ketone in Eq. (1). The aluminum enolate was observed by ^{1}H NMR, with the ^{t}Bu group observed upfield at -1.46 ppm, and the ketone was formed after addition of acid. The reaction was accelerated by irradiation with visible light or, less effectively, accelerated in the dark by the addition of 1-methylimidazole.[34,40] Enolates could also be formed by the conjugate addition of vinyl ketones to Al(TPP)SR, or by hydrogen abstraction from saturated ketones by Al(TPP)NEt_2 [Eqs. (2,3)], although irradiation was not required for these reactions. The high Lewis acidity of aluminum is illustrated by the fact that only O-bound enolates were observed. Stereochemical aspects (*cis/trans* isomerism) of enolate formation were also considered, although the products were characterized only by NMR spectroscopy,[34] and there is scope for this area to be further advanced by a structural study. Al(TPP)Et is active for the living polymerization of cyclosiloxanes. Surprisingly, unlike the polymerization reactions reported here, visible light irradiation was not reported to be required for this reaction.[41]

Et, Al; ^{t}Bu, O; → ^{t}Bu, O, Et, Al; H^+ → ^{t}Bu, O, Et (1)

SPh, Al; ^{t}Bu, O; → ^{t}Bu, O, SPh, Al (2)

NEt_2, Al; Ph, O; → Ph, O, Al + $HNEt_2$ (3)

Another very important visible light-initiated reaction of alkyl aluminum porphyrins is their 1,4-addition to alkyl methacrylates to produce ester enolate species [Eq. (4)]. This enolate then acts as the active species in the subsequent polymerization of the acrylate monomer. For example, Al(TPP)Me acts as a photocatalyst to produce polymethylmethacrylate with a narrow molecular weight distribution in a living polymerization process [Eq. (4)].[42] Visible light is essential for both the initiation step (addition of methylmethacrylate to Al(TPP)Me) and the propagation step (addition of the monomer to the Al(TPP) enolate), although acceleration occurs to different extents. The reaction does not proceed in the dark. The actual mechanistic steps in the monomer addition steps have not been elucidated, and it has been proposed that the polymerization proceeds via a concerted mechanism where approach of the methacrylate monomer to the aluminum center and conjugate addition of the methyl or enolate group simultaneously take place.[42]

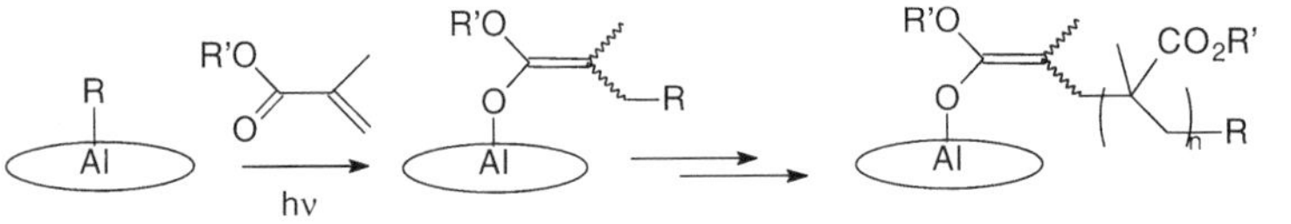

(4)

C. *Alkylaluminum Porphyrins as Initiators for Living Polymerization Processes*

The use of aluminum porphyrins as catalysts for living polymerization reactions is now a sophisticated area. Two recent reviews are available,[43] so only an outline of the process will be given here. The alkylaluminum porphyrins simply act as initiators, undergoing a visible light-activated addition reaction to the first methacrylate monomer, producing an aluminum enolate which then becomes the growing species, adding to further monomers at the Al—O bond. In fact, both aluminum porphyrin enolate and aluminum porphyrin thiolate complexes such as Al(TPP)SPr are more efficient initiators than Al(TPP)R (R = Me, Et) and do not require activation by visible light.[44] Block copolymers can be prepared, by adding epoxide or lactone monomers to the aluminum porphyrin enolate prepolymer. In each case, monomer addition occurs at the aluminum end of the growing chain.[45,46] Catalytic polymerization of lactone or epoxide monomers alone can be initiated by aluminum porphyrins bearing chloride, alkoxide, or thiolate axial ligands, but these reactions do not involve alkyl aluminum porphyrin species.[31,47–50]

The methyl methacrylate polymerization reactions initiated by Al(TPP)Me can be further accelerated by addition of a sterically crowded organoaluminum Lewis acid, for example, $MeAl(O\text{-}2,4\text{-}{}^tBu_2C_6H_3)_2$,[51] which activates the monomer and hence accelerates chain growth of the enolate polymer. The combination of the bulky Lewis acid and the sterically demanding aluminum porphyrin enolate prevents a side reaction from occurring between the nucleophilic aluminum enolate and the Lewis acid, which would inhibit the polymerization, as shown schematically in Fig. 4. For example, if $AlMe_3$ is added as the Lewis acid to the growing Al(TPP)(enolate) polymer, the result is formation of Al(TPP)Me and Me_2Al (enolate), and polymerization is suppressed.[47,52,53] Steric factors can be tuned by varying both the substituents on the aluminum Lewis acid and on the periphery of the porphyrin.[54]

Further evidence from the mechanism of the polymerization reactions arose from using two initiators with different reactivity, Al(TPP)Me and Al(EtioP)Me (EtioP = etioporphyin). A unimodal polymer molecular weight distribution

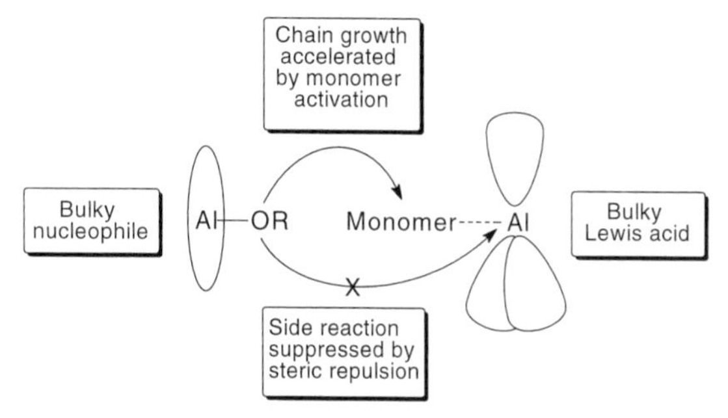

FIG. 4. Schematic illustration of "high-speed living polymerization" of methacrylate esters accelerated by steric separation of the aluminum porphyrin nucleophile and bulky Lewis acid.[52,53]

resulted, and an acyclic transition-state polymerization mechanism involving the participation of two aluminum porphyrin molecules was proposed.[55] Methacrylonitrile can also be polymerized under these conditions.[56] Overall, under optimum conditions, these Lewis acid-assisted high-speed living polymerizations using aluminum porphyrin initiators can give high-molecular-weight ($M_n = 10^5$–10^6) poly(methylmethacrylate) polymers with a narrow molecular weight distribution. Either Al(TPP)R (R = Me, Et) with irradiation by visible light, or Al(TPP)SR (R = alkyl) complexes without irradiation are the most effective initiators.

D. *Activation of CO_2*

Aluminum porphyrins first came to attention with the discovery that the simple alkyl complex Al(TPP)Et was capable of activating CO_2 under atmospheric pressure.[25] Both irradiation with visible light and addition of 1-methylimidazole were required for the reaction, which was proposed to proceed by initial coordination of the base to aluminum. The aluminum porphyrin containing direct product of CO_2 insertion was not isolated, but was proposed on the basis of IR data to be (TPP)AlOC(O)Et, which was then treated with HCl gas, presumably liberating propanoic acid, subsequently isolated as the butyl or methyl ester after reaction with 1-butanol or diazomethane, respectively [Eq. (5)]. Insertion of CO_2 into the Al—C bond of an ethylaluminum phthalocyanine complex has also been reported.[57]

Et–Al + 1-methylimidazole (N-Me) → Et–Al–N(imidazole)Me —CO_2→ EtC(O)O–Al–N(imidazole)Me (5)

Aluminum porphyrins with alkoxide,[30,58,59] carboxylate,[32] or enolate[60] can also activate CO_2, some catalytically. For example, Al(TPP)OMe (prepared from Al(TPP)Et with methanol) can bring about the catalytic formation of cyclic carbonate or polycarbonate from CO_2 and epoxide [Eq. (6)],[30,58,59] and Al(TPP)OAc catalyzes the formation of carbamic esters from CO_2, dialkylamines, and epoxide.[32] Neither of the reactions requires activation by visible light, in contrast to the reactions involving the alkylaluminum precursors. Another key difference is that the ethyl group in Al(TPP)Et remains in the propionate product after CO_2 insertion, whereas the methoxide or acetate precursors in the other reactions do not, indicating that quite different mechanisms are possibly operating in these processes. Most of this chemistry has been followed via spectroscopic (IR and ^{1}H NMR) observation of the aluminum porphyrin species, and by organic product analysis, and relatively little is known about the details of the CO_2 activation steps.

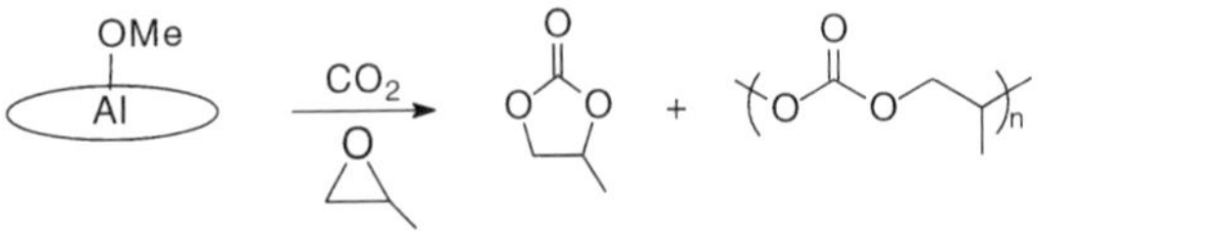

OMe Al $\xrightarrow{CO_2}$ (epoxide) cyclic carbonate + polycarbonate (n) (6)

Both CO_2 activation and enolate formation are combined in the preparation of malonic acid derivatives. The reaction of CO_2 with methacrylic esters or methacrylonitrile and under visible light irradiation produced the corresponding aluminum porphyrin malonate complex. When diethylzinc was added to this system, Al(TPP)Et could be regenerated by axial ligand exchange reactions, and the malonic acid derivatives were formed catalytically with respect to the aluminum porphyrins in a one-pot photosynthetic route (Scheme 1).[61] The first step in this

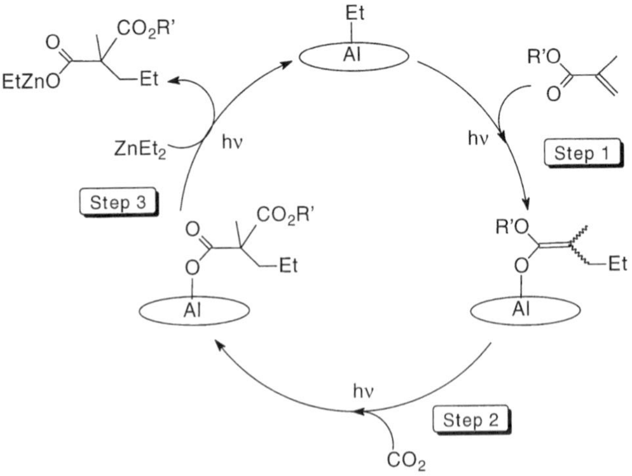

SCHEME 1

process is proposed to be addition of Al(TPP)Et to the monomer to produce an aluminum enolate species (as observed in the polymerization reactions). However, this enolate reacts more rapidly with CO_2 than with further monomers, giving an aluminum malonate derivative. The choice of $ZnEt_2$ to regenerate Al(TPP)Et from the malonate is important, as it reacts with neither CO_2 nor with the monomer. Irradiation is essential for steps 1 and 3 in Scheme 1, and although it is not essential for step 2 it does result in acceleration of this step.[61]

IV

GALLIUM, INDIUM, AND THALLIUM

These elements comprise the longest established family of organometallic main group porphyrin complexes, with extensive development of the chemistry during the late 1980s. One of the motivations for this development was the use of Group 13 complexes as models for iron porphyrin complexes. Comparison of iron porphyrins with porphyrin complexes in which the central metal does not undergo redox reactions helps to isolate ring-centered from metal-centered processes. More recently, Group 13 phthalocyanine complexes have shown promise as materials for nonlinear optics, and this is becoming a new growth area.

A. *Synthesis, Spectroscopy, Structure, and Electrochemistry*

Organoelement porphyrin complexes of gallium, indium, and thallium, M(Por)R, have been prepared from reactions of the chloride complex M(Por)Cl with lithium (LiR) or Grignard (RMgX) reagents. One exception is the use of $RTl(OAc)_2$ to metallate H_2OEP or H_2TPP, forming the methyl or norbornenyl complexes, Tl(Por)R.[62] The range of σ-bonded groups in the porphyrin complexes include alkyl, aryl, fluorophenyl, vinyl, acetylide, and allyl, as shown in Table II.[25,26,62–76] Aryl indium porphycene (EtioPc) and phthalocyanine ($^tBu_4Pc$, R_8Pc) complexes are also featured in Table II.[72–74] The stability of the complexes varies depending on the metal and the axial R group. For example, the alkyl and aryl gallium porphyrin complexes can be purified by chromatography so long as light is excluded,[63] whereas the allyl complexes are more reactive and cannot be chromatographed.[67]

The alkyl and aryl complexes show hyper-type UV-visible spectra with a split Soret band, although the spectra of the gallium and indium acetylide complexes more closely resemble the normal UV-visible spectra observed for the ionic complexes M(Por)Cl or $M(Por)ClO_4$, and the fluorophenyl indium complexes do in fact show normal UV-visible spectra. The position of the Soret band for complexes of a given porphyrin ligand allows an order of decreasing electron-donating ability of the axial ligand to be determined as follows: $^tBu > {}^iPr > {}^nBu > Et > Me > p\text{-}C_6H_4OMe > C_6H_5 > CH{=}CHPh > C{\equiv}CPh > C_6F_4H \approx C_6F_5 > Cl$.[64,70,75,77]

A similar order of electron-donating ability can also be determined from the chemical shift of the methine (meso) CH peak in the In(OEP)R complexes.[77] In fact the sensitivity of the porphyrin proton chemical shifts to the nature of the axial ligand increases from gallium, for which the porphyrin proton resonances of Ga(Por)R do not depend on the electron-donating ability of R,[64] to In(Por)R for which a correlation can be observed, to Tl(Por)R where the porphyrin electron density is very sensitive to the nature of R (alkyl, aryl, or ionic).[75] A characteristic of the ^{1}H NMR spectra of all the σ-bonded complexes is the upfield shift of the axial ligand protons in the ^{1}H NMR spectra, with the most pronounced shift observed for the protons closest to the porphyrin ring. This is nicely illustrated for the gallium complex Ga(OEP)^{n}Bu in which the α, β, γ, and δ protons of the n-butyl group appear at -6.15, -3.79, -1.79, and -0.80 ppm, respectively.[63,203,205] Tl coupling is observed in the thallium alkyl complexes, with ^{205}Tl—^{1}H ($^2J = 715$ Hz) and ^{205}Tl—^{13}C ($^1J = 5835$ Hz) observed for Tl(OEP)Me.[62,76]

The organo-gallium, indium, and thallium TPP and OEP complexes have been extensively studied by electrochemistry and spectroelectrochemistry.[64,70,75,77] One (OEP) or two (TPP) reversible reductions are observed in CH_2Cl_2 for the gallium and indium complexes, and from the electrochemical potentials and spectroscopic data these are confirmed to be porphyrin ring-centered reductions. More interesting behavior is observed upon electrooxidation in CH_2Cl_2, which for each complex gives an oxidized complex, $[M(Por)R]^{\cdot +}$, whose stability is very much dependent on both R and metal. For alkyl and aryl gallium and indium complexes the oxidation step is followed by rapid M—C bond cleavage to form an ionic porphyrin species in solution, $[M(Por)]^{\cdot +}$, which in turn undergoes two reversible ring-centered oxidations [Eq. (7)]. The most unstable oxidized $[Ga(Por)R]^{\cdot +}$ and $[In(Por)R]^{\cdot +}$ complexes are those in which the R group has the most σ-bonded character. Thus the most rapid M—C bond cleavage is seen for the alkyl groups and slower cleavage is seen for aryl, vinyl, and acetylide groups. The one-electron oxidation processes which form the most stable products are reversible, as seen for the indium acetylide complexes In(Por)C≡CR in benzonitrile and the fluorophenyl complexes in CH_2Cl_2.[70,77] The oxidized complexes containing the strongly electron-withdrawing fluorophenyl groups are sufficiently stable so that they can be characterized spectroscopically, confirming that the site of oxidation is the porphyrin ring.[70]

$$\mathrm{M(Por)R} \underset{e^-}{\overset{-e^-}{\rightleftharpoons}} [\mathrm{M(Por)(R)}]^{\cdot +} \longrightarrow [\mathrm{M(Por)}]^+ + \mathrm{R}^\bullet \qquad (7)$$

The thallium complexes show somewhat different electrochemical behavior, and reversible oxidations are observed for both σ-alkyl and σ-aryl thallium porphyrins, indicating that the oxidized complexes have a more stable metal—carbon bond than the gallium or indium analogs. Spectroelectrochemistry revealed that the first oxidation is porphyrin ring-centered. The first reduction is reversible and ring centered, but a slow chemical process follows the second reduction. This represents the only evidence seen in the Group 13 porphyrin complexes for a redox process involving the central metal. In this case it is possible that Tl(I) is formed and is followed by demetallation.[75]

Overall, the UV-visible, NMR, and electrochemical data for the range of σ-bonded complexes of gallium, indium, and thallium porphyrins show that there is a trend from pure σ-bonded character of the M—C bond in the alkyl complexes to the much more ionic character observed for the fluorophenyl and acetylide complexes. These more ionic complexes show NMR and UV-visible characteristics more like those of the ionic M(Por)Cl or M(Por)ClO_4 complexes.[64,65]

One of the motivations for studying Group 13 porphyrin complexes has been their use as electroinactive models for trivalent transition metal porphyrin complexes.[65,67,71–73] The electrochemistry of the indium and gallium complexes In(Por)R and Ga(Por)R can be compared to that of the iron and cobalt complexes, Fe(Por)R and Co(Por)R, which show rather different behavior. One-electron reduction of the cobalt complexes results in Co—C cleavage, as does reduction of the fluorophenyl iron complexes Fe(Por)C_6F_5 and Fe(Por)C_6F_4H. Stable reduction products are observed for the other σ-bonded iron complexes, and also for the gallium and indium σ-bonded complexes, although reduction occurs at the metal center for iron, giving $[Fe^{II}(Por)R]^-$ anions, whereas for the gallium and indium complexes both one- and two-electron reductions are ring centered. Electrochemical oxidations of the gallium and indium alkyl and aryl complexes give products which are unstable and rapid metal—carbon bond cleavage ensues. This contrasts with the one-electron oxidized iron and cobalt complexes which undergo rapid migration of the σ-bonded group from the metal to the porphyrin nitrogen, with concomitant formal reduction of the metal to the Fe(II) or Co(II) oxidation state. This kind of migration is not observed for gallium and indium, presumably because of the lack of a stable +II oxidation state.[64] The difference between the iron and indium complexes is further illustrated by the difference in the reduction potentials between the In(TPP)R and In(TPP)ClO_4 complexes, which is small ($\leq$0.17 V) compared to the corresponding difference, up to 1.0 V, between the iron complexes Fe(TPP)R and Fe(TPP)ClO_4. This reflects the fact that the site of reduction in the indium complexes is the porphyrin ring, and there is little involvement of the central metal or axial ligand, whereas in the iron complex reduction involves the metal center, which is very sensitive to changes in the axial ligand.[77]

Group 13 phthalocyanine complexes are interesting because of their optical properties, and several indium phthalocyanine derivatives have recently been reported. The development of phthalocyanine chemistry has tended to lag behind that of porphyrin chemistry, in part because of solubility problems. Incorporation of alkyl groups on the periphery of the phthalocyanine ligands ($^tBu_4Pc$ and R_8Pc, R = n-pentyl) has helped to overcome these. Aryl indium phthalocyanine complexes were prepared, In($^tBu_4Pc$)Ar and In(R_8Pc)Ar, in which the aryl ligands are

CF_3- or F-substituted (see Table II). These were chosen because of their greater stability relative to the more electron-rich alkyl or aryl groups. Although the simple phenyl complexes were also reported, they had the lowest stability of the series.[74] The indium complex In(^{t}Bu$_4$Pc)Cl has one of the largest positive nonlinear absorption coefficients among phthalocyanine complexes, and it has been used to manufacture optical limiting devices.[78] For this reason the harmonic-generation capabilities and nonlinear absorption, refraction, and optical-limiting properties of the arylindium derivatives have been explored.[79]

There is currently significant growth in the area of metal complexes of expanded, contracted, and isomeric porphyrins.[11] A phenyl indium complex of the porphyrin isomer porphycene, In(EtioPc)C_6H_5, has been reported, including a crystal structure and electrochemical study.[72,73] Corroles are tetrapyrrole macrocycles similar to porphyrins but in which one *meso*-carbon is replaced by a direct bond between two α-pyrrolic carbon atoms (Fig. 1). A corrole macrocycle bears a 3− charge in its deprotonated form, and transition metal corrole complexes are stabilized in higher oxidation states than their porphyrin counterparts. Preliminary evidence has been reported for an indium corrole complex,[80] and more recently, gallium corrole complexes have been prepared. Because the 3− charge on the ligand matches the 3+ charge on the metal, there are no additional anionic axial ligands, and the complexes have the simple formula M(corrole), although neutral pyridine is coordinated to the gallium complex. Gallium corroles are likely to occupy the same niche in corrole chemistry as do zinc porphyrins in porphyrin chemistry, serving as the "proto-metallacorrole."[81–83]

Selected data for structurally characterized Group 13 organometallic porphyrin complexes are shown in Table III and representative examples are shown in Fig. 5.[26,27,65–67,73,84] With so few structural data it is difficult to draw conclusions, but the small number of examples do illustrate the five-coordinate, square pyramidal geometry with the metal atom displaced above the mean N_4 plane. This displacement increases with the size of the central atom, and is ca. 0.5, 0.6, 0.8, and 1.0 Å for Al, Ga, In, and Tl, respectively. The gallium alkyl, vinyl, and acetylide complexes form a useful series containing sp^3, sp^2, and *sp* hybridized carbon atoms bonded to gallium. As expected, both Ga—C and average Ga—N bonds shorten as the *s* character in the Ga—C bond increases.[65–67]

B. *Activation of SO_2 and CO_2*

Most of the studies on organo-gallium, indium, and thallium porphyrin complexes have focused on synthesis and properties of the complexes, and rather little attention has been devoted to reaction chemistry. Two areas which have received some attention are the insertion of small molecules (SO_2 or CO_2) into the metal—carbon bonds and photochemical metal—carbon bond cleavage. The

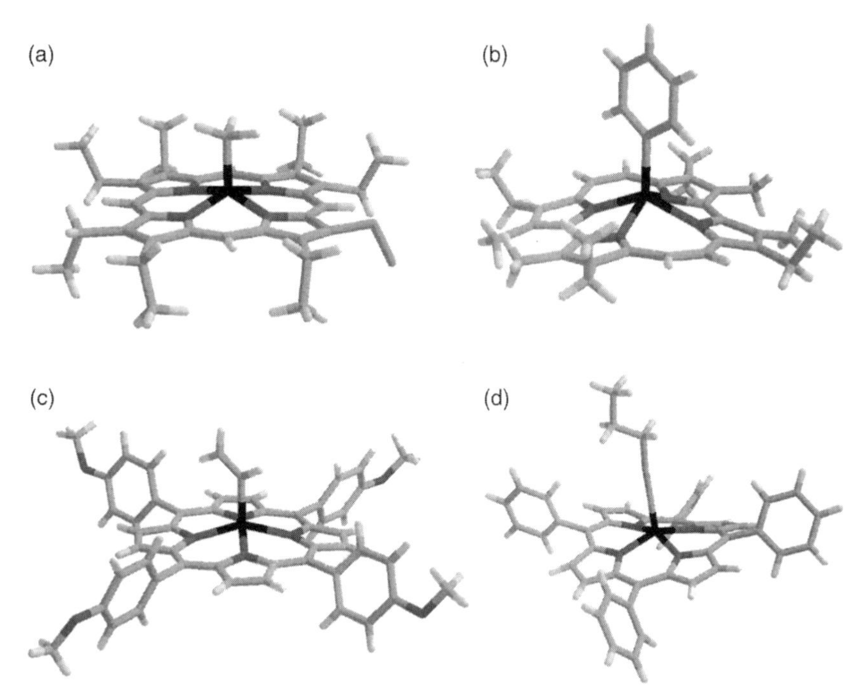

FIG. 5. Molecular structures of selected Group 13 organometallic porphyrin complexes: (a) Al(OEP)(CH_3),[26] (b) In(EtioPc)(C_6H_5),[73] (c) Ga(TAP)(CH=CH_2) (TAP = tetraanisylporphyrin),[67] (d) Ga(TPP)(C≡CPr).[65]

gallium and indium OEP and TPP complexes Ga(Por)Me and In(Por)R (R = Me, ^{t}Bu, Ph) will insert SO_2 into the metal—carbon bond under mild conditions to give sulfinato complexes, M(Por)OS(O)R, which can then be air oxidized to the sulfonato complexes, M(Por)OS$(O)_2$R. The sulfonato complexes can be independently prepared from the reaction of Ga(Por)Cl with $MeSO_3H$, or by controlled air oxidation of In(Por)SR (R = Me, ^{t}Bu) to give the sulfinato complexes at 0°C, and then the sulfonato complexes at 50–60°C (Scheme 2). Metal—O rather than metal—S bonding is confirmed for the sulfonato complexes by X-ray crystal structures of Ga(OEP)OS$(O)_2$Me and In(TPP)OS$(O)_2$Me, and proposed for the sulfinato complexes on the basis of IR data.[85,86]

The CO_2 activation reactions seen for aluminum porphyrins are also observed for In(Por)Me (Por = OEP, TPP), which will insert CO_2 in the presence of pyridine and under irradiation by visible light to give the acetato complex In(Por)OC(O)Me.[87] The indium acetato product has been characterized by X-ray crystallography, whereas in the aluminum complex it was observed only by spectroscopy. An alternative synthesis of the acetato complex is by treatment of In(Por)Cl by alumina and water, followed by acetic acid. For the indium and

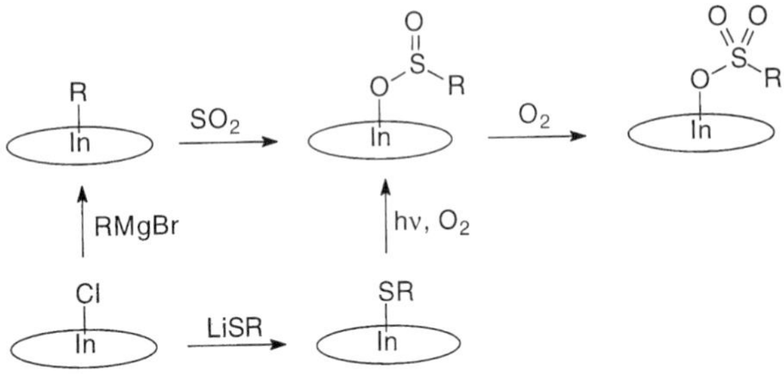

SCHEME 2

aluminum CO_2 insertion reactions, both irradiation and an added base (pyridine 1-methylimidazole, respectively) are essential for the reaction to proceed [Eq. (8)]. Addition of pyridine to In(Por)Me in benzene or CH_2Cl_2 yielded no evidence (by UV-visible spectroscopy) for formation of a six-coordinate pyridine adduct. However, irradiation of In(Por)Me in benzene/pyridine resulted in reduction to form radical species which were observed by ESR. A tentative mechanism was proposed, in which pyridine coordinates to the initially formed reduced radical complex, followed by addition of the indium radical to CO_2 [Eq. (9)].[87] Evidence for coordination of pyridine in the reduced species $[In(Por)(py)Me]^-$ comes from studies on the electrochemical reduction of In(Por)Me in pyridine.[77]

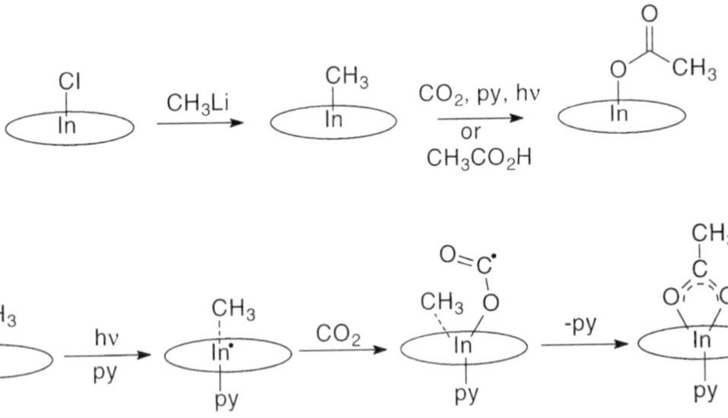

(8)

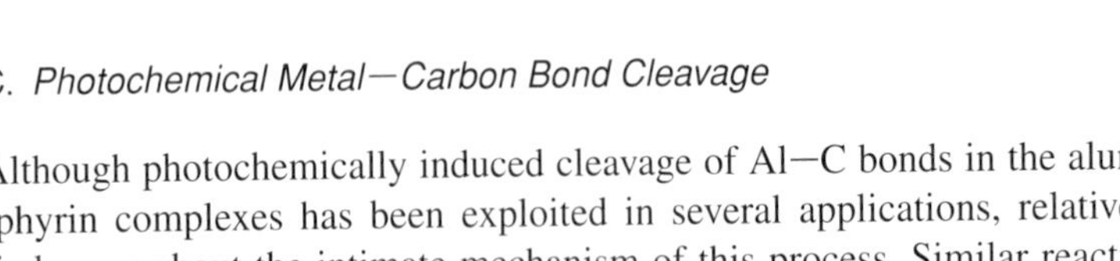

(9)

C. *Photochemical Metal—Carbon Bond Cleavage*

Although photochemically induced cleavage of Al—C bonds in the aluminum porphyrin complexes has been exploited in several applications, relatively little is known about the intimate mechanism of this process. Similar reactivity is observed for the organo-gallium and indium porphyrins, and for these elements several detailed studies on the photoreactivity have been carried out. Photolysis ($\lambda < 400$ nm) of In(TPP)Et in 2-methyltetrahydrofuran at room temperature led to homolytic In—C bond cleavage. This is well known for metal(III) transition metal porphyrins, and leads to formation of M^{II}(Por) and R· products.[7] However, in the indium case the In(TPP) photoproduct was proposed to be a zwitterionic In(III) porphyrin radical anion, formulated as $(In^{III})^+(TPP\cdot^-)$. This species carries no net charge, and comprises an In(III) cation and a $(TPP\cdot^-)^{3-}$ anion. The species exhibits an ESR spectrum which disappears on aeration, and UV-visible spectroscopic data are consistent with a porphyrin radical anion.[88] In—C bond homolysis in In(Por)R gives R· and a radical $[M^{III}(Por)]\cdot$ in which the electron locates on the porphyrin ring resulting in the zwitterion $(In^{III})^+(TPP\cdot^-)$ rather than at the metal center to give M^{II}(Por) as is observed in transition metal porphyrin complexes.

Laser flash photolysis of In(TPP)Et in benzene provides evidence that the first step is formation of a transient photoexcited triplet state, ^{3}[In(TPP)Et]*, which then undergoes bond homolysis. The reaction was quenched by the addition of ferrocene, but quantum yields improved when a strong electron acceptor (2,4,7-trinitro-9-fluorenone) was added.[89] Addition of pyridine also increased the quantum yield, explained by facile homolysis of the In—C bond assisted by the coordination of a pyridine ligand in the excited triplet state complex ^{3}[In(TPP)(py)Et] [Eq. (10)].[90] The lowest excited triplet state of In(TPP)Et was considered to be a charge transfer triplet state formed by transfer of an electron from an In—C σ-bonding orbital to the porphyrin LUMO. This weakens the In—C bond and homolysis ensues, although rapid recombination of transient In^{II}(TPP) and Et· can reform In(TPP)Et, resulting in a low quantum yield. Axially coordinated pyridine raises the energy of the indium 5s orbital producing the zwitterion $(In^{III})^+(TPP\bar{\cdot})$-(py) which is more stable than the In^{II}(TPP) transient, thus suppressing recombination and raising the quantum yield.[90]

$$\mathrm{In(TPP)(C_2H_5)} \xrightarrow[\text{py}]{h\nu} {}^3[\mathrm{In(TPP)(C_2H_5)}]^* \longrightarrow (\mathrm{In^{III}})^+(\mathrm{TPP}\bar{\cdot})(\mathrm{Py}) + {}^\bullet\mathrm{C_2H_5} \qquad (10)$$

These initial studies focused only on In(TPP)Et, and a more wide-ranging study looked at the photoreactivity of M(Por)R for both gallium and indium OEP and TPP complexes, where R = Me, Et, ^{n}Bu, ^{i}Pr, ^{t}Bu, CH=CHPh and C≡CPh.[91] Photodissociation was observed for all the alkyl axial ligands, but not for the vinyl or acetylide complexes. This was attributed to the electron-withdrawing capabilities of these groups resulting in the more ionic character of the M—C bond, in line with earlier spectroscopic and electrochemical studies. Laser photolysis confirmed that the photoreaction occurs via a triplet state originating from the porphyrin macrocycle, and that the zwitterionic indium photoproduct $(In^{III})^+(Por\cdot^-)$ is formed. For example, for a given metal (In or Ga) and porphyrin ligand (OEP or TPP) the UV-visible spectrum after irradiation was the same, independent of the R group. A comparison of the photochemical and electrochemical reduction of In(TPP)Cl and In(TPP)Et showed that the ESR spectrum of electrochemically reduced In(TPP)Cl and photochemically reduced In(TPP)Et are almost identical, corresponding to $(In^{III})^{+}(TPP\cdot^{-})$, whereas electrochemically reduced In(TPP)Et had quite a different ESR spectrum arising from $In(TPP\cdot^{-})Et$. Higher quantum yields were observed for more polar solvents (for example, DMF relative to benzene), consistent with formation of the polar, zwitterionic photoproduct.[91]

The fate of the alkyl radical and the further chemical reactivity of the metal porphyrin complex formed in the photoinduced homolysis have been addressed using gallium alkyl complexes, Ga(TPP)R, where R = cyclopentylmethyl or 5-hexenyl. These were investigated so that the well-known cyclization of the 5-hexenyl radical could be used to monitor radical formation during photolysis [Eq. (11)].[66] Photolysis of 5-hexenyl complex $Ga(TPP)CH_2(CH_2)_3CH{=}CH_2$ in benzene for 1 h resulted in a 37% yield of the cyclopentylmethyl complex $Ga(TPP)CH_2(\textit{cyclo}\text{-}C_5H_9)$, observed by ^1H NMR spectroscopy. The isomerization results from homolysis of the Ga—C bond to form the 5-hexenyl radical which then undergoes rearrangement and recombination. In the presence of a radical trap such as nitrosobenzene the alkyl radicals are trapped to form stable nitroxide radicals observed by ESR spectroscopy. Photolysis of either the Ga(TPP)Et or the 5-hexenyl complex in benzene in the presence of O_2 led to formation of the alkylperoxide complexes Ga(TPP)OOEt or $Ga(TPP)(OO(CH_2)_4CH{=}CH_2)$. Ga(TPP)OOEt could be confirmed by its independent preparation from Ga(TPP)Cl and HOOEt. The lack of isomerization in the photolytic formation of the 5-hexenylperoxide product results from the fact that reaction of the 5-hexenyl radical with O_2 to form the peroxy radical is faster than the cyclization reaction. The photochemical reactions of these gallium alkyl complexes are summarized in Scheme 3.[66]

$$\cdot CH_2CH_2CH_2CH_2CH{=}CH_2 \longrightarrow \cdot CH_2\text{-}cyclo\text{-}C_5H_9 \tag{11}$$

Ga(CH₂R) —hν→ Ga• + •CH₂R —→ RCH₂(Ph)NO•

Ga• —O₂→ Ga• + •OOCH₂R —→ Ga(OOCH₂R)

Ga• + •CH₂R' —→ Ga(CH₂R')

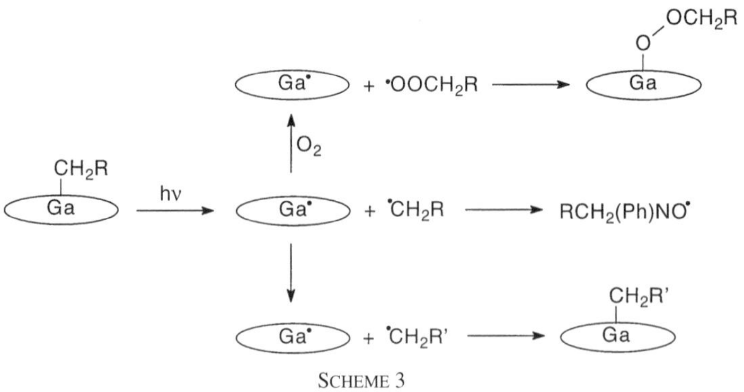

SCHEME 3

V

SILICON, GERMANIUM, AND TIN

Simple porphyrin coordination complexes of the form $M(Por)X_2$ have been known for a relatively long time for germanium and tin but to a lesser extent for silicon, in part because of the very high hydrolytic reactivity of $Si(Por)Cl_2$.[92] Organo-germanium porphyrin chemistry was developed in the 1980s, but it was not until the 1990s that the corresponding chemistry of silicon and tin was developed. The main difficulty in preparing and handling the Group 14 organoelement porphyrins is their photosensitivity, which is the most pronounced for tin and results in ready cleavage of the Sn—C bonds.

Organogermanium porphyrins comprise the very earliest examples of main group organometallic porphyrins, with a report from 1973 detailing their potential as NMR shift reagents.[93,94] Organosilicon phthalocyanine complexes have been known for even longer, and were first reported in the 1960s.[95] Tin and germanium porphyrins have demonstrated significant biological activity, with a report in 1983 that a dimethyl germanium porphyrin complex showed antineoplastic activity against three types of solid tumors in mice.[96] Tin porphyrins have been shown to act as competitive inhibitors of heme oxygenase,[97] and have been investigated for the suppression of neonatal jaundice.[98]

A. *Synthesis, Spectroscopy, Structure, and Electrochemistry*

1. *Synthesis*

The range of organo-silicon, germanium, tin, and lead porphyrin complexes reported to date are given in Table IV.[93,96,99–113] Some mono- and dialkyltin complexes, $Sn(Por)R_2$ or $Sn(Por)R(OH)$, which have been characterized only by NMR spectroscopy have been omitted from the table.[114] The complexes are six-coordinate, with two *trans* axial ligands except for two exceptions which exhibit *cis* geometry. The general classes of complex are $M(Por)R_2$ and $M(Por)RX$, where X is an anionic ligand. Four general routes have been described for the preparation of these complexes, shown in Eqs. (12–15): (i) reaction of $M(Por)X_2$ ($X =$ halide or OH) with RMgBr or RLi, (ii) oxidative addition of RX to divalent M(Por), (iii) direct metallation of $Li_2(Por)$ with R_2MX_2, and (iv) photochemical cleavage of one axial ligand in $M(Por)R_2$ to give $M(Por)RX$.

The diorganoelement complexes $M(Por)R_2$ are formed from $Si(Por)F_2$ or $M(Por)Cl_2$ ($M =$ Si, Ge, Sn) or $Sn(Por)(OH)_2$ with alkyl Grignard or aryl lithium reagents.[93,99–105,114] Two early studies looked in detail at the mechanism of the reactions of alkyl Grignard reagents with $Sn(Por)X_2$ ($X =$ Cl, OH) and $Ge(Por)Cl_2$.[100,114] The key intermediate is a radical anion, $[M(Por)X_2]^{\cdot -}$, formed

TABLE IV

SILICON, GERMANIUM, TIN, AND LEAD PORPHYRIN COMPLEXES, $M(Por)R_2$ AND M(Por)XY

Porphyrin	$M(Por)R_2$	M(Por)XY		Reference
	R	X	Y	
Silicon				
OEP	Me, Ph			106
OEP		Ph	OH	106
TPP	Me, ^{n}Pr, CH_2SiMe_3, $CH{=}CH_2$, Ph, C≡CPh			101, 102
Pc		Ph	Cl, OH, OR (dendrimer)	107
$Pc(OR)_8$, PcR_8[a]	Me			108
$Pc(OR)_8$, PcR_8[a]		Me	Cl, OH, OTs, $OSiR_3$, OMe, OEt, O^{i}Pr, OTf	108, 109
Germanium				
TPP	Me, Et, Pr, ^{i}Pr, Bu, ^{i}Bu, n-octyl, CH_2SiMe_3, CH_2Ph, ferrocenyl, Ph			93, 96, 99 100, 103, 110
OEP	Me, CH_2Ph, ferrocenyl			103, 110, 111
TPP, OEP		Ph	Cl, OH, ClO_4	103
TPP, OEP		Ferrocenyl	Ph	110
Tin				
OEP, TTP, TMP		Me	I	112
TPP, TBPP,[b] OEP	*cis*-Ph_2			104
TPP, TTP	*trans*-Ph_2			104
TTP		Ph	Cl	104
TTP	Et, $CH_2{}^tBu$, *cis*-Me_2, *trans*-Me_2, C≡CPh, C≡$CSiMe_3$			105
TTP		Me	Br	105
TTP		C≡CPh	OMe	105
OEC		Ph	—	113

[a] $Pc(OR)_8$, PcR_8: R = C_5H_{11}.
[b] TBPP = dianion of tetra-(4-^{t}Bu-phenyl)porphyrin.

by electron transfer from the Grignard reagent, which then reacts either at the metal center or at the periphery of the porphyrin to give alkylated products. Diethylzinc has been used with $Sn(TTP)Cl_2$ to produce $Sn(TTP)Et_2$.[105]

One startling result is that the stereochemistry of the tin porphyrin complexes can be affected by the choice of synthetic route. The lithiated porphyrins $Li_2(Por)(L)_n$[12] have not been much utilized for the preparation of main group porphyrins. Direct metallation of $Li(TPP)(OEt_2)_2$ using Ph_2SnCl_2 produced the relatively unusual *cis* arrangement of the two phenyl groups in the products, *cis*-$Sn(TPP)Ph_2$. The more conventional reaction of *trans*-$Sn(TPP)Cl_2$ with $MgPh_2$ produced the stereoisomer *trans*-$Sn(TPP)Ph_2$.[104] The methyl derivatives *cis*- and *trans*-$Sn(TTP)Me_2$ have been produced by similar means, although the *trans* isomer is extremely photosensitive and of limited stability even in the dark. There are also solvent effects, illustrated by the formation of *cis*-$Sn(TTP)Me_2$ from Me_2SnBr_2 with $Li_2(TTP)(THF)_2$ in toluene, whereas the same reaction in CH_2Cl_2 gives *trans*-Sn(TTP)(Me)Br.[105]

$$X\text{–}M(Por)\text{–}X \xrightarrow[\text{RLi}]{\text{RMgBr or}} R\text{–}M(Por)\text{–}R \quad (12)$$

$$M(Por) \xrightarrow{\text{RX}} R\text{–}M(Por)\text{–}X \quad (13)$$

$$Li_2Por(THF)_2 \xrightarrow{R_2MX_2} cis\text{-}M(Por)R_2 \quad (14)$$

$$R\text{–}M(Por)\text{–}R \xrightarrow[X^-]{h\nu} R\text{–}M(Por)\text{–}X \quad (15)$$

Tin represents one of the few main group elements for which the availability of stable complexes in the lower Sn(II)[115] oxidation state has been utilized for the synthesis of organometallic derivatives. Oxidative addition of methyl iodide to a selection of Sn(Por) complexes produced the tin(IV) derivatives Sn(Por)(Me)I. The iodide ligand is only weakly coordinating and readily ionizes in polar solvents to form the solvated cations $[Sn(Por)(Me)(S)]^+$.[112] Similar chemistry has been observed for tin complexes of the related tetramethyltetraazadibenzo[14]annulene ($tmtaa^{2-}$)macrocycle, in which oxidative addition of alkyl iodides to Sn(tmtaa) produced cationic, five-coordinate complexes, $[Sn(tmtaa)R]^+$ (R = Me, Et, Pr, Bu).[116]

A pronounced characteristic of the dialkyl and diaryl Group 14 porphyrin compounds $M(Por)R_2$ is their sensitivity to light. Photochemical reactions in the absence of oxygen lead either to cleavage of one metal—carbon in the more stable complexes (usually where R = Ph) to give M(Por)(R)X or to complete decomposition in the case of the more labile tin alkyl complexes. This route has been used for the preparation of the unsymmetrically substituted complexes M(Por)(Ph)X (M = Si, Ge; X = Cl, OH).[103,104,106] The germanium ferrocenyl complexes $Ge(Por)Fc_2$[93,110,111] and the tin alkynyl complexes $Sn(TTP)(C{\equiv}CR)_2$ (R = Ph, $SiMe_3$)[105] are much less sensitive to both light and air. This stability has been attributed to electronic interactions between the ferrocenyl and porphyrin ligands in the germanium complexes. The tin examples reflect the trend observed for the Group 15 complexes where the lability of the metal—carbon bond is related to the basicity of the axial ligand, and the more electron-withdrawing alkynyl ligands are less reactive. The higher ionic character of the Sn—alkynyl bond is illustrated by the formation of Sn(TTP)(C≡CPh)(OMe) by methanolysis of $Sn(TTP)(C{\equiv}CPh)_2$ in a process which does not require irradiation.[105]

One example of a tin porphycene has been reported, but as yet no organometallic derivatives have been reported.[117] A small number of tin corrole complexes are known including one organotin example, Sn(OEC)Ph, prepared from the reaction of Sn(OEC)Cl with PhMgBr. A crystal structure of Sn(OEC)Ph shows it to have both shorter Sn—N and Sn—C bonds than $Sn(TPP)Ph_2$, with the tin atom displaced 0.722 Å above the N_4 plane of the domed macrocycle (Fig. 6). The complex undergoes reversible one-electron electrochemical oxidation and reduction at the corrole ring, and also two further ring oxidations which have no counterpart in tin porphyrin complexes.[80,113,118]

2. *Structural and Spectroscopic Features*

The M(IV) complexes, $M(Por)R_2$ and M(Por)RX, show normal type UV-visible spectra, in contrast to the hyper spectra observed for the Group 13 complexes M(Por)X. Hyperporphyrin spectra are observed for the lower valent tin(II) porphyrins.[115,119] UV-visible absorptions for $M(Por)R_2$ and M(Por)RX are shifted to longer wavelengths compared to $M(Por)X_2$. For example, the shift of the Soret band can be observed in the series $Ge(OEP)Ph_2$ (442 nm), $Ge(OEP)(Ph)ClO_4$ (411 nm), and $Ge(OEP)(ClO_4)_2$ (404 nm). The germanium porphyrin complexes $Ge(Por)R_2$ are like Ga(Por)R but unlike In(Por)R in that the chemical shifts of the protons on the periphery of the porphyrin are not sensitive to the nature of the axial ligands R.[103]

Only a small number of organometallic Group 14 complexes have been structurally characterized, with selected data given in Table V and representative structures shown in Fig. 6. As expected, the M—C bond length increases on going from silicon to germanium to tin. A useful series exists for silicon, $Si(TPP)R_2$, containing alkyl, vinyl, alkynyl, and phenyl ligands.[102] Comparing these, it is apparent that a similar trend is observed to the series of gallium complexes, with the Si—C bond length shortening from 1.929(6) to 1.819(2) Å on going from sp^3 to sp^2 to sp hybridized carbon. The longest Si—C bonds, 1.943(4) and 1.950(3) Å, are observed for the phenyl complex $Si(TPP)Ph_2$, probably as a result of steric repulsion between the *ortho* protons of the phenyl group and the porphyrin ligand. This arises because of the short bonds between carbon and the small element silicon, and the six-coordination which constrains the Si atom to lie in the center of the N_4 plane. The porphyrin ring in $Si(TPP)Ph_2$ is ruffled so as to minimize the steric interaction.[102] Nonplanar distortions of the porphyrin ligand in silicon complexes are also featured in phosphorus complexes, and both are discussed in a later section.

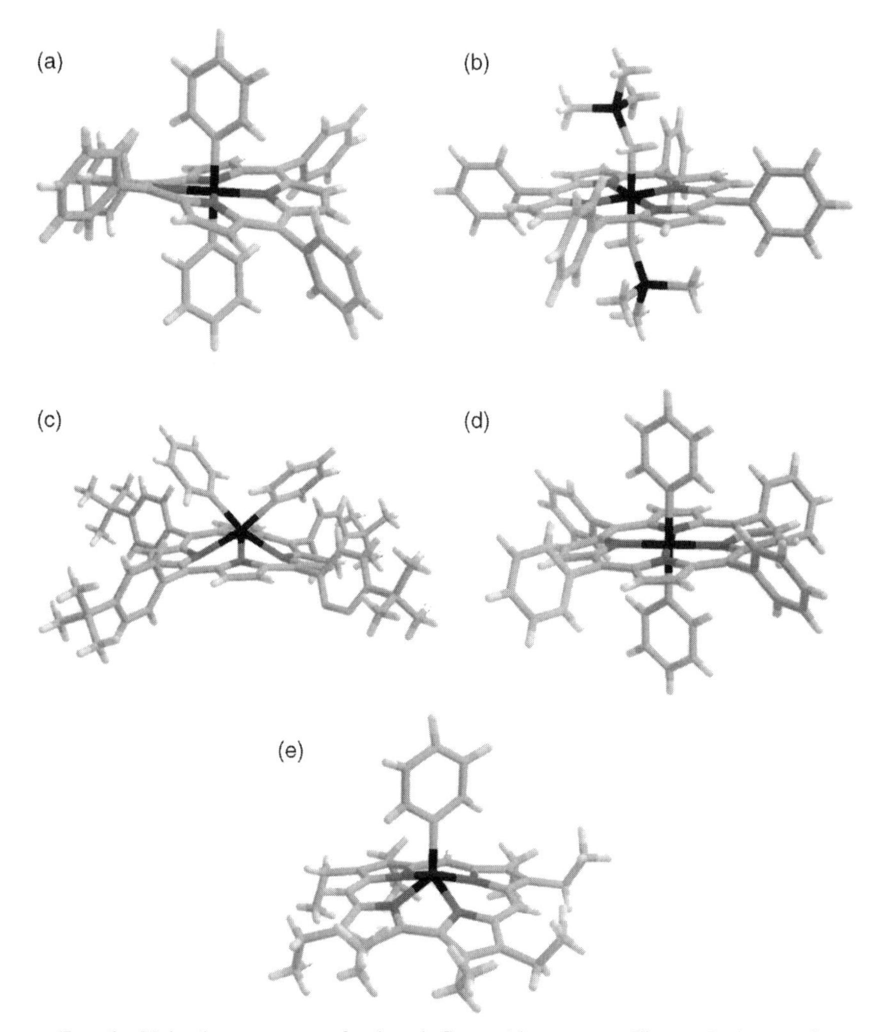

FIG. 6. Molecular structures of selected Group 14 organometallic porphyrin complexes: (a) $Si(TPP)(C_6H_5)_2$,[102] (b) $Si(TPP)(CH_2SiMe_3)_2$,[102] (c) *cis*-$Sn(TBPP)(C_6H_5)_2$ (TBPP = tetra(4-^{t}Bu-phenyl)porphyrin),[104] (d) *trans*-$Sn(TPP)(C_6H_5)_2$,[104] (e) $Sn(OEC)(C_6H_5)$.[113]

The Sn—C bond lengths in $Sn(TTP)(C{\equiv}CPh)_2$ are slightly shorter than those in the two $Sn(TPP)Ph_2$ isomers.[105] The *cis* and *trans* isomers of $Sn(Por)Ph_2$ offer a useful comparison.[104] The tin atom is displaced 1.11 Å out of the N_4 plane in the *cis* isomer, longer even than the out-of-plane distance of 1.018 Å found in the tin(II) porphyrin complex Sn(OEP).[115] The tin(II) porphyrin contains a much larger tin ion but no axial ligands. *cis*-$Sn(TBPP)Ph_2$ exhibits two long (averaging 2.354 Å) and two short (2.177 Å) Sn—N bond lengths and a porphyrin ring that is

TABLE V

SELECTED DATA FOR STRUCTURALLY CHARACTERIZED GROUP 14 PORPHYRIN COMPLEXES

	M—C bond length/Å	$M—N_{av}$ bond length/Å	$M—N_4$ plane/Å	Other		Reference
$Si(TPP)(CH_2SiMe_3)_2$	1.929(6)	2.01				102
$Si(TPP)(C_6H_5)_2$	1.943(4)	1.97				102
	1.950(3)					
$Si(TPP)(CH{=}CH_2)_2$	1.93(2)	2.01				102
$Si(TPP)(C{\equiv}CPh)_2$	1.819(2)	1.98				102
$Ge(T\text{-}3,5\text{-}{}^tBu_2PP)(CH_3)_2$	1.99(3)	2.03				96
cis-$Sn(TBPP)(C_6H_5)_2$	2.210(7)	2.367(6)	1.11	C—Sn—C	98.7°	104
	2.193(7)	2.341(6)				
		2.171(5)				
		2.183(5)				
trans-$Sn(TPP)(C_6H_5)_2$	2.196(4)	2.134(4)				104
	2.212(4)					
$Sn(TPP)(C{\equiv}CPh)_2$	2.167(2)	2.117		C≡C	1.197(3) Å	105
				Sn—C≡C	178.5(3)°	
				C≡C—C	170.1(2)°	
$Sn(OEC)C_6H_5$	2.105(7)	2.067	0.723			113

severely distorted, both ruffled and domed. These features might arise from steric interactions between the phenyl rings and the ligand. Despite the difference in geometry, however, the Sn—C bond lengths in the *cis* and *trans* complexes are almost identical, close to 2.20 Å.[104]

3. *Electrochemistry*

Sn(Por)(Me)I ionizes in polar solvents (S) to $[Sn(Por)(Me)(S)]^+$, and the cations show two reversible electrochemical reductions.[112] Similar reversible reductions at the porphyrin ring are observed for Ge(OEP)(Ph)X (X = Cl, OH, ClO_4).[103] Electrochemical data for $Sn(Por)R_2$ are not available presumably because of the high photosensitivity and lability of these compounds. The electrochemistry of silicon and germanium porphyrins $Si(OEP)R_2$ (R = Me, Ph)[106] and $Ge(Por)R_2$ (Por = OEP, TPP; R = Me, Ph, CH_2Ph)[103] is broadly similar. The complexes undergo reversible one-electron reduction at the porphyrin ring, with the phenyl complexes being more easily reduced than their alkyl counterparts. The first one-electron oxidation is irreversible and is followed by M—C bond cleavage of one R group to give M(Por)(R)X, where X = Cl, OH, or ClO_4 depending on the solvent and supporting electrolyte. Further oxidation (for X = Cl or OH) results in oxidation of the X^- anion and formation of $M(Por)(R)ClO_4$. The oxidative metal—carbon bond cleavage in the the silicon and germanium complexes is reminiscent of that observed in the organo-gallium and indium porphyrin complexes.

The bis-ferrocenyl (Fc) germanium OEP and TPP complexes Ge(Por)Fc_2 comprise a special category on account of the electroactive axial ligands. Monoferrocenyl complexes can be prepared from the reaction of Ge(Por)(Ph)Cl with FcLi.[93,110,111] The ferrocenyl complexes Ge(Por)Fc_2 can be viewed as examples of bridged ferrocenyl compounds Fc—E—Fc in which the bridging group E in this case is the large Ge(Por) group rather than a small fragment like Se or CH_2. Four oxidations are seen for Ge(OEP)Fc_2, the first two of which are ferrocene oxidations and the second two are porphyrin ring oxidations. The electronic properties of the ferrocene group are apparent from the spectral changes of the porphyrin ligand. On going from the neutral to the one- and two-electron-oxidized complexes the Soret band is progressively shifted to shorter wavelengths and decreases in intensity. The shift to shorter wavelengths is consistent with a more electron-withdrawing ligand; in other words, the oxidized Fc^+ ligand is a stronger acceptor than is neutral Fc. The decrease in intensity of the Soret band arises from the delocalization of the charge on the porphyrin a_{1u} and a_{2u} orbitals. This charge delocalization means that the Ge porphyrin-bridged biferrocene complexes are very easily oxidized (at a more negative potential) than simple bridged biferrocenes. Two oxidations are seen for the monoferrocenyl complexes Ge(Por)(Fc)Ph, the first of which occurs at the ferrocene group and the second of which, at the porphyrin ring, is followed by Ge—Ph bond cleavage.[110,111]

B. *Photochemical Metal—Carbon Bond Cleavage*

Lability of the metal—carbon bonds even under ambient light is a characteristic of the Group 14 organometallic porphyrin complexes. This photoreactivity is dependent on the nature of the axial ligands. The silicon vinyl and alkynyl,[120,121] germanium ferrocenyl,[110,111] and tin alkynyl[105] complexes, like their gallium and indium vinyl and alkynyl counterparts,[91] are much more stable to both light and oxygen as a consequence of the greater electron-withdrawing capability of the ligands. The photolability of the silicon and germanium dialkyl and diaryl complexes has been specifically addressed in two studies.[120,122] Steady-state irradiation of Ge(TPP)R_2 (R = CH_2Ph or Ph) in THF resulted in photocleavage of one Ge—C bond to give a product formulated as a zwitterionic complex, Ge^+(TPP $\cdot^-$)(R), characterized by UV-visible and ESR spectroscopy. In $CDCl_3$ solution the product is Ge(TPP)(R)Cl. The alkyl germanium zwitterion is similar to the In^+(Por $\cdot^-$) species formed from In—C bond photocleavage in In(Por)R.[88,91] The quantum yield was lower for the phenyl complex than the benzyl complex, consistent with a stronger Ge—C bond in the former. The ferrocenyl complexes, Ge(TPP)Fc_2 and Ge(TPP)(Fc)Ph, were photostable under the same conditions, with no Ge—C bond homolysis observed even after several hours irradiation. The initially formed photoexcited state was investigated by laser flash photolysis, and for all the complexes

(Ge(TPP)R_2, Ge(TPP)(Fc)Ph, and Ge(TPP)Fc_2) the spectrum after 2 μs was consistent with a triplet excited state, although this decayed much faster for the ferrocenyl complexes. Addition of ferrocene to Ge(TPP)R_2 also quenches triplet lifetimes. A similar situation was observed for the indium complexes In(Por)R, and the triplet-state quenching was attributed to an energy transfer process from the excited-state triplet to ferrocene. In the case of the germanium porphyrins, the longer-lived triplet state in Ge(TPP)R_2 is responsible for the Ge—C bond homolysis, and both inter- and intramolecular quenching by ferrocene is observed.[122]

The photochemistry of organosilicon porphyrin complexes has been investigated in the presence of nitroxyl ($R_2NO\cdot$) radical scavengers.[120] Irradiation of Si(TPP)Pr_2 in C_6D_6 in the presence of TEMPO (tetramethypiperidinyl oxide) led at first (after 4 min) to Si(TPP)(Pr)(TEMPO) and finally (after 20 min) to quantitative formation of the bis(nitroxy)silicon porphyrin complex Si(TPP)$(TEMPO)_2$, both observed by ^{1}H NMR. The propyl radical was trapped as TEMPO-Pr. The silicon phenyl, vinyl, and alkynyl complexes did not react under these conditions, again in line with their stronger Si—C bonds. Photolysis of Si(TPP)Pr_2 in C_6D_6 with no added nitroxyl radical also resulted in homolysis of both Si—C bonds, and the EPR-active silicon product was formulated as a silicon diradical, which remained stable in solution for up to 50 days. This remarkable radical does not react with TEMPO in the dark but does so when irradiated by visible light, resulting in Si(TPP)$(TEMPO)_2$. The explanation posed for this "photoswitchable radical" is that in the dark the electrons in the silicon diradical are partially delocalized over the porphyrin ring and hence lack reactivity towards TEMPO. When excited by visible light the delocalization is "switched off" resulting in a silicon-centered diradical which does react with TEMPO. These transformations are summarized in Scheme 4.[120]

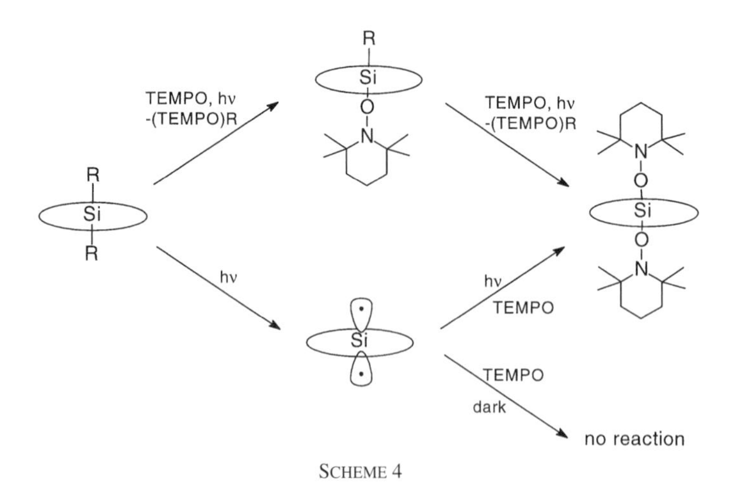

SCHEME 4

C. *Activation of O_2*

Simple photocleavage of one M—C bond in the Group 14 porphyrins requires the absence of oxygen. The reason for this is that irradiation in the presence of oxygen results in the formation of stable peroxo compounds. This was noted in early investigations of alkylgermanium porphyrins, and stepwise insertion of two O_2 molecules into the Ge—C bonds of Ge(Por)R_2 to give Ge(TPP)(R)(OOR) and Ge(TPP)$(OOR)_2$ could be observed by NMR spectroscopy.[100,123] This early NMR study was not able to readily distinguish between alkylperoxide (OOR) ligands and their potential alkoxide (OR) ligand hydrolysis products. This was followed up in a later study which began by unambiguously establishing the spectroscopic properties of Ge(TPP)Et_2, Ge(TPP)$(OEt)_2$, and Ge(TPP)$(OOEt)_2$.[124] The ethoxide and ethylperoxide complexes were independently prepared from the reactions of Ge(TPP)Cl_2 with LiOEt, and of Ge(TPP)$(OH)_2$ with EtOOH, and both were structurally characterized. Reactivity differences are apparent, with the ethyl and ethylperoxide but not the ethoxide complexes exhibiting photsensitivity, while the ethoxide and ethylperoxide but not the ethyl species are hydrolytically sensitive.

Ge(TPP)Et_2 does not react with oxygen in the dark, but when irradiated with long-wavelength visible light ($\lambda > 638$ nm) conversion to Ge(TPP)(Et)(OOEt) can be observed by NMR spectroscopy. When higher-energy radiation is used ($\lambda > 498$ nm) the photoproduct is Ge(TPP)$(OOEt)_2$. The need for higher-energy light results from the systematic shift in the electronic absorption peaks to shorter wavelength accompanying the stepwise transformation from Ge—Et to Ge—OOEt bonds. Controlling the wavelength in the irradiation step allows a selective pathway for preparing mixed ligand complexes such as Ge(TPP)(Et)(OOEt). The ethylperoxide ligand in this complex can be hydrolyzed to produce Ge(TPP)(Et)(OH), and the same product forms upon oxygen atom abstraction from Ge(TPP)(Et)(OOEt) by PPh_3 in the presence of H_2O. These transformations are summarized in Scheme 5.[124] Photochemically induced insertion of O_2 into the Si—C bonds of Si(TPP)Et_2 is observed under similar conditions, giving stepwise formation of Si(TPP)(Et)(OOEt) and Si(TPP)$(OOEt)_2$.[121]

The stability of the germanium and silicon porphyrin alkylperoxide complexes contrasts with the lability of the corresponding iron porphyrin alkylperoxide complexes. For example, Fe(Por)(OOEt) was formed from the reaction of Fe(Por)Et with O_2 and has been characterized only by spectroscopy as it decomposes to Fe(Por)OH and CH_3CHO above $-80°C$. The key to the difference in stability is likely to be the availability of higher oxidation states in the iron porphyrins as the decomposition of Fe(Por)(OOR) may proceed through ferryl (Fe^{IV}(Por)=O or $Fe^{IV}(Por^{\cdot +})$=O intermediates. Many metalloporphyrin complexes are bleached by peroxides, but the germanium porphyrins remain intact even in the presence of excess alkylhydroperoxide. This suggests that high oxidation state intermediates

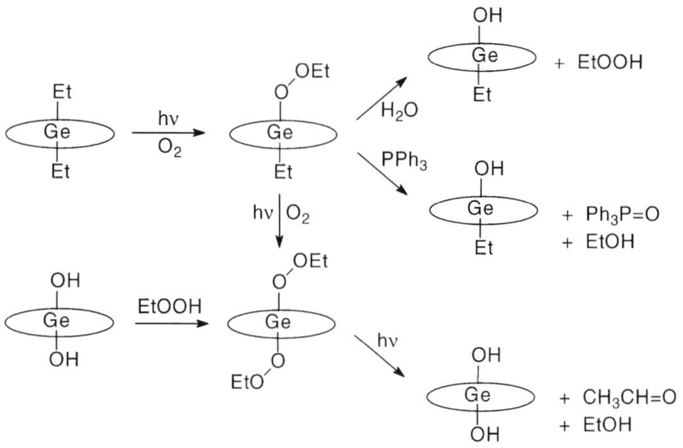

SCHEME 5

formed by a reaction of the peroxides at the metal center are involved in the bleaching process rather than direct attack of the peroxides at the porphyrin ligand.[125]

D. *Group 14 Phthalocyanines*

Group 14 phthalocyanines have been utilized in a range of applications based on their potential for energy capture and their optical properties. Very few of these applications involve organometallic derivatives, although very recently a small number of examples involving organosilicon groups have been reported. In porphyrin chemistry the free base macrocycle is first synthesized and the coordinating element is inserted in a subsequent step. In contrast, phthalocyanine complexes are often prepared by assembly and condensation of four isoindoline units around a central element which serves as a template. Silicon phthalocyanines were first reported as early as 1966, and several examples were prepared from the reaction of $MeSiCl_3$ or $PhSiCl_3$ with 1,3-diiminoisoindoline to give Si(Pc)(R)Cl (R = Me, Et, Pr, Ph) and subsequent substitution of the chloro group to give Si(Pc)(R)X (X = OH, F, OPh and $OSiR'_3$).[95,126] The reactions of either Si(Pc)(octyl)Cl (prepared using (n-octyl)$SiCl_3$) or $Si(Pc)Cl_2$ with aryl Grigard reagents has produced a family of diorganosilicon complexes comprising of Si(Pc)(octyl)(Ar), $Si(Pc)(octyl)_2$, or $Si(Pc)Ar_2$. The electrophilic cleavage of the silicon carbon bonds in these compounds by N-bromosuccinimide, halogens, and copper(II) halides was investigated.[127] Alkynyl Grignard reagents with $M(Pc)Cl_2$ gave the *bis*-alkynyl silicon, germanium, and tin derivatives, $M(Pc)(C{\equiv}CR)_2$ (R = H, Me, ^{t}Bu, Ph). The ease of purification and the stability of the compounds decreased down the series, with the tin compounds being most problematical due to reduction to form Sn(Pc).[128] The alkynyl chemistry was extended to a novel acteylene-bridged polymer by treating $Ge(Pc)Cl_2$ with the ethynediyl di-Grignard reagent BrMgC≡CMgBr.[129]

An important area of research in silicon phthalocyanine chemistry has been the preparation of conducting polymers through Si—O—Si links.[130] In one recent example, silicon phthalocyanine complexes have been appended to a dendrimer framework through Si—O—C(triazine) bonds formed by reaction of the salt $Na^+[Si(Pc)(Ph)O^-]$ with a dichlorotriazine derivative.[107]

Phthalocyanine chemistry has been hampered for a long time by limited solubility, and organosilicon derivatives have been prepared using a macrocycle with eight n-pentyl or n-pentyloxy groups appended to the periphery (R_8Pc or $(RO)_8Pc$, respectively). These were designed specifically to form self-assembled silicon phthalocyanine monolayers. The disubstituted isoindolines were refluxed with $MeSiCl_3$ to give $Si(R_8Pc)(Me)Cl$ or the corresponding $(RO)_8Pc$ complex. Derivatives of these were prepared by replacing the very hydrolytically sensitive chloride by OH, OR (R = Me, Et, ^{i}Pr), OSO_2(*p*-tol), OSO_2CF_3, or OCH_2CH_2SH. The dimethyl complex $Si(R_8Pc)(Me)_2$ was prepared from $Si(R_8Pc)(Me)Cl$ with MeMgI although it proved difficult to isolate. Like their porphyrin counterparts the complexes are highly light-sensitive. Dehydration of $Si(R_8Pc)(Me)(OH)$ in a sealed tube produced the μ-oxo dinuclear complex $MeSi(R_8Pc)$—O—$OSi(R_8Pc)(Me)$, and the methyl groups in this complex were replaced by OH or OCH_2CH_2SH groups by photolysis in benzene containing H_2O or $HOCH_2CH_2SH$. The silicon phthalocyanine complexes with appended thiol groups are useful for their ability to form thin films on gold surfaces.[108,109]

VI

PHOSPHORUS, ARSENIC, AND ANTIMONY

Porphyrin complexes containing the Group 15 elements were developed relatively late, with first phosphorus(V) and then antimony(V) porphyrins established as cationic, six-coordinate complexes of the form $[M(Por)(X)_2]^+$, where X = halide, hydroxide, or alkoxide. The compounds were prepared from either pentavalent reagents (PCl_5, $POCl_3$, $SbCl_5$) or from trivalent halides (PCl_3, $SbCl_3$) under oxidative conditions.[131] Arsenic porphyrin complexes proved much more difficult to prepare. The pentahalides such as PCl_5 and $SbCl_5$ are readily accessible but the arsenic congener $AsCl_5$ decomposes above −50°C, and the use of $AsCl_3$ with air or H_2O_2 as an oxidant resulted in demetallation. The first well—characterized arsenic porphyrin was not reported until 1996 and was prepared using a non-nucleophilic oxidant such as pyridinium tribromide.[132] Another problem that had dogged Group 15 porphyrin chemistry was the inability to prepare organosubstituted derivatives from the reactions of $[M(Por)(X)_2]^+$ with organolithium or Grignard reagents, which led usually to demetallated or decomposed products.[133] However, in 1994, Akiba reported that the use of trialkylaluminum reagents, particularly $AlMe_3$, overcame this problem, resulting in the first alkylantimony porphyrin complexes,[134] followed soon by the phosphorus and arsenic analogs.[132,135] Since then, organoelement porphyrin derivatives containing phosphorus, arsenic, and antimony have flourished, all reported from the research group of Akiba. These developments have been reviewed recently[136] and the main advances will be summarized here. Bismuth porphyrins have been the last to be developed and the first structurally characterized example appeared only very recently.[137] In contrast to the lighter Group 15 elements bismuth porphyrins are found only in the +3 oxidation state and to date no organometallic examples have been reported.

Over the last 10 years, phosphorus porphyrins are finding increasing applications especially in extended arrays. Advantages are the high oxidation state central element which can form complexes of the type $[P(Por)X_2]^+$ which bear a permanent positive charge but are not redox active at the central element. The six-coordinate complexes can be linked into polymers utilizing the ready substitution chemistry of the P—X bonds.[138] Antimony(V) porphyrins are finding applications based on their photosensitizing properties.[139]

A. *Synthesis, Spectroscopy, Structure, and Electrochemistry*

All of the organoelement Group 15 porphyrins contain the element in the +5 oxidation state. Most of the complexes are cationic, of the form $[M(Por)R_2]^+$ or $[M(Por)(R)X]^+$, and are generally isolated as halide, ClO_4^-, or PF_6^- salts. The nature of the counterion does not affect the chemistry of the cationic complexes, and in the discussion that follows the counterions have been omitted from formulae for the sake of simplicity. One hallmark of the organoelement Group 15 porphyrin complexes is the inertness of the E—C bonds compared to their Groups 13 and 14 counterparts. The compounds are generally stable to air, moisture, and light, and can be chromatographed and recrystallized with relative ease. A listing of organoelement Group 15 porphyrin complexes is given in Table VI.[35,132,134–136,140–146]

Organophosphorus porphyrins can be prepared by two general routes.[135] The first involves stirring the readily available dichloro complex $[P(OEP)Cl_2]^+$ with AlR_3 (R = Me, Et) in CH_2Cl_2 at room temperature, giving $[P(OEP)R_2]^+$ (R = Me, Et). Alternatively, the reaction of $RPCl_2$ with the free base porphyrin with air-oxidation during workup gave $[P(OEP)(R)(OH)]^+$ (R = Me, Et, Ph). A special feature of these complexes is the acid–base chemistry of the hydroxo group proton, which can be removed by treatment with base to give the neutral P=O

TABLE VI

PHOSPHORUS, ARSENIC, AND ANTIMONY PORPHYRIN COMPLEXES, $[M(Por)(X)(Y)]^+$ AND M(Por)(X)(=O)

Porphyrin	X	Y	Reference
Neutral porphyrin complexes M(Por)(R)(=O)			
Phosphorus			
OEP	Me, Et, Ph	=O	135, 136
Arsenic			
OEP	Me, Et, Ph	=O	132, 140
Cationic porphyrin complexes $[M(Por)(X)(Y)]^+$			
Phosphorus			
OEP	Me	Me, Et	135
	Me	OH, F	136
	Et	OH, NEt_2	135, 136
	Ph	OH, O^nPr	135, 136
OEP	Me	OAl(OEP)	35
	Me	OOH	136
OETPP	Me	Me, F, OH	141
Arsenic			
TPP	Me	Me	142
OEP	Me	Me, Et	140
	Et	Et	140
	Me, Et, Ph	OH	132, 140
	Me	Cl, OMe, OEt, O^iPr, NH^nBu, NH-*p*-tol	140
	Me	OOH	136
OEP	Me	OAl(OEP)	35
Antimony			
OEP	Me	Me	134
	Me, Et	OH	134
	Me	OAl(OEP)	35
TPP	Me	Me, Et, ^iBu	143, 144
	Me, Et	OH, Cl	136, 143
	Me	F, Br	143
	Me	OMe, OEt, O-*p*-tol	143, 144
	Me	OC(O)-*m*-C_6H_4Cl	144
	Me	OO^tBu, OOC(O)-*m*-C_6H_4Cl, OOH	136, 144
	Me	NH-*p*-tol, $NHCH_2Ph$, S-*p*-tol	144
Neutral corrole complexes M(OEC)(X)(Y)			
P(OEC)	H	H	145
	Me	Me	145
	Ph	Ph	145
Cationic corrole complexes $[M(OEC)(X)]^+$			
P(OEC)	Me		145
As(OEC)	Me		146

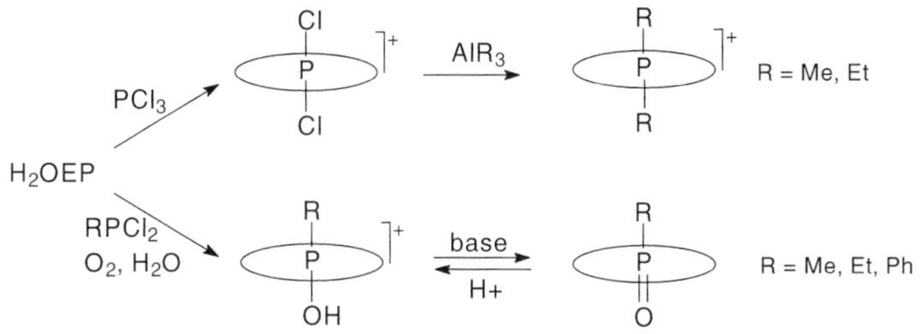

SCHEME 6

bonded complexes P(OEP)(R)(=O). The presence of the P=O double bond is confirmed by the short P—O bond length observed in the molecular structure of P(OEP)(Et)(=O). Treatment with dilute HCl reforms the cationic hydroxo complex. These transformations are summarized in Scheme 6.[135] Substitution of the hydroxo ligand in $[P(OEP)(R)(OH)]^+$ has produced a range of mixed ligand complexes $[P(OEP)(R)(X)]^+$, where R = Me, Et, or Ph, and X = halide, OR, or NR_2 (see Table VI).[136] An important consequence of the small size of the phosphorus atom is the prevalence of nonplanar distortions of the porphyrin ring (see the following for further discussion). Such distortions can also be introduced by using the sterically encumbered porphyrin OETPP, and three phosphorus complexes containing this ligand have been reported, $[P(OETPP)(Me)(X)]^+$ (X = Me, F, OH).[141]

The difficulties in accessing arsenic porphyrin chemistry were outlined earlier. Briefly, reaction of $AsCl_3$ with $H_2(Por)$ gives a poorly characterized arsenic(III) species which is oxidized using pyridinium tribromide. Hydrolysis during the workup forms $[As(Por)(OH)_2]^+$, followed by treatment with oxalyl bromide to give $[As(Por)Br_2]^+$ which serves as a precursor to $[As(Por)R_2]^+$ through reactions with AlR_3. Limiting the reaction time gives the unsymmetrical complexes $[As(OEP)(R)(OH)]^+$ (R = Me, Et), from which can be prepared $[As(OEP)(R)(X)]^+$ (X = Cl, Et, OR or NHR) through further substitution reactions (Table VI, Scheme 7).[140] Like its phosphorus analog, $[As(OEP)(Ph)(OH)]^+$ is prepared using $PhAsCl_2$ with H_2OEP. The hydroxo complexes also show acid–base behavior and the neutral complexes As(OEP)(R)(=O) (R = Me, Ph, Et) can be isolated.[132,140,142]

Several symmetrical $[Sb(Por)R_2]^+$ and a wide range of unsymmetrical $[Sb(Por)(R)(X)]^+$ organoantimony porphyrin complexes are known. These are listed in Table VI, and their chemistry is summarized in Scheme 8. Several differences are evident for antimony porphyrins compared to their phosphorus and arsenic relatives. The diorgano complexes $[Sb(Por)R_2]^+$ are prepared from the reaction of AlR_3 with the antimony(III) precursor Sb(Por)Br, whereas reaction of the antimony(V) precursor $[Sb(TPP)(OMe)(OH)]^+$ with AlR_3 gives $[Sb(TPP)(R)(OH)]^+$. The chloroantimony complex $[Sb(TPP)(R)(Cl)]^+$ is less reactive than its arsenic counterpart, reacting only with methanol to give $[Sb(TPP)(R)(OMe)]^+$, and the more labile bromo analog $[Sb(TPP)(R)(Br)]^+$ was used for further substitution reactions. Finally, the acid–base chemistry observed for

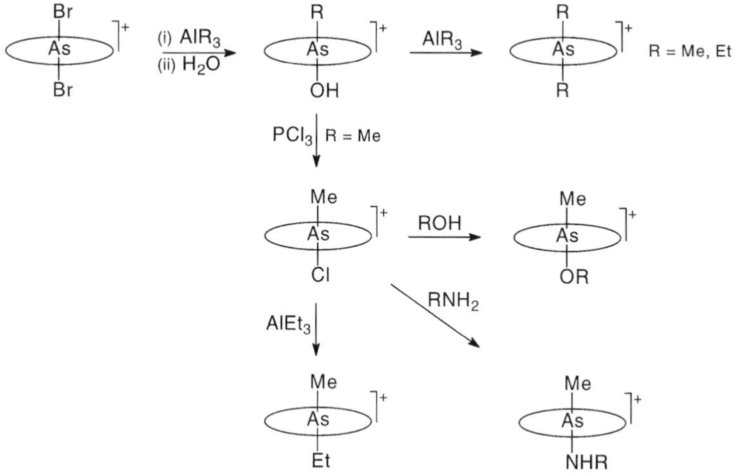

SCHEME 7

$[P(OEP)(R)(OH)]^+$ and $[P(OEP)(R)(OH)]^+$ does not occur for the corresponding antimony hydroxo complexes.[134,143,144]

All three hydroxo species $[M(OEP)(Me)(OH)]^+$ (M = P, As, Sb]) are sufficiently acidic to react with the aluminum porphyrin complex Al(OEP)Me, which is known to eliminate methane on reaction with protic reagents. Three novel binuclear

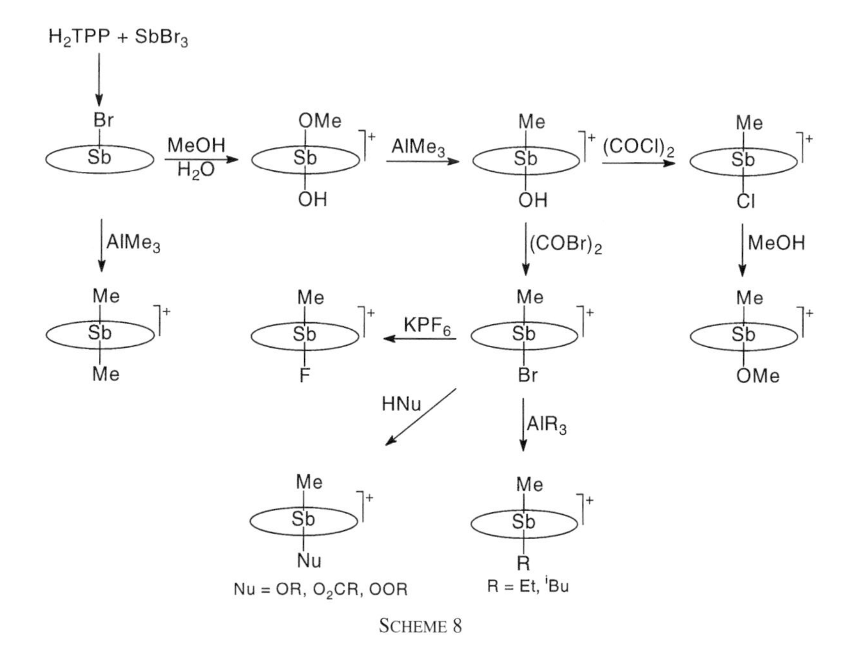

SCHEME 8

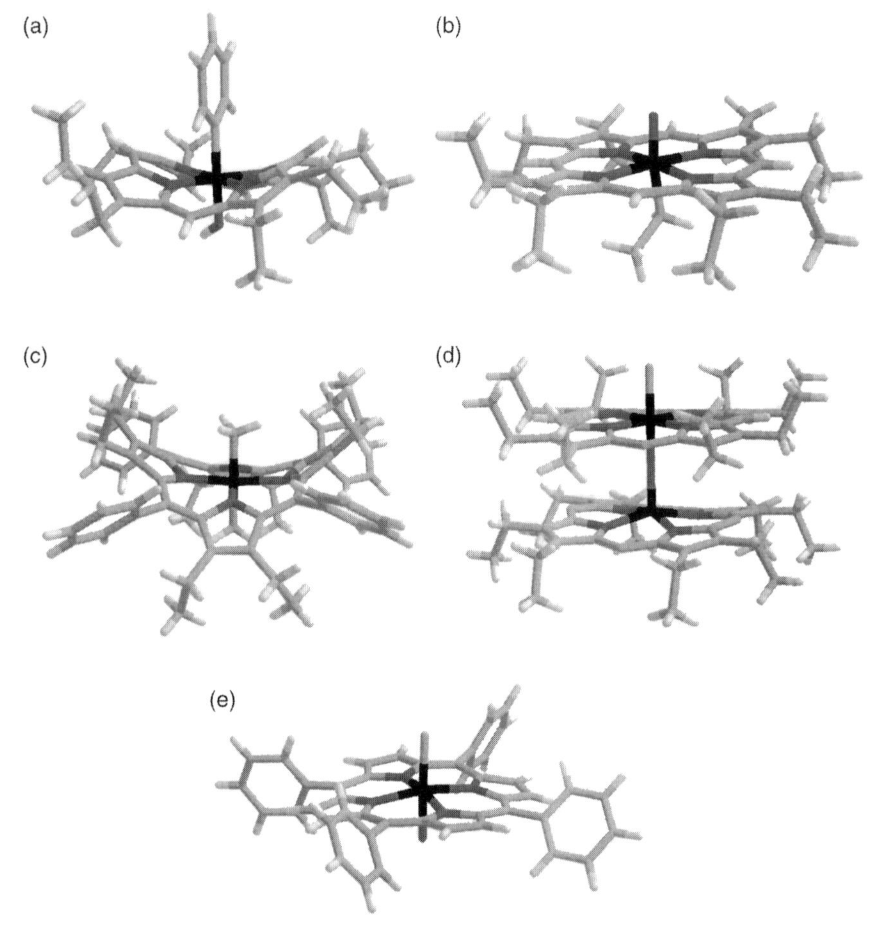

FIG. 7. Molecular structures of selected Group 15 organometallic porphyrin cations and complexes: (a) $[P(OEP)(C_6H_5)(OH)]PF_6$,[135] (b) $P(OEP)(C_2H_5)({=}O)$,[135] (c) $[P(OETPP)(CH_3)_2]PF_6$,[141] (d) $[(OEP)Al{-}O{-}As(OEP)(CH_3)]ClO_4$,[35] (e) $[Sb(TPP)(CH_3)(F)]PF_6$.[134]

μ-oxo complexes resulted, MeM(OEP)—O—Al(OEP), and the arsenic complex has been structurally characterized (Fig. 7). The rates of the reactions decreased in the order P > As > Sb, reflecting the relative difference in the acidity of the M—OH group.[35]

Metalloporphyrins containing low valent main group elements show hyper-type UV-visible spectra and the corresponding higher valent complexes show normal spectra.[131,147] Despite this, the phosphorus complexes $[P(OEP)R_2]^+$ show spectra with distinctly hyper character, with two extra Soret bands apparent. The extra Soret band in hyper complexes has been assigned to charge transfer from the low valent central element to the porphyrin ligand, and the observation of this band in the formally phosphorus(V) porphyrin complexes was attributed to the electron-rich alkyl ligands.[135] This effect is not observed for the oragnoarsenic porphyrins which show normal UV-visible spectra.[140] In general, the bands for the organo-substituted Group 15 porphyrin complexes are shifted to longer wavelength relative to the corresponding OH- or halide-substituted complexes, as seen for Groups 13 and 14 organometallic derivatives.

The ^{1}H NMR spectra show the usual upfield shifts for the axial ligands induced by the porphyrin ring current. The chemical shift of the axial methyl group in the series of antimony complexes $[Sb(TPP)(Me)(X)]^+$ shows a progressive downfield shift as the electronegativity of the axial ligand X increases. This is indicated by a linear relationship in a plot of the values of δ_{Me} versus σ_1 (the Hammet constant for inductive effects) for 14 compounds. No correlation is seen for the chemical shift of the porphyrin β-pyrrolic protons indicating that electron density at the porphyrin core is not greatly affected by the axial substituents.[144]

Electrochemical data have been collected for a selection of the antimony OEP and TPP complexes including $[Sb(Por)Me_2]^+$ and $[Sb(Por)(R)(OH)]^+$ (R = Me, Et). The complexes show one-electron oxidations and reductions at the porphyrin rings. Spectroelectrochemistry indicated that small amounts of antimony(III) products may be formed through a chemical reaction following the first reduction.[143]

Photocleavage of the element—carbon bonds features prominently in Groups 13 and 14 organoelement porphyrins but is not observed for Group 15 complexes. This photolytic stability has been recently exploited in a very interesting application in which phosphorous etioporphyrin complexes P(EtioPor)(R)(=O), where R = Ph or Et, act as photocatalysts for the hydration of benzonitrile.[148] For example, irradiation of P(EtioPor)(Ph)(=O) in MeOH with 10 equivalents of benzonitrile in the presence of aqueous base and oxygen gives a 95% yield of benzamide after 25 h. The base is consumed over the course of the reaction. Irradiation is essential and the reaction could be stopped and started by turning the light off and on. The phosphorous porphyrin is proposed to act as a photosensitizer for electron transfer from OH^- to O_2. This kind of photosensitization has also been observed for antimony porphyrins.[139] Addition of DMSO as an oxygen acceptor dramatically accelerates the reaction. Overall, the organophosphorus porphyrin sensitizer is useful because of its resistance to photochemical decomposition, allowing hydration of a nitrile to an amide under mild conditions without the use of peroxy compounds.[148]

One further recent preliminary report describes the preparation of reactive hydroperoxide compounds $[M(OEP)(Me)(OOH)]^+$ (M = P, As, Sb) and their oxygen atom transfer chemistry with triphenylphosphine,[136] indicating that further interesting chemical applications of organoelement Group 15 porphyrins might be expected.

The trianionic octaethylcorrole ligand forms complexes with a different net charge than complexes of the dianionic porphyrin ligand. The chemistry of Group 15 corrole complexes is very new and still under development. Arsenic, antimony, and bismuth corrole complexes containing the central element in the +3 oxidation state have the simple formulation M(OEC). As(OEC) reacts with methyl iodide to produce the cationic complex $[As(OEC)(Me)]^{+}I^{-}$, but similar chemistry has not been reported for the heavier elements.[146] Phosphorus corroles bearing oxygen axial donors have been reported with a variety of formulations, including [P(EMC)(OH)]Cl,[149] $P(tfpc)(OH)_2$,[118] and P(OEC)(=O).[145] The last of these was prepared from the reaction of H_3OEC with PCl_3, with oxidation to phosphorus(V) occurring on contact with H_2O and air during the workup. The compound reacts with $LiAlH_4$ to give the only main group porphyrinoid hydride complex to be reported so far, $P(OEC)H_2$. The ^1H NMR spectrum of this unusual compound shows a $^{31}P-^{1}H$ coupling constant of 921 Hz. P(OEC)(=O) can be converted to organophosphorus derivatives $P(OEC)Ph_2$, $[P(OEC)(Me)]^+$, and $P(OEC)Me_2$ by reaction with PhMgBr or MeMgI. Electrochemistry of $P(OEC)Ph_2$ and $P(OEC)Me_2$ shows a reversible one-electron oxidation and reduction at the corrole ring in each case.[145] Both $[As(OEC)(Me)]^+$ and $[P(OEC)(Me)]^+$show a reversible one-electron oxidation, but for both complexes the first reduction is irreversible and results in loss of the methyl group and formation of an As(III) or P(III) product.[145,146]

Selected data for structurally characterized Group 15 porphyrin complexes are given in Table VII, and representative structures are shown in Fig. 7. The focus for the structural studies on the phosphorus porphyrins has been to elucidate the relationship between the electronic properties of the axial ligands and the degree of ruffling (nonplanar distortion) of the porphyrin ligand. As expected, both the mean E—C and E—N bond lengths increase from phosphorus to arsenic to antimony. The Me and OH groups in $[Sb(TPP)(Me)(OH)]PF_6$ were disordered,[143] whereas in the arsenic analog, $[As(OEP)(Me)(OH)]ClO_4$, the perchlorate counterion is hydrogen-bonded to the OH group allowing the two groups to be distinguished crystallographically.[140] The unsymmetrical antimony complexes $[Sb(TPP)(Me)(X)]^+$ show a displacement of the Sb atom out of the N_4 plane toward the carbon atom, and the degree of displacement becomes larger as the X group becomes more electronegative.[144]

B. *Ruffling Deformations in Groups 14 and 15 Porphyrins*

The porphyrin ligand is sufficiently flexible to adopt nonplanar conformations in response to steric or electronic effects induced by the central metal, the axial ligands, or substitution at the porphyrin periphery. Coordination of the small, nonmetallic Si(IV), Ge(IV), P(V), and As(V) ions to porphyrins offer an ideal opportunity to study not only the direct effect on this phenomenon of

TABLE VII

SELECTED DATA FOR STRUCTURALLY CHARACTERIZED GROUP 15 PORPHYRIN COMPLEXES

	M—C bond length/Å	M—N_{av} bond length/Å	M—N_4 plane/Å	Other		Reference
$[P(OEP)(C_6H_5)(O^nPr)]ClO_4$		1.869(7)				136
$[P(OEP)(CH_3)(F)]PF_6$		1.85(1)				136
$[P(OEP)(C_6H_5)(OH)]PF_6$	1.865(7)	1.873(5)		P—O	1.636 Å	135
$[P(OEP)(CH_2CH_3)(OH)]ClO_4$	1.971(6)	1.884(3)		P—O	1.635 Å	135
$[P(OEP)(CH_2CH_3)(NEt_2)]ClO_4$		1.92(1)				136
$[P(OEP)(CH_2CH_3)_2]PF_6$		1.947(3)				136
$[P(OEP)(CH_3)_2]PF_6$	1.863(8) 1.864(8)	1.974(6)				135
$P(OEP)(=O)(CH_2CH_3)$	1.84(1) 2.00(1)			P=O	1.487(8) Å	135
$[P(OETPP)(CH_3)(OH)]ClO_4$	1.854(4)	1.849(4)		P—O	1.657(3) Å	141
$[(OETPP)(CH_3)_2]PF_6$	1.85(1) 1.82(1)	1.94(1)				141
$[P(OETPP)(CH_3)(F)]PF_6$	1.83(2)	1.810(8)		P—F	1.648(9)	141
$[CH_3(OEP)As\text{-}O\text{-}Al(OEP)]ClO_4$	1.971(4)	As 2.031 Al 2.011		As—O Al—O	1.690(3) Å 1.731(3) Å	35
$[As(OEP)(CH_3)(OH)]ClO_4$	1.870(6)	2.004		As—O	1.826(6)	132, 140
$[As(TPP)(CH_3)_2]PF_6$	1.953(3)	2.062				142
$[Sb(OEP)(CH_3)_2]PF_6$	2.121(7) 2.061(9)	2.106		C—Sb—C	178.8(3)	134
$[Sb(TPP)(CH_3)F]PF_6$	2.115(6)	2.086	0.201 toward C	Sb—F	1.928(3)	143
$[Sb(TPP)(CH_3)(O_2CAr)]PF_6$ Ar = $m\text{-}C_6H_4Cl$	2.13(1)	2.09	0.182 toward C	Sb—O	2.040(9)	144

the size of the central element but also the electronic effect of axial ligands of differing electronegativities. Structural data are available for phosphorus, arsenic, silicon, and germanium porphyrins in which the axial ligands vary from the very electronegative fluoride ligand to the much more electron-donating alkyl ligands.

One of the two most common nonplanar deformations of the porphyrin ligand is the saddle conformation in which the pyrrole C_β—C_β bonds are displaced alternately above and below the mean 24-atom plane. The other is the ruffled conformation in which the C_{meso} carbon atoms are displaced alternately above and below the mean 24-atom plane with concomitant twisting of the pyrrole rings. There are various measures for the extent of ruffling; for example, the $C_{meso} \cdots C_{meso}$ cross-ring distance decreases and the mean displacement of C_{meso} from the mean 24-atom plane increases as the structures become more ruffled. The average M—N distance is also significant, with a fairly sharp threshold dividing the ruffled from the planar structures determined to be 1.95 Å for a series of phosphorus porphyrin complexes[136] or 2.00–2.02 Å when a larger number of elements are considered.[150]

An extensive and consistent series of structural data has very recently become available for phosphorus OEP complexes, for which structural data for over 10 complexes with different sets of axial ligands can be compared. Table VIII lists these complexes, together with the average M—N bond length and Δr, a parameter derived from the root mean squares of displacements for the entire

TABLE VIII

PHOSPHORUS, ARSENIC, AND SILICON AND GERMANIUM PORPHYRINS: M—N BOND LENGTH AND DEGREE OF RUFFLING[a]

$[E(Por)(X)(Y)]^{n+}$					
$[E(Por)]^{n+}$	X	Y	M—N bond length/Å	Δr/Å[b]	Deformation
$[P(OEP)]^+$	F	OH	1.848(5)	0.539	Ruffled
$[P(OEP)]^+$	Cl	Cl	1.840(8)	0.518	Ruffled
$[P(OEP)]^+$	O^nPr	Ph	1.869(7)	0.505	Ruffled
$[P(OEP)]^+$	F	Me	1.85(1)	0.499	Ruffled
$[P(OEP)]^+$	OH	Ph	1.877(5)	0.485	Ruffled
$[P(OEP)]^+$	OH	Et	1.892(5)	0.462	Ruffled
$[P(OEP)]^+$	NEt_2	Et	1.92(1)	0.381	Ruffled
$[P(OEP)]^+$	Et	Et	1.947(3)	0.266	Ruffled
$[P(OEP)]^+$	Me	Me	1.990(8)	0.128	Planar
$[P(OEP)]^+$	=O	Et	2.01(1)	0.073	Planar
$[P(OETPP)]^+$	Me	Me	1.94(1)	0.767	Saddle
$[P(OETPP)]^+$	OH	Me	1.849(4)	0.638	Ruffled
$[P(OETPP)]^+$	F	Me	1.810(8)	0.690	Ruffled
$[As(TPP)]^+$	F	F	1.927	0.427	Ruffled
$[As(OEP)]^+$	F	F	1.966	0.024	Planar
$[As(OEP)]^+$	OH	Me	2.009(2)	0.086	Planar
$[As(TPP)]^+$	Me	Me	2.062(2)	0.035	Planar
Si(TPP)	OTf	OTf	1.870	0.468	Ruffled
Si(TTP)	F	F	1.918	0.371	Ruffled
Si(TPP)	Ph	Ph	1.947	0.288	Ruffled
Si(TPP)	C≡CPh	C≡CPh	1.98	0.037	Planar (waved)
Si(TPP)	CH_2SiMe_3	CH_2SiMe_3	2.01	0.023	Planar
Ge(TPP)	OAc	OAc	1.963	0.025	Planar
Ge(OEP)	F	F	1.986	0.265	Ruffled
Ge(TPP)	OMe	OMe	2.032	0.032	Planar
Ge(TPP)	Cl	Cl	2.019	0.021	Planar

[a] References 92, 101, 136, and 150, and references therein.

[b] Δr, a parameter derived from the root mean squares of displacements for the entire 24-atom core.

24-atom core.[135,136,140] Data for silicon, germanium, and arsenic complexes are also included.[92,101,150] Overall, these data show clearly that as the electronegativity of the axial ligand increases, the M—N bond length contracts and the extent of ruffling increases. These conclusions are supported by a recent density functional theory study on silicon, germanium, phosphorus, and arsenic porphyrins. The calculations show that, in addition to electronegativity effects, the *ortho* hydrogens of axial phenyl groups can contribute to ruffling.[150] From the data in Table VIII the ruffling threshold appears to lie between an M—N bond distance of 1.95 and 2.00 Å. Complexes with M—N bond lengths less than 1.95 Å are clearly ruffled, and those above 2.00 Å are planar, with not quite such a clear-cut correlation between 1.95 and 2.00 Å.

Some interesting fine detail is apparent comparing specific pairs of complexes and from the structures shown in Fig. 7. For example, $[P(OEP)(Et)(OH)]^+$ is markedly ruffled, whereas P(OEP)(Et)(=O) is planar, despite the similarity in donor atoms, and indicates that the doubly bonded oxygen atom has good electron donor properties.[135] The ruffled porphyrin in $[P(OEP)(Et)(OH)]^+$ compares with the planar porphyrin in $[As(OEP)(Me)(OH)]^+$ showing that replacing phosphorus by the slightly larger arsenic atom relieves some of the strain.[132,135] The porphyrin ligand can influence the geometry, as shown by ruffled $[As(TPP)F_2]^+$ (As—N = 1.927 Å) versus planar $[As(OEP)F_2]^+$ (As—N = 1.966 Å).[142] The sterically crowded OETPP ligand is already predisposed to nonplanar porphyrin distortions. However, the three phosphorus complexes show a similar trend with a transition from the less strained saddle-shaped conformation to the more strained ruffled geometry as the electronegativity of the axial ligands increases.[141]

VII

MAIN-GROUP PORPHYRIN COMPLEXES CONTAINING METAL—METAL BONDS

One further category of main group organometallic complexes that will be considered in this review comprises main group porphyrins which bear one or more axial bonds to an organotransition metal fragment.[15] These complexes will be discussed for two reasons. First, they are organometallic species by virtue of the transition—metal carbon bonds in the axial ligands. Second, using the isolobal analogy, any transition metal ML_n fragment which is isolobal with CH_3 could conceptually replace the methyl groups in $In(Por)CH_3$ or $Ge(Por)(CH_3)_2$. Main group porphyrin organometallic complexes are limited to σ-bonded carbon ligands. Carbene-type ligands can be envisaged for Group 14 porphyrin complexes, which would have the form $Ge(Por)(=CR_2)$. Again, using the isolobal analogy,

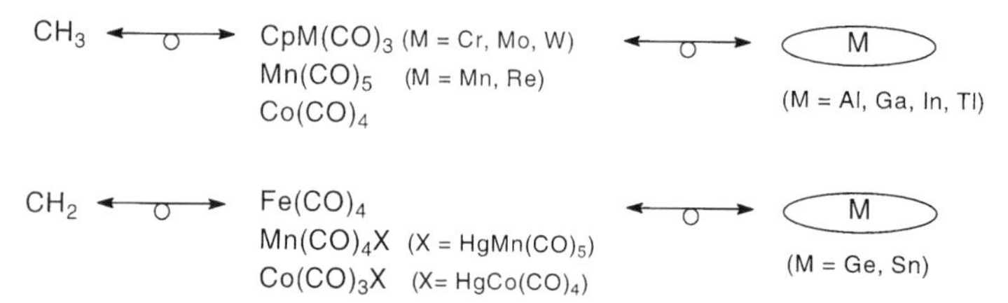

FIG. 8. Isolobal relationship between methyl and methylene fragments, transition metal groups, and main group porphyrins.

any transition metal ML_n fragment which is isolobal with CH_2 could conceptually form a carbenoid complex with a Ge(Por) fragment, Indeed, In(Por) and Ge(Por) are themselves isolobal with CH_3 and CH_2, respectively, and the analogy can be extended to diporphyrin complexes (Por)M-M'(Por'). Several of these types of complexes have been realized for Groups 13 and 14 porphyrin complexes, and the isolobal relationships are shown in Fig. 8. The complete list of complexes reported to date is given in Table IX,[151–173] data for structurally characterised examples in Table X, and representative structures are shown in Fig. 9. Most of this chemistry was reported during the 1980s, and there have been few significant developments since then,[174] so only an outline is given in the following.

The complete set of 20 indium and thallium complexes In(Por)ML_n where Por = OEP and TPP, and where ML_n = M(Cp)$(CO)_3$ (M = Cr, Mo, W), $Mn(CO)_5$, and $Co(CO)_4$ have been reported. The references in Table IX give details of the syntheses and spectroscopic and electrochemical studies. Three different synthetic routes are possible, shown for the indium complexes in Eqs. (16–18).[154] The most general route involves reaction of the ML_n^- anion with In(Por)Cl [Eq. (16)]. An alternative is photochemical cleavage both of the In—C bond in In(Por)CH_3 and the M—M bond in the dimer $L_nM—ML_n$ [Eq. (17)]. The complexes show hyper-type UV-visible spectra and bands shifted to longer wavelength than In(Por)Cl, indicative of increased charge density on the In(Por) fragment, and the metallate anions are deduced from spectral data to be more electron donating than all of the alkyl and aryl σ-bonded ligands with the exception of the t-butyl ligand. IR data for the CO stretching bands were useful for investigating electronic effects in detail. Very weak or nonexistent molecular ion peaks are observed in the mass spectra, indicating that the In—M bonds are relatively weak. This is borne out by the electrochemical data, for which one-electron oxidative and reductive processes are both followed by In—M bond cleavage.[154] The complexes are also somewhat light sensitive, especially when handled in chlorinated solvents, in which decomposition to In(Por)Cl occurs.[157] The thallium complexes exhibit more stable Tl—M bonds than do their indium counterparts, with reversible one-electron oxidations to give radical cations with

TABLE IX

METAL—METAL BONDED PORPHYRIN COMPLEXES (Por)M—M′L_n

Group 13 complexes (Por)M—M′L_n			
M	Porphyrin	M′L_n	Reference
In	OEP, TPP	$CrCp(CO)_3$ $MoCp(CO)_3$ $WCp(CO)_3$ $Mn(CO)_5$ $Co(CO)_4$	151–156
In	TBPP,[a] PFPP[b]	$MoCp(CO)_3$ $Mn(CO)_5$ $Co(CO)_4$	157
Tl	OEP, TPP	$CrCp(CO)_3$ $MoCp(CO)_3$ $WCp(CO)_3$ $Mn(CO)_5$ $Co(CO)_4$	156, 158, 159
Al	OEP	$Re(CO)_5$	160
Ga, In, Tl	OEP, TPP	$Re(CO)_5$	160
Group 14 complexes (Por)M—M′L_n			
Ge, Sn	OEP, TPP, TTP	$Fe(CO)_4$	161–164
Sn	TPP	$Mn(CO)_4$—Hg—$Mn(CO)_5$ $Co(CO)_3$—Hg—$Co(CO)_4$	153, 165
Sn	DecPTriP,[c] PalPTriP[d]	$Fe(CO)_4$	166
$[Sn]^+$	TTP	$Re(CO)_5$ [(TTP)Sn—$Re(CO)_5$]BF_4	167

Trinuclear groups 13 and 14 complexes (PorM)—M′L_n—M(Por)			
M(Por)	M′L_n	Complex	Reference
In(OEP), In(TPP)	$Fe(CO)_4$	(Por)In—$Fe(CO)_4$—In(Por)	151, 168
$[Sn(TTP)]^+$	$Re(CO)_4$	[(TTP)Sn—$Re(CO)_4$—Sn(TTP)]BF_4	167
Sn(TPP)	$Re(CO)_3$	$(CO)_3$Re—Sn(TPP)—$Re(CO)_3$	169
Sn(TPP)	$CRe(CO)_3$	$(CO)_3$ReC—Sn(TPP)—$CRe(CO)_3$	170

Dinuclear group 13 diporphyrin complexes (Por)M—M′(Por′)				
M	Por	M′	Por′	Reference
In	OEP	Rh	OEP	171
In	OEP, TPP	Rh	OEP, TPP	172
Tl	OEP, TPP	Rh	OEP, TPP	173

[a]TBPP = dianion of tetrakis(3,5-di-t-butylphenyl)porphyrin.
[b]PFPP = dianion of tetrakis(2,3,4,5,6-pentafluorophenyl)porphyrin.
[c]DecPTriP = dianion of (4-N-decanoylaminophenyl)triphenylporphyrin.
[d]PalPTriP = dianion of (4-N-palmitoylaminophenyl)triphenylporphyrin.

TABLE X

STRUCTURAL DATA FOR METAL—METAL BONDED PORPHYRIN COMPLEXES

	M—M′ bond length/Å	M—M′ bond length/Å	M—N_{av} bond length/Å	M—N_4 plane/Å	Other		Reference
(OEP)InMn$(CO)_5$	In—Mn	2.705(1)	2.193	0.744	In—M—C	88.8(5)°	154
					Mn—CO_{eq}	1.839(4) Å	
					Mn—CO_{ax}	1.809(6)Å	
(OEP)TlMn$(CO)_5$	Tl—Mn	2.649(1)	2.263	0.939	Tl—Mn—C	84.3(9)°	158, 159
					Mn—$CO_{eq(av)}$	1.847(8)	
					Mn—CO_{ax}	1.834(8)	
(OEP)GeFe$(CO)_4$	Ge—Fe	2.370(2)	2.078		Ge—Fe—C_{ax}	178.9(4)°	164
					Ge—Fe—C_{eq}	88.7°	
(OEP)SnFe$(CO)_4$	Sn—Fe	2.492(1)	2.187	0.818(9)	Sn—Fe—C_{ax}	1.791(1)°	161
(TPP)Sn—$Mn^1(CO)_4$—Hg—$Mn^2(CO)_5$	Sn—Mn	2.554(3)	2.18(1)	0.85	Mn^1—Hg	2.582(3) Å	153, 165
					Hg—Mn^2	2.659(3) Å	
					Sn—Mn—Hg	94.54°	
					Mn—Hg—Mn	164.4(1)°	
(OEP)In—Rh(OEP)	In—Rh	2.584(2)	In 2.202(2)	In 0.83	Por · · · Por	3.41 Å	171
			Rh 2.036(2)	Rh 0.01			
$(CO)_3$Re≡C—Sn(TPP)—C≡Re$(CO)_3$	Sn—C	2.14	2.07(2)		Re≡C	1.75 Å	170
					Sn—C≡Re	138.5(8)	

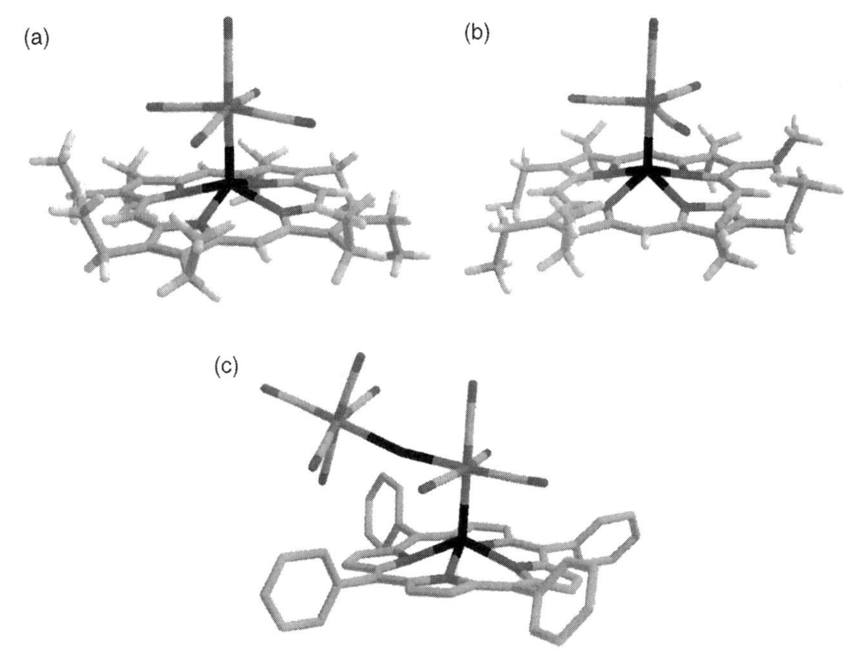

FIG. 9. Molecular structures of selected metal—metal bonded main group porphyrin complexes: (a) (OEP)InMn(CO)$_5$,[154] (b) (OEP)SnFe(CO)$_4$,[101] (c) (TPP)SnMn(CO)$_4$HgMn(CO)$_5$.[153]

appreciable lifetimes observed for some complexes. This parallels the greater stability of both the neutral and oxidized σ-bonded Tl—C bonded porphyrin complexes relative to their indium congeners.[158]

The trinuclear indium OEP and TPP complexes (Por)In—Fe(CO)$_4$—In(Por) can be prepared from In(Por)Cl with $Fe(CO)_4^{2-}$ or $HFe(CO)_4^-$, or from In(TPP)Me

$$\text{(Por)In–Cl} + [\text{ML}_n]^- \xrightarrow[\text{dark}]{\text{THF}} \text{(Por)In–ML}_n \quad (16)$$

$[\text{ML}_n]^-$ = $[\text{Mn(CO)}_5]^-$, $[\text{Cp(CO)}_4]^-$, $[\text{CpM(CO)}_3]^-$ (M = Cr, Mo, W)

$$\text{(Por)In–CH}_3 + \text{L}_n\text{M-ML}_n \xrightarrow[h\nu]{\text{THF}} \text{(Por)In–ML}_n \quad (17)$$

$\text{L}_n\text{M-ML}_n$ = $Co_2(CO)_8$, $[CpMo(CO)_3]_2$

$$\text{(Por)In–Cl} + Co_2(CO)_8 \longrightarrow \text{(Por)In–Co(CO)}_4 \quad (18)$$

with $Fe_2(CO)_9$. They show characteristics very similar to their dinuclear counterparts. Electrochemical data indicate that the two In(Por) units are equivalent and there is no interaction across the $Fe(CO)_4$ bridge.[168]

Examples of aluminum and gallium metal—metal bonded complexes are few, but the series of complexes $M(Por)Re(CO)_5$ for all the Group 13 elements Al, Ga, In, and Tl does provide an opportunity to make comparisons down the group.[160] The stability of the Group 15 metal—metal bonded complexes appears to parallel that of the σ-bonded complexes M(Por)R, the most stable of which are observed for thallium, and the least stable for aluminium and gallium. In addition, the In—M and Tl—M bond strengths increase with the nucleophilicity of the ML_n^- ion, so the more nucleophilic $Re(CO)_5^-$ ion is suited to the stabilization of the gallium and aluminum complexes. The indium and thallium examples of $M(Por)Re(CO)_5$ were prepared at ambient temperature, while the aluminum and gallium reactions were performed at $-70°C$. Electrochemical data for the series show oxidations followed by rapid M—M′ bond cleavage for Al, Ga and In, but formation of a stable radical cation for thallium. The oxidations show a correlation between electrode potentials and the axial substituent, suggesting that the electron is abstracted from an orbital with some M—M′ σ-bond character. Reductions, on the other hand, are insensitive to the nature of the axial substituent, and ring-centered one-electron reductions are followed in each case by M—M′ bond cleavage.[160]

The carbenoid Group 14 complexes contain Ge or Sn and TPP, TTP, or OEP, and are prepared from $M(Por)Cl_2$ with $Na_2Fe(CO)_4$. The UV-visible spectra of the complexes show two Soret bands indicative of the hyper-type spectra observed for low valent Ge(II) or Sn(II) porphyrin species. In contrast to the Group 15 examples, the germanium and tin complexes were stable after electroreduction, and two reversible one-electron processes could be observed. The one-electron oxidation processes occurred at the porphyrin ring and were followed by cleavage of the M—M′ bond.[163] Two cationic tin—rhenium complexes $[(TTP)Sn{-}Re(CO)_5]BF_4$ and $[(TTP)Sn{-}Re(CO)_4{-}Sn(TTP)]BF_4$,[167] were prepared during an attempt to repeat the reactions claimed to give the other two trinuclear tin-rhenium complexes shown in Table IX. The formulation of these two last species is doubtful.[169,170]

Structural data for $(OEP)MMn(CO)_5$ (M = In, Tl)[154,158,159] and (OEP)MFe-$(CO)_4$ (M = Ge, Sn)[161,164] allow for ready comparisons between like pairs of complex. A surprising feature is the that the Tl—Mn bond is shorter than the In—Mn bond, attributed to the thallium atom being displaced further out of the porphyrin plane and larger orbital overlap between the thallium atom and mangnese atoms.[158] The Groups 13 and 14 complexes have local C_{4v} and C_{3v} symmetry at the $Mn(CO)_5$ and $Fe(CO)_4$ sites, respectively, with different $M'{-}CO_{axial}$ and $M'{-}CO_{equatorial}$ bond lengths. In each case the $M'{-}CO_{eq}$ bonds are tilted slightly toward the main group element. In the indium complex the $Mn{-}CO_{ax}$ bond (1.839(4) Å) is longer than the $Mn{-}CO_{eq}$ bonds (1.809(6) Å) suggesting that the CO ligand has a larger *trans* influence than the In(OEP) unit.[154]

There is little doubt that it is appropriate to formulate the Group 13 metal–metal bonded complexes as M(III) species with electron-rich anionic metallate ions as the axial ligands. For the Group 14 complexes, however, two oxidation state formalisms are possible, either M(IV) porphyrins with Fe(-II) $Fe(CO)_4^{2-}$ donors, or M(II) porphyrins with zerovalent $Fe(CO)_4$ acceptors. The assignment of M(II) and Fe(0) oxidation states is consistent with IR and Mössbauer data, and with the hyper UV-visible spectra.[161,163] The same is true for the cationic tin—rhenium complexes $[(TTP)Sn{-}Re(CO)_5]BF_4$ and $[(TTP)Sn{-}Re(CO)_4{-}Sn(TTP)]BF_4$,[167] for which Sn(IV)/Re(−I) or Sn(II)/Re(+I) formalisms are possible, and again spectroscopic and structural data are consistent with Sn(II). In summary, the carbenoid germanium and tin porphyrin complexes behave as (Por)M: fragments acting as donors toward the electron-accepting transition metal fragment.

Metal—metal bonded porphyrin dimers are better known for transition metal than for main group porphyrins, with examples such as $[Mo(Por)]_2$, $[Ru(Por)]_2$, $[Os(Por)]_2$, and $[Rh(Por)]_2$ having proved to be pivotal species in organotransition metal porphyrin chemistry.[115,175] Homonuclear porphyrin dimers are almost completely unknown for main group elements, and the family of In—Rh and Tl—Rh dinuclear complexes (Por)M—Rh(Por) comprise the only heterobimetallic examples (Table IX).[171–173] The prototypical example, (OEP)In—Rh(OEP), was prepared from In(OEP)Cl and $[Rh(OEP)]^-$, and has been characterized structurally (Table X, Fig. 9).[171] A recent report claiming the preparation of a disilicon diphthalocyanine, $[Si(Pc)]_2$, from the reaction of $Cl_3Si{-}SiCl_3$ with diiminoisoindoline[176] has been shown to be in error.[177]

References

(1) Smith, K. M., Ed. *Porphyrins and Metalloporphyrins;* Elsevier: New York, 1975.

(2) Dolphin, D., Ed. *The Porphyrins;* Academic: New York, 1979; Vols. 1–7.

(3) Brothers, P. J.; Collman, J. P. *Acc. Chem. Res.* **1986,** *19,* 209.

(4) Guilard, R.; Lecomte, C.; Kadish, K. M. *Struct. Bond.* **1987,** *64,* 205.

(5) Guilard, R.; Kadish, K. M. *Chem. Rev.* **1988,** *88,* 1121.

(6) Guilard, R.; Van Caemelbecke, E.; Tabard, A.; Kadish, K. M. In *The Porphyrin Handbook;* Kadish, K. M.; Smith, K. M.; Guilard, R., Eds.; Academic: San Diego, 1999; Vol. 3, Ch 21.

(7) Brothers, P. J. *Adv. Organomet. Chem.* **2000,** *46,* 223–321.

(8) Belcher, W. J.; Boyd, P. D. W.; Brothers, P. J.; Liddell, M. J.; Rickard, C. E. F. *J. Am. Chem. Soc.* **1994,** *116,* 8416–8417.

(9) Belcher, W. J.; Breede, M.; Brothers, P. J.; Rickard, C. E. F. *Angew. Chem. Int. Ed.* **1998,** *37,* 1112–1114.

(10) Woo, L. K.; Grubisha, D. S.; Guzei, I. A. In *Book of Abstracts; 219th ACS National Meeting, San Francisco, CA, March 26–30, 2000,* 2000; pp. INOR-662.

(11) (a) Sessler, J. L.; Weghorn, S. J. *Expanded, Contracted and Isomeric Porphyrins;* Elsevier: Oxford, 1997. (b) Kadish, K. M.; Smith, K. M.; Guilard, R., Eds. *The Porphyrin Handbook;* Academic: San Diego, 1999; Vol. 2.
(12) (a) Arnold, J. J. *Chem. Soc. Chem. Commun.* **1990,** 976–978. (b) Arnold, J.; Dawson, D. Y.; Hoffman, C. G. *J. Am. Chem. Soc.* **1993,** *115,* 2707–2713.
(13) Arnold, J. In Kadish, K. M.; Smith, K. M.; Guilard, R., Eds. *The Porphyrin Handbook;* Academic: San Diego, 1999; Vol. 3, Ch 17.
(14) Brand, H.; Arnold, J. *Coord. Chem. Rev.* **1995,** *140,* 137–168.
(15) Barbe, J. M.; Guilard, R. In Kadish K. M.; Smith, K. M.; Guilard, R., Eds. *The Porphyrin Handbook;* Academic: San Diego, 1999; Vol. 3, Ch 19.
(16) (a) Scheidt, W. R.; Lee, Y. J. *Struct. Bond.* **1987,** *64,* 1. (b) Scheidt, W. R. In *The Porphyrin Handbook;* Kadish, K. M.; Smith, K. M.; Guilard, R., Eds.; Academic: San Diego, 1999; Vol. 3, Ch 16.
(17) Lemke, F. R.; Lorenz, C. R. *Recent. Res. Dev. Electroanal. Chem.* **1999,** *1,* 73–89.
(18) (a) Power, P. P. *Chem. Rev.* **1999,** *99,* 3463–3503. (b) Power, P. P. *J. Chem. Soc. Dalton Trans.* **1998,** 2939–2951. (c) Okazaki, R.; West, R. *Adv. Organomet. Chem.* **1996,** *39,* 231–273. (d) Brothers, P. J.; Power, P. P. *Adv. Organomet. Chem.* **1996,** *39,* 1–69.
(19) (a) Belcher, W. J. unpublished results. (b) Weiss, A.; Siebert, W.; Brothers, P. J.; Pritzkow, H. unpublished results.
(20) Rauschnabel, J.; Hanack, M. *Tetrahedron Lett.* **1995,** *36,* 1629–1632.
(21) Geyer, M.; Plenzig, F.; Rauschnabel, J.; Hanack, M.; Del Rey, B.; Sastre, A.; Torres, T. *Synthesis* **1996,** 1139–1151.
(22) Hanack, M.; Geyer, M. *J. Chem. Soc. Chem. Commun.* **1994,** 2253–2254.
(23) Potz, R.; Goldner, M.; Huckstadt, H.; Cornelissen, U.; Tutass, A.; Homborg, H. Z. *Anorg. Allg. Chem.* **2000,** *626,* 588–596.
(24) (a) Kobayashi, N.; Ishizaki, T.; Ishii, K.; Konami, H. *J. Am. Chem. Soc.* **1999,** *121,* 9096–9110. (b) Kobayashi, N. *J. Porphyrins Phthalocyanines* **1999,** *3,* 453–467.
(25) Inoue, S.; Takeda, N. *Bull. Chem. Soc. Jpn.* **1977,** *50,* 984–986.
(26) Guilard, R.; Zrineh, A.; Tabard, A.; Endo, A.; Han, B. C.; Lecomte, C.; Souhassou, M.; Habbou, A.; Ferhat, M.; Kadish, K. M. *Inorg. Chem.* **1990,** *29,* 4476–4482.
(27) Guilard, R.; Pichon-Pesme, V.; Lachekar, H.; Lecomte, C.; Aukauloo, A. M.; Boulas, P. L.; Kadish, K. M. *J. Porphyrins Phthalocyanines* **1997,** *1,* 109–119.
(28) Sugimoto, H.; Aida, T.; Inoue, S. *J. Chem. Soc. Chem. Commun.* **1995,** 1411–1412.
(29) Arai, T.; Inoue, S. *Tetrahedron* **1990,** *46,* 749–760.
(30) Takeda, N.; Inoue, S. *Bull. Chem. Soc. Jpn.* **1978,** *51,* 3564–3567.
(31) Asano, S.; Aida, T.; Inoue, S. *Macromolecules* **1985,** *18,* 2057–2061.
(32) Kojima, F.; Aida, T.; Inoue, S. *J. Am. Chem. Soc.* **1986,** *108,* 391–395.
(33) Kubo, H.; Aida, T.; Inoue, S.; Okamoto, Y. *J. Chem. Soc. Chem. Commun.* **1988,** 1015–1017.
(34) Arai, T.; Murayama, H.; Inoue, S. *J. Org. Chem.* **1989,** *54,* 414–420.
(35) Yamamoto, G.; Nadano, R.; Satoh, W.; Yamamoto, Y.; Akiba, K.-Y. *Chem. Commun.* **1997,** 1325–1326.
(36) Murayama, H.; Inoue, S.; Ohkatsu, Y. *Chem. Lett.* **1983,** 381–384.
(37) Hirai, Y.; Murayama, H.; Aida, T.; Inoue, S. *J. Am. Chem. Soc.* **1988,** *110,* 7387–7390.
(38) Tero-Kubota, S.; Hoshino, N.; Kato, M.; Goedken, V. L.; Ito, T. *J. Chem. Soc. Chem. Commun.* **1985,** 959–961.
(39) Rohmer, M. M. *Chem. Phys. Lett.* **1989,** *157,* 207–210.
(40) Murayama, H.; Inoue, S. *Chem. Lett.* **1985,** 1377–1380.
(41) Yoshinaga, K.; Iida, Y. *Chem. Lett.* **1991,** 1057–1058.
(42) Kuroki, M.; Aida, T.; Inoue, S. *J. Am. Chem. Soc.* **1987,** *109,* 4737–4738.

(43) (a) Sugimoto, H.; Aida, T.; Inoue, S. *Bull. Chem. Soc. Jpn.* **1995,** *68,* 1239–1246. (b) Aida, T.; Inoue, S. *Acc. Chem. Res.* **1996,** *29,* 39–48.
(44) Hosokawa, Y.; Kuroki, M.; Aida, T.; Inoue, S. *Macromolecules* **1991,** *24,* 824–829.
(45) Kuroki, M.; Nashimoto, S.; Aida, T.; Inoue, S. *Macromolecules* **1988,** *21,* 3114–3115.
(46) Takeuchi, D.; Watanabe, Y.; Aida, T.; Inoue, S. *Macromolecules* **1995,** *28,* 651–652.
(47) Watanabe, Y.; Yasuda, T.; Aida, T.; Inoue, S. *Macromolecules* **1992,** *25,* 1396–1400.
(48) Akatsuka, M.; Aida, T.; Inoue, S. *Macromolecules* **1994,** *27,* 2820–2825.
(49) Yasuda, T.; Aida, T.; Inoue, S. *Macromolecules* **1983,** *16,* 1792–1796.
(50) Endo, M.; Aida, T.; Inoue, S. *Macromolecules* **1987,** *20,* 2982–2988.
(51) Kuroki, M.; Watanabe, T.; Aida, T.; Inoue, S. *J. Am. Chem. Soc.* **1991,** *113,* 5903–5904.
(52) Adachi, T.; Sugimoto, H.; Aida, T.; Inoue, S. *Macromolecules* **1993,** *26,* 1238–1243.
(53) (a) Adachi, T.; Sugimoto, H.; Aida, T.; Inoue, S. *Macromolecules* **1992,** *25,* 2280–2281. (b) Sugimoto, H.; Kuroki, M.; Watanabe, T.; Kawamura, C.; Aida, T.; Inoue, S. *Macromolecules* **1993,** *26,* 3403–3410.
(54) Sugimoto, H.; Aida, T.; Inoue, S. *Macromolecules* **1994,** *27,* 3672–3674.
(55) Aida, T.; Sugimoto, H.; Kuroki, M.; Inoue, S. *J. Phys. Org. Chem.* **1995,** *8,* 249–257.
(56) Sugimoto, H.; Saika, M.; Hosokawa, Y.; Aida, T.; Inoue, S. *Macromolecules* **1996,** *29,* 3359–3369.
(57) Kasuga, K.; Moriwaki, N.; Handa, M. *Inorg. Chim. Acta* **1996,** *244,* 137–139.
(58) Aida, T.; Inoue, S. *J. Am. Chem. Soc.* **1983,** *105,* 1304–1309.
(59) Sugimoto, H.; Inoue, S. *Pure Appl. Chem.* **1998,** *70,* 2365–2369.
(60) Hirai, Y.; Aida, T.; Inoue, S. *J. Am. Chem. Soc.* **1989,** *111,* 3062–3063.
(61) Komatsu, M.; Aida, T.; Inoue, S. *J. Am. Chem. Soc.* **1991,** *113,* 8492–8498.
(62) (a) Henrick, K.; Matthews, R. W.; Tasker, P. A. *Inorg. Chem.* **1977,** *16,* 3293–3298. (b) Brady, F.; Henrick, K.; Matthews, R. W. *J. Organomet. Chem.* **1981,** *210,* 281–288.
(63) Coutsolelos, A.; Guilard, R. *J. Organomet. Chem.* **1983,** *253,* 273–282.
(64) Kadish, K. M.; Boisselier-Cocolios, B.; Coutsolelos, A.; Mitaine, P.; Guilard, R. *Inorg. Chem.* **1985,** *24,* 4521–4528.
(65) Balch, A. L.; Latos-Grazynski, L.; Noll, B. C.; Phillips, S. L. *Inorg. Chem.* **1993,** *32,* 1124–1129.
(66) Balch, A. L.; Hart, R. L.; Parkin, S. *Inorg. Chim. Acta* **1993,** *205,* 137–143.
(67) Arasasingham, R. D.; Balch, A. L.; Olmstead, M. M.; Phillips, S. L. *Inorg. Chim. Acta* **1997,** *263,* 161–170.
(68) Guilard, R.; Cocolios, P.; Fournari, P. *J. Organomet. Chem.* **1977,** *129,* C11–C13.
(69) Cocolios, P.; Guilard, R.; Fournari, P. *J. Organomet. Chem.* **1979,** *179,* 311–322.
(70) Tabard, A.; Guilard, R.; Kadish, K. M. *Inorg. Chem.* **1986,** *25,* 4277–4285.
(71) Kadish, K. M.; Van Caemelbecke, E.; D'Souza, F.; Medforth, C. J.; Smith, K. M.; Tabard, A.; Guilard, R. *Inorg. Chem.* **1995,** *34,* 2984–2989.
(72) Kadish, K. M.; D'Souza, F.; Van Caemelbecke, E.; Boulas, P.; Vogel, E.; Aukauloo, A. M.; Guilard, R. *Inorg. Chem.* **1994,** *33,* 4474–4479.
(73) Kadish, K. M.; Tabard, A.; Van Caemelbecke, E.; Aukauloo, A. M.; Richard, P.; Guilard, R. *Inorg. Chem.* **1998,** *37,* 6168–6175.
(74) Hanack, M.; Heckmann, H. *Eur. J. Inorg. Chem.* **1998,** 367–373.
(75) Kadish, K. M.; Tabard, A.; Zrineh, A.; Ferhat, M.; Guilard, R. *Inorg. Chem.* **1987,** *26,* 2459–2466.
(76) Tang, S.-S.; Sheu, M.-T.; Lin, Y.-H.; Liu, I.-C.; Chen, J.-H.; Wang, S.-S. *Polyhedron* **1995,** *14,* 301–305.
(77) Kadish, K. M.; Boisselier-Cocolios, B.; Cocolios, P.; Guilard, R. *Inorg. Chem.* **1985,** *24,* 2139–2147.
(78) Perry, J. W.; Mansour, K.; Lee, I. Y. S.; Wu, X. L.; Bedworth, P. V.; Chen, C. T.; Ng, D.; Marder, S. R.; Miles, P.; *et. al. Science (Washington, D.C.)* **1996,** *273,* 1533–1536.
(79) (a) Shirk, J. S.; Pong, R. G. S.; Flom, S. R.; Heckmann, H.; Hanack, M. *J. Phys. Chem. A* **2000,**

104, 1438–1449. (b) Rojo, G.; Martin, G.; Agullo-Lopez, F.; Torres, T.; Heckmann, H.; Hanack, M. *J. Phys. Chem. B* **2000,** *104,* 7066–7070.
(80) Paolesse, R.; Licoccia, S.; Boschi, T. *Inorg. Chim. Acta* **1990,** *178,* 9–12.
(81) Simkhovich, L.; Goldberg, I.; Gross, Z. *J. Inorg. Biochem.* **2000,** *80,* 235–238.
(82) Bendix, J.; Dmochowski, I. J.; Gray, H. B.; Mahammed, A.; Simkhovich, L.; Gross, Z. *Angew. Chem. Int. Ed.* **2000,** *39,* 4048–4051.
(83) Ghosh, A.; Wondimagegn, T.; Parusel, A. B. J. *J. Am. Chem. Soc.* **2000,** *122,* 5100–5104.
(84) Lecomte, C.; Protas, J.; Cocolios, P.; Guilard, R. *Acta Crystallogr., Sect. B* **1980,** *B36,* 2769–2771.
(85) (a) Guilard, R.; Cocolios, P.; Fournari, P.; Lecomte, C.; Protas, J. *J. Organomet. Chem.* **1979,** *168,* C49–C51. (b) Cocolios, P.; Fournari, P.; Guilard, R.; Lecomte, C.; Protas, J.; Boubel, J. C. *J. Chem. Soc. Dalton Trans.* **1980,** 2081–2089.
(86) Boukhris, A.; Lecomte, C.; Coutsolelos, A.; Guilard, R. *J. Organomet. Chem.* **1986,** *303,* 151–165.
(87) Cocolios, P.; Guilard, R.; Bayeul, D.; Lecomte, C. *Inorg. Chem.* **1985,** *24,* 2058–2062.
(88) Hoshino, M.; Yamaji, M.; Hama, Y. *Chem. Phys. Lett.* **1986,** *125,* 369–372.
(89) (a) Hoshino, M.; Ida, H.; Yasufuku, K.; Tanaka, K. *J. Phys. Chem.* **1986,** *90,* 3984–3987. (b) Yamaji, M.; Hama, Y.; Arai, S.; Hoshino, M. *Inorg. Chem.* **1987,** *26,* 4375–4378.
(90) Hoshino, M.; Hirai, T. *J. Phys. Chem.* **1987,** *91,* 4510–4514.
(91) Kadish, K. M.; Maiya, G. B.; Xu, Q. Y. *Inorg. Chem.* **1989,** *28,* 2518–2523.
(92) Kane, K. M.; Lemke, F. R.; Petersen, J. L. *Inorg. Chem.* **1995,** *34,* 4085–4091.
(93) Kenney, M. E.; Maskasky, J. E. *J. Am. Chem. Soc.* **1971,** *93,* 2060–2062.
(94) Maskasky, J. E.; Kenney, M. E. *J. Am. Chem. Soc.* **1973,** *95,* 1443–1448.
(95) (a) Esposito, J. N.; Lloyd, J. E.; Kenney, M. E. *Inorg. Chem.* **1966,** *5,* 1979–1984. (b) Kane, A. R.; Yalman, R. G.; Kenney, M. E. *Inorg. Chem.* **1968,** *7,* 2588–2592.
(96) Miyamoto, T. K.; Sugita, N.; Matsumoto, Y.; Sasaki, Y.; Konno, M. *Chem. Lett.* **1983,** 1695–1698.
(97) Philippova, T. O.; Galkin, B. N.; Golovenko, N. Y.; Zhilina, Z. I.; Vodzinskii, S. V. *J. Porphyrins Phthalocyanines* **2000,** *4,* 243–247.
(98) Vremen, H. J.; Cipkala, D. A.; Stevenson, D. K. *Can. J. Physiol. Pharmacol.* **1996,** *74,* 278–285.
(99) Cloutour, C.; Lafargue, D.; Richards, J. A.; Pommier, J. C. *J. Organomet. Chem.* **1977,** *137,* 157–163.
(100) Cloutour, C.; Lafargue, D.; Pommier, J. C. *J. Organomet. Chem.* **1978,** *161,* 327–334.
(101) Kane, K. M.; Lemke, F. R.; Petersen, J. L. *Inorg. Chem.* **1997,** *36,* 1354–1359.
(102) Zheng, J.-Y.; Konishi, K.; Aida, T. *Inorg. Chem.* **1998,** *37,* 2591–2594.
(103) Kadish, K. M.; Xu, Q. Y.; Barbe, J. M.; Anderson, J. E.; Wang, E.; Guilard, R. *J. Am. Chem. Soc.* **1987,** *109,* 7705–7714.
(104) Dawson, D. Y.; Sangalang, J. C.; Arnold, J. *J. Am. Chem. Soc.* **1996,** *118,* 6082–6083.
(105) Chen, J.; Woo, L. K. *Inorg. Chem.* **1998,** *37,* 3269–3275.
(106) Kadish, K. M.; Xu, Q. Y.; Barbe, J. M.; Guilard, R. *Inorg. Chem.* **1988,** *27,* 1191–1198.
(107) Kraus, G. A.; Louw, S. V. *J. Org. Chem.* **1998,** *63,* 7520–7521.
(108) Li, Z.; Lieberman, M. *Inorg. Chem.* **2001,** *40,* 932–939.
(109) Li, Z.; Lieberman, M. *Supramol. Sci.* **1998,** *5,* 485–489.
(110) Xu, Q. Y.; Barbe, J. M.; Kadish, K. M. *Inorg. Chem.* **1988,** *27,* 2373–2378.
(111) Kadish, K. M.; Xu, Q. Y.; Barbe, J. M. *Inorg. Chem.* **1987,** *26,* 2565–2566.
(112) Kadish, K. M.; Dubois, D.; Koeller, S.; Barbe, J. M.; Guilard, R. *Inorg. Chem.* **1992,** *31,* 3292–3294.
(113) Kadish, K. M.; Will, S.; Adamian, V. A.; Walther, B.; Erben, C.; Ou, Z.; Guo, N.; Vogel, E. *Inorg. Chem.* **1998,** *37,* 4573–4577.
(114) Cloutour, C.; Debaig-Valade, C.; Gacherieu, C.; Pommier, J. C. *J. Organomet. Chem.* **1984,** *269,* 239–247.

(115) Barbe, J. M.; Ratti, C.; Richard, P.; Lecomte, C.; Gerardin, R.; Guilard, R. *Inorg. Chem.* **1990,** *29,* 4126–4130.
(116) Belcher, W. J.; Brothers, P. J.; Meredith, A. P.; Rickard, C. E. F.; Ware, D. C. *J. Chem. Soc., Dalton Trans.* **1999,** 2833–2836.
(117) Guldi, D. M.; Neta, P.; Vogel, E. *J. Phys. Chem.* **1996,** *100,* 4097–4103.
(118) Simkhovich, L.; Mahammed, A.; Goldberg, I.; Gross, Z. *Chem. Eur. J.* **2001,** *7,* 1041–1055.
(119) Kadish, K. M.; Dubois, D.; Barbe, J. M.; Guilard, R. *Inorg. Chem.* **1991,** *30,* 4498–4501.
(120) Zheng, J.-Y.; Konishi, K.; Aida, T. *J. Am. Chem. Soc.* **1998,** *120,* 9838–9843.
(121) Zheng, J.-Y.; Konishi, K.; Aida, T. *Chem. Lett.* **1998,** 453–454.
(122) Maiya, G. B.; Barbe, J. M.; Kadish, K. M. *Inorg. Chem.* **1989,** *28,* 2524–2527.
(123) Cloutour, C.; Lafargue, D.; Pommier, J. C. *J. Organomet. Chem.* **1980,** *190,* 35–42.
(124) Balch, A. L.; Cornman, C. R.; Olmstead, M. M. *J. Am. Chem. Soc.* **1990,** *112,* 2963–2969.
(125) Balch, A. L. *Inorg. Chim. Acta* **1992,** *198–200,* 297–307.
(126) Kane, A. R.; Sullivan, J. F.; Kenny, D. H.; Kenney, M. E. *Inorg. Chem.* **1970,** *9,* 1445–1448.
(127) Tamao, K.; Akita, M.; Kato, H.; Kumada, M. *J. Organomet. Chem.* **1988,** *341,* 165–175.
(128) (a) Hanack, M.; Mitulla, K.; Pawlowski, G.; Subramanian, L. R. *Angew. Chem. Int. Ed. Engl.* **1979,** *18,* 322–323. (b) Hanack, M.; Mitulla, K.; Pawlowski, G.; Subramanian, L. R. *J. Organomet. Chem.* **1981,** *204,* 315–325.
(129) Hanack, M.; Metz, J.; Pawlowski, G. *Chem. Ber.* **1982,** *115,* 2836–2853.
(130) (a) Dirk, C. W.; Inabe, T.; Schoch, K. F. Jr.; Marks, T. J. *J. Am. Chem. Soc.* **1983,** *105,* 1539–1550. (b) DeWulf, D. W.; Leland, J. K.; Wheeler, B. L.; Bard, A. J.; Batzel, D. A.; Dininny, D. R.; Kenney, M. *Inorg. Chem.* **1987,** *26,* 266–270.
(131) Sayer, P.; Gouterman, M.; Connell, C. R. *Acc. Chem. Res.* **1982,** *15,* 73–79.
(132) Satoh, W.; Nadano, R.; Yamamoto, Y.; Akiba, K.-Y. *Chem. Commun.* **1996,** 2451–2452.
(133) Barbour, T.; Belcher, W. J.; Brothers, P. J.; Rickard, C. E. F.; Ware, D. C. *Inorg. Chem.* **1992,** *31,* 746–754.
(134) Akiba, K.; Onzuka, Y.; Itagaki, M.; Hirota, H.; Yamamoto, Y. *Organometallics* **1994,** *13,* 2800–2803.
(135) Yamamoto, Y.; Nadano, R.; Itagaki, M.; Akiba, K.-Y. *J. Am. Chem. Soc.* **1995,** *117,* 8287–8288.
(136) Yamamoto, Y.; Akiba, K.-Y. *J. Organomet. Chem.* **2000,** *611,* 200–209.
(137) (a) Michaudet, L.; Fasseur, D.; Guilard, R.; Ou, Z.; Kadish, K. M.; Dahaoui, S.; Lecomte, C. *J. Porphyrins Phthalocyanines* **2000,** *4,* 261–270. (b) Michaudet, L.; Richard, P.; Boitrel, B. *Chem. Commun.* **2000,** *17,* 1589–1590. (c) Brothers, P. J.; Belcher, W. J.; Al Salim, N. I.; Rickard, C. E. F., unpublished results.
(138) (a) Reddy, D. R.; Maiya, B. G. *Chem. Commun.* **2001,** 117–118. (b) Giribabu, L.; Rao, T. A.; Maiya, B. G. *Inorg. Chem.* **1999,** *38,* 4971–4980. (c) Segawa, H.; Kunimoto, K.; Susumu, K.; Taniguchi, M.; Shimidzu, T. *J. Am. Chem. Soc.* **1994,** *116,* 11193–11194.
(139) Shiragami, T.; Kubomura, K.; Ishibashi, D.; Inoue, H. *J. Am. Chem. Soc.* **1996,** *118,* 6311–6312.
(140) Satoh, W.; Nadano, R.; Yamamoto, G.; Yamamoto, Y.; Akiba, K.-Y. *Organometallics* **1997,** *16,* 3664–3671.
(141) Yamamoto, A.; Satoh, W.; Yamamoto, Y.; Akiba, K.-Y. *Chem. Commun.* **1999,** 147–148.
(142) Yasumoto, M.; Satoh, W.; Nadano, R.; Yamamoto, Y.; Akiba, K.-Y. *Chem. Lett.* **1999,** 791–792.
(143) Kadish, K. M.; Autret, M.; Ou, Z.; Akiba, K.-Y.; Masumoto, S.; Wada, R.; Yamamoto, Y. *Inorg. Chem.* **1996,** *35,* 5564–5569.
(144) Satoh, W.; Masumoto, S.; Shimizu, M.; Yamamoto, Y.; Akiba, K.-Y. *Bull. Chem. Soc. Jpn.* **1999,** *72,* 459–463.
(145) Kadish, K. M.; Ou, Z.; Adamian, V. A.; Guilard, R.; Gros, C. P.; Erben, C.; Will, S.; Vogel, E. *Inorg. Chem.* **2000,** *39,* 5675–5682.
(146) Kadish, K. M.; Erben, C.; Ou, Z.; Adamian, V. A.; Will, S.; Vogel, E. *Inorg. Chem.* **2000,** *39,* 3312–3319.

(147) Knör, G.; Vogler, A. *Inorg. Chem.* **1994,** *33,* 314–318.
(148) Igarashi, T.; Konishi, K.; Aida, T. *Chem. Lett.* **1998,** 1039–1040.
(149) Paolesse, R.; Boschi, T.; Licoccia, S.; Khoury, R. G.; Smith, K. M. *Chem. Commun.* **1998,** 1119–1120.
(150) Vangberg, T.; Ghosh, A. *J. Am. Chem. Soc.* **1999,** *121,* 12154–12160.
(151) Cocolios, P.; Moise, C.; Guilard, R. *J. Organomet. Chem.* **1982,** *228,* C43–C46.
(152) Cocolios, P.; Chang, D.; Vittori, O.; Guilard, R.; Moise, C.; Kadish, K. M. *J. Am. Chem. Soc.* **1984,** *106,* 5724–5726.
(153) Onaka, S.; Kondo, Y.; Yamashita, M.; Tatematsu, Y.; Kato, Y.; Goto, M.; Ito, T. *Inorg. Chem.* **1985,** *24,* 1070–1076.
(154) Guilard, R.; Mitaine, P.; Moise, C.; Lecomte, C.; Boukhris, A.; Swistak, C.; Tabard, A.; Lacombe, D.; Cornillon, J. L.; Kadish, K. M. *Inorg. Chem.* **1987,** *26,* 2467–2476.
(155) Goral, J.; Proniewicz, L. M.; Nakamoto, K.; Kato, Y.; Onaka, S. *Inorg. Chim. Acta* **1988,** *148,* 169–175.
(156) Guilard, R.; Tabard, A.; Zrineh, A.; Ferhat, M. *J. Organomet. Chem.* **1990,** *389,* 315–324.
(157) Takagi, S.; Kato, Y.; Furuta, H.; Onaka, S.; Miyamoto, T. K. *J. Organomet. Chem.* **1992,** *429,* 287–299.
(158) Guilard, R.; Zrineh, A.; Ferhat, M.; Tabard, A.; Mitaine, P.; Swistak, C.; Richard, P.; Lecomte, C.; Kadish, K. M. *Inorg. Chem.* **1988,** *27,* 697–705.
(159) Richard, P.; Zrineh, A.; Guilard, R.; Habbou, A.; Lecomte, C. *Acta Crystallogr. Sect. C: Cryst. Struct. Commun.* **1989,** *C45,* 1224–1226.
(160) Guilard, R.; Zrineh, A.; Tabard, A.; Courthaudon, L.; Han, B.; Ferhat, M.; Kadish, K. M. *J. Organomet. Chem.* **1991,** *401,* 227–243.
(161) Barbe, J. M.; Guilard, R.; Lecomte, C.; Gerardin, R. *Polyhedron* **1984,** *3,* 889–894.
(162) Kadish, K. M.; Boisselier-Cocolios, B.; Swistak, C.; Barbe, J. M.; Guilard, R. *Inorg. Chem.* **1986,** *25,* 121–122.
(163) Kadish, K. M.; Swistak, C.; Boisselier-Cocolios, B.; Barbe, J. M.; Guilard, R. *Inorg. Chem.* **1986,** *25,* 4336–4343.
(164) Habbou, A.; Lecomte, C.; Barbe, J. M. *Acta Crystallogr. Sect. C: Cryst. Struct. Commun.* **1992,** *C48,* 921–923.
(165) Onaka, S.; Kondo, Y.; Toriumi, K.; Ito, T. *Chem. Lett.* **1980,** 1605–1608.
(166) Guilard, R.; Barbe, J. M.; Fahim, M. *New J. Chem.* **1992,** *16,* 815–820.
(167) Guilard, R.; Fahim, M.; Zaegel, F.; Barbe, J.-M.; d'Souza, F.; Atmani, A.; Adamian, V. A.; Kadish, K. M. *Inorg. Chim. Acta* **1996,** *252,* 375–382.
(168) Guilard, R.; Mitaine, P.; Moise, C.; Cocolios, P.; Kadish, K. M. *New J. Chem.* **1988,** *12,* 699–705.
(169) Kato, S.; Noda, I.; Mizuta, M.; Itoh, Y. *Angew. Chem. Int. Ed. Engl.* **1979,** *18,* 82–83.
(170) Noda, I.; Kato, S.; Mizuta, M.; Yasuoka, N.; Kasei, N. *Angew. Chem. Int. Ed. Engl.* **1979,** *18,* 83.
(171) Jones, N. L.; Carroll, P. J.; Wayland, B. B. *Organometallics* **1986,** *5,* 33–37.
(172) (a) Lux, D.; Daphnomili, D.; Coutsolelos, A. G. *Polyhedron* **1994,** *13,* 2367–2377. (b) Coutsolelos, A. G.; Lux, D.; Mikros, E.; *Polyhedron* **1996,** *15,* 705–715.
(173) Daphnomili, D.; Scheidt, W. R.; Zajicek, J.; Coutsolelos, A. G. *Inorg. Chem.* **1998,** *37,* 3675–3681.
(174) Guilard, R.; Kadish, K. M. *Comments Inorg. Chem.* **1988,** *7,* 287–305.
(175) Collman, J. P.; Arnold, H. J. *Acc. Chem. Res.* **1993,** *26,* 586.
(176) Kobayashi, N.; Furuya, F.; Yug, G.-C. *J. Porphyrins Phthalocyanines* **1999,** *3,* 433–438.
(177) Li, J.; Subramanian, L. R.; Hanack, M. *J. Porphyrins Phthalocyanines* **2000,** *4,* 739–741.

Index

R

S

T

U

V

W

Y

Z

Cumulative List of Contributors for Volumes 1–36

Cumulative Index for Volumes 37–48